Ecotoxicology Modeling

Emerging Topics in Ecotoxicology

Principles, Approaches and Perspectives

Volume 2

Series Editor

Lee R. Shugart
L.R. Shugart and Associates, Oak Ridge, TN, USA

For other volumes published in this series, go to
http://www.springer.com/series/7360

James Devillers
Editor

Ecotoxicology Modeling

Springer

Editor
James Devillers
Centre de Traitement de
l'Information Scientifique (CTIS)
3 chemin de la Gravière
69140 Rillieux La Pape
France
j.devillers@ctis.fr

ISSN 1868-1344 e-ISSN 1868-1352
ISBN 978-1-4614-2934-0 e-ISBN 978-1-4419-0197-2
DOI 10.1007/978-1-4419-0197-2
Springer Dordrecht Heidelberg London New York

Printed on acid-free paper

Springer is part of Springer Science+Business Media (www.springer.com)

Preface

The term "ecotoxicology" was first coined in the early 1970s by the French toxicologist, Professor René Truhaut, who defined it as a branch of toxicology concerned with the study of the adverse effects caused by the natural and synthetic pollutants to the biota in the aquatic and terrestrial ecosystems. Inherent in this concept is the investigation of how and to what level the wildlife species and humans are exposed to pollutants but also the study of the manner in which chemicals are released into the environment, are transported between the different compartments of the biosphere, and are transformed from abiotic and biotic processes. Ecotoxicology is therefore multidisciplinary in essence being strongly rooted in toxicology, environmental chemistry, pharmacology, and ecology. Modeling is also undoubtedly a component of this multifaceted discipline. Indeed, the fate and effects of chemicals in the environment are governed by complex phenomena and modeling approaches have proved to be particularly suited not only to better understand these phenomena but also to simulate them in the frame of predictive hazard and risk assessment schemes.

The vocation of this book is not to catalog all the types of models and approaches that can be used in ecotoxicology. To date, the task should be quite impossible due to the huge number of available linear and nonlinear methods, the possibility to hybridize them, and so on.

Nevertheless, the book provides a clear overview of the main modeling approaches that can be used in ecotoxicology. Limiting the mathematical descriptions to a minimum, it presents numerous case studies to enable the reader to understand the interest but also the limitations of models in ecotoxicology.

I am extremely grateful to the contributors for accepting to participate in this book and for preparing valuable contributions. To ensure the scientific quality and clarity of the book, each chapter was sent to two referees for review. I would like to thank all the referees for their useful comments. Finally, I would like to thank Melinda Paul and all the publication teams at Springer for making the publication of this book possible.

Rillieux La Pape
France

James Devillers

Contents

Contributors

Jon A. Arnot Canadian Environmental Modeling Centre, Trent University, Peterborough, ON, Canada K9J 7B8

Jan Baas Faculty of Earth and Life Sciences, Vrije Universiteit, de Boelelaan 1085, 1081 HV Amsterdam, The Netherlands

Elise Billoir Université de Lyon, F-69000 Lyon, Université Lyon 1, CNRS, UMR5558, Laboratoire de Biométrie et Biologie Evolutive, F-69622 Villeurbanne, France

Antonio Bispo ADEME, 20 Avenue du Grésillé, BP 90406, 49004 Angers Cedex 01, France

Daniel Bontje Faculty of Earth and Life Sciences, Vrije Universiteit, de Boelelaan 1085, 1081 HV Amsterdam, The Netherlands

Mieke Broerse Faculty of Earth and Life Sciences, Vrije Universiteit, de Boelelaan 1085, 1081 HV Amsterdam, The Netherlands

Anne-Marie Charissou IPL Santé Environnement Durables Est, Rue Lucien Cuenot, ZI Saint Jacques II, 54320 Maxeville, France

Sandrine Charles Université de Lyon, F-69000 Lyon, Université Lyon 1, CNRS, UMR5558, Laboratoire de Biométrie et Biologie Evolutive, F-69622 Villeurbanne, France, scharles@biomserv.univ-lyon1.fr

Arnaud Chaumot CEMAGREF, F-69000, Lyon, UR "Biologie des Ecosystèmes Aquatiques", Laboratoire d'écotoxicologie, 3bis quai Chauveau, CP 220, F-69336 Lyon Cedex 9, France

Trine Dalkvist Department of Wildlife Ecology and Biodiversity, National Environmental Research Institute, University of Aarhus, Grenåvej 14, DK-8410 Rønde, Denmark
Centre for Integrated Population Ecology, Roskilde University, PO Box 260, 4000 Roskilde, Denmark

James Devillers CTIS, 3 Chemin de la Gravière, 69140 Rillieux La Pape, France, j.devillers@ctis.fr

Valery E. Forbes Centre for Integrated Population Ecology, Roskilde University, PO Box 260, 4000 Roskilde, Denmark
Department of Environmental, Social and Spatial Change, Roskilde University, PO Box 260, 4000 Roskilde, Denmark

Volker Grimm Department of Ecological Modelling, Helmholtz Centre for Environmental Research – UFZ, Permoserstr. 15, 04318 Leipzig, Germany

Tjalling Jager Faculty of Earth and Life Sciences, Vrije Universiteit, de Boelelaan 1085, 1081 HV Amsterdam, The Netherlands

Sven E. Jørgensen Institute A, Section for Environmental Chemistry, Copenhagen University, University Park 2, 2100 Copenhagen Ø, Denmark, msijapan@hotmail.com

Sebastiaan A.L.M. Kooijman Faculty of Earth and Life Sciences, Vrije Universiteit, de Boelelaan 1085, 1081 HV Amsterdam, The Netherlands, bas.kooijman@falw.vu.nl

Kannan Krishnan Département de santé environnementale et santé au travail, 2375 chemin de la Cote Ste Catherine, Pavillon Marguerite d'Youville, Université de Montréal, Montreal, QC, Canada, kannan.krishnan@umontreal.ca

Sunil A. Kulkarni Existing Substances Division, Health Canada, Ottawa, ON, Canada K1A 0K9, sunil_kulkarni@hc-sc.gc.ca

Christelle Lopes Université de Lyon, F-69000 Lyon, Université Lyon 1, CNRS, UMR5558, Laboratoire de Biométrie et Biologie Evolutive, F-69622 Villeurbanne, France

Donald Mackay Canadian Environmental Modeling Centre, Trent University, Peterborough, ON, Canada K9J 7B8, dmackay@trentu.ca

Anne Merle ADEME, 20 Avenue du Grésillé, BP 90406 49004 Angers Cedex 01, France

Pascal Pandard INERIS, Parc Technologique ALATA, BP n° 2, 60550 Verneuil en Halatte, France

Grace Patlewicz DuPont Haskell Global Centers for Health and Environmental Sciences, 1090 Elkton Road, Newark, DE 19711, USA, patlewig@hotmail.com

Willie J.G.M. Peijnenburg Laboratory for Ecological Risk Assessment, National Institute of Public Health and the Environment, PO Box 1, 3720 BA Bilthoven, The Netherlands, WJGM.Peijnenburg@rivm.nl

Thomas Peyret Département de santé environnementale et santé au travail, 2375 chemin de la Cote Ste Catherine, Pavillon Marguerite d'Youville, Université de Montréal, Montreal, QC, Canada

Lüsa Reid Canadian Environmental Modeling Centre, Trent University, Peterborough, ON, Canada K9J 7B8

David W. Roberts School of Pharmacy and Chemistry, Liverpool John Moores University, Byrom Street, Liverpool L3 3AF, UK

Richard M. Sibly Centre for Integrated Population Ecology, Roskilde University, PO Box 260, 4000 Roskilde, Denmark
School of Biological Sciences, University of Reading, Whiteknights, Reading RG6 6PS, UK

Eric Thybaud INERIS, Parc Technologique ALATA, BP n° 2, 60550 Verneuil en Halatte, France

Christopher J. Topping Department of Wildlife Ecology and Biodiversity, National Environmental Research Institute, University of Aarhus, Grenåvej 14, DK-8410 Rønde
Centre for Integrated Population Ecology, Roskilde University, PO Box 260, 4000 Roskilde, Denmark, cjt@dmu.dk

Stefan Trapp DTU Environment, Department of Environmental Engineering, Technical University of Denmark, Miljoevej, Building 113, DK-2800 Kgs. Lyngby, Denmark, stt@env.dtu.dk

Cees A.M. van Gestel Faculty of Earth and Life Sciences, Vrije Universiteit, de Boelelaan 1085, 1081 HV Amsterdam, The Netherlands

Martina G. Vijver Institute of Environmental Sciences (CML), Leiden University, Leiden, The Netherlands

Eva Webster Canadian Environmental Modeling Centre, Trent University, Peterborough, ON, Canada K9J 7B8

Jiping Zhu Exposure and Biomonitoring Division, Health Canada, Ottawa, ON, Canada K1A 0K9

Artificial Neural Network Modeling of the Environmental Fate and Ecotoxicity of Chemicals

James Devillers

Abstract An artificial neural network (ANN) includes nonlinear computational elements called neurons, which are linked by weighted connections. Typically, a neuron receives an input information and performs a weighted summation, which is propagated by an activation function to other neurons through the ANN. Numerous ANN paradigms have been proposed for pattern classification, clustering, function approximation, prediction, optimization, and control. In this chapter, an attempt is made to review the main applications of ANNs in ecotoxicology. Our goal was not to catalog all the models in the field but only to show the diversity of the situations in which these nonlinear tools have proved their interest for modeling the environmental fate and effects of chemicals.

Keywords Artificial neural network · QSPR · QSAR · Environmental contamination · Nonlinear methods

1 Introduction

The last decade has witnessed a surge interest in the use of artificial neural networks (ANNs) for modeling complex tasks in a variety of fields including data mining, speech, image recognition, finance, business, drug design, and so on [1–6]. The *raison d'être* of these powerful tools is to exploit the imprecision and uncertainty of real-world problems for deriving valuable and robust models.

The concepts of ANNs are directly inspired by neurobiology. Thus, the cerebral cortex contains about 100 billion neurons, which are special cells processing information. A biological neuron receives signals from other neurons through its dendrites and transmits information generated by its soma along its axon. In the brain, each neuron is connected to 1,000–11,000 other neurons via synapses in

J. Devillers (✉)
CTIS, 3 Chemin de la Gravière, 69140 Rillieux La Pape, France
e-mail: j.devillers@ctis.fr

J. Devillers (ed.), *Ecotoxicology Modeling*, Emerging Topics in Ecotoxicology: Principles, Approaches and Perspectives 2, DOI 10.1007/978-1-4419-0197-2_1,

which neurotransmitters inducing different activities are released. The human brain contains approximately 10^{14}–10^{15} interconnections [7–9]. Consequently, the brain can be viewed as a nonlinear and highly parallel biological device characterized by robustness and fault tolerance. It can learn, handle imprecise, fuzzy, and noisy information, and can generalize from past and/or new experiences [10, 11]. ANNs can be defined as weighted directed graphs with connected nodes called neurons that attempt to mimic some of the basic characteristics of the human brain [11]. Consequently, it is not surprising to see that now these nonlinear statistical tools are widely used in numerous technical and scientific domains to process complex information. After a brief overview of the characteristics of ANNs, this chapter will review the main applications of ANNs for modeling the toxicity and ecotoxicity of chemicals as well as their environmental fate. Their advantages and limitations will be also stressed.

2 Characteristics of ANNs

A precise definition of learning is difficult to formulate but the fundamental questions that neurobehaviorists try to answer are: How do we learn? Which is the most efficient process for learning? How much and how fast can we learn? In a neurocomputing context, a learning process can be viewed as a method for updating the architecture as well as the connection weights of an ANN to optimize its efficiency to perform a specific task. The three main learning paradigms are the following: supervised, unsupervised (or self-organized), and reinforcement. Each category includes numerous algorithms. Supervised is the most commonly employed learning paradigm to develop classification and prediction applications. The algorithm takes the difference between the observed and calculated output and uses that information to adjust the weights in the network so that next time, the prediction will be closer to the correct answer (Fig. 1) [1]. Unsupervised learning is used when

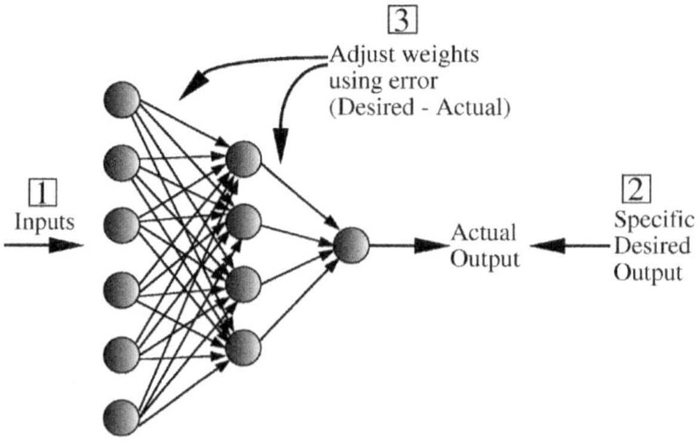

Fig. 1 Supervised learning paradigm (adapted from [1])

we want to perform a clustering of the input data. ANNs that are trained using this learning process are called self-organizing neural networks because they receive no direction on what the desired output should be. Indeed, when presented with a series of inputs, the outputs self-organize by initially competing to recognize the input information and then cooperating to adjust their connection weights. Over time, the network evolves so that each output unit is sensitive to and will recognize inputs from a specific portion of the input space (Fig. 2) [1]. Reinforcement learning attempts to learn the input–output mapping through trial and error with a view to maximizing a performance index called the reinforcement signal (Fig. 3). Reinforcement learning is particularly suited to solve difficult temporal (time-dependent) problems [1].

ANNs are also characterized by their connection topology. The arrangement of neurons and their interconnections can have an important impact on the modeling capabilities of the ANNs. Generally, ANNs are organized into layers of neurons. Data can flow between the neurons in these layers in two different ways. In feedforward networks, no loops occur while in recurrent networks feedback connections are found.

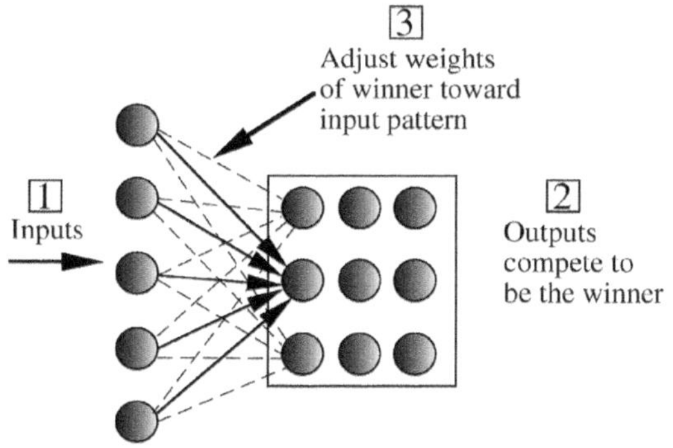

Fig. 2 Unsupervised learning paradigm (adapted from [1])

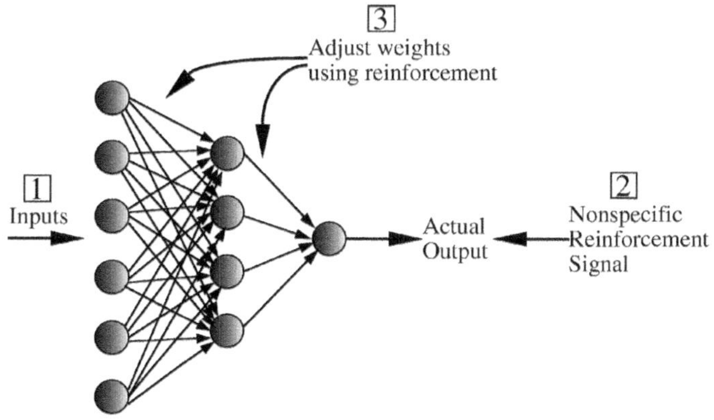

Fig. 3 Reinforcement learning paradigm (adapted from [1])

Table 1 Taxonomy of the main types of ANNs (adapted from [1, 11])

ANN paradigm	Architecture	Learning type
Multilayer perceptron[a]	Feedforward	Supervised
Radial basis function network	Feedforward	Hybrid
Probabilistic neural network	Feedforward	Supervised
Kohonen self-organizing map	Recurrent	Unsupervised
Learning vector quantization	Competitive	Supervised
ART networks	Recurrent	Supervised/unsupervised

[a] Mostly three layers

The description of the different ANN paradigms is beyond the scope of this chapter and the interested readers are invited to consult the rich body of literature on this topic (see e.g., [12–17]). However, Table 1 summarizes the main characteristics of the different types of ANNs cited in the following sections. It is also beyond the scope of this chapter to provide information on computer tools that can be used for deriving ANN models. However, it is noteworthy that a list of freeware, shareware, and commercial ANN software can be found in Devillers and Doré [18].

3 Use of ANNs in Quantitative Structure–Property Relationship (QSPR) Modeling

Knowing the physicochemical properties of xenobiotics is a prerequisite to estimate their bioactivity, bioavailability, transport, and distribution between the different compartments of the biosphere [19–22]. Unfortunately, there are very limited or no experimental physicochemical data available for most of the chemicals susceptible to contaminate the aquatic and terrestrial ecosystems. Consequently, for the many compounds without experimental data, the only alternative to using actual measurements is to approximate values by means of estimation models, which are generically termed quantitative structure–property relationships (QSPRs). The ingredients necessary to derive a QSPR model are given in Fig. 4. Although most of the QSPR models have been derived from simple contribution methods and regression analysis [23–27], attempts have been made to use ANNs for modeling the intrinsic physicochemical properties of organic molecules as well as their environmental degradation parameters linked to transformation process. These models are discussed in the following sections.

3.1 Boiling Point

The normal boiling point (BP), corresponding to the temperature at which a substance presents a vapor pressure (VP) of 760 mmHg, depends on a number of

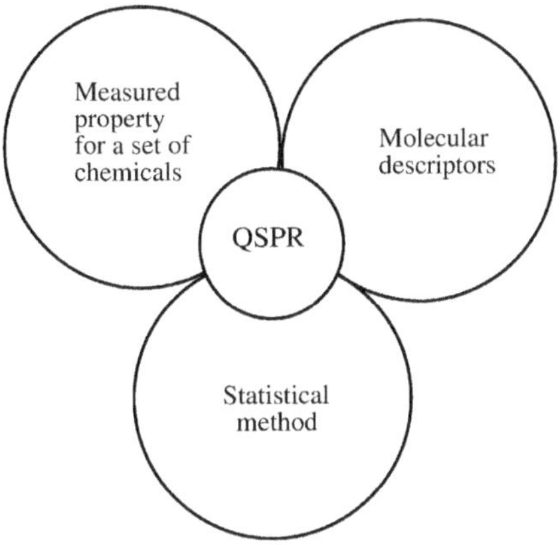

Fig. 4 Ingredients for deriving a QSPR model

molecular properties that control the ability of a molecule to escape from the surface of a liquid into the vapor phase. These properties are molecular size, polar and hydrogen bonding forces, and entropic factors such as flexibility and orientation [27]. Different types of ANNs have been used for computing BP models. Thus, a radial basis function (RBF) network was used by Lohninger [28] for predicting the BPs of 185 ethers, peroxides, acetals, and their sulfur analogs. Molecules were described by two sets of three topological and structural descriptors yielding the design of two models, both including 20 hidden neurons and cross-validated from a leave-25%-out procedure. Both models outperformed regressions models obtained under the same conditions. Cherquaoui and coworkers [29] used the same data set of 185 molecules but their ANN was a three-layer perceptron (TLP) trained by the backpropagation algorithm, and the chemical structures were characterized by embedding frequencies. The ANN presented 20 input neurons and a bias, from 3 to 8 hidden neurons and a bias, and an output neuron. Their selected 20/5/1 (input/hidden/output) TLP after 4,000 iterations presented good statistics but undoubtedly this model presented a problem of overtraining, and it is noteworthy that the number of connections within the ANN is high. At that time, other TLP models allowing the estimation of BPs of chlorofluorocarbons with 1, 1–2, or 1–4 carbon atoms ($n = 15$, 62, and 276, respectively) as well as of halomethanes with up to four different halogen atoms ($n = 48$) were also proposed [30]. Egolf and coworkers [31] used a TLP trained by the Broyden–Fletcher–Goldfarb–Shanno (BFGS) quasi-Newton optimization method for deriving a model allowing for the prediction of the BP of industrial chemicals. A database of 298 structurally diverse chemicals was first split into a learning set (LS), a cross-validation set (CVS), and an external testing set (ETS) of 241, 27, and 30 chemicals, respectively. It is noteworthy that the CVS is used to monitor the ANN. Topological, geometrical, and electronic descriptors were generated for characterizing the molecules. The best configuration was a 8/3/1 ANN yielding RMS error values of 11.18, 9.17, and 10.69 K for the LS, CVS, and ETS, respectively. The same methodology was applied to a larger database [32]. The selected 6/5/1 ANN gave RMS error values of 5.7 K for the training and CVSs of 267 and 29 chemicals, respectively. The network model was validated with a 15-member external prediction set. The RMS error of prediction was 7.1 K. This was substantially better than the 8.5 K error obtained from a regression model derived under the same conditions and with the same descriptors. E-state indices [33] for 19 atom types were used [34] as inputs neurons of a TLP trained by the backpropagation algorithm for predicting the BPs of chemicals from a LS and ETS of 268 and 30 compounds, respectively. The best model included five neurons on the hidden layer. It produced a mean absolute error of 3.86 and 4.57 K for the LS and ETS, respectively. These authors experienced the same strategy on a larger database of 372 chemicals but only including alkanes, alcohols, and (poly)chloroalkanes [35]. The interest of the TLP and a fuzzy ARTMAP ANN was tested by Espinosa et al. [36] from a limited database including 140 alkanes, 144 alkenes, and 43 alkynes. Even if this kind of study allows us to compare methods and/or descriptors, it is obvious that ANNs show their full interest when models are derived from large sets of molecules from which, it is not easy to relate the structure of the molecules to a property (or activity) under study from classical linear methods. Thus, an interesting approach based on the use of a TLP and descriptors calculated using AM1 and PM3 semiempirical quantum-chemical methods was used by Chalk and coworkers [37] for deriving models from a database of 6,629 experimental BPs. The LS and ETS included 6,000 and 629 chemicals, respectively. The best results were obtained with a 18/10/1 ANN architecture. Ten separate ANNs with random starting weights were then trained with different LSs and ETSs, chosen such that each chemical appeared only once in an ETS. The standard deviations (means of the results for 10 nets) for the LS and ETS were 16.54 and 19.02 K with the AM1 approach and 18.33 and 20.27 K with the PM3 approach.

3.2 Vapor Pressure

The VP determines the potential of a chemical to volatilize from its condensed or dissolved phases and to therefore exist as a gas [38]. VP strongly depends on the temperature as expressed in the classical Clausius–Clapeyron equation [24]. As previously seen, the BP of a chemical can be easily derived from its VP. Numerous methods can be used for estimating the VPs of chemicals, and among them, some are based on the use of ANNs. Thus, different regression and ANN models were tested by Liang and Gallagher [39] from a set of 479 chemicals described by various descriptors encoding the structure and physicochemical properties of the molecules. Standard errors of 0.534 and 0.522 (log units, Torr) were obtained for the regression models with seven independent variables and a 7/5/1 ANN. However, the interest of the results is very limited because of total lack of information on the conditions in which the models were derived. More reliable models were designed by McClelland and Jurs [40]. TLP models were developed to relate the structural characteristics of 420 diverse organic compounds to their VP at 25°C expressed as log (VP in Pascals). The log (VP) values ranged over eight orders of magnitude from −1.34 to 6.68 log units. The database was split into a learning set (LS), a CVS, and an ETS of 290, 65, and 65 chemicals, respectively. A 8/3/1 TLP trained by a BFGS optimization algorithm and only including topological descriptors yielded RMS errors of 0.26, 0.29, and 0.37 for the LS, CVS, and ETS, respectively (log units, Pa). An alternative 10/4/1 TLP containing a lager selection of descriptor types (e.g., quantum mechanical descriptors) resulted in improved performance with RMS errors of 0.19, 0.24, and 0.33 for the LS, CVS, and ETS, respectively [40]. In the same way, Beck and coworkers [41] derived a 10/8/1 TLP trained by the backpropagation algorithm for estimating the log VP at 25°C. Descriptors derived from quantum mechanical calculations were used for describing the 551 chemicals constituting the learning and testing sets. The leave-one-out (LOO) cross validation gave a standard deviation of 0.37 log units (Torr) and a maximum absolute error of 1.65. A temperature-dependent model based on a TLP trained by the backpropagation algorithm and descriptors calculated using AM1 semiemperical MO-theory was proposed by Chalk et al. [42]. A data set of 8,542 measurements at various temperatures for a total of 2,349 molecules was divided into a training set of 7,681 measurements and an external validation set of 861 measurements in such a manner that the validation set spans the full range of VPs. The standard deviation of the error (log units, Torr) for the learning, LOO cross-validation, and validation sets obtained with the selected 27/15/1 TLP was equal to 0.32, 0.46, and 0.33, respectively. Yaffe and Cohen [43] also computed a temperature-dependent QSPR model for VP of aliphatic, aromatic, and polycyclic aromatic hydrocarbons, ranging from 4 to 12 carbon atoms using a TLP trained by the backpropagation algorithm with connectivity indices [44], molecular weight, and temperature as input parameters in the ANN. The database of 274 molecules included 7,613 vapor pressure–temperature data. It was split into a learning set (LS), a CVS, and an ETS of 5,330, 754, and 1,529 chemicals, respectively. The best model was a 7/29/1 TLP yielding average absolute VP errors of 11.6% (0.051 log units or 34 kPa), 8.2% (0.036 log units or 23.2 kPa), 9.2% (0.039 log units or 26.8 kPa), and 10.7% (0.046 log P units or 31.1 kPa) for the training, test, validation, and overall sets, respectively.

3.3 Water Solubility

Of the various parameters affecting the fate and transport of organic chemicals in the ecosystems, water solubility is one of the most important. Highly soluble chemicals are easily and rapidly distributed in the environment. These chemicals tend to have relatively low adsorption coefficients in soils and sediments and also negligible bioconcentration factors in living species. They tend to be more readily biodegradable by microorganisms. The water solubility of chemicals also influences their photolysis, hydrolysis, oxidation, and volatilization [23]. A quite large number of estimation methods have been proposed for modeling the water solubility of organic chemicals, and some of them are based on the use of ANNs. Thus, a database of water solubility values for 157 substituted aromatic hydrocarbons described from structural fragments was randomly split into a LS, a CVS, and an ETS of 95, 31, and 31 chemicals, respectively [45]. A TLP trained by the backpropagation algorithm was used as statistical engine. The best model was a 9/11/1 ANN (learning rate 0.35, 276 cycles) yielding a mean square error (MSE) of 0.21 from 40 randomly selected test data sets. For comparison purpose, the MSE obtained with a regression analysis was 0.25. A rather similar approach was used by Sutter and Jurs [46] from solubility data for 140 organic compounds presenting diverse structures, which were divided into a LS, a CVS, and an ETS of 116, 11, and 13 chemicals, respectively. Chemicals were described by means of 144 descriptors encoding topological and/or physicochemical properties. This pool of descriptors was reduced to nine that were used for deriving a regression model from a LS of 127 (116 + 11) chemicals. An RMS error of 0.321 log units was found. However, four chemicals were detected as outlier, and their removal from the regression model allowed to obtain an RMS error of 0.277 log units. A 9/3/1 TLP including the nine descriptors as input neurons was then derived. It gave RMS errors of 0.217, 0.282, and 0.222 log units for the LS ($n = 112$), CVS ($n = 11$), and ETS ($n = 13$), respectively. It is noteworthy that another 9/3/1 TLP model was computed by Sutter and Jurs [46] after exclusion of the polychlorinated biphenyls (PCBs). In that case, RMS error values of 0.145, 0.151, and 0.166 log units were obtained for the LS ($n = 94$), CVS ($n = 13$), and ETS ($n = 13$), respectively. Other TLP models for predicting the aqueous solubility of chemicals were proposed by Mitchell and Jurs [47], McElroy and Jurs [48], and Huuskonen et al. [49] from databases of limited sizes. Yaffe and coworkers [50] used a heterogeneous set of 515 organic compounds with their solubility data for comparing the performances of a TLP and fuzzy ARTMAP ANNs. The first ANN model derived from a large diverse set of aqueous solubility data was proposed by Huuskonen [51]. A database of 1,297 chemicals with their aqueous solubility values was split into a TS and an ETS of 884 and 413 chemicals, respectively. Another testing set (ETS+) of 21 chemicals was also considered. All the chemicals were encoded from the 30 following topological indices: 24 atom-type electrotopological state indices [33], path 1 simple and valence connectivity indices [44], flexibility index, the number of H-bond acceptors, and indicators of aromaticity and for aliphatic hydrocarbons. A 30/12/1 TLP trained by the backpropagation algorithm yielded standard deviation values of 0.47, 0.60, and 0.63 for the LS, ETS, and ETS+, respectively. A regression analysis performed under the same conditions gave standard deviation values of 0.67, 0.71, and 0.88 for the LS, ETS, and ETS+, respectively [51]. Liu and So [52] tried to derive an ANN with fewer connections but presenting similar performances by using a LS and an ETS of 1,033 and 258 chemicals, respectively. A 7/2/1 TLP with the 1-octanol/walter partition coefficient (log P), topological polar surface area (TPSA), molecular weight, and four topological indices as input neurons gave standard deviation values of 0.70 and 0.71 in log units for the LS and ETS, respectively. An interesting hybrid model was proposed by Hansen and coworkers [53] developed for the prediction of pH-dependent aqueous solubility of chemicals. It used a TLP ANN trained from 4,548 solubility values and a commercial software tool for estimating the acid/base dissociation coefficients.

It is important to note that the aqueous solubility estimations obtained from QSPR models have to be used with caution. Thus, for example, QSPR models generally calculate solubility in pure water at 25°C while it is well-known that the varying temperatures found in the environment change the solubility of chemicals. The degree of salinity of the aquatic ecosystems also influences the solubility of the chemicals in these media.

3.4 Henry's Law Constant

The Henry's law constant (Hc) of a chemical is defined as the ratio of its concentration in air to its concentration in water when these two phases are in contact and equilibrium distribution of the chemical is achieved [25]. Hc is of first importance for assessing the environmental distribution of chemicals. The different methods allowing to calculate this parameter have been reviewed by Dearden and Schüürmann [54]. Among them, two studies deal with the use of ANNs for modeling the Hc of chemicals at 25°C. A database of 357 organic chemicals with their log H values ranged from −7.08 to 2.32 was used by English and Carroll [55] for deriving their ANN models. Chemicals were described by 29 descriptors including topological indices, physicochemical properties, and atomic and group contributions. The best results were obtained in 3,000 cycles with a 10/3/1 TLP. The standard errors for the LS ($n = 261$), CVS ($n = 42$), and ETS ($n = 54$) were equal to 0.202, 0.157, and 0.237 log units, respectively. Comparatively, the standard errors obtained with a regression analysis, performed according to the same conditions, were 0.262 and 0.285 log units for the LS ($n = 303$) and ETS ($n = 54$), respectively.

Experimental Hc at 25°C for a diverse set of 495 chemicals were collected by Yaffe et al. [56]. The log H values ranged from −6.72 to 2.87. Six physicochemical descriptors (heat of formation, dipole moments, ionization potential, average polarizability) and the second-order valence molecular connectivity index were used as input parameters for a fuzzy ARTMAP ANN and a TLP ANN trained by the back-propagation algorithm. The average absolute error values obtained with the fuzzy ARTMAP ANN were 0.01 and 0.13 for the LS ($n = 421$) and ETS ($n = 74$). The selected 7/17/1 TLP yielded average absolute error values of 0.29, 0.28, and 0.27 for the LS ($n = 331$), validation set ($n = 421$) and ETS ($n = 74$).

3.5 Octanol/Water Partition Coefficient

In 1872, Berthelot [57] undertook the study of partitioning as a purely physicochemical phenomenon. He was the first to collect the evidence proving that the ratio of the concentrations of small solutes when distributed between water and an immiscible solvent (e.g., ether) remained constant even when the solvent ratios varied widely [58]. In 1891, Nernst [59] put this type of equilibrium on a firmer thermodynamic basis. About a decade later, Meyer [60] and Overton [61], who showed that the narcotic action of simple chemicals was reflected rather closely by their oil–water partition coefficients, initiated the use of this physicochemical property for deriving structure–activity relationships. In the first part of the twentieth century, many different organic solvent/water systems were tested to derive structure–activity relationships. However, in 1962–1964, the 1-octanol was adopted as solvent of choice after the pioneering works of Hansch and coworkers in quantitative structure–activity relationships (QSARs) [62, 63] demonstrating that the 1-octanol/water partition coefficient (Kow) could provide a rationalization for the interaction of organic chemicals with living organisms or for biological processes occurring in organisms [64]. Kow is simply defined as the ratio of a chemical's concentration in the octanol phase to its concentration in the aqueous phase of a two-phase octanol/water system. Values of Kow are thus unitless and are expressed in a logarithmic form (i.e., log Kow or log P) when used in pharmaceutical and environmental modeling. There are numerous methods available for the experimental measurement of log P as well as for its estimation from contribution methods or from linear and nonlinear QSPRs [23, 24, 58, 64, 65]. Different ANN models for log P have been derived from a limited number of chemicals (see e.g., [66–68]). A database of 1,870 log P values for structurally diverse chemicals was used by Huuskonen and coworkers [69] for deriving a log P model based on atom-type electrotopological state indices [33] and a TLP. It was split into a LS and an ETS of 1,754 and 116 molecules, respectively. The best configuration included the molecular weight and 38 electrotopological state indices as input neurons, five hidden neurons, and bias neurons. Averaged results of 200 ANN simulations were used to calculate the final outputs. With this strategy, RMS (LOO) values of 0.46 and 0.41 were obtained for the LS and ETS, respectively. This model was further refined from an extended LS and is now called ALOGPS [70, 71]. A log P model was designed by Devillers and coworkers [72–74] from a TLP trained by the backpropagation algorithm using 7,200 log P values for the learning process. Experimental log P values were retrieved from original publications or unpublished results. The log P values of the LS ranged between −3.7 and 9.95 with a mean of 2.13 and a standard deviation of 1.65. Molecules were described by means of autocorrelation descriptors [75, 76] encoding lipophilicity (H) defined according to Rekker and Mannhold [65], molecular refractivity (MR), and H-bonding donor (HBD) and H-bonding acceptor (HBA) abilities. Prior to calculations, data were scaled with a classical min/max equation. The optimal architecture and set of parameters for the neural network model were determined by means of a trial and error procedure. The different training exercises were monitored with a validation set of 200 molecules presenting a high structural diversity but not deviating too much from the chemical structures included in the training set. This procedure showed that a neural network model with 35 input neurons (i.e., H_0 to H_{14}, MR_0 to MR_{14}, HBA_0 to HBA_3, and HBD_0) was necessary to correctly describe the molecules and model the 7,200 experimental log P values. The hidden layer consisted of 32 neurons. It was found that a learning rate of 0.5 and a momentum term of 0.9 always gave good neural network generalization within ca. 5,500 cycles. A composite network constituted of four configurations was selected as final model (RMS $= 0.37$, $r = 0.97$) because it allowed to obtain the best simulation results on an ETS of 519 chemicals (RMS $= 0.39$, $r = 0.98$). It has been shown that this model competed favorably with other log P models [77,78] and was particularly suited for estimating the log P values of pesticides [79]. It is noteworthy that a commercial version of this model called AUTOLOGP^{TM} is available [80, 81].

3.6 Degradation Parameters

Biodegradation is an important mechanism for eliminating xenobiotics by biotransforming them into simple organic and inorganic products. Two types of biodegradation can be distinguished. The primary biodegradation denotes a simple transformation not leading to a complete mineralization. The biodegradation products are specifically measured from chromatographic methods, and the results are expressed by means of kinetic parameters such as biodegradation rate constant (k) and half-life ($T_{1/2}$). The ultimate (or total) biodegradation totally converts chemicals into simple molecules such as CO_2 and H_2O. Biodegradation tests are time consuming, expensive, and their results are difficult to interpret because they depend on numerous parameters linked to the experimental conditions such as the nature and concentration of the inoculum, cultivation, and adaptation of the microbial culture, concentration of the test substance [82–84]. Because ANNs are particularly suited for modeling noisy data, they have been successfully used to model biodegradation processes [85]. Thus, for example, 47 molecules presenting a high degree of heterogeneity were described in a qualitative way for their biodegradability (i.e., $0 = $ weak, $1 = $ high) from a survey made by 22 experts in microbial degradation [86]. They were encoded from 11 Boolean descriptors representing structural features associated with persistent or degradable chemicals. These descriptors are listed in Table 2. A TLP trained by the backpropagation algorithm was used as statistical engine to find a relationship between the structure of the

Table 2 Boolean descriptors[a] used as input neurons in a TLP designed for predicting the biodegradability of chemicals

N°	Descriptor
1	Heterocycle N
2	Ester, amide, anhydride
3	$\geq$2 Cl
4	Bicyclic alkane
5	Only C, H, N, O
6	NO_2
7	$\geq$2 cycles
8	Epoxide
9	Primary or aromatic OH
10	Molecular weight <235.9 g mol^{-1}[b]
11	O bound to C

[a] 1 when present in the molecule and 0 otherwise
[b] Mean calculated from the training set

molecules and their biodegradation potential. The learning phase yielded 100% of good classification (i.e., 47/47) with a 11/4/1 ANN in 500 cycles. The predictive power of this model was estimated from two ETSs. With the former ETS, 78% of good classifications (i.e., 18/23) were obtained while with the latter, 94% (i.e., 16/17) of the chemicals were correctly classified. The use of Boolean descriptors as input neurons in a TLP especially for modeling a complex property can induce problems of overfitting. To avoid this drawback without losing the interest of fragment descriptors, the usefulness of correspondence factor analysis [87] for reducing the dimensionality of a data matrix was tested. Thus, a CFA was used to scale the 47×11 Boolean matrix and the CFA factors were directly introduced as inputs in the ANN. Same results were obtained also in 500 cycles with only the first seven factors (87.9% of the total inertia). It is noteworthy that an intercommunicating hybrid system including this ANN model and a genetic algorithm [88] was then constructed for designing molecules with specific biodegradability characteristics [89].

TLPs with structural descriptors [90, 91] or autocorrelation descriptors [92] were used for modeling the biodegradability of other sets of aliphatic and aromatic chemicals. The field half-lives of 110 pesticides were modeled using a TLP trained by the backpropagation algorithm [93]. Because periodicities in agricultural calendars are measured in days, weeks, and months (i.e., seasons), the field half-lives ($T_{1/2}$) of pesticides were divided into the three following classes: Class 1 (encoded 100 in the ANN output) contained pesticides with $T_{1/2} \leq 10$ days, class 2 (encoded 010) included pesticides with 10 days $< T_{1/2} \leq 30$ days, and class 3 (encoded 001) included pesticides with 30 days $< T_{1/2} \leq 90$ days. Molecules were described by means of the frequency of 17 structural fragments. Different scaling transformations were tested but the best results were obtained with a CFA, which also allowed a reduction of the dimensionality of the descriptor matrix. The optimal results were obtained by using the first 12 factors (95.8% of the total inertia) as input neurons and seven neurons for the hidden layer. With this configuration, 95.5% of correct classifications were obtained with the LS. The performances of the selected ANN model were tested from an ETS of 13 pesticides representing the three classes of field half-lives. The testing phase with CFA gave 84.6% of correction predictions. A discriminant factor analysis at three classes was performed for comparison purposes. In that case, 60% and 53.8% of good classifications were obtained for the LS and ETS, respectively [93].

4 Use of ANNs in Quantitative Structure–Activity Relationship (QSAR) Modeling

The knowledge about systematic relationships between the structure of chemicals and their biological activity dates back to the prime infancy of the modern pharmacology and toxicology. Thus, for example, Cros [94] stressed, in the last page of his thesis published in 1863, an empirical relationship between the number of carbon and hydrogen atoms in a series of alcohols and their solubility in water and toxicity. Until about the middle of the twentieth century, most of these structure–activity relationships were only qualitative. The dramatic change resulted from the systematic use, in the early 1960s, of linear regression analysis for correlating biological activities of congeneric series of molecules with their physicochemical properties or some of their structural features encoded by means of Boolean descriptors (i.e., 0/1). These contributions started the development of two QSAR methodologies later termed Hansch analysis [62,63] and Free-Wilson analysis [95], respectively.

Nowadays, regression analysis remains the most widely used statistical tool for deriving QSARs, even if most of the basic statistical assumptions for its correct use are often not satisfied with numerous data sets [96]. In addition, the choice of regression analysis can also be annoying because a postulate is made that only linear relationships exist between the variables involved in the modeling process, while generally it is not true. Since about one decade, ANNs have become the focus of much attention in QSAR to find complex relationships between the structure of molecules and their toxicity. These models have been derived on various organisms such as the marine luminescent bacterium *Vibrio fischeri* (formerly known as *Photobacterium phosphoreum*) [97,98], the freshwater protozoan *Tetrahymena pyriformis* [99–110], the waterflea *Daphnia magna* [111], the freshwater amphipod *Gammarus fasciatus* [112], the midge *Chironomus riparius* [113], the fathead minnow *Pimephales promelas* [114–122], the rainbow trout *Oncorhynchus mykiss* [123], the bluegill *Lepomis macrochirus* [124], and the honey bee *Apis mellifera* [125, 126]. All these models were recently analyzed [127]. Consequently, only the main characteristics of some of them are presented in Table 3.

It is interesting to note that due to their high flexibility and their ability to find complex relationships between variables, ANNs can be used to derive QSARs from sets of variables encoding, as usual, the structure and physicochemical properties

Table 3 Selected ANN QSAR models derived from noncongeneric data sets

Species[a]	n	Learning/testing	ANN	Reference
V.f.	747	454/150 + 143	TLP	[97]
V.f.	1,308	1,068/240	TLP	[98]
T.p.	825	600/150 + 75	Probabilistic	[108]
T.p.	1,084	1,000/84	Probabilistic	[109]
T.p.	1,371	914/457	TLP	[110]
D.m.	776	700/76	Probabilistic	[111]
P.p.	865	80–20%	Probabilistic	[117]
P.p.	886	800/86	Probabilistic	[118]
P.p.	562	392/170	BP + Fuzzy-ANN	[120]
P.p.	551	LOO + 80 − 20% + 541/10	Counterpropagation	[121]
P.p.	569	484/85	Reg-TLP	[122]

[a] *V.f. Vibrio fischeri, T.p. Tetrahymena pyriformis, D.m. Daphnia magna, P.p. Pimephales promelas, P.r. Poecilia reticulata*

of the molecules but also the experimental conditions in which the different tests are performed such as the time of exposure [98] or the temperature, pH, hardness of the medium, and size of the organisms [112, 123, 124]. In the same way, due to their pure nonlinear nature, ANNs can be used in synergy with another statistical tool, especially regression analysis. Devillers [122] showed that this kind of modeling approach was particularly interesting in the common situation in which the toxicity of molecules mainly depended on their log P. In that case, in a first step, a classical regression equation with log P is derived. The residuals obtained with this simple linear equation are then modeled from a TLP including different molecular descriptors as input neurons. Finally, results produced by the linear and nonlinear QSAR models are both considered for calculating the toxicity values, which are then compared with the initial toxicity data.

5 Use of ANNs for Modeling Environmental Contaminations

5.1 Air Pollution

There is a large body of evidence suggesting that exposure to air pollution, even at the levels commonly achieved nowadays in the industrial countries, leads to adverse health effects. In particular, exposure to pollutants such as particulate matter and ozone has been found to be associated with increases in hospital admissions for cardiovascular and respiratory diseases and to the incidence of cancers [128]. Air pollution not only affects the quality of the air we breathe, but it also directly and indirectly impacts the biotopes and the biocenoses constituting the aquatic and terrestrial ecosystems. For the evaluation of air pollution events in a particular geographical area, it is crucial to have a powerful mapping technique allowing to perform typologies, compare sampling sites, and so on. The Kohonen self-organizing map (KSOM) [16] is particularly suited to perform these tasks. Thus, for example, Ferré-Huguet and coworkers [129] used a KSOM to assess the environmental impact and human health risks of polychlorinated dibenzo-p-dioxins and dibenzofurans in the vicinity of a new hazardous waste incinerator in Spain 4 years after regular operation of the facility. More specifically, KSOM, which was a 48 (8×6) rectangular grid, was applied to soil and herbage samples to establish pattern similarities among the samples as well as to identify hot spots near the plant. Lee and coworkers [130] used a KSOM of 150 (15×10) output neurons to examine the influence of urbanization on the assembly patterns of 52 breeding birds in 367 sites.

Undoubtedly KSOM offers an interesting tool for data compression of p multivariate samples defined in an n-dimensional space into v clusters (loaded neurons). This data reduction to a few clusters provides an optimal data structure display. However, in KSOM, the problem is that information about the correct distance between the neurons disappears during the projection onto the 1, 2, or 3D array of nodes. To overcome this problem, a minimum spanning tree (MST) [131] can be calculated between the loaded neurons of a trained KSOM to visualize the shortest distances between them. The hybridization of the KSOM and MST algorithms constitutes the basis of the 3MAP algorithm designed and used by Wienke for locating fine airborne particle sources [132–135]. It is noteworthy that because there remains information not represented, about the correct distances between all the loaded neurons, a nonlinear mapping (NLM) [136] performed on these loaded neurons can be used to visualize all the distances separating them. The hybridization of the KSOM, MST, and NLM algorithms constitutes the basis of the N2M algorithm [137, 138] (Fig. 5). A rather similar hybridization approach in combination with a multilayer perceptron (MLP) was used by Kolehmainen and coworkers [139] to forecast urban air quality. Hourly airborne pollutant and meteorological averages collected during the years 1995–1997 were analyzed to identify air quality episodes having typical and the most probable combinations of air pollutants and meteorological variables. This modeling was performed from KSOM, NLM, and fuzzy distance metrics. Several overlapping MLPs were then applied to the clustered data, each representing a pollution episode.

KSOM is not the unique ANN clustering technique that was used to visualize air pollution events. Thus, Owega and coworkers [140] used cluster analysis and an adaptive resonance theory (ART-2a) [141] ANN to classify back trajectories of air

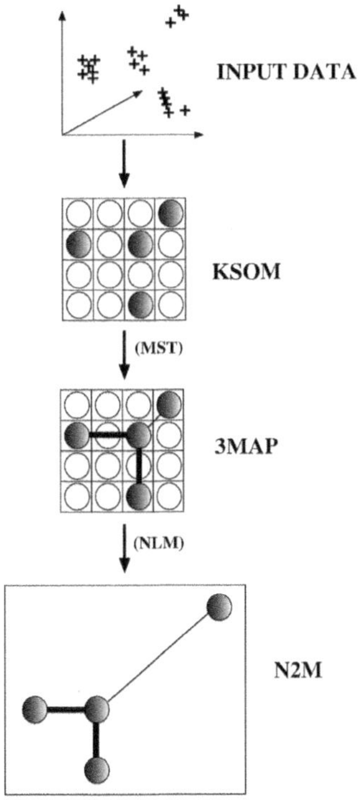

Fig. 5 N2M algorithm flow diagram (adapted from [140])

masses arriving in Toronto (Canada) into distinct transport patterns. Spencer and coworkers [142] also used an ART-2a ANN to analyze ambient aerosol particles in Riverside (California).

Numerous MLPs have been used alone or in combination or in competition with other statistical approaches for estimating various atmospheric pollution events. Some examples are given in Table 4 [143–150].

Table 4 Examples of MLP models designed for estimating atmospheric pollution events

Atmospheric pollution event	Reference
Hourly levels up to 8 h ahead for SO_2, CO, NO_2, NO, and O_3 and six locations in the area of Bilbao (Spain)	[143]
SO_2, PM10[a], CO levels for the next 3 days in Istanbul (Turkey)	[144]
SO_2 concentrations in Istanbul (Turkey)	[145]
NO_2 concentrations at three sites in Kolkata (India)	[146]
Hourly concentrations of NO_2 at a traffic station in Helsinki (Finland)	[147]
Ozone concentrations	[148–155]
PM10 concentrations in the city of Thessaloniki (Greece)	[156]
PM10 concentrations in the urban area of Volos (Greece)	[157]
PM2.5 concentrations in downtown Santiago (Chile)	[158]
PM2.5 concentrations on the US-Mexico border	[159]
Lead concentrations in grasses from urban descriptors in Athens (Greece)	[160]
NO_2 dispersion from vehicular exhaust emissions	[161]
Mercury speciation in combustion flue gases	[162]
Benzene concentrations in a street canyon	[163]
Benzene concentrations with an electronic nose	[164]
Odor thresholds for chemicals of environmental and industrial concern	[165]

[a]Particulate matter with aerodynamic diameter less than 10 microns (PM10) or less than 2.5 microns (PM2.5)

5.2 Aquatic Contaminations

The worldwide environmental problem of eutrophication in lenthic ecosystems is caused by an unbalanced increase in the nutrient inflow due to the human activities. Indeed, when the nutrient concentration increases under high-temperature conditions in a lake during the summertime, certain microalgae can overgrow yielding the production of blooms, which can cause water discolorations, mortality in fish and invertebrates as well as in humans because of the production of harmful toxins [166]. It is obvious that these deleterious effects could be prevented or at least minimized if the algal blooms could be predicted in an early stage. Different ANNs have been used to reach this goal. Thus, Recknagel and coworkers [167] used a TLP trained by the backpropagation algorithm for modeling algal bloom in three lakes and a river. The lakes, located in Japan and Finland, were of different characteristics including a variety of nutrient levels, light and temperature conditions, depth and water retention time. The river was located in Australia. Four different ANNs were computed. Different parameters such as concentration in nitrate, water temperature, concentration in chlorophyll a, and concentration in dissolved oxygen were used as input neurons. The dominating algal species (in number of cells/mL or mg/L for the Finnish lake) were considered as output neurons. One or two hidden layers having a maximum of 20 neurons per layer were used to distribute the information within the networks. The ANNs were trained for 500,000 cycles with measured input and output data from 6 to 10 years. For the validation of model predictions, data of 2 independent years were used for each ANN model. More realistic and optimized models were proposed by Lee and coworkers [168] for predicting the algal bloom dynamics for two bays in the eutrophic coastal waters of Hong Kong. A TLP was also used as statistical engine. Biweekly water quality data were tested as input neurons. Concentration in chlorophyll-a or cell concentration of *Skeletonema* were used as output neurons in each ANN model. Data collected in different years were used to train (3,000 cycles) and test the two ANN models. Different combinations of parameters were tested as inputs but in both cases, the best results were obtained by only using the time-lagged chlorophyll-a or log (*Skeletonema* (cells/l) as input neurons. This work clearly suggested that the algal concentration in the eutrophic subtropical coastal waters was mainly dependent on the antecedent algal concentrations in the previous 1–2 weeks.

Oh and coworkers [169] used a KSOM for patterning algal communities and then a TLP for identifying important factors causing algal blooms in Daechung reservoir (Korea). Thirty-nine samples were used for KSOM analysis. The patterns of the sample communities were investigated on the basis of community abundance data (Cyanophyceae, Chlorophyceae, Bacillariophyceae, and others) in percentages for 1999 and 2003. The best arrangement of the output layer of 24 (6×4) neurons was a hexagonal lattice. Interestingly, a hierarchical cluster analysis, based on Ward algorithm and using the Euclidean distance, was performed on the KSOM units. Analysis of the results showed that the clustering was based on the phytoplankton communities and sampling time. A TLP was used to predict the chlorophyll-a concentration and abundance of Cyanophyceae from environmental factors including the total nitrogen, total dissolved nitrogen, total particulate nitrogen, total phosphorus, total dissolved phosphorus, total particulate phosphorus, temperature, DO, pH, conductivity, turbidity, Secchi depth, precipitation, and daily irradiance. Data were collected from 54 samples over 3 years. Gradient descent optimization was used for error reduction. The best models for chlorophyll-a concentration and abundance of Cyanophyceae were 14/3/1 and 14/6/1 TLPs. The predictive performances of the models were not estimated from an ETS. Conversely, a sensitivity analysis was performed to determine the most influential variables. Results showed that they were different for the two TLP ANNs.

Lenthic and lotic ecosystems are also contaminated by numerous xenobiotics resulting from agricultural and industrial activities. Thus, pesticides are used to control weeds, insects, and other organisms in a wide variety of agricultural and nonagricultural settings yielding their release into the environment including the aquatic compartment. Among the collection of models available for predicting the environmental fate and effects of pesticides, some of them are based on nonlinear methods, especially the ANNs. Thus, for example, Kim and coworkers [170] coupled wavelet analysis and a TLP trained by the backpropagation algorithm for modeling the movement behavior of *Chironomus samoensis* larvae in response to treatments of carbofuran at 0.1 mg/L in seminatural conditions. Various ANN paradigms have been also used for modeling the contamination of groundwater by pesticides and other anthropic pollutants [171–176].

Samecka-Cymerman and coworkers [177] used a KSOM to perform a typology of three species of aquatic bryophytes (*Fontinalis antipyretica*, *Platyhypnidium riparioides*, *Scapania undulata*) according to their concentration in Al, Be, Ca, Cd, Co, Cr, Cu, Fe, K, Mg, Mn, Ni, Pb, and Zn. The sampling sites were divided into three groups depending on the type of rock basement of the stream. Sampling sites in group one consisted of granites and gneisses ($n = 21$), those in group two of sandstones ($n = 5$), and those in group three of limestones and dolomites ($n = 26$). The output layer of 5×5 neurons visualized by hexagonal cells showed that the bryophytes were clustered according to their sampling origin. There was no difference between the bryophytes from the three types of rock in terms of concentrations in Be, Fe, K, Co, and Cu. Conversely, bryophytes growing in streams flowing through granites/gneisses contained significantly higher concentrations of Cd and Pb, while bryophytes from streams flowing through sandstones contained significantly higher concentrations of Cr. Bryophytes from group three were characterized by high concentrations in Ca and Mg. These results were confirmed from a PCA.

Last, it is noteworthy that ANNs have been used in the areas of wastewater treatment and analyses [178–180].

5.3 *Soil and Sediment Contaminations*

Soils and sediments can be contaminated by various pollutants released into the environment from a number of anthropogenic sources. ANNs have shown their interest for characterizing and/or quantifying these contaminations. Thus, for example, in

Winter 2002, 24 soil and 12 wild chard (*Beta vulgaris*) samples were collected by Nadal et al. [181] in Tarragona County (Catalonia, Spain). Soil sampling points were chosen as follows: 15 in the industrial complex (8 in the vicinity of chemical industries and 7 near petroleum refineries), 5 in Tarragona downtown and its residential area, and 4 in presumably unpolluted zones. The number of wild chard samples collected from industrial, residential, and unpolluted areas were 6, 3, and 3, respectively. The samples were analyzed for their concentrations in As, Cd, Cr, Hg, Mn, Pb, and V. In chard samples, significant differences between areas were only found for vanadium (V). Regarding the soil samples, the differences and concentrations between the three zones were higher. A KSOM was successfully used to perform their typology according their differences in metal concentrations. The same type of methodology based on KSOM was applied by Arias and coworkers [182] for evaluating the pollution level in Cu, Mn, Ni, Cr, Pb, and Zn of the sediments dredged from the dry dock of a former shipyard in the Bilbao estuary (Bizkaia, Spain). KSOM was compared with different cluster analysis algorithms to classify 407 samples of various origins contaminated by polychlorinated dibenzodioxins and polychlorinated dibenzofurans [183].

Other ANN paradigms were used to model soil and sediment contaminations. Thus, for example, Kanevski [184] tested the usefulness of general regression ANNs, based on kernel statistical estimators for predicting the soil contamination in Cs137 in Western part of Briansk region following Chernobyl accident.

6 Conclusion

On the basis of a computing model similar to the underlying structure of a mammalian brain, ANNs share the brain's ability to learn or adapt in responses to external inputs. When exposed to a stream of training data, they can uncover previously unknown relationships and learn complex mappings in the data. Under these conditions, ANNs provide interesting alternatives to well-established linear methods commonly used in ecotoxicology modeling. In this chapter, different ANN models computed for predicting the environmental fate and effects of chemicals are presented. Our goal was not to catalog all the models in the field but only to show the diversity of the situations in which these nonlinear tools have proved their interest. Their correct use requires to have some practical experience for architecture and parameter setting as well as to interpret the modeling results. They also need to respect some rules dealing with the size of the data sets, the constitution of learning and testing sets, and so on. Despite these limitations, it is obvious that their use in ecotoxicology modeling will continue to grow, especially in combination with other linear and nonlinear statistical methods to create powerful hybrid systems.

References

1. Bigus JP (1996) Data mining with neural networks. Solving business problems – From application development to decision support. McGraw-Hill, New York
2. Bengio Y (1996) Neural networks for speech and sequence recognition. International Thomson Computer Press, London
3. Zupan J, Gasteiger J (1993) Neural networks for chemists. An introduction. VCH, Weinheim
4. Devillers J (1996) Neural networks in QSAR and drug design. Academic Press, London
5. Guegan JF, Lek S (2000) Artificial neuronal networks: Application to ecology and evolution. Springer, New York
6. Christopoulos C, Georgiopoulos M (2001) Applications of neural networks in electromagnetics. Artech House Publishers, London
7. Lisboa PJG, Taylor MJ (1993) Techniques and applications of neural networks. Ellis Horwood, London
8. Wade N (1998) The science Times book of the brain. Lyons Press, New York
9. Müller B, Reinhardt J, Strickland MT (1995) Neural networks. An introduction. Springer, Berlin
10. Pal SK, Srimani PK (1996) Neurocomputing. Motivation, models, and hybridization. Computer March: 24–28
11. Jain AK, Mao J, Mohiuddin KM (1996) Artificial neural networks: A tutorial. Computer March: 31–44
12. Wasserman PD (1989) Neural computing: Theory and practice. Van Nostrand Reinhold, New York
13. Pao YH (1989) Adaptive pattern recognition and neural networks. Addison-Wesley Publishing Company, Reading
14. Eberhart RC, Dobbins RW (1990) Neural network PC tools. A practical guide. Academic Press, San Diego
15. Wasserman PD (1993) Advanced methods in neural computing. Van Nostrand Reinhold, New York
16. Kohonen T (1995) Self-organizing maps. Springer, Berlin
17. Fiesler E, Beale R (1997) Handbook of neural computation. IOP Publishing Ltd, Bristol
18. Devillers J, Doré JC (2002) e-statistics for deriving QSAR models. SAR QSAR Environ Res 13: 409–416
19. Domine D, Devillers J, Chastrette M, Karcher W (1992) Multivariate structure-property relationships (MSPR) of pesticides. Pestic Sci 35: 73–82
20. Samiullah Y. (1990) Prediction of the environmental fate of chemicals. Elsevier, London
21. Mackay D, Di Guardo A, Hickie B, Webster E (1997) Environmental modelling: Progress and prospects. SAR QSAR Environ Res 6: 1–17
22. Hemond HF, Fechner EJ (1994) Chemical fate and transport in the environment. Academic Press, San Diego
23. Lyman WJ, Reehl WF, Rosenblatt DH (1990) Handbook of chemical property estimation methods. American Chemical Society, Washington, DC
24. Reinhard M, Drefahl A (1999) Handbook for estimating physicochemical properties of organic compounds. Wiley, New York
25. Boethling RS, Howard PH, Meylan WM (2004) Finding and estimating property data for environmental assessment. Environ Toxicol Chem 23: 2290–2308
26. Cronin MTD, Livingstone DJ (2004) Calculation of physicochemical properties. In: Cronin MTD, Livingstone DJ (eds) Predicting chemical toxicity and fate, CRC Press, Boca Raton
27. Dearden JC (2003) Quantitative structure-property relationships for prediction of boiling point, vapor pressure, and melting point. Environ Toxicol Chem 22: 1696–1709
28. Lohninger H (1993) Evaluation of neural networks based on radial basis functions and their application to the prediction of boiling points from structural parameters. J Chem Inf Comput Sci 33: 736–744

29. Cherqaoui D, Villemin D, Mesbah A, Cense JM, Kvasnicka V (1994) Use of a neural network to determine the normal boiling points of acyclic ethers, peroxides, acetals and their sulfur analogues. J Chem Soc Faraday Trans 90: 2015–2019
30. Balaban AT, Basak SC, Colburn T, Grunwald GD (1994) Correlation between structure and normal boiling points of haloalkanes C_1–C_4 using neural networks. J Chem Inf Comput Sci 34: 1118–1121
31. Egolf LM, Wessel MD, Jurs PC (1994) Prediction of boiling points and critical temperatures of industrially important organic compounds from molecular structure. J Chem Inf Comput Sci 34, 947–956
32. Wessel MD, Jurs PC (1995) Prediction of normal boiling points of hydrocarbons from molecular structure. J Chem Inf Comput Sci 35: 68–76
33. Kier LB, Hall LH (1999) The electrotopological state; Structure modeling for QSAR and database analysis. In: Devillers J, Balaban AT (eds) Topological indices and related descriptors in QSAR and QSPR, Gordon and Breach Publishers, Amsterdam
34. Hall LH, Story CT (1996) Boiling point and critical temperature of a heterogeneous data set: QSAR with atom type electrotopological state indices using artificial neural networks. J Chem Inf Comput Sci 36: 1004–1014
35. Hall LH, Story CT (1997) Boiling point of a set of alkanes, alcohols and chloroalkanes: QSAR with atom type electrotopological state indices using artificial neural networks. SAR QSAR Environ Res 6: 139–161
36. Espinosa G, Yaffe D, Cohen Y, Arenas A, Giralt F (2000) Neural network based quantitative structural property relations (QSPRs) for predicting boiling points of aliphatic hydrocarbons. J Chem Inf Comput Sci 40: 859–879
37. Chalk AJ, Beck B, Clark T (2001) A quantum mechanical/neural net model for boiling points with error estimation. J Chem Inf Comput Sci 41: 457–462.
38. Anonymous (1998) QSARs in the assessment of the environmental fate and effects of chemicals. Technical report no. 74, ECETOC, Brussels
39. Liang C, Gallagher DA (1998) QSPR prediction of vapor pressure from solely theoretically-derived descriptors. J Chem Inf Comput Sci 38: 321–324
40. McClelland HE, Jurs PC (2000) Quantitative structure-property relationships for the prediction of vapor pressures of organic compounds from molecular structures. J Chem Inf Comput Sci 40: 967–975
41. Beck B, Breindl A, Clark T (2000) QM/NN QSPR models with error estimation: Vapor pressure and log P. J Chem Inf Comput Sci 40: 1046–1051
42. Chalk AJ, Beck B, Clark T (2001) A temperature-dependent quantum mechanical/neural net model for vapor pressure. J Chem Inf Comput Sci 41: 1053–1059
43. Yaffe D, Cohen Y (2001) Neural network based temperature-dependent quantitative structure property relations (QSPRs) for predicting vapor pressure of hydrocarbons. J Chem Inf Comput Sci 41: 463–477
44. Hall LH, Kier LB (1999) Molecular connectivity Chi indices for database analysis and structure-property modeling. In: Devillers J, Balaban AT (eds) Topological indices and related descriptors in QSAR and QSPR, Gordon and Breach Publishers, Amsterdam
45. Chow H, Chen H, Ng T, Myrdal P, Yalkowsky SH (1995) Using backpropagation networks for the estimation of aqueous activity coefficients of aromatic organic compounds. J Chem Inf Comput Sci 35: 723–728
46. Sutter JM, Jurs PC (1996) Prediction of aqueous solubility for a diverse set of heteroatom-containing organic compounds using a quantitative structure-property relationship. J Chem Inf Comput Sci 36: 100–107
47. Mitchell BE, Jurs PC (1998) Prediction of aqueous solubility of organic compounds from molecular structure. J Chem Inf Comput Sci 38: 489–496
48. McElroy NR, Jurs PC (2001) Prediction of aqueous solubility of heteroatom-containing organic compounds from molecular structure. J Chem Inf Comput Sci 41: 1237–1247
49. Huuskonen J, Salo M, Taskinen J (1998) Aqueous solubility prediction of drugs based on molecular topology and neural network modeling. J Chem Inf Comput Sci 38: 450–456

50. Yaffe D, Cohen Y, Espinosa G, Arenas A, Giralt F (2001) A fuzzy ARTMAP based on quantitative structure-property relationships (QSPRs) for predicting aqueous solubility of organic compounds. J Chem Inf Comput Sci 41: 1177–1207
51. Huuskonen J (2000) Estimation of aqueous solubility for a diverse set of organic compounds based on molecular topology. J Chem Inf Comput Sci 40: 773–777
52. Liu R, So SS (2001) Development of quantitative structure-property relationship models for the early ADME evaluation in drug discovery. 1. Aqueous solubility. J Chem Inf Comput Sci 41: 1633–1639
53. Hansen NT, Kouskoumvekaki I, Jørgensen FS, Brunak S, Jonsdottir SO (2006) Prediction of pH-dependent aqueous solubility of druglike molecules. J Chem Inf Comput Sci 46: 2601–2609
54. Dearden JC, Schüürmann G (2003) Quantitative structure-property relationships for predicting Henry's law constant from molecular structure. Environ Toxicol Chem 22: 1755–1770
55. English NJ, Caroll DG (2001) Prediction of Henry's law constants by a quantitative structure property relationship and neural networks. J Chem Inf Comput Sci 41: 1150–1161
56. Yaffe D, Cohen Y, Espinosa G, Arenas A, Giralt F (2003) A fuzzy ARTMAP-based quantitative structure-property relationship (QSPR) for the Henry's law constant of organic compounds. J Chem Inf Comput Sci 43: 85–112
57. Berthelot M (1872) Sur les lois qui président au partage d'un corps entre deux dissolvants (Théorie). Ann Chim Phys 26: 408–417
58. Hansch C, Leo A (1995) Exploring QSAR. Fundamentals and applications in chemistry and biology. American Chemical Society, Washington
59. Nernst W (1891) Verteilung eines Stoffes zwischen zwei Lösungsmitteln und zwischen Lösungsmittel und Dampfraum. Z Phys Chem 8: 110–139
60. Meyer H (1899) Zur Theorie der Alkoholnarkose. Arch Exp Pathol Pharmakol 42: 109–118
61. Overton E. (1901) Studien über die Narkose. Gustav Fischer, Jena
62. Hansch C, Maloney PP, Fujita T, Muir RM (1962) Correlation of biological activity of phenoxyacetic acids with Hammett substituent constants and partition coefficients. Nature 194: 178–180
63. Hansch C, Fujita T (1964) ρ-σ-π Analysis. A method for the correlation of biological activity and chemical structure. J Am Chem Soc 86: 1616–1626
64. Sangster J (1997) Octanol-water partition coefficients: Fundamentals and physical chemistry. Wiley, Chichester
65. Rekker RF, Mannhold R (1992) Calculation of drug lipophilicity. The hydrophobic fragmental constant approach. VCH, Weinheim
66. Schaper KJ, Samitier MLR (1997) Calculation of octanol/water partition coefficients (log P) using artificial neural networks and connection matrices. Quant Struct Act Relat 16: 224–230
67. Bodor N, Ming-Ju H, Harget A (1994) Neural network studies. III: Prediction of partition coefficients. J Molec Struct Theochem 309: 259–266
68. Yaffe D, Cohen Y, Espinosa G, Arenas A, Giralt F (2002) Fuzzy ARTMAP and back-propagation neural networks based quantitative structure-property relationships (QSPRs) for octanol-water partition coefficient of organic compounds. J Chem Inf Comput Sci 42: 162–183
69. Huuskonen JJ, Livingstone DJ, Tetko IV (2000) Neural network modeling for estimation of partition coefficient based on atom-type electrotopological states indices. J Chem Inf Comput Sci 40: 947–955
70. Tetko IV, Tanchuk VY, Villa AEP (2001) Prediction of n-octanol/water partition coefficients from PHYSPROP database using artificial neural networks and E-states indices. J Chem Inf Comput Sci 41: 1407–1421
71. Tetko IV, Tanchuk VY (2002) Application of associative neural networks for prediction of lipophilicity in ALOGPS 2.1 program. J Chem Inf Comput Sci 42: 1136–1145
72. Devillers J, Domine D, Guillon C, Bintein S, Karcher W (1997) Prediction of partition coefficients (log P_{oct}) using autocorrelation descriptors. SAR QSAR Environ Res 7: 151–172
73. Devillers J, Domine D, Guillon C (1998) Autocorrelation modeling of lipophilicity with a back-propagation neural network. Eur J Med Chem 33, 659–664

74. Devillers J, Domine D, Guillon C, Karcher W (1998) Simulating lipophilicity of organic molecules with a back-propagation neural network. J Pharm Sci 87, 1086–1090
75. Broto P, Devillers J (1990) Autocorrelation of properties distributed on molecular graphs. In: Karcher W, Devillers J (eds) Practical applications of quantitative structure-activity relationships (QSAR) in environmental chemistry and toxicology, Kluwer, Dordrecht
76. Devillers J (1999) Autocorrelation descriptors for modeling (eco)toxicological endpoints. In: Devillers J, Balaban AT (eds) Topological indices and related descriptors in QSAR and QSPR, Gordon and Breach Publishers, Amsterdam
77. Devillers J, Domine D (1997) Comparison of reliability of log P values calculated from a group contribution approach and from the autocorrelation method. SAR QSAR Environ Res 7: 195–232
78. Devillers J (2000) EVA/PLS versus autocorrelation/neural network estimation of partition coefficients. Pespect Drug Discov Design 19: 117–131
79. Devillers J (1999). Calculation of octanol/water partition coefficients for pesticides. A comparative study. SAR QSAR Environ Res 10: 249–262
80. Domine D, Devillers J (1998) A computer tool for simulating lipophilicity of organic molecules. Sci Comput Autom 15: 55–63
81. Devillers J (1999) AUTOLOGP^{TM}: A computer tool for simulating *n*-octanol-water partition coefficients. Analusis 27 23–29
82. Pitter P, Chudoba J (1990) Biodegradability of organic substances in the aquatic environment. CRC Press, Boca Raton
83. Kuenemann P, Vasseur P, Devillers J (1990) Structure-biodegradability relationships. In: Karcher W, Devillers J (eds) Practical applications of quantitative structure-activity relationships (QSAR) in environmental chemistry and toxicology, Kluwer, Dordrecht
84. Vasseur P, Kuenemann P, Devillers J (1993) Quantitative structure-biodegradability relationships for predictive purposes. In: Calamari D (ed) Chemical exposure predictions, Lewis Publishers, Boca Raton
85. Devillers J (1996). On the necessity of multivariate statistical tools for modeling biodegradation. In: Ford MG, Greenwood R, Brooks CT, Franke R (eds) Bioactive compound design: Possibilities for industrial use. BIOS Scientific Publishers, Oxford
86. Cambon B, Devillers J (1993) New trends in structure-biodegradability relationships. Quant Struct Act Relat 12 49–56
87. Devillers J, Karcher W (1990) Correspondence factor analysis as a tool in environmental SAR and QSAR studies. In: Karcher W, Devillers J (eds) Practical applications of quantitative structure-activity relationships (QSAR) in environmental chemistry and toxicology, Kluwer, Dordrecht
88. Devillers J (1996) Genetic algorithms in molecular modeling. Academic Press, London
89. Devillers J (1996) Designing molecules with specific properties from intercommunicating hybrid systems. J Chem Inf Comput Sci 36: 1061–1066
90. Devillers J (1993) Neural modelling of the biodegradability of benzene derivatives. SAR QSAR Environ Res 1: 161–167
91. Tabak HH, Govind R (1993) Prediction of biodegradation kinetics using a nonlinear group contribution method. Environ Toxicol Chem 12: 251–260
92. Devillers J, Domine D, Boethling RS (1996) Use of a backpropagation neural network and autocorrelation descriptors for predicting the biodegradation of organic chemicals. In: Devillers J (ed) Neural networks in QSAR and drug design, Academic Press, London
93. Domine D, Devillers J, Chastrette M, Karcher W (1993) Estimating pesticide field half-lives from a backpropagation neural network. SAR QSAR Environ Res 1: 211–219
94. Cros AFA (1863) Action de l'alcool amylique sur l'organisme. Thesis, University of Strasbourg, Strasbourg
95. Free SM, Wilson JW (1964) A mathematical contribution to structure-activity studies. J Med Chem 1: 395–399
96. Devillers J, Lipnick RL (1990) Practical applications of regression analysis in environmental QSAR studies. In: Karcher W, Devillers J (eds) Practical applications of quantitative structure-activity relationships (QSAR) in environmental chemistry and toxicology, Kluwer, Dordrecht

97. Devillers J, Bintein S, Domine D, Karcher W (1995) A general QSAR model for predicting the toxicity of organic chemicals to luminescent bacteria (Microtox® test). SAR QSAR Environ Res 4: 29–38
98. Devillers J, Domine D (1999) A noncongeneric model for predicting toxicity of organic molecules to *Vibrio fischeri*. SAR QSAR Environ Res 10: 61–70
99. Xu L, Ball JW, Dixon SL, Jurs PC (1994) Quantitative structure-activity relationships for toxicity of phenols using regression analysis and computational neural networks. Environ Toxicol Chem 13: 841–851
100. Serra JR, Jurs PC, Kaiser KLE (2001) Linear regression and computational neural network prediction of *Tetrahymena* acute toxicity of aromatic compounds from molecular structure. Chem Res Toxicol 14: 1535–1545
101. Burden FR, Winkler DA (2000) A Quantitative structure-activity relationships model for the acute toxicity of substituted benzenes to *Tetrahymena pyriformis* using Bayesian-regularized neural networks. Chem Res Toxicol 13: 436–440
102. Winkler D, Burden F (2003) Toxicity modelling using Bayesian neural nets and automatic relevance determination. In: Ford M, Livingstone D, Dearden J, van de Waterbeemd H (eds) EuroQSAR 2002. Designing drugs and crop protectants: Processes, problems and solutions, Blackwell publishing, Malden
103. Devillers J (2004) Linear versus nonlinear QSAR modeling of the toxicity of phenol derivatives to *Tetrahymena pyriformis*. SAR QSAR Environ Res 15: 237–249
104. Yao XJ, Panaye A, Doucet JP, Zhang RS, Chen HF, Liu MC, Hu ZD, Fan BT (2004) Comparative study of QSAR/QSPR correlations using support vector machines, radial basis function neural networks, and multiple linear regression. J Chem Inf Comput Sci 44: 1257–1266
105. Ren S (2003) Modeling the toxicity of aromatic compounds to *Tetrahymena pyriformis*: The response surface methodology with nonlinear methods. J Chem Inf Comput Sci 43: 1679–1687
106. Panaye A, Fan BT, Doucet JP, Yao XJ, Zhang RS, Liu MC, Hu ZD (2006) Quantitative structure-toxicity relationships (QSTRs): A comparative study of various non linear methods. General regression neural network, radial basis function neural network and support vector machine in predicting toxicity of nitro- and cyano-aromatics to *Tetrahymena pyriformis*. SAR QSAR Environ Res 17: 75–91
107. Novic M, Vracko M (2003) Artificial neural networks in molecular-structures-property studies. In: Leardi R (ed) Nature-inspired methods in chemometrics: Genetic algorithms and artificial neural networks, Elsevier, Amsterdam
108. Niculescu SP, Kaiser KLE Schultz TW (2000) Modeling the toxicity of chemicals to *Tetrahymena pyriformis* using molecular fragment descriptors and probabilistic neural networks. Arch Environ Contam Toxicol 39: 289–298
109. Kaiser KLE, Niculescu SP, Schultz TW (2002) Probabilistic neural network modeling of the toxicity of chemicals to *Tetrahymena pyriformis* with molecular fragment descriptors. SAR QSAR Environ Res 13: 57–67
110. Kahn I, Sild S, Maran U (2007) Modeling the toxicity of chemicals to *Tetrahymena pyriformis* using heuristic multilinear regression and heuristic back-propagation neural networks. J Chem Inf Model 47: 2271–2279
111. Kaiser KLE, Niculescu SP (2001) Modeling acute toxicity of chemicals to *Daphnia magna*: A probabilistic neural network approach. Environ Toxicol Chem 20: 420–431
112. Devillers J (2003) A QSAR model for predicting the acute toxicity of pesticides to gammarids. In: Leardi R (ed) Nature-inspired methods in chemometrics: Genetic algorithms and artificial neural networks, Elsevier, Amsterdam
113. Devillers J (2000) Prediction of toxicity of organophosphorus insecticides against the midge, *Chironomus riparius*, via a QSAR neural network model integrating environmental variables. Toxicol Meth 10: 69–79
114. Kaiser KLE, Niculescu SP, Schüürmann G (1997) Feed forward backpropagation neural networks and their use in predicting the acute toxicity of chemicals to the fathead minnow. Water Qual Res J Canada 32: 637–657

115. Kaiser KLE, Niculescu SP, McKinnon MB (1997) On simple linear regression, multiple linear regression, and elementary probabilistic neural network with Gaussian kernel's performance in modeling toxicity values to fathead minnow based on Microtox data, octanol/water partition coefficient, and various structural descriptors for a 419-compound dataset In: Chen F, Schüürmann G, Proceedings of the 7th international workshop on QSAR in environmental sciences, SETAC Press
116. Eldred DV, Weikel CL, Jurs PC, Kaiser KLE (1999) Prediction of fathead minnow acute toxicity of organic compounds from molecular structure. Chem Res Toxicol 12: 670–678
117. Kaiser KLE, Niculescu SP (1999) Using probabilistic neural networks to model the toxicity of chemicals to the fathead minnow (*Pimephales promelas*): A study based on 865 compounds. Chemosphere 38: 3237–3245
118. Niculescu SP, Atkinson A, Hammond G, Lewis M (2004) Using fragment chemistry data mining and probabilistic neural networks in screening chemicals for acute toxicity to the fathead minnow. SAR QSAR Environ Res 15: 293–309
119. Espinosa G, Arenas A, Giralt F (2002) An integrated SOM-fuzzy ARTMAP neural system for the evaluation of toxicity. J Chem Inf Comput Sci 42: 343–359
120. Mazzatorta P, Benfenati E, Neagu CD, Gini G (2003) Tuning neural and fuzzy-neural networks for toxicity modeling. J Chem Inf Comput Sci 43: 513–518
121. Vracko M, Bandelj V, Barbieri P, Benfenati E, Chaudry Q, Cronin M, Devillers J, Gallegos A, Gini G, Gramatica P, Helma C, Mazzatorta P, Neagu D, Netzeva T, Pavan M, Patlewicz G, Randic M, Tsakovska I, Worth A (2006) Validation of counter propagation neural network models for predictive toxicology according to the OECD principles: A case study. SAR QSAR Environ Res 17: 265–284
122. Devillers J (2005) A new strategy for using supervised artificial neural networks in QSAR. SAR QSAR Environ Res 16: 433–442
123. Devillers J, Flatin J (2000) A general QSAR model for predicting the acute toxicity of pesticides to *Oncorhynchus mykiss*. SAR QSAR Environ Res 11: 25–43
124. Devillers J. (2001) A general QSAR model for predicting the acute toxicity of pesticides to *Lepomis macrochirus*. SAR QSAR Environ Res 11: 397–417
125. Devillers J, Pham-Delègue MH, Decourtye A, Budzinski H, Cluzeau S, Maurin G (2002) Structure-toxicity modeling of pesticides to honey bees. SAR QSAR Environ Res 13: 641–648
126. Devillers J, Pham-Delègue MH, Decourtye A, Budzinski H, Cluzeau S, Maurin G (2003) Modeling the acute toxicity of pesticides to *Apis mellifera*. Bull Insect 56: 103–109
127. Devillers J (2008) Artificial neural network modeling in environmental toxicology. In: Livingstone D (ed) Artificial neural networks: Methods and protocols, Humana Press, New York
128. WHO (2003) Health aspects of air pollution with particulate matter, ozone and nitrogen dioxide, Germany, Bonn
129. Ferré-Huguet N, Nadal M, Schuhmacher M, Domingo JL (2006) Environmental impact and human health risks of polychlorinated dibenzo-p-dioxins and dibenzofurans in the vicinity of a new hazardous waste incinerator: A case study. Environ Sci Technol 40: 61–66
130. Lee J, Kwak IS, Lee E, Kim KA (2007) Classification of breeding bird communities along an urbanization gradient using an unsupervised artificial neural network. Ecol Model 203: 62–71
131. Devillers J, Doré JC (1989) Heuristic potency of the minimum spanning tree (MST) method in toxicology. Ecotoxicol Environ Safety 17: 227–235
132. Wienke D, Hopke PK (1994) Visual neural mapping technique for locating fine airborne particles sources. Environ Sci Technol 28: 1015–1022
133. Wienke D, Gao N, Hopke PK (1994) Multiple site receptor modeling with a minimal spanning tree combined with a neural network. Environ Sci Technol 28: 1023–1030
134. Wienke D, Hopke PK (1994) Projection of Prim's minimal spanning tree into a Kohonen neural network for identification of airborne particle sources by their multielement trace patterns. Anal Chim Acta 291: 1–18
135. Wienke D, Xie Y, Hopke PK (1995) Classification of airborne particles by analytical SEM imaging and a modified Kohonen neural network (3MAP). Anal Chim Acta 310: 1–14

136. Domine D, Devillers J, Chastrette M, Karcher W (1993) Non-linear mapping for structure-activity and structure-property modelling. J Chemom 7: 227–242
137. Domine D, Devillers J, Wienke D, Buydens L (1996) Test series selection from nonlinear neural mapping. Quant Struct Act Relat 15: 395–402
138. Domine D, Wienke D, Devillers J, Buydens L (1996) A new nonlinear neural mapping technique for visual exploration of QSAR data. In: Devillers J (ed) Neural networks in QSAR and drug design, Academic Press, London
139. Kolehmainen M, Martikainen H, Hiltunen T, Ruuskanen J (2000) Forecasting air quality parameters using hybrid neural network modelling. Environ Monit Ass 65: 277–286
140. Owega S, Khan BUZ, Evans GJ, Jervis RE, Fila M (2006) Identification of long-range aerosol transport patterns to Toronto via classification of back trajectories by cluster analysis and neural network techniques. Chemom Int Lab Syst 83: 26–33
141. Wienke D, Domine D, Buydens L, Devillers J (1996) Adaptive resonance theory based neural networks explored for pattern recognition analysis of QSAR data. In: Devillers J (ed) Neural networks in QSAR and drug design, Academic Press, London
142. Spencer MT, Shields LG, Prather KA (2007) Simultaneous measurement of the effective density and chemical composition of ambient aerosol particles. Environ Sci Technol 41: 1303–1309
143. Ibarra-Berastegi G, Elias A, Barona A, Saenz J, Ezcurra A, Diaz de Argandonia J (2008) From diagnosis to prognosis for forecasting air pollution using neural networks: Air pollution monitoring in Bilbao. Environ Model Soft 23: 622–637
144. Kurt A, Gulbagci B, Karaca F, Alagha O (2008) An online air pollution forecasting system using neural networks. Environ Int 34: 592–598
145. Sahin U, Ucan ON, Bayat C, Oztorun N (2005) Modeling of SO_2 distribution in Istanbul using artificial neural networks. Environ Model Ass 10: 135–142
146. Chelani AB, Singh RN, Devotta S (2005) Nonlinear dynamical characterization and prediction of ambient nitrogen dioxide concentration. Water Air Soil Pollut 166: 121–138
147. Niska H, Hiltunen T, Karppinen A, Ruuskanen J, Kolehmainen M (2004) Evolving the neural network model for forecasting air pollution time series. Eng Appl Artif Int 17: 159–167
148. Gardner MW, Dorling SR (2000) Statistical surface ozone models: An improved methodology to account for non-linear behaviour. Atmos Environ 34: 21–34
149. Balaguer Ballester E, Camps i Valls G, Carrasco-Rodriguez JL, Soria Olivas E, del Valle-Tascon S (2002) Effective 1-day ahead prediction of hourly surface ozone concentrations in eastern Spain using linear models and neural networks. Ecol Model 156: 27–41
150. Abdul-Wahab SA, Al-Alawi SM (2002) Assessment and prediction of tropospheric ozone concentration levels using artificial neural networks. Environ Model Soft 17: 219–228
151. Corani G (2005) Air quality prediction in Milan: Feed-forward neural networks, pruned neural networks and lazy learning. Ecol Model 185: 513–529
152. Sousa SIV, Martins FG, Pereira MC, Alvim-Ferraz MCM (2006) Prediction of ozone concentrations in Oporto city with statistical approaches. Chemosphere 64: 1141–1149
153. Lu HC, Hsieh JC, Chang TS (2006) Prediction of daily maximum ozone concentrations from meteorological conditions using a two-stage neural network. Atmos Res 81: 124–139
154. Karatzas KD, Kaltsatos S (2007) Air pollution modeling with the aid of computational intelligence methods in Thessaloniki, Greece. Simulat Model Pract Theor 15: 1310–1319
155. Salazar-Ruiz E, Ordieres JB, Vergara EP, Capuz-Rizo SF (2008) Development and comparative analysis of tropospheric ozone prediction models using linear and artificial intelligence-based models in Mexicalli, Baja California (Mexico) and Calexico, California (US). Environ Model Soft 23: 1056–1069
156. Slini T, Kaprara A, Karatzas K, Moussiopoulos N (2006) PM_{10} forecasting for Thessaloniki, Greece. Environ Model Soft 21: 559–565
157. Papanastasiou DK, Melas D, Kioutsioukis I (2007) Development and assessment of neural network and multiple regression models in order to predict PM10 levels in a medium-sized Mediterranean city. Water Air Soil Pollut 182: 325–334
158. Perez P, Reyes J (2001) Prediction of particulate air pollution using neural techniques. Neural Comput Applic 10: 165–171

159. Ordieres JB, Vergara EP, Capuz RS, Salazar RE (2005) Neural network prediction model for fine particulate matter (PM2.5) on the US-Mexico border in El Paso (Texas) and Ciudad Juárez (Chihuahua). Environ Model Soft 20: 547–559
160. Dimopoulos I, Chronopoulos J, Chronopoulou-Sereli A, Lek S (1999) Neural network models to study relationships between lead concentration in grasses and permanent urban descriptors in Athens city (Greece). Ecol Model 120: 157–165
161. Nagendra SMS, Khare M (2006) Artificial neural network approach for modelling nitrogen dioxide dispersion from vehicular exhaust emissions. Ecol Model 190: 99–115
162. Jensen RR, Karki S, Salehfar H (2004) Artificial neural network-based estimation of mercury speciation in combustion flue gases. Fuel Process Technol 85: 451–462
163. Karakitsios SP, Papaloukas CL, Kassomenos PA, Pilidis GA (2006) Assessment and prediction of benzene concentrations in a street canyon using artificial neural networks and deterministic models. Their response to "what if" scenarios. Ecol Model 193: 253–270
164. De Vito S, Massera E, Piga M, Martinotto L, Di Francia G (2008) On field calibration of an electronic nose for benzene estimation in an urban pollution monitoring scenario. Sensors Actuators B 129: 750–757
165. Devillers J, Guillon C, Domine D (1996) A neural structure-odor threshold model for chemicals of environmental and industrial concern. In: Devillers J (ed) Neural networks in QSAR and drug design, Academic Press, London
166. Guyot M, Doré JC, Devillers J (2004) Typology of secondary cyanobacterial metabolites from minimum spanning tree analysis. SAR QSAR Environ Res 15: 101–114
167. Recknagel F, French M, Harkonen P, Yabunaka KI (1997) Artificial neural network approach for modelling and prediction of algal blooms. Ecol Model 96: 11–28
168. Lee JHW, Huang Y, Dickman M, Jayawardena AW (2003) Neural network modelling of coastal algal blooms. Ecol Model 159: 179–201
169. Oh HM, Ahn CY, Lee JW, Chon TS, Choi KH, Park YS (2007) Community patterning and identification of predominant factors in algal bloom in Deachung reservoir (Korea) using artificial neural networks. Ecol Model 203: 109–118
170. Kim CK, Kwak IS, Cha EY, Chon TS (2006) Implementation of wavelets and artificial neural networks to detection of toxic response behavior of chironomids (Chironomidae: Diptera) for water quality monitoring. Ecol Model 195: 61–71
171. Ray C, Klindworth KK (2000) Neural networks for agrichemical vulnerability assessment of rural private wells. J Hydrol Eng 5: 162–171
172. Mishra A, Ray C, Kolpin DW (2004) Use of qualitative and quantitative information in neural networks for assessing agricultural chemical contamination of domestic wells. J Hydrol Eng 9: 502–511
173. Sahoo GB, Ray C, Wade HF (2005) Pesticide prediction in ground water in North Carolina domestic wells using artificial neural networks. Ecol Model 183: 29–46
174. Sahoo GB, Ray C, Mehnert E, Keefer DA (2006) Application of artificial neural networks to assess pesticide contamination in shallow groundwater. Sci Total Environ 367: 234–251
175. Stenemo F, Lindahl AML, Gärdenäs A, Jarvis N (2007) Meta-modeling of the pesticide fate model MACRO for groundwater exposure assessments using artificial neural networks. J Contam Hydrol 93: 270–283
176. El Tabach E, Lancelot L, Shahrour I, Najjar Y (2007) Use of artificial neural network simulation metamodelling to assess groundwater contamination in a road project. Math Comput Model 45: 766–776
177. Samecka-Cymerman A, Stankiewicz A, Kolon K, Kempers AJ (2007) Self-organizing feature map (neural networks) as a tool in classification of the relation between chemical composition of aquatic bryophytes and types of streambeds in the Tatra national park in Poland. Chemosphere 67: 954–960
178. Gagné F, Blaise C (1997) Predicting the toxicity of complex mixtures using artificial neural networks. Chemosphere 35: 1343–1363
179. Pigram GM, Macdonald TR (2001) Use of neural network models to predict industrial bioreactor effluent quality. Environ Sci Technol 35: 157–162

180. Lopez Garcia H, Machon Gonzalez I (2004) Self-organizing map and clustering for wastewater treatment monitoring. Eng Appl Art Int 17: 215–225
181. Nadal M, Schuhmacher M, Domingo JL (2004) Metal pollution of soils and vegetation in an area with petrochemical industry. Sci Total Environ 321: 59–69
182. Arias R, Barona A, Ibarra-Berastegi G, Aranguiz I, Elias A (2008) Assessment of metal contamination in degded sediments using fractionation and self-organizing maps. J Hazad Mat 151: 78–85
183. Götz R, Lauer R (2003) Analysis of sources of dioxin contamination in sediments and soils using multivariate statistical methods and neural networks. Environ Sci Technol 37: 5559–5565
184. Kanevski MF (1999) Spatial predictions of soil contamination using general regression neural networks. Int J Syst Res Inf Syst 8: 241–256

(Q)SAR Models for Genotoxicity Assessment

Sunil A. Kulkarni and Jiping Zhu

Abstract Assessment of genotoxicity of chemicals is one of the utmost priorities of a regulatory agency since it is indicative of its potential carcinogenic properties. A major challenge to the regulatory agencies today is how to assess the genotoxicity of the large proportion of existing and new substances that have otherwise very little or no information on their genotoxicity potential given the high costs and large time-scales associated with experimental testing. (Quantitative) structure–activity relationships ((Q)SAR)-based methodologies have the potential to serve as rapid and reliable genotoxicity screening tools. Such tools are very useful for regulatory agencies to assess the safety of chemicals, whereas for drug or new chemical manufacturers these aid in providing an insight into the potential genotoxic/mutagenic properties of their novel molecules. To assess genotoxicity of diverse groups of chemicals, various methods ranging from traditional linear modeling techniques to modern machine learning algorithms have been applied by researchers to develop a large variety of (Q)SAR models. This chapter provides an overview of some of the existing (Q)SAR models that have the potential to be integrated in a regulatory framework for nonempirical genotoxicity assessment.

Keywords Genotoxicity · (Q)SAR · Risk assessment · Modeling · Regulation

1 Introduction

The human population is continually exposed to a variety of chemicals typically through food, air, water, and use of consumer products. Exposure to many of these chemicals is of great concern for human health owing to their potential to exert genotoxic/mutagenic effects. Such chemicals are termed as genotoxic since they are capable of altering the genes/chromosomes in some way that can cause point mutation, chromosomal aberration, or DNA damage. A prior knowledge of genotoxicity

S.A. Kulkarni (✉)
Existing Substances Division, Health Canada, Ottawa, ON, K1A 0K9, Canada
e-mail: sunil_kulkarni@hc-sc.gc.ca

J. Devillers (ed.), *Ecotoxicology Modeling*, Emerging Topics in Ecotoxicology: Principles, Approaches and Perspectives 2, DOI 10.1007/978-1-4419-0197-2_2,

of a substance is indicative, in many cases, of its carcinogenic and teratogenic potential. Therefore, assessment of genotoxicity of in-commerce chemicals is one of the essential components of a sound regulatory policy. For a drug manufacturer, a prior knowledge on genotoxicity of a chemical in early stages of product development saves a huge amount of resources.

Since direct genotoxicity tests on humans are not possible for a number of reasons, rodent models are generally used for experimental testing. However, on the one hand, one reliable in vivo testing of chronic effects such as genotoxicity for one single chemical requires huge amounts of resources in terms of money, time, and animals. In addition, there are sensitive issues pertaining to ethical and humane treatment of laboratory animals during testing. On the other hand, in vitro genotoxicity tests such as bacterial tests and chromosomal aberration assays are comparatively quicker and relatively less expensive, but these too could be time-consuming and expensive when testing vast numbers of chemicals [1].

In this context, in silico methods such as (quantitative) structure–activity relationships ((Q)SARs), which are data and/or knowledge driven, have the potential to offer fairly reliable prediction results on genotoxicity. The in silico method serves as a screening tool and has several advantages when compared with traditional in vivo and in vitro tests. First, being a mathematical and statistical tool, it is capable of processing a large amount of information in a short time using minimum resources. Second, these models in some cases are able to simulate real-life biological systems and also provide mechanistic basis for the genotoxicity. Third, in today's fast-paced world thousands of new chemicals in the form of pesticides, drugs, fine chemicals, and dyes are synthesized every year in different parts of the world. Typically in countries where regulatory laws are stringent every chemical, either produced or imported, has to be screened for its genotoxic properties and prioritized for further experimental testing. Given the fact that there is still a large number of existing chemicals that as yet do not have adequate information on their genotoxic/mutagenic status, experimental testing based approach alone cannot meet the demand for assessing the genotoxicity of chemicals. As a screening tool, positive hits on genotoxic (Q)SAR models help to flag the substances as having potential hazard to human health and thus, subject to further investigation using in vitro and/or in vivo tests. Fourth, the experimental genotoxicity screening to predict genotoxicity potential of novel molecules that are still under planning and as yet not synthesized is not practicable. Finally, the high levels of uncertainty associated with genotoxicity testing calls for dependable theoretical methods to identify suspect experimental results [2].

The term "genotoxic" generally applies to chemicals or processes, which alter the structure, information content, or segregation of DNA, whereas the term "mutagenic" is used to designate chemicals that are able to make heritable changes to the genetic material. In this chapter, the terms "genotoxic" and "mutagenic" have been used to indicate a toxic response elicited by the interaction of a chemical with the DNA, and therefore, we have treated them as one category for addressing various (Q)SAR models.

2 Elements of (Q)SAR Models for Genotoxicity

A typical (Q)SAR model for prediction of genotoxicity comprises certain essential elements including training set chemicals, descriptors (e.g., structural, physicochemical), genotoxic/mutagenic endpoints, and a methodology. These components are interconnected and each plays an essential role in the model. The usefulness of these elements and their importance in the genotoxicity-based (Q)SAR model are briefly outlined in the following subsections.

2.1 Training Set Chemicals and Descriptors

A training set is a set of chemicals that has data on a genotoxicity endpoint and is used to develop a (Q)SAR model. Such data-rich chemicals could be gathered from publicly available sources such as Carcinogenic Potency Databases (CPDB), United States Food and Drug Administration (USFDA), Distributed Structure-Searchable Toxicity (DSSTox) Database, Genetox, National Toxicology Program or from proprietary databases (e.g., from a pharmaceutical industry). Alternatively, the training set could be developed in-house by generating one's own empirical genotoxicity data. Commercial (Q)SAR models use public and/or proprietary databases. One important aspect of a training set pertains to its composition with respect to diversity of molecular structures since it decides the scope of a model. Models based on a congeneric set of chemicals with a focus on a specific functional group or structural characteristic have a narrow applicability compared with ones that have a wide range of chemicals.

Genotoxicity has been successfully modeled using global and/or fragment-based descriptors. Global descriptors, on the one hand, tend to encode information about the whole molecule, do not change from point to point in space, and help to identify trends and tendencies within a chemical dataset. The fragment-based descriptors, on the other hand, are localized and vary with the nature of functional groups/bonds. A descriptor, therefore, is the final result of a logic and mathematical procedure, which transforms chemical information encoded within a symbolic representation of a molecule into a useful number or the result of some standardized experiment [3]. The descriptors can either be determined experimentally or computed using software programs such as DRAGON [4], ADRIANA.code [5], PowerMV [6]. Some of the descriptors that are commonly used to model genotoxicity include hydrophobicity (generally measured as the logarithm of n-octanol-water partition coefficient (log Kow)), quantum-mechanical (e.g., highest occupied molecular orbital (HOMO), lowest unoccupied molecular orbital (LUMO), geometry-optimised bond length, etc.), thermodynamic, steric, bulkiness, topological, Taft constant σ^*, Hammett constant σ, and STERIMOL. In addition to these, it is possible to compute a very large number of other descriptors using one of the programs mentioned above.

The choice of descriptors generally depends on factors including the genotoxic effect to be predicted, nature of training set chemicals, and availability of computational resources. From a practical standpoint, it is worth noting that there is always a trade-off between the level of approximation as far as the information from molecular descriptor is concerned and the demand on resources. For instance, the computationally intensive ab initio molecular descriptors result in a heavy demand for computer resources and can make models built on it unsuitable for rapid screening of large numbers of chemicals. Descriptors such as the E-state (electrotopological) indices that encode similar molecular information as the conventional descriptors are relatively easy to compute [7].

Generally techniques such as genetic algorithm (GA), generalized simulated annealing (GSA), and fuzzy logic are used by model developers to provide a high-level reasoning capability and/or in finding optimal subsets of descriptors from a large descriptor population. A brief description of some of the molecular descriptors commonly used to model genotoxicity is presented in Table 1.

2.2 Genotoxicity Endpoints

The molecular mechanisms of genotoxicity include DNA intercalation by aromatic ring of a chemical, gene mutation, DNA adduct formation and strand break, and unscheduled DNA synthesis. Indirect-acting chemicals that exert genotoxicity upon metabolic activation are generally found to cause chromosomal aberrations, micronuclei formation, sister chromatid exchanges, and cell death [8, 9]. Several in vitro and in vivo tests and testing protocols have been developed to measure one of these genotoxic/mutagenic effects. The most commonly used in vitro genotoxic endpoint is the Ames test, a microbial assay that uses various strains of the bacteria *Salmonella typhimurium*, where the strains are sensitive to specific types of mutation [10]. To make the in vitro tests metabolically relevant for higher organisms (humans), the rat liver enzymes (S9) are generally added to the *microbial assay* especially when assessing indirect-acting substances [10]. Some of the common genotoxicity endpoints (or tests) used to develop (Q)SAR models have been summarized in Table 2.

2.3 Model Development Methodologies

In the process of (Q)SAR model development, it is the methodology that ultimately establishes the relationship between the structural characteristics of a chemical and the genotoxic outcome, and paves the way for interpretation of the underlying genotoxic mechanisms. These methodologies include the classical linear regression techniques [31–33], machine learning techniques such as inductive logic programming [34], k-nearest neighbors [35], decision trees [36], neural networks [37], pattern recognition [38], decision forests [39], support vector machines [40], random forest [2], linear discriminant analysis [41], molecular modeling methods that employ three-dimensional (3D) techniques that model relevant

Table 1 Common molecular descriptors used for modeling genotoxicity/mutagenicity

Molecular descriptor	Description/interpretation
Hydrophobicity (log Kow or log P)	Measures affinity of a molecule or a moiety for a lipophilic environment and tells about penetration, distribution, bioaccumulation
Hansch-Fujita π constant	Describes the contribution of a substituent to the lipophilicity of a compound
Structural (general)	Size of aromatic ring system, presence/absence/orientation/number of a specific functional group, number of ring heteroatoms
Taft steric parameter (Es)	A relative reaction parameter encoding the reaction rate retardation due to the size of a substituent group
Molar refractivity (MR)	Quantifies bulkiness, steric hindrance, size and polarizability of a fragment or molecule. Higher value prevents enzymatic access to molecule and thus, formation of reactive intermediate
STERIMOL (L, B1, B2, B3, B4)	Length and width of molecule/substituent
Topological	Characterize structures according to molecular size, degree of branching, and overall shape
Electrotopological state (E-state) indices	Encode electronic, connectivity and topological information for each skeletal atom in a molecule. Also, take into account the structure of the entire molecule, and contain some form of shape information
Geometrical	Encode information on overall size and shape, molecular surface area, molecular volume (require accurate 3D geometries)
Three-dimensional	Encode spatial relationships between atoms, ring centroids, and planes. Considers conformational flexibility
Quantum-mechanical – HOMO	Higher HOMO value means chemical is easier to oxidize hence, should be readily bioactivated. Mutagenicity should increase with increasing HOMO
Quantum-mechanical – LUMO	Mutagenic potency increases as LUMO decreases
Hammett sigma constant (σ)	Electronic substituent descriptor reflecting the electron-donating or accepting properties of a substituent. Positive values indicate electron withdrawal by substituent and negative values indicate electron release to an aromatic ring
Thermodynamic	Enthalpy of reaction (ΔH)
Electrostatic	Molecular polarizability, dipole moment, average ionization energy

biochemical events for toxicity [42], rule-based and knowledge-based expert systems that mimic human reasoning about toxicological phenomena such as deductive estimation of risk from existing knowledge (DEREK) [43], HazardExpert [44] and prediction of activity spectra for biologically active substances (PASS) [45], highly populated, structurally diverse data-driven global model systems such as computer-automated structure evaluation of toxicity (CASETOX) [46] and toxicity prediction by komputer-assisted technology (TOPKAT) [47], as well as the hybrid methods that apply modern machine learning approaches, account for metabolic activation, and integrate expert knowledge with quantitative aspects of modeling using molecular descriptors [48,49].

Table 2 Common endpoints/tests used to model genotoxicity

Name of test	Brief description	Reference
Gene mutation		
Salmonella typhimurium reverse mutation assay (Ames test)	Measures his− → his+ reversion induced by chemicals which cause base changes or frameshift mutations in the genome of *Salmonella*. Strains TA98, 1537 and 1538 are more prone to capture frameshift mutations whereas TA100, 1535 are specific for base-pair substitution mutations. Highly sensitive, effective, rapid screening test. Limitations being the bacterial nature and positive testing of several chemicals not known to be carcinogenic	[10, 11]
Escherichia coli	*E. coli* strains WP2 and WP2 uvrA are used to detect base-pair mutations	[12, 13]
Phage T7-inactivation test	It is based on the determination of the inactivation of bacteriophage T7	[14]
Chinese hamster lung (V79) cell mutagenicity assay	Primarily measures gene mutation at hypoxanthine-guanine phosphoribosyl transferase (HGPT) or the Na+/K+ ATPase locus	[15]
In vitro mouse lymphoma tk locus assay	Quantifies genetic alterations affecting expression of the thymidine kinase (*TK*) gene (tk). Able to detect a wide range of genetic damages, including gene mutations, larger scale chromosomal changes, recombination, aneuploidy, and others	[16, 17]
Induction of chromosomal aberrations		
Micronucleus test in mouse lymphoma (L5178Y) cells	Sensitive and specific to determine clastogenic or aneugenic potential of a test compound; Excellent correlation between chromosomal aberration and micronucleus data in vitro, in primary cells	[18, 19]
Sister chromatid exchange (SCE)	Detects reciprocal exchanges of DNA between two sister chromatids of a duplicating chromosome. Represent the interchange of DNA replication products at apparently homologous loci. Provides a measurement of effects at the highest level of genetic organization. Possible both in vitro and in vivo	[20]
Mammalian micronucleus test in bone marrow	In vivo test for measuring damage to chromosomes or the mitotic apparatus of erythroblasts by analysis of erythrocytes as sampled in bone marrow and/or peripheral blood cells of animals, usually rodents	[21, 22]
Somatic mutation and recombination test (SMART) in *Drosophila melanogaster*	Good capacity for describing mutagenic and carcinogenic activity of various chemical compounds	[23, 24]
Induction of chromosomal malsegregation in *Aspergillus nidulans* (mold)	It measures the aneugenic potential of a chemical	[25]

(continued)

Table 2 (continued)

Name of test	Brief description	Reference
DNA damage		
SOS (salt-overly-sensitive) Chromotest	SOS response is an inducible DNA repair system that allows bacteria to survive sudden increases in DNA damage. The test measures induction of a lac Z reporter gene in response to DNA damage caused by a broad spectrum of genotoxic substances	[26, 27]
Comet assay (single cell gel (SCG) electrophoresis)	Detects DNA strand breaks. The DNA damage is indicated by the extent of migration of DNA fragments when subjected to electrophoresis. Can be conducted in both in vitro and in vivo systems. It is rapid, simple to perform, and requires only a small amount of test substance	[28]
Umu test	It is based on the ability of DNA damaging agents to induce the expression of the *umuC* gene, which is also a SOS response gene. By the umu-test, using the single tester strain, many types of DNA-damaging agents can be detected for which the Ames test requires several tester strains	[29]
Unscheduled DNA synthesis (UDS)	Measures the DNA repair synthesis after excision and removal of a stretch of DNA containing the region of damage induced by chemical and physical agents. It may also be measured in in vivo systems	[30]

3 Overview of Important Genotoxicity-Based (Q)SAR Studies

A vast number of genotoxicity-based SAR and QSAR studies have been carried out and published for a diverse group of chemicals applying different kinds of techniques. The techniques range from classical multiple regression analysis to machine learning (neural networks/pattern recognition/statistical learning methods) to 2D and 3D QSAR methodologies (e.g., comparative molecular field analysis (CoMFA)). Table 3 provides a summary of some of the important studies carried out over the last three decades. Even though a significant number of studies have been carried out on aromatic amines alone, a wide variety of chemical classes have been investigated for (Q)SAR model development. Gene mutation in *Salmonella* strains has been the most commonly used genotoxic endpoint. Although the frequent choice of methodology in the development of QSAR models is multiple linear regressions (MLR) involving smaller datasets, classification/pattern recognition algorithms are some of the preferred methods for models involving large and structurally diverse datasets. From Table 3 it can be observed that MLR was a preferred method in the earlier days, whereas modern day modelers clearly seem to prefer machine learning algorithms. For example, Debnath and coworkers [33] established a relationship using MLR method, $\log \text{TA98} = 1.08(\pm 0.26)\ \log\ P + 1.28(\pm 0.64)\ \text{HOMO} - 0.73(\pm 0.41)\ \text{LUMO} + 1.46(\pm 0.56)\ I_{\text{L}} + 7.20(\pm 5.4))$ with $n = 88$, $r = 0.898$, $s = 0.860$ (n is the number of chemicals in the training set, r is correlation coefficient, s is standard error) for the development of Ames

Table 3 Summary of representative genotoxicity-based (Q)SAR studies

Chemical class	Parameters	Genotoxic test	Training set(s)	Method	References
Triazenes and Platinum amines	Log Kow and Hammett constant	Ames TA92	17	MLR	[31]
O-phenylenediamine platinum dichloride	Hammett constant	Ames TA92	13	LR	[32]
5-nitroimidazoles	Steric, H-bonding	Ames TA100	20	MLR	[51]
N-nitrosomethylaniline compounds	Hammett's constant, hydrophobic substituent constant, dammy, STERIMOL	Ames TA100	16	MLR	[52]
Alkyl alkane sulfonates	Hammett constant and Es	Ames TA100	12	MLR	[53]
Chalcones (aromatic ketones)	Hydrophilicity, steric, electronic	Ames TA98, TA100	31	MLR	[54]
Aromatic amines and acetamides	Lipophilicity, position and nature of amine group	Ames TA98, TA100	19	MLR	[55]
Aromatic and heteroaromatic nitro compounds	Lipophilicity, quantum chemical	Ames TA98	188	MLR	[56]
Halogenated aliphatics	Physchem, log Kow, van der Waals, ionization potential	Gene mutation using Chinese hamster cells	58	MLR	[57]
Polycyclic aromatic nitro compounds	Lipophilicity, quantum chemical	*Escherichia coli*	15	MLR	[58]
Aromatic and heteroaromatic amines	Lipophilicity, quantum chemical	Ames TA100	106, 117	MLR	[33]
Aromatic and heteroaromatic amines	Lipophilicity, quantum chemical	Ames TA98	7, 67, 88	MLR	[59]

Quinolines	Lipophilicity, quantum chemical	Ames TA100	21	MLR	[60]
Heteroaromatic amines	Nitrenium ion stability	Ames TA98, TA100	28	MLR	[61]
Heteroaromatic amines	Nitrenium ion stability	Ames TA98, TA100, TA1538	14, 13, 6	MLR	[62]
Chlorofuranones (MX compounds)	Quantum-chemical	Ames TA100	17	MLR	[63]
Nitrofurans	Electronic, hydrophobic, steric	SOS Chromotest	40	MLR	[64]
Methane sulfonic esters	Log Kow, reaction rate	Ames TA100	15	MLR	[65]
Propylene oxides	Electronic, steric, physicochemical	Ames TA100, TA1535	17	MLR	[66]
N-methylbenzimidazoles	Log Kow, positional	Ames TA98	30	MLR	[67]
Nitro-, amino- and aminonitro-carbazoles	Log Kow, oxidative potential, quantum chemical, retention volume	Ames TA98	8,6,6	MLR	[68]
Aromatic and heterocyclic amines	Structural, quantum chemical	Ames TA98, TA1538	165	MLR	[69]
Quinolines	Lipophilicity, steric	Ames TA100	13	MLR	[70]
Chlorinated hydrocarbons	Quantum chemical, log Kow, MR, electronic	Micronucleus test and Comet assay	8	MLR	[71]
Aromatic and heteroaromatic amines	Electrotopological state indices	Ames TA98	95	MLR	[7]
Aromatic and heterocyclic primary amines	Structural, quantum chemical, and hydropathic	Ames TA98, TA1538	80	MLR	[72]
N-acyloxy-*N*-alkoxyamides	Log Kow, steric, pKa	Ames TA100	41	MLR	[73]

(continued)

Table 3 (continued)

Chemical class	Parameters	Genotoxic test	Training set(s)	Method	References
Benzidine analogues	Physicochemical	Ames TA98, TA100	11	MLR	[74]
Aromatic amines	Quantum chemical and lipophilicity	Ames TA98 and TA100	198	MLR	[75]
Nitroarenes	Topological	Ames TA100	48	MLR	[76]
Halogenated aliphatics	Physchem, 3D, quantum-mechanical	*Drosophila* wing spot	17	MLR	[77]
Quinolones	Lipophilicity, quantum chemical	SOS/umu bioassay	15	MLR	[78]
2-amino-trimethylimidazopyridine isomers (TMIP)	Structural, quantum chemical, hydropathic	Ames TA98	12	MLR	[79]
Benzazoles	Hydrophobic, electronic, steric, structural	*Bacillus subtilis* rec-assay	16	MLR	[80]
Aminoazo dyes	Quantum-chemical	Ames TA98	43	MLR	[81]
N-acyloxy-*N*-alkoxyamides	Log Kow, structural	Ames TA100	26	MLR/SAR	[82]
p-alkoxynitrosobenzenes and *p*-phenoxynitrosobenzene	Steric and electronic	Ames TA100	–	SAR	[83]
Allyl and propenyl benzenes	Quantum-mechanical	UDS	11	SAR	[84]
Styrene oxides	Hydrophobicity, substituent effects	*E. coli* WP2 and TA100	6	SAR	[85]
Glycidyl ethers	Hydrophobicity, substituent effects	Ames test	7	SAR	[86]
4-nitro-(imidazoles and pyrazoles)	Nature of substituents, hydrophobicity	Ames TA98 TA100	13	SAR	[87]
Hydrazine compounds	Quantum chemical and lipophilicity	Ames TA100	12	PLS	[88]

Hetroaromatic triazenes	Quantum chemical topology	Ames TA92	23	GA + PLS and PCs + PLS	[89]
Halogenated hydroxyfuranones	Quantum chemical topology	Ames TA100	24	GA + PLS and PCs + PLS	[89]
Aliphatic *N*-nitrosamines	Molecular connectivity, geometry, sigma charge	Gene mutation using Chinese Hamster lung cells (V79)	21	PR	[38]
Acridine	Several	Ames TA1537	40	PR	[90]
Nitroarenes	Relational description of atom/bond	Ames TA98	230	ILP (PROGOL)	[34]
Polycyclic aromatic hydrocarbons	Topological, geometric, electronic, polar surface area	SOS Chromotest	277	Classification	[41]
Thiophene	Numerical	SOS Chromotest	140	Classification (LDA, kNN, PNN)	[37]
Diverse organics	Topological, geometrical, electronic	Chromosomal aberration	383	Classification (kNN, LDA, PNN, SVM)	[35]
Aminoazo dyes and metabolites	Quantum-chemical	Ames TA98	74	MLR and ANN	[91]
Diverse organics	Topological	Ames test	2963	TLP	[39]
Diverse organics	Quantum mechanical, thermodynamic, steric, structural, electronic, Hydrophobicity	Ames TA100	1196	Probabilistic classification (COREPA)	[48]

(continued)

Table 3 (continued)

Chemical class	Parameters	Genotoxic test	Training set(s)	Method	References
Simple and α, β-unsaturated aldehydes	Log Kow, MR, LUMO, ΔH, partial charges	Ames TA100 and SOS Chromotest	36	MLR and LDA	[92]
Pharmaceuticals	Topological, quantum-chemical, geometrical	Ames test, cytogenetics, mouse lymphoma assay, SOS Chromotest	860	ML (SVM, PNN, kNN, DT)	[36]
Diverse organics	Several descriptors	Ames test	4337	MEV + SVM	[40]
Diverse organics	HOMO, LUMO, MW, log Kow	Ames TA1535 pSK1002 Umu test	82	ANN	[93]
Diverse organics	Several descriptors	Chromosomal aberrations	650	Classification	[50]
PAHs (cyclopentaphenanthrenes and chrysenes)	Holistic molecular descriptors	Ames TA98, TA100	32	Classification	[94]
Diverse organics	MOLMAP descriptors	Ames test	4083	RF, CART	[2]
Diverse organics	Electronic, quantum mechanical, log Kow, electronegativity, dipole moment, volume polarizability	Chromosomal aberrations	538	Probabilistic classification	[49]
Diverse organics	Quantum mechanical, thermodynamic, steric, structural, electronic, hydrophobicity	Ames all strains	2844	Probabilistic classification	[95]
Polycyclic aromatic hydrocarbons	Several descriptors (holistic and fragment-based)	Human B-lymphoblastoid (h1A1v2) cells	70	kNN, CART	[96]
Aromatic and heteroaromatic amines	Topological, quantum chemical, log Kow	Ames TA100	73	QMSA	[97]

Aromatic and heteroaromatic amines	Topostructural, topochemical, geometric, quantum chemical, log Kow	Ames test	95	Hierarchy-based	[98]
Diverse organics	Structural alerts, linear fragments	Ames test	1447	Inductive database (similarity-based)	[99]
Aliphatic and heterocyclic compounds	Number and ratio of elements, side chains, bonding position, and microenvironment of side chains	Ames test	–	Expert system	[100]
Nitroaromatics	Quantum-mechanical, steric, hydrophobicity	Ames TA98, TA100	358	PLS/CoMFA	[101]
Nitroarenes	2D and 3D descriptors, hydrophobicity, steric, solubility, ionization constants	Ames TA98	197	2D (HQSAR and GFA), 3D (CoMFA and CoMSIA)	[42]

MLR multiple linear regression, *LR* linear regression, *SLM* statistical learning method, *ANN* artificial neural network, *TLP* three-layer perceptron, *GA* genetic algorithm, *SAR* structure–activity relationship, *LDA* linear discriminant analyses, *HQSAR* hologram QSAR, *GFA* genetic function approximation, *PLS* partial least squares, *CoMFA* comparative molecular field analysis, *CoMSIA* comparative molecular similarity indices analysis, *UDS* unscheduled DNA synthesis, *SOS* salt-overly sensitive, *MW* molecular weight, *HOMO* highest occupied molecular orbital, *LUMO* lowest unoccupied molecular orbital, *Kow* *n*-octanol-water partition coefficient, *PNN* probabilistic neural networks, *SVM* support vector machines, *kNN* k-nearest neighbors, *ML* machine learning, *QMSA* quantitative molecular similarity analysis, *PR* pattern recognition, *Physchem* physico-chemical, *MEV* molecular electrophilicity vector, *RF* random forest, *CART* classification and regression tree, *MOLMAP* molecular map of atom-level properties, *COREPA* Common reactivity patterns mutagenicity QSAR on a set of aromatic amines. Studies based on machine learning methods do not generate such equations but instead compute a sensitivity factor that shows how well the classifier was able to segregate genotoxic from nongenotoxic chemicals [35, 50]. In the following subsections, we will describe in brief some of the important studies that were carried out by various researchers from which the relationship between structures and genotoxic endpoints are established.

3.1 QSARs Based on Classical Techniques

The work of Debnath and coworkers [60, 65], which mainly focused on different types of aromatic amines and nitroarenes demonstrated the importance of hydrophobicity and molecular orbital energies in predicting their mutagenicity. These studies considered not only the *S. typhimurium* strains but also other bacterial species such as *Escherichia coli*. However, the QSARs developed by Debnath and coworkers were found to be unsuitable for differentiating the inactives from the actives [75]. According to Benigni and coworkers, first it was necessary to separately investigate the structure–activity relationships (SARs) for discrimination between positive and negative chemicals, and the SARs for the potency of the positive chemicals. Second, it was necessary to investigate the degree of homogeneity (congenericity) of apparently similar chemicals to assess and describe the various mechanisms of action that may be elicited by the chemicals [102]. It was found that hydrophobicity alone had no discriminating power in the *Salmonella* strains TA98 and TA100, which was at odds with the major role played in the modulation of potency within the group of active compounds. Discriminant functions separating mutagenic from nonmutagenic amines were based mainly on electronic and steric hindrance factors. Since no satisfactory models existed for discriminating between mutagenic and nonmutagenic aromatic amines, a new set of QSAR models focusing on the mechanistic aspects were built using data (mostly homocyclic aromatic amines) taken from Debnath's studies [75]. The models for strains TA98 and TA100 correctly reclassified 89.2% and 87.4% of the compounds, respectively.

The genotoxicity of these arylamines is generally thought to be due to their abilities to covalently modify nucleic acid base sites following metabolic activation via N-oxidation to the highly reactive arylnitrenium ions [103]. In the backdrop of the significant success of simple molecular orbital procedures in correlating the genetic toxicities of the polycyclic aromatic hydrocarbons (PAHs) and the landmark development of the "bay region theory," Ford and Herman reexamined the role played by stabilities of the aromatic nitrenium ions in the genotoxicity of polycyclic aromatic amines [61]. They used modern molecular orbital techniques to study the effects. However, limited linear correlations were obtained between energetics of nitrenium ion formation and mutagenicities in the TA98 and TA100 strains of *Salmonella*. On similar lines Ford and Griffin also reported approximately linear correlation between relative stabilities of nitrenium ion and mutagenicity of a variety of heteroaromatic amines present in cooked foods [62].

Using a hierarchical approach, class-specific QSAR models were developed for the prediction of mutagenicity for a set of 95 aromatic and heteroaromatic amines [98]. The hierarchical approach was designed such that the simplest molecular descriptors, the topostructural, which encoded limited chemical information, was first used and gradually the complexity was increased by adding topochemical, geometric, and finally quantum chemical parameters. Log Kow was also added to the set of independent variables. It was found that the topological parameters such as topostructural and topochemical indices explained the majority of the variance, and that the inclusion of log Kow, geometric, and quantum chemical parameters did not result in significantly improved predictive models.

Cash [7] carried out a QSAR model development study by correlating the measured mutagenicities of a set of aromatic and heteroaromatic amines with easily calculated sums of E-state indices and principal components derived from them. The correlations obtained were comparable to previously published models that used log Kow, HOMO, and LUMO, or a selection of topological and geometric variables, but they were not as good as a model that relied on SCF/AM1 (self-consistent field/Austin model 1) energy optimization. However, the model performed poorly when subjected to external validation [104]. This indicated the incapability of E-state indices to adequately describe key positional features that influence the stability of reactive intermediates of aromatic amines that are so crucial in their mutagenic action.

A QSAR methodology using new structural factors and quantum chemical Hückel and ab initio calculations on a diverse set of 80 chemicals consisting of aromatic and heterocyclic amines was proposed by Hatch and coworkers [72]. The results indicated that the main determinant of mutagenic potency was the extent of the aromatic π-electron system with minor contributions made by both the dipole moment and the calculated stability of the nitrenium ion.

A series of 17 mono-substituted propylene oxides were investigated for their mutagenic effects in *Salmonella* strains TA100 and TA1535 using two different assays (liquid suspension assay (LSA) and standard plate incorporation assay (PIA)) [66]. The relative mutagenicity was found to differ not only in the two assays but also in the two strains. The assay differences (greater mutagenicity for propylene oxide in LSA than in the PIA) were attributed to epoxide stability, whereas the observed variations between strains were due to the response of the error-prone repair system to the stronger alkylating agents that is found only in the TA100 strain.

A structure–mutagenicity relationship study on a set of 31 para-monosubstituted chalcones and their corresponding oxides using *Salmonella* strains TA98 and TA100 revealed how the latter group of chemicals exhibited varying sensitivities toward the two strains. Additionally, the importance of increasing hydrophilicity (as indicated by the Hansch π parameter) and resonance electronic contributions when compared with steric terms in explaining their mutagenicity was also observed [54]. The chalcone oxides were found to be more mutagenic with TA100 than TA98 since the former strain was particularly sensitive to alkylating agents that tend to interact covalently with *S. typhimurium* DNA. Epoxidation, in general, was found to increase the mutagenic activity of the respective chalcone. Benzoyl (4′) substituted chalcones and their oxides with an electron-withdrawing substituent (e.g., nitro, fluoro) usually had higher activity than their phenyl (4) substituted counterparts, whereas the converse was the case with electron-donating substituents (e.g., acetamido, methoxy).

A QSAR study on a set of 12 alkylated hydrazine compounds used CoMFA methodology and partial least squares (PLS) statistics to determine the factors governing their mutagenic potency to *Salmonella* strain TA100 [88]. Six different QSAR equations with different descriptors were developed. Energy of the lowest unoccupied molecular orbital (E_{LUMO}) together with log Kow were found to be the most important descriptors in determining the mutagenic activity.

A structure–activity study on a series of 15 quinoline congeners was carried out to demonstrate how two simultaneously occurring biological processes, mutagenicity and cytotoxicity in *Salmonella* strain TA100 could be modeled especially when the hydrophobicity plays an important part in the process [70]. It was found that for the mutagenic potency, both hydrophobic and steric interactions were important.

Tafazoli and coworkers developed genotoxicity-based QSAR model using a small number of aliphatic chlorinated hydrocarbons [71]. The QSAR analysis highlighted that the toxicity of the tested chemicals was influenced by different parameters, like lipophilicity (log Kow), electron donor ability (charge), and longest carbon–chlorine ($\mathrm{LB_{C-Cl}}$) bond length. In addition, steric parameters, such as molar refractivity (MR) and $\mathrm{LB_{C-Cl}}$, and electronic parameters, such as $\mathrm{E_{LUMO}}$ that indicated electrophilicity, were predominant factors discriminating genotoxins from non-genotoxins in the presence but not in the absence of metabolising enzymes (S9).

Bonin and coworkers obtained QSARs for N-acyloxy-N-alkoxyamides, which are a class of direct-acting mutagens not requiring metabolic activation [82]. Structural factors were found to have more influence on binding and reactivity toward DNA when compared with hydrophobicity, which plays a more important role in case of indirect-acting mutagens [67].

A number of QSAR analyses were performed on a set of simple and α, β-unsaturated aldehydes to predict their genotoxicity [92]. One of the interesting findings was related to the parameter ΔH (reaction enthalpy), which accounts for the ease of adduct formation between α,β-unsaturated aldehyde and the DNA bases. In general, the lower the ΔH, the higher is the probability of adduct formation. Out of a group of descriptors used for the analyses the authors found that the descriptors most relevant in explaining the segregation of these chemicals into active and inactive ones were hydrophobicity (log Kow) and bulkiness (MR) accounting for a cumulative 62% variance. The other two descriptors, LUMO (related to the reactivity of the chemicals) and ΔH taken together were able to explain a further 20% of variance.

3.2 (Q)SARs Based on Machine Learning Methods

A pattern recognition analysis using Automated Data Analysis by Pattern-recognition Technique (ADAPT) software was applied to a set of 21 aliphatic

N-nitrosamines to classify them into nonmutagenic and mutagenic groups [38]. The authors considered molecular connectivity, geometry, and sigma charge on Nitrogen as the molecular descriptors and applied linear learning machine and iterative least squares algorithms. Using sets of four, three, and two descriptors, they could achieve 100%, 90.5% and 87.5% correct classification, respectively.

A set of aromatic and heterocyclic nitro compounds was used to develop a SAR based on relational description that considers substructures in a chemical and their associations using inductive logic programming (ILP) named PROGOL algorithm [34]. This approach allowed the use of a rich representation of chemical structure and had the advantage that it did not need the use of any indicator variables handcrafted by an expert, and was based on easily comprehensible rules. The SAR obtained was found to be significantly more accurate than linear regression, quadratic regression, and back-propagation methods.

Several predictive binary classification models have been presented that directly link the genetic toxicity of a series of 140 thiophene derivatives with information derived from molecular structures [37]. Genotoxicity was measured using a salt-overly sensitive (SOS) Chromotest. Maximal SOS induction factor (IMAX) values were recorded for each of the 140 compounds both in the presence and absence of S9 rat liver homogenate. Compounds were classified as genotoxic if IMAX $\geq$ 1.5 and non-genotoxic if IMAX $<$ 1.5 for both tests. The molecular structures were represented by numerical descriptors that encoded the topological, geometric, electronic, and polar surface area properties of the thiophene derivatives. Classification models such as linear discriminant analysis (LDA), *k*-nearest neighbor (kNN), and probabilistic neural network (PNN) were used in conjunction with either genetic algorithm (GA) or a generalized simulated annealing (GSA) to find optimal subsets of descriptors for each classifier. One noteworthy point was that these models were not necessarily representative of cause–effect relationships and in many cases were simply correlations between an observed genotoxicity indicator and a set of descriptor without an intuitive chemical sense.

Li and coworkers tested several statistical learning methods (SLMs) that included support vector machines (SVM), PNN, kNN, and C4.5 decision tree (DT) for a dataset of 860 pharmaceuticals and a set of 199 molecular descriptors (143 topological, 31 quantum chemical, and 25 geometrical) [36]. The resulting genotoxicity prediction systems were able to predict accurately up to 77.8% for genotoxic and 92.7% for non-genotoxic chemicals.

A recent study on a large and diverse group of chemicals integrated the novel molecular representation method molecular electrophilicity vector (MEV) and SVM to develop a mutagenicity prediction model [40]. The MEV was devised to characterize the electrophilicity and topology of a molecule, accounting for both direct and indirect mechanisms of genotoxicity. The model exhibited a superior efficiency in data fitting with a concordance rate of 91.86%. Sensitivity and specificity were found to be 93.63 and 89.67%, respectively. For the external validation set, a prediction accuracy of 84.80% was obtained. The authors also used the same dataset to evaluate the performance of TOPKAT model and found that it could correctly identify true negatives with a specificity of 85.10%. The sensitivity and overall accuracy was found to be much lower at 77.32 and 80.81%, respectively.

To account for genotoxicity of chemicals that exert their effect after metabolic activation, an approach based on the common reactivity patterns (COREPA) was used to delineate the structural requirements for eliciting mutagenicity in terms of ranges of descriptors associated with three-dimensional molecular structures [48]. Out of a total set of 1196 chemicals, a model was developed using 148 chemicals that tested positive in TA100 strain without rat liver enzymes (S9) and 188 chemicals that tested positive in TA100 strain with S9. A decision tree was developed by first comparing the reactivity profile of chemicals that were positive in TA100 without S9 to the reactivity profile of the remaining 1048 chemicals. A series of hierarchically-ordered metabolic transformations were used to develop an S9 metabolism simulator to identify the chemicals that are positive only in the presence of metabolic activation. The 1,048 chemicals were then passed through the simulator, and the potential metabolites were screened through the decision tree to identify reactive mutagens. This model correctly identified 77% of the metabolically activated chemicals in the training set.

A binary classification of structural chromosome aberrations (clastogenic and non-clastogenic) for a diverse set of 383 organic compounds was proposed by Serra and coworkers [35]. Using topological, geometrical, and electronic descriptors, the classification schemes such as kNN and SVM were applied to generate predictions on an external prediction set. In vitro chromosomal aberration assay with Chinese hamster lung cells was considered as the endpoint. The overall classification success rate for a kNN classifier built with six topological descriptors was found to be 81.2% for the training set and 86.5% for the external prediction set, whereas the same was found to be 99.7% and 92.1% for a three-descriptor SVM model, respectively. To demonstrate that the models were not built on chance correlation, the authors also performed scrambling experiments. The main advantage of developing such in silico inductive classification methods was that no prior knowledge of mechanism of action was needed.

An approach that considered a diverse set consisting of both pharmaceuticals and industrial chemicals and based on machine learning classification methods has been proposed for the prediction of the chromosome-damaging potential of chemicals as assessed in the in vitro chromosome aberration (CA) test [50]. Using the publicly available CA-test results of more than 650 chemical substances, half of which were drug-like chemicals, two different computational models were developed. The first model, which uses the (Q)SAR tool MCASE (MULTICASE), gave a limited performance (53%) for the assessment of a chromosome-damaging potential (sensitivity), whereas for CA-test negative compounds, it correctly predicted with a specificity of 75%. The second model, constructed with a machine learning approach, generated a classification model based on 14 molecular descriptors, obtained after feature selection. The performance of the second model was found to be superior to the first one, primarily because of an improved sensitivity which might suggest that the more complex molecular descriptors in combination with statistical learning approaches are better suited to model the complex nature of mechanisms leading to a positive effect in the CA-test.

Shoji and Kawakami applied a molecular diversity-based ANN approach to estimate the genotoxicity using Umu test and systemic toxicity data on a highly diverse heterogeneous dataset of 82 environmental chemicals [93]. Using HOMO, LUMO, molecular weight, and log Kow as descriptors, the ANN approach was able to account for approximately 94% of the variation in the genotoxicity results. The authors noted that the predictive power was higher than the traditional regression approach, even though the detailed biochemical mechanisms responsible for genotoxicity were not clearly delineated by the ANN.

Votano and coworkers presented a structure–genotoxicity study by applying three different methods, ANN (three layer perceptron (TLP)), kNN, and Decision Forest (DF), to 3,363 diverse compounds that included more than 300 drugs [39]. All models were developed using the same initial set of 148 topological indices: molecular connectivity chi indices, electrotopological state indices (atom-type, bond-type, and single-atom E-State), as well as binary indicators. The three models yielded an average training/test concordance value of 88%, with a low percentage of false positive and negatives. When subjected to external validation using a dataset of 400 compounds not used in the development of the QSAR models, it gave an average concordance of 82%. This value increased to 92% upon removal of less reliable outcomes as determined by a reliability criterion used within each model. The ANN model showed the best performance in predicting drug compounds, yielding a 97% concordance after the removal of less reliable predictions. An interesting finding was that 14 of the most important descriptors related directly to known toxicophores involved in potent genotoxic responses in *S. typhimurium*.

QSAR studies carried out on mutagenicity of aminoazo dyes reported that when the structural diversity of the compounds in the original training dataset was enhanced, it required several new descriptors to formulate a credible model [81, 91]. The initial study by Garg and coworkers [81] considered a more homogeneous group of aminoazo derivatives containing only a single azo linkage between two aromatic rings, whereas the study by Sztandera and coworkers [91] included in addition, their *N*-hydroxy and ester metabolites, and also those containing two azo linkages and sulfonic acid groups to make it diverse. Both studies employed MLR and ANN methodologies to build the QSAR models.

Lazy structure–activity relationship (lazar) is a tool for the prediction of toxic properties including mutagenicity of chemical structures [99]. For a given query chemical, it first searches a database with chemical structures and experimental data (training set) for structures that are similar to it (nearest neighbors), and by using a modified kNN algorithm it computes a prediction from the experimental measurements of the neighbors/structurally similar analogues. Leave-one-out and external validation experiments indicated that *Salmonella* mutagenicity using lazar could be predicted with 85% accuracy for compounds within the applicability domain of the training set chemicals (3,895 chemicals from the Chemical Potency Database). One of the advantages of lazar is that it provides the rationales in the form of structural features and similar compounds for the prediction, in addition to a reliable confidence index that indicates if a query structure falls within the applicability domain of the training database.

By applying the machine learning classifier random forest to a large dataset of 4,083 organic chemicals and using Molecular Map of Atom-level Properties (MOLMAP) descriptors of bond properties, Zhang and Aires-de-Sousa were not only able to associate meaningful probabilities to the mutagenicity predictions but also to explain the predictions in terms of similarities between query structures and compounds in the training set [2]. For an external validation set of 472 compounds, they could achieve an error percentage as low as 15%.

A recent study by Papa and coworkers used human B-lymphoblastoid cells as an alternative to Ames bacterial test to model mutagenicity of 70 polycyclic aromatic hydrocarbons [96]. They applied QSAR classification methods such as kNN and CART (classification and regression tree), using hundreds of theoretical descriptors (1D, 2D, 3D, quantum-chemical) that were selected by GA. An interesting finding of the study was that the GA did not select classical parameters such as log Kow or quantum-chemical descriptors but instead it selected descriptors of dimension and shape (S1K). Both classifiers performed equally well giving an overall accuracy of about 80% for classifying actives and inactives and a pretty low percentage of false negatives and false positives.

3.3 Miscellaneous (Q)SARs

A study developed and compared a series of methods for prediction of mutagenic potency in *S. typhimurium* strain TA100 for 73 aromatic and heteroaromatic amines, using quantitative molecular similarity analysis (QMSA) [97]. The first method called the atom pair (AP) was based on topological descriptors. Out of the remaining ones that were based on Euclidean distance (ED) within an *n*-dimensional space, two were derived from physicochemical and electronic parameters, whereas the last one was derived from a combination of both topological and physicochemical parameters. The similarity spaces were based on parameters that quantify size, shape, branching, and bonding patterns in molecular architecture in addition to those that adequately reflect different aspects of mechanism of action of the chemicals at the molecular level. The study found that the AP-based similarity method was almost as effective as the property-based (PROP) method (ED within 3-dimensional space consisting of original physicochemical properties) in predicting mutagenicity.

Chroust and coworkers developed a QSAR approach to predict genotoxicity of saturated and unsaturated aliphatic halogenated compounds using Wing Spot test of *Drosophila melanogaster* [77]. They applied principal component analysis (PCA) for the selection of training set and the descriptors. In all, 28 descriptors complemented by 55 3D and 24 quantum-mechanical descriptors were used to develop the QSAR model for the set of 17 halogenated compounds by means of PLS. Nucleophilic superdelocalizability calculated by quantum mechanics was found to be a good parameter for predicting both toxicity and genotoxicity effects of halogenated aliphatic compounds.

A recent study on nitroarenes demonstrated how the use of modern techniques such as 2D and 3D QSAR could highlight certain subtleties (e.g., alignment of molecules when interacting with DNA, fragment-wise color-coding that describes the most and least contributing mutagenic fragment, etc.) involved in the interactions of toxic chemicals with receptors that are normally not captured by traditional

QSAR methodologies [42]. The authors compared four different methodologies, two 2D QSAR methods, Hologram QSAR (HQSAR) and Genetic Function Approximation (GFA), and two 3D QSAR methods, CoMFA and Comparative Molecular Similarity Indices Analysis (CoMSIA). They found that certain classes of nitroarenes particularly those with pyrene and biphenyl structures were predicted better by 3D QSAR methods, whereas some others such as nitrobenzenes and indazoles were well-predicted by 2D methods.

Debnath and coworkers demonstrated how a judicious use of classical QSAR and CoMFA could complement each other and enhance the applicability of structure–activity relationship technique [64]. They applied both of these techniques to elucidate the mechanisms of genotoxicity as indicated by SOS-inducing potential of nitrofuran derivatives on *Escherichia coli* PQ37. Three important factors, namely, electronic (qc2), hydrophobic (log P), and steric were found to be contributing toward the genotoxic activity of these compounds. qc2, the charge on the c2 atom attached to the NO_2 group, was found to support a furan ring opening mechanism in explaining the genotoxicity. The study also demonstrated the potential of CoMFA analysis to unravel the steric features of the molecules through contour maps.

For a diverse set of 358 nitroaromatics tested using *Salmonella* strains TA98 and TA100, four CoMFA models were developed [101]. These models not only agreed with the postulated mechanisms of mutagenicity but also could explain over 70% of the corresponding mutagenic variance. The use of CoMFA methodology in this study also aided the elucidation of genotoxicity mechanisms.

A study reported the application of a novel method called quantum topological molecular similarity (QTMS) to two different sets of mutagenic compounds, triazenes, and hydroxyfuranones (MX) derivatives [89]. The method generated bond descriptors from contemporary geometry-optimized ab initio wave functions and used chemometric analysis to generate QSARs. The QTMS selects the bonds directly involved in the activity, and therefore is able to highlight the active center of the mutagens thereby providing mechanistic information on their genotoxicity. For instance, the study indicated that the triazene hydroxylation pathway involving direct hydrogen abstraction from the methyl group was strongly disfavored, whereas for the MX derivatives it underlined the central role played by C_α in the $C_\beta = C_\alpha - C = O$ system and did not exclude lactone ring opening.

The machine learning models have been criticized for their inability to either lend support or propose or even contradict a particular mechanism of action [49]. To account for mechanistic basis, a novel 2-step approach that combined the Ames model for bacterial mutagenicity and another model accounting for additional mechanisms that led to chromosomal aberrations (CA) has been proposed [49]. A set of 497 chemicals, for which data on induced CA without S9 activation were available, was used to derive a CA model that did not account for metabolic activation of chemicals, whereas another set comprising 162 chemicals, for which data on induced CA in the presence of S9 activation were available, was used for modeling CA with metabolic activation of chemicals. The approach was unique because it integrated the tissue metabolism simulator (TIMES) that facilitates consideration of both parent chemicals as well as their active metabolites to determine their activity toward the nucleic acid. The alerting groups associated with different mechanisms were defined by specific structural boundaries as well as by 2D and 3D parameter ranges describing effects of bioavailability and reactivity of alerts that were conditioned by the rest of the molecules. Moreover, the role of each alert has been justified by an interaction mechanism(s) identified in the literature or introduced by experts. The performance of the model without metabolic activation was characterized by sensitivity and specificity values of 77 and 82%, respectively, whereas for the model coupled with the metabolic simulator the sensitivity reached 75% but the specificity dropped to 56%.

3.4 Expert Systems and Commercial Programs

Using number and ratio of elements, side chains, bonding position, and microenvironment of side chains as the structural factors of chemicals, an expert system to predict their mutagenicity was developed by Nakadate [100]. Eight rules that were analyzed by discriminant analysis were used to predict the mutagenicity of aliphatic and heterocyclic chemicals (which were not used to make the rules) with 90% accuracy.

Regulatory agencies and pharmaceutical industries often use commercial genotoxicity prediction programs such as TOPKAT, CASETOX and DEREK. As mentioned earlier, the first two are data-driven models, whereas the last one is a rule-based expert system. CASETOX is a statistically driven substructure/fragment-based system that does not use prior knowledge of mechanism of action or structural alerts but develops its own rules dynamically during model development, relating test chemical structures with the endpoint under investigation. The CASETOX Ames mutagenicity (all strains combined) model uses around 5,864 training set chemicals whose genotoxicity was determined by Ames tests of different bacterial strains, while TOPKAT mutagenicity model contains a training set of about 1,865 chemicals. Additionally, the CASETOX, on the one hand, has other genotoxicity modules with limited training sets built into it that can predict chromosomal aberrations, induction of micronuclei, unscheduled DNA synthesis, mouse lymphoma mutation, and *Drosophila* somatic mutation. DEREK, on the other hand, codifies existing knowledge derived from human experts into generalized rules. Unlike CASETOX or TOPKAT, DEREK does not have a chemical database built into it but instead has a set of knowledge-based rules and structural alerts embedded in it. These three programs are among the most popular ones that are commercially available. Therefore, these have been subjected to frequent scrutiny to investigate their efficiency. Cariello and coworkers carried out a study involving 400 chemicals with known Ames tests results to compare TOPKAT and DEREK to determine their abilities in predicting bacterial mutagenicity [105]. The overall concordance of the TOPKAT program was found to be higher than DEREK. TOPKAT fared more poorly than DEREK in the critical Ames-positive category, where 60% of the chemicals were incorrectly predicted by TOPKAT as negative but were mutagenic in the Ames test. For DEREK, 54% of the Ames-positive molecules had no structural alerts and were predicted to be nonmutagenic. Another study compared the validation results of their in-house built models with those of TOPKAT, CASETOX, and DEREK and demonstrated that the in-house built models performed substantially better than the commercial ones [39]. A comparative evaluation of commercial programs CASETOX, TOPKAT, and DEREK for the prediction of Ames test mutagenicity using a set of 614 compounds (520 drug candidates and 94 industrial) revealed that all three programs predicted with similar level of accuracy for both types of chemicals at greater than 80% [106]. However, higher confidence could only be assigned to the prediction of nonmutagens and not to the prediction of mutagens since the accuracy was primarily driven by the specificity values of a dataset that was heavily weighted with nonmutagens.

The MULTICASE methodology has also been applied to develop a chromosomal aberration (CA) prediction model using a dataset of 233 chemicals (current version 1.9 has 805 chemicals) and it now forms a part of the genotoxicity suite of models in the CASETOX program [107]. Using an internal validation strategy, the observed sensitivity and specificity (i.e., the correct prediction of positives and negatives, respectively) of the model were found to be 53 and 71%, respectively [108].

MDL QSAR is another commercial program available that predicts mutagenicity of organic chemicals [109]. It comprises models that are developed using atom-type E-state indices and nonparametric discriminant analysis. The mutagenicity models included are *S. typhimurium* gene mutation tests (all strains combined TA97, TA98, TA100, TA1535, TA1536, TA1537, and TA1538) containing 3228 chemicals, *E. coli* gene mutation tests (WP2, WP100, and polA) containing 472 chemicals and a composite microbial mutation model that combines the first two with a *Bacillus subtilis* rec spot test study results, making a total of 3338 chemicals. External test sets of 1,444 and 1,485 compounds were used to validate the *Salmonella* and the composite microbial mutagenesis models, respectively. The average specificity, sensitivity, positive predictivity, concordance, and coverage of both models were found to be 76%, 81%, 73%, 78%, and 98%, respectively.

4 Key Considerations in Regulatory Decision-Making Based on (Q)SAR

Over the past three decades, a number of genotoxicity-based (Q)SAR models have been built with a specific rationale and purpose, and, therefore, some of them may not necessarily be developed according to the needs of regulatory assessment. The benefits of using these models for regulatory decision-making may only be realized once all the underlying factors including its reliability, uncertainty, and predictivity have been properly assessed. Eleven important considerations in this regard should be taken into account. First, the Organization for Economic Cooperation and Development (OECD) has recommended certain "Principles for (Q)SAR Validation," which serve as a compliance guideline for a (Q)SAR model to be considered for regulatory applications [110]. Accordingly, a genotoxicity (Q)SAR model should be associated with a defined endpoint, an unambiguous algorithm, a defined domain of applicability, appropriate measures of goodness-of-fit, robustness and predictivity, and a mechanistic interpretation. Second, since (Q)SAR model is based on a finite set of information (its domain of applicability (DA)), therefore, predictions outside of this domain may not be completely reliable. Hence, for a proper interpretation of a model's prediction, it is important to know, if practicable, where the query chemical lies with respect to the model's DA [111–113]. Third, human exposures to a wide variety of chemicals generally occur at a low concentration level. However, chemicals (or their metabolites) may induce damage to DNA only at certain concentrations above a defined threshold, or at extreme or nonphysiological conditions that are not present in exposed humans [114]. Thus, conclusions drawn from predictions obtained from (Q)SAR models that are developed using studies at high concentrations may not reflect responses at a more biologically relevant dose. Fourth, in vitro tests may not be capable of differentiating between genders and also replicating genotoxic biotransformation processes as the in vivo systems do. Therefore, predictions obtained from (Q)SAR models that use in vivo data should be preferred. Fifth, a (Q)SAR model may perform well on internal validation but its real strength lies in its ability to predict chemicals outside its training set. For example, some of the models developed by Debnath and coworkers [59] and Cash [7] to predict the mutagenic potency of amines exhibited good statistics from internal validation with the training set but lacked predictivity for chemicals not used in the development of the models [104, 115–117]. Sixth, when using a rule-based expert system, it is important to consider the steric and electronic environment surrounding a structural alert fragment (labeled as toxicophore) in a given molecule. It may sometimes diminish or enhance its genotoxic potency or create a toxic fragment that has not been previously identified. This, in turn, may lead to a false prediction. Therefore, an expert opinion should be taken while interpreting information based on structural alerts. Seventh, when using (Q)SAR models based on classifiers (e.g., kNN, DT, RF, CART), the most important parameter to judge their performance is the percentage of correctly classified mutagenic compounds (sensitivity) [96]. A high sensitivity means that the classifier will correctly assign active molecules to the mutagen category (true positive) rather than classifying an inactive one as mutagenic (false positive). Eighth, for the effective application of (Q)SAR models, the descriptors must have a physicochemical interpretation that conforms with a known mechanism of biological action for the endpoint of interest. This is sometimes an issue with machine learning methods that generally consider a large number of descriptors in model development. For such models, it is difficult to establish a scientific correlation between the descriptors and the genotoxic mechanism and tend to fall short when subjected to external validation [104]. Ninth, no matter how good a (Q)SAR model performs statistically, it is important to ensure that the laboratory conditions used to derive the genotoxicity values were comparable because the quality of datasets eventually determines the quality of predictions. Tenth, it is critical to remember that in vitro mammalian cell genotoxicity tests have often resulted in an extremely high false-positive rate when compared with in vivo carcinogenicity in rodents [118]. (Q)SAR predictions based on false-positive genotoxicity data, therefore, will lead to unreliable results. Lastly, to minimize the uncertainty associated with the use of a single (Q)SAR model, a "battery" of endpoints covering a variety of test systems, species, and (Q)SAR methodologies should generally be considered in any risk assessment [119, 120]. Consistency in predictions across a "battery" of endpoints presents a trend and could help establish with reasonable certainty if a query chemical possesses genotoxic properties.

5 Conclusion

A genotoxic event is indicative of potential adverse human health effects including carcinogenicity, chromosomal aberrations, sister chromatid exchanges, and/or direct damage to DNA. Many chemicals are capable of causing these types of genetic alterations and therefore, screening of chemicals for their likely genotoxicity/mutagenicity is one of the important priorities of regulatory agencies. In silico (Q)SAR methodologies have the potential to serve as rapid and reliable screening tools capable of identifying chemicals likely to exert genotoxic effects. At the same time, these are in tune with the principle of 3Rs (replacement, reduction, replacement of animal testing of all chemicals of regulatory concern) and are highly cost-effective. Genotoxicity (Q)SAR models are currently being integrated into emerging data-gap filling applications such as the OECD's QSAR Application Toolbox [121].

A large proportion of early genotoxicity models, the so-called traditional QSARs, were developed using congeneric chemicals and multiple linear regression analysis. They performed extremely well on internal validation and were useful in providing insight into mechanistic basis. However, being built using small datasets their utility was limited. The advent of fast-paced personal computers saw a new era of highly sophisticated techniques such as chemical data mining, modern machine learning algorithms, programs to compute complex molecular descriptors, and three-dimensional technologies. These gave a whole new dimension to the (Q)SAR model development process. The new generation of genotoxicity models uses a very large training set that runs into few thousands, contains diverse chemical population, and has the capability of rapidly processing thousands of molecular descriptors. Commercial data-driven global models such as CASETOX that combines traditional and modern techniques or the rule-based expert systems, such as DEREK that typically combine mechanistic hypothesis, expert judgment, and empirical observation, are some of the most up-to-date programs and considered reliable. Finally, there is the class of hybrid models that employ newer machine learning algorithms, account for metabolism, and integrate the best features of the rule-based and the quantitative systems which hold a lot of promise.

Acknowledgments The authors thank Dr. Bhaja, K. Padhi, and Dr. Guosheng Chen for critical review of this chapter, and Dr. Eeva Leinala and Christine Norman of Health Canada for their support and encouragement.

References

1. Hayashi M, Kamata E, Hirose A, Takahashi M, Morita T, Ema M (2005) In silico assessment of chemical mutagenesis in comparison with results of Salmonella microsome assay on 909 chemicals. Mutat Res 588:129–135
2. Zhang Q-Y, Aires-de-Sousa J (2007) Random forest prediction of mutagenicity from empirical physicochemical descriptors. J Chem Inf Model 47:1–8
3. Todeschini R, Consonni V (2002) Handbook of molecular descriptors. Wiley-VCH, Germany
4. Todeschini R, Consonni V, Mauri A, Pavan (2005) M DRAGON Version 5.3, Telete srl, Milan, Italy
5. ADRIANA.Code. http://www.molecular-networks.com/software/adrianacode/index.html (accessed March 2008)
6. Liu K, Feng J, Young SS (2005) PowerMV: A software environment for molecular viewing, descriptor generation, data analysis and hit evaluation. J Chem Inf Model 45:515–522
7. Cash GG (2001) Prediction of the genotoxicity of aromatic and heteroaromatic amines using electrotopological state indices. Mutat Res 491:31–37
8. Bolzan AD, Bianchi MS (2002) Genotoxicity of streptozotocin. Mutat Res 512:121–134
9. Snyder RD, Pearl GS, Mandakas G, Choy WN, Goodsaid F, Rosenblum IY (2004) Assessment of the sensitivity of the computational programs DEREK, TOPKAT, and MCASE in the prediction of the genotoxicity of pharmaceutical molecules. Environ Mol Mutagen 43: 143–158
10. Maron DM, Ames BN (1983) Revised methods for the *Salmonella* mutagenicity test. Mutat Res 113:173–215
11. Ames BN, McCann J, Yamasaki E (1975) Method for detecting carcinogens and mutagens with the *Salmonella*/mammalian microsome test. Mutat Res 31:347–364
12. Williams GM, Kroes R, Waaijers HW, van de Poll KW (1980) The predictive value of short-term screening tests in carcinogenicity evaluation. Elsevier, Amsterdam
13. Mohn GR, Ellenberger J, van Bladeren PJ (1980) Evaluation and relevance of *Escherichia coli* test systems for detecting and for characterizing chemical carcinogens and mutagens. In: Williams GM, Kroes R, Waaijers HW, van de Poll KW (eds) Applied methods in oncology, Vol 3. Elsevier, Amsterdam
14. Rontó G, Tarján I, Gáspár S (1986) Phage T7-inactivation test. A possibility of quantitative mutagenicity screening. Physiol Chem Phys Med NMR 18:275–285
15. Bradley MO, Bhuyan B, Francis MC, Langenbach R, Peterson A, Huberman E (1981) Mutagenesis by chemical agents in V79 Chinese hamster cells: a review and analysis of the literature. Mutat Res 87:81–142
16. Clive D, Spector JFS (1975) Laboratory procedure for assessing specific locus mutations at the Tk locus in cultured L5178Y mouse lymphoma cells. Mutat Res 31:17–29
17. Clive D, Johnson KO, Spector JF, Batson AG, Brown MM (1979) Validation and characterization of the L5178Y/TK$+/-$ mouse lymphoma mutagen assay system. Mutat Res 59:61–108
18. Clive D, McCuen R, Spector JFS, Piper C, Mavourkin KH (1983) Specific gene mutations in L5178Y cells in culture. Mutat Res 115:225–251
19. Fenech M, Morley AA (1985) Measurement of micronuclei in lymphocytes. Mutat Res 147: 29–36
20. Latt SA, Allen J, Bloom SE, Carrano A, Falke E, Kram D, Schneider E, Schreck R, Tice R, Whitfield B, Wolff S (1981) Sister chromatid exchanges: A report of the Gene-Tox Program. Mutat Res 87:17–62
21. Schmid W (1975) The Micronucleus test. Mutat Res 31:9–15
22. Heddle JA, Hite M, Kirkhart B, Mavournin K, MacGregor JT, Newell GW, Salamone MF (1983) The induction of micronuclei as a measure of genotoxicity. A report of the U.S. Environmental Protection Agency Gene-Tox Program. Mutat Res 123:61–118
23. Graf U, Juon H, Katz AJ, Frei HI, Wuergler FW (1983) A pilot study on a new Drosophila spot test. Mutat Res 120:233–239

24. Rodriguez-Arnaiz R, Tellez GO (2002) Structure–activity relationships of several anisidine and dibenzanthracene isomers in the w/w+ somatic assay of *Drosophila melanogaster*. Mutat Res 514:193–200
25. Morpurgo GD, Bellincampi G, Gualandi L, Baldineili, Crescenzi OS (1979) Analysis of mitotic non-disjunction with *Aspergillus nidulans*. Environ Health Perspect 31:81–95
26. Hofnung M, Quillardet P (1988) The SOS Chromotest, a colorimetric assay based on the primary cellular responses to genotoxic agents. Ann NY Acad Sci 534:817–825.
27. Quillardet P, Hofnung M (1993) The SOS Chromotest: A review. Mutat Res 297:235–279
28. Speit G, Hartmann A (1999) The comet assay (single-cell gel test): a sensitive genotoxicity test for the detection of DNA damage and repair. DNA Repair Protocols 113:203–212
29. Oda Y, Nakamuro S, Oki T, Kato T, Shinagawa H. (1985) Evaluation of the new system (umu-test) for the detection of environmental mutagens and carcinogens. Mutat Res 147:219–229
30. Williams GM (1977) Detection of chemical carcinogens by unscheduled DNA synthesis in rat liver primary cell cultures. Cancer Res 37:1845–1851
31. Venger BH, Hansch C, Hatheway GJ, Amrein YU (1979) Ames test of 1-(X-phenyl)-3,3-dialkyltriazenes. A quantitative structure-activity study. J Med Chem 22:473–476
32. Hansch C, Venger BH, Panthananickal A (1980) Mutagenicity of substituted (*o*-phenylenediamine) platinum dichloride in the Ames test. A quantitative structure-activity analysis. J Med Chem 23:459–461
33. Debnath AK, de Compadre RLL, Shusterman AJ, Hansch C (1992) Quantitative structure-activity relationship investigation of the role of hydrophobicity in regulating mutagenicity in the Ames test, 2. Mutagenicity of aromatic and heteroaromatic nitro compounds in *Salmonella typhimurium* TAI00. Environ Mol Mutagen 19:53–70
34. King RD, Muggleton SH, Srinivasan A, Sternberg MJE (1996) Structure-activity relationships derived by machine learning: The use of atoms and their bond connectivities to predict mutagenicity by inductive logic programming. Proc Natl Acad Sci USA 93:438–442
35. Serra JR, Thompson ED, Jurs PC (2003) Development of binary classification of structural chromosome aberrations for a diverse set of organic compounds from molecular structure. Chem Res Toxicol 16:153–163
36. Li H, Ung CY, Yap CW, Xue Y, Li ZR, Cao ZW, Chen YZ (2005) Prediction of genotoxicity of chemical compounds by statistical learning methods. Chem Res Toxicol 18:1071–1080
37. Mosier PD, Jurs PC, Custer LL, Durham SK, Pearl GM (2003) Predicting the genotoxicity of thiophene derivatives from molecular structure. Chem Res Toxicol 16:721–732
38. Nesnow S, Langenbach R, Mass MJ (1985) Pattern recognition analysis of a set of mutagenic aliphatic N-nitrosamines. Environ Health Perspect 61:345–349
39. Votano JR, Parham M, Hall LH, Kier LB, Oloff S, Tropsha A, Xie Q, Tong W (2004) Three new consensus QSAR models for the prediction of Ames genotoxicity, Mutagenesis 19: 365–377
40. Zheng M, Liu Z, Xue C, Zhu W, Chen K, Luo Z, Jiang H (2006) Mutagenic probability estimation of chemical compounds by a novel molecular electrophilicity vector and support vector machine. Bioinformatics 22:2099–2106
41. He L, Jurs PC, Custer LL, Durham SK, Pearl GM (2003) Predicting the genotoxicity of polycyclic aromatic compounds from molecular structure with different classifiers. Chem Res Toxicol 16:1567–1580
42. Nair PC, Sobhia ME (2008) Comparative QSTR studies for predicting mutagenicity of nitro compounds. J Mol Graph Model 26:916–934
43. Greene N, Judson PN, Langowski JJ, Marchant CA (1999) Knowledge-based expert systems for toxicity and metabolism prediction: DEREK, StAR and METEOR. SAR QSAR Environ Res 10:299–314
44. CompuDrug Inc., HazardExpert. http://www.compudrug.com/(accessed March 2008).
45. Poroikov VV, Filimonov DA, Ihlenfeldt WD, Gloriozova TA, Lagunin AA, Borodina YV, Stepanchikova AV, Nicklaus MC (2003). PASS biological activity spectrum predictions in the enhanced open NCI database browser. J Chem Inf Comput Sci 43:228–236

46. Rosenkranz HS, Cunningham AR, Zhang YP, Claycamp HG, Macina OT, Sussman NB, Grant SG, Klopman G (1999) Development, characterization and application of predictive-toxicology models. SAR QSAR Environ Res 10:277–298
47. Enslein K, Gombar VK, Blake BW (1994). Use of SAR computer-assisted prediction of carcinogenicity and mutagenicity of chemicals by the TOPKAT program. Mutat Res 305:47–61
48. Mekenyan O, Dimitrov S, Serafimova R, Thompson E, Kotov S, Dimitrova N, Walker JD (2004) Identification of the structural requirements for mutagenicity by incorporating molecular flexibility and metabolic activation of chemicals I: TA100 model. Chem Res Toxicol 17:753–766
49. Mekenyan O, Todorov M, Serafimova R, Stoeva S, Aptula A, Finking R, Jacob E (2007) Identifying the structural requirements for chromosomal aberration by incorporating molecular flexibility and metabolic activation of chemicals. Chem Res Toxicol 20:1927–1941
50. Rothfuss A, Steger-Hartmann T, Heinrich N, Wichard J (2006) Computational Prediction of the chromosome-damaging potential of chemicals. Chem Res Toxicol 19:1313–1319
51. Biagi GL, Barbaro AM, Guerra MC, Cantelli Forti G, Aicardi G, Borea PA. (1983) Quantitative relationship between structure and mutagenic activity in a series of 5-nitroimidazoles. Teratog Carcinog Mutagen 3:429–438
52. Jinno K (1984) The quantitative structure-activity relationship approach to the mutagenicity of *N*-nitrosomethylaniline compounds. Mutat Res 141:141–143
53. Hakura A, Ninomiya S, Kohda K, Kawazoe Y (1984) Studies on chemical carcinogens and mutagens. XXVI. Chemical properties and mutagenicity of alkyl alkanesulfonates on *Salmonella typhimurium* TA100. Chem Pharm Bull (Tokyo) 32:3626–3635
54. Rashid KA, Mullin CA, Mumma RO (1986) Structure-mutagenicity relationships of chalcones and their oxides in the *Salmonella* assay. Mutat Res 169:71–79
55. Trieff NM, Biagi GL, Ramanujam VMS, Connor TH, Cantelli-Forti G, Guerra MC, Bunce HB III, Legator MS (1989) Aromatic amines and acetamides in *Salmonella typhimurium* TA98 and TA100: A QSAR relation study. J Mol Toxicol 2:53–65
56. Debnath AK, de Compadre RLL, Debnath G, Shusterman AJ, Hansch C (1991) Structure-activity relationship of mutagenic aromatic and heteroaromatic nitro compounds. Correlation with molecular orbital energies and hydrophobicity. J Med Chem 34:786–797
57. Eriksson L, Jonsson J, Hellberg S, Lindgren F, Sjostrom M, Wold S (1991) A strategy for ranking environmentally occurring chemicals. Part V: The development of two genotoxicity QSARs for halogenated aliphatics. Environ Tox Chem 10:585–596
58. Debnath AK, Hansch C (1992) Structure-activity relationship of genotoxic nitropolycyclic aromatic compounds: Further evidence for the importance of hydrophobicity and molecular orbital energies in genetic toxicity. Environ Mol Mutagen 20:140–144
59. Debnath AK, Debnath G, Shusterman AJ, Hansch C (1992) A QSAR investigation of the role of hydrophobicity in regulating mutagenicity in the Ames test, 1. Mutagenicity of aromatic and heteroaromatic amines in *Salmonella typhimurium* TA98 and TA100. Environ Mol Mutagen 19:37–52
60. Debnath AK, de Compadre RLL, Hansch C (1992) Mutagenicity of quinolines in *Salmonella typhimurium* TA100, A QSAR study based on hydrophobicity and molecular orbital determinants. Mutat Res 280:55–65
61. Ford GP, Hermans PS (1992) Relative stabilities of nitrenium ions derived from polycyclic aromatic amines – Relationship to mutagenicity. Chem Biol Interact 81:1–18
62. Ford GP, Griffin GR (1992) Relative stabilities of nitrenium ions derived from heterocyclic amine food carcinogens: Relationship to mutagenicity. Chem Biol Interact 81:19–33
63. Tuppurainen K, Lötjönen S, Laatikainen R, Vartiainen T (1992) Structural and electronic properties of MX compounds related to TA100 mutagenicity. A semi-empirical molecular orbital QSAR study. Mutat Res 266:181–188
64. Debnath AK, Hansch C, Kim KH, Martin YC (1993) Mechanistic interpretation of the genotoxicity of nitrofurans (antibacterial agents) using quantitative structure-activity relationships and comparative molecular field analysis. J Med Chem 36:1007–1016
65. Debnath AK, Hansch C (1993) The importance of hydrophobicity in the mutagenicity of methanesulfonic acid esters with S. typhimurium. Chem Res Toxicol 6:310–312

66. Hooberman BH, Chakraborty PK, Sinsheimer JE (1993) Quantitative structure–activity relationships for the mutagenicity of propylene oxides with *Salmonella*. Mutat Res 299:85–93
67. Debnath AK, Shusterman AJ, Lopez de Compadre RL, Hansch C (1994) The importance of the hydrophobic interaction in the mutagenicity of organic compounds. Mutat Res 305:63–72
68. Andre V, Boissart C, Sichel F, Gauduchon P, Le Talaer JY, Lancelot JC, Robba M, Mercier C, Chemtob S, Raoult E, Tallec A (1995) Mutagenicity of nitro- and amino-substituted carbazoles in *Salmonella typhimurium*. II. Ortho-aminonitro derivatives of 9H-carbazole. Mutat Res 345:11–25
69. Hatch FT, Colvin ME (1997) Quantitative structure-activity (QSAR) relationships of mutagenic aromatic and heterocyclic amines. Mutat Res 376:87–96
70. Smith CJ, Hansch C, Morton MJ (1997) QSAR treatment of multiple toxicities: the mutagenicity and cytotoxicity of quinolines. Mutat Res 379:167–175
71. Tafazoli M, Baeten A, Geerlings P, Kirsch-Volders M (1998) In vitro mutagenicity and genotoxicity study of a number of short-chain chlorinated hydrocarbons using the micronucleus test and the alkaline single cell gel electrophoresis technique (Comet assay) in human lymphocytes: a structure-activity relationship (QSAR) analysis of the genotoxic and cytotoxic potential, Mutagenesis 13:115–126
72. Hatch FT, Knize MG, Colvin ME (2001) Extended quantitative structure-activity relationships for 80 aromatic and heterocyclic amines: structural, electronic, and hydropathic factors affecting mutagenic potency. Environ Mol Mutagen 38:268–291
73. Andrews LE, Bonin AM, Fransson LE, Gillson AM, Glover SA (2006) The role of steric effects in the direct mutagenicity of *N*-acyloxy-*N*-alkoxyamides. Mutat Res 605:51–62
74. Chung KT, Chen SC, Wong TY, Li YS, Wei CI, Chou MW (2000) Mutagenicity studies of benzidine and its analogs: Structure–activity relationships. Toxicol Sci 56:351–356
75. Benigni R, Bossa C, Netzeva T, Rodomonte A, Tsakovska I (2007) Mechanistic QSAR of aromatic amines: New models for discriminating between homocyclic mutagens and nonmutagens, and validation of models for carcinogens. Environ Mol Mutagen 48:754–771
76. Gramatica P, Pilutti P, Papa E (2007) Approaches for externally validated QSAR modeling of nitrated polycyclic aromatic hydrocarbon mutagenicity. SAR QSAR Environ Res 18:169–178
77. Chroust K, Pavlova M, Prokop Z, Mendel J, Bozkova K, Kubat Z, Zajickova V, Damborsky J (2007) Quantitative structure–activity relationships for toxicity and genotoxicity of halogenated aliphatic compounds: Wing spot test of *Drosophila melanogaster*. Chemosphere 67:152–159
78. Hu J, Wang W, Zhu Z, Chang H, Pan F, Lin B (2007) Quantitative structure–activity relationship model for prediction of genotoxic potential for quinolone antibacterials. Environ Sci Technol 41:4806–4812
79. Knize MG, Hatch FT, Tanga MJ, Lau EY, Colvin ME (2006) A QSAR for the mutagenic potencies of twelve 2-amino-trimethylimidazopyridine isomers: structural, quantum chemical, and hydropathic factors. Environ Mol Mutagen 47:132–146
80. Tekiner-Gulbas B, Temiz-Arpaci O, Oksuzoglu E, Eroglu H, Yildiz I, Diril N, Aki-Sener E, Yalcin I (2007) QSAR of genotoxic active benzazoles. SAR QSAR Environ Res 18:251–263
81. Garg A, Bhat KL, Bock CW (2002) Mutagenicity of aminoazobenzene dyes and related structures: A QSAR/QPAR investigation. Dyes Pigments 55:35–52
82. Bonin AM, Banks TM, Campbell JJ, Glover SA, Hammond GP, Prakash AS, Rowbottom CA (2001) Mutagenicity of electrophilic *N*-acyloxy-*N*-alkoxyamides. Mutat Res 494:115–134
83. Gupta RL, Dey DK, Juneja TR (1985) Structure-mutagenicity relationships within a series of para-alkoxynitrosobenzenes. Toxicol Lett 28:125–132
84. Tsai RS, Carrupt PA, Testa B, Caldwell J (1994) Structure-genotoxicity relationships of allylbenzenes and propenylbenzenes: A quantum chemical study. Chem Res Toxicol 7:73–76
85. Sugiura K, Goto M (1981) Mutagenicities of styrene oxide derivatives on bacterial test systems: Relationship between mutagenic potencies and chemical reactivity. Chem-Biol Interac 35:71–91
86. Sugiura K, Goto M (1983) Mutagenicities of glycidyl ethers for *Salmonella typhimurium*: Relationship between mutagenic potencies and chemical reactivity. Chem-Biol Interac, 45:153–169

87. Hrelia P, Fimognari C, Maffei F, Brighenti B, Garuti L, Burnelli S, Cantelli-Forti G (1998) Synthesis, metabolism and structure–mutagenicity relationships of novel 4-nitro-imidazoles and pyrazoles/in *Salmonella typhimurium*. Mutat Res 397:293–301
88. Poso A, Wright A, Gynther J (1995) An empirical and theoretical study on mechanisms of mutagenic activity of hydrazine compounds. Mutat Res 332:63–71
89. Popelier PL, Smith PJ, Chaudry UA (2004) Quantitative structure-activity relationships of mutagenic activity from quantum topological descriptors: Triazenes and halogenated hydroxyfuranones (mutagen-X) derivatives. J Comput Aided Mol Des 18:709–718
90. Henry DR, Lavine BK, Jurs PC (1987) Electronic factors and acridine frameshift mutagenicity – A pattern recognition study. Mutat Res 179:115–121
91. Sztandera L, Garg A, Hayik S, Bhat KL, Bock CW (2003) Mutagenicity of aminoazo dyes and their reductive-cleavage metabolites: a QSAR/QPAR investigation. Dyes Pigments 59: 117–133
92. Benigni R, Conti L, Crebelli R, Rodomonte A, Vari MR (2005) Simple and alpha, beta-unsaturated aldehydes: Correct prediction of genotoxic activity through structure-activity relationship models. Environ Mol Mutagen 46:268–280
93. Shoji R, Kawakami M (2006) Prediction of genotoxicity of various environmental pollutants by artificial neural network simulation. Mol Divers 10:101–108
94. Gramatica P, Papa E, Marrocchi A, Minuti L, Taticchi A (2007) Quantitative structure–activity relationship modeling of polycyclic aromatic hydrocarbon mutagenicity by classification methods based on holistic theoretical molecular descriptors. Ecotoxicol Environ Safety 66:353–361
95. Serafimova R, Todorov M, Pavlov T, Kotov S, Jacob E, Aptula A, Mekenyan O (2007) Identification of the structural requirements for mutagenicity, by incorporating molecular flexibility and metabolic activation of chemicals. II. General Ames mutagenicity model. Chem Res Toxicol 20:662–676
96. Papa E, Pilutti P, Gramatica P (2008) Prediction of PAH mutagenicity in human cells by QSAR classification. SAR QSAR Environ Res 19:115–127
97. Basak SC, Grunwald GD (1995) Predicting mutagenicity of chemicals using topological and quantum chemical parameters: A similarity based study. Chemosphere 31:2529–2546
98. Basak SC, Mills DR, Balaban AT, Gute BD (2001) Prediction of mutagenicity of aromatic and heteroaromatic amines from structure: a hierarchical QSAR approach. J Chem Inf Comput Sci 41:671–678
99. Helma C (2005) lazar: Lazy Structure – Activity Relationships for toxicity prediction. In: Helma C (ed) Predictive Toxicology, Taylor and Francis, Boca Raton, FL
100. Nakadate M (1998) Toxicity prediction of chemicals based on structure activity relationships. Toxicol Lett 102–103:627–629
101. Fan M, Byrd C, Compadre CM, Compadre RL (1998) Comparison of CoMFA models for *Salmonella typhimurium* TA98, TA100, TA98 + S9 and TA100 + S9 mutagenicity of nitroaromatics. SAR QSAR Environ Res 9:187–215
102. Benigni R, Passerini L, Gallo G, Giorgi F, Cotta-Ramusino M (1998) QSAR models for discriminating between mutagenic and nonmutagenic aromatic and heteroaromatic amines. Environ Mol Mutagen 32:75–83
103. Kadlubar FF, Beland FA (1985) Chemical properties of ultimate carcinogenic metabolites of arylamines and arylamides. In: Harvey RG (ed) Polycyclic hydrocarbons and carcinogenesis, American Chemical Society, Washington, DC
104. Cash GG, Anderson B, Mayo K, Bogaczyk S, Tunkel J (2005). Predicting genotoxicity of aromatic and heteroaromatic amines using electrotopological state indices. Mutat Res 585:170–183
105. Cariello NF, Wilson JD, Britt BH, Wedd DJ, Burlinson B, Gombar V (2002) Comparison of the computer programs DEREK and TOPKAT to predict bacterial mutagenicity. Mutagenesis 17:321–329
106. White AC, Mueller RA, Gallavan RH, Aaron S, Wilson AGE (2003) A multiple in silico program approach for the prediction of mutagenicity from chemical structure. Mutat Res 539:77–89

107. Rosenkranz HS, Ennever FK, Dimayuga M, Klopman G (1990) Significant differences in the structural basis of the induction of sister chromatid exchanges and chromosomal aberrations in Chinese hamster ovary cells. Environ Mol Mutagen 16:149–177
108. Rosenkranz HS (2004) SAR modeling of genotoxic phenomena: the consequence on predictive performance of deviation from a unity ratio of genotoxicants/non-genotoxicants. Mutat Res 559:67–71
109. Contrera JF, Matthews EJ, Kruhlak NL, Benz RD (2005) In silico screening of chemicals for bacterial mutagenicity using electrotopological E-state indices and MDL QSAR software. Regul Toxicol Pharmacol 43:313–323
110. OECD (2007) Guidance document on the validation of (Quantitative) structure-activity relationship [(Q)SAR] models. Series on Testing and Assessment No. 69, Paris, (ENV/JM/MONO(2007)2). Available at http://www.oecd.org/dataoecd/55/35/38130292.pdf
111. Kulkarni SA, Zhu J (2008) Integrated approach to assess the domain of applicability of some commercial (Q)SAR models. SAR QSAR Env Res 19:39–54
112. Dimitrov S, Dimitrova G, Pavlov T, Dimitrova N, Patlewicz G, Niemela J, Mekenyan O (2005) A stepwise approach for defining the applicability domain of SAR and QSAR models. J Chem Inf Model 45:839–849
113. Schultz TW, Hewitt M, Netzeva TI, Cronin MTD (2007) Assessing applicability domains of toxicological QSARs: Definition, confidence in predicted values, and the role of mechanisms of action. QSAR Comb Sci 26:238–254
114. Kirkland DJ, Aardema M, Banduhn N, Carmichael P, Fautz R, Meunier JR, Pfuhler S (2007) In vitro approaches to develop weight of evidence (WoE) and mode of action (MoA) discussions with positive in vitro genotoxicity results. Mutagenesis 22:161–175
115. Gramatica P, Consonni V, Pavan M (2003) Prediction of aromatic amines mutagenicity from theoretical molecular descriptors. SAR QSAR Environ Res 14:237–250
116. Glende C, Schmitt H, Erdinger L, Engelhardt G, Boche G (2001) Transformation of mutagenic aromatic amines into nonmutagenic species by alkyl substituents. Part I. Alkylation ortho to the amino function. Mutat Res 498:19–37
117. Glende C, Klein M, Schmitt H, Erdinger L, Boche G (2002) Transformation of mutagenic aromatic amines into non-mutagenic species by alkyl substituents. Part II. Alkylation far away from the amino function. Mutat Res 515:15–38
118. Kirkland D, Aardema M, Henderson L, Mueller L (2005) Evaluation of the ability of a battery of 3 in vitro genotoxicity tests to discriminate rodent carcinogens and non-carcinogens. I. Sensitivity, specificity and relative predictivity. Mutat Res 584:1–256
119. Cahill PA, Knight AW, Billinton N, Barker MG, Walsh L, Keenan PO, Williams CV, Tweats DJ, Walmsley RM (2004) The GreenScreen genotoxicity assay: a screening validation programme. Mutagenesis 19:105–119
120. Matthews EJ, Kruhlak NL, Cimino MC, Benz RD, Contrera JF (2006) An analysis of genetic toxicity, reproductive and developmental toxicity, and carcinogenicity data: II. Identification of genotoxicants, reprotoxicants, and carcinogens using in silico methods. Regul Toxicol Pharmacol 44:97–110
121. QSAR Application Toolbox. OECD. www.oecd.org/document/23/0,3343, en_2649_ 34379_ 33957015_1_1_1_1,00.html (accessed April 2008)

Chemistry Based Nonanimal Predictive Modeling for Skin Sensitization

David W. Roberts and Grace Patlewicz

Abstract Skin sensitization is a significant environmental and occupational health concern. The possibility of workers and consumers becoming sensitized is a major problem for individuals, employers and marketing certain products. Consequently, there exists an important need to accurately identify chemicals that have the potential to cause skin sensitization. Under Registration, Evaluation, Authorization and Restriction of Chemicals (REACH), the sensitizing potential needs to be assessed for substances manufactured or imported at levels of 1 ton or greater per annum. Assessment of skin sensitization has traditionally relied on animal testing, but there are now strong pressures to reduce and ultimately eliminate the use of animals for this purpose. Building on research dating back at least 7 decades, a quite detailed mechanistic understanding of skin sensitization, both in terms of the underlying biological processes and the underlying chemistry, has been developed, with significant advances being made in the present century. This chapter presents an overview of the current biological and chemical mechanistic understanding, and reviews recent progress in nonanimal predictive modeling for skin sensitization, with an emphasis on how the understanding of these mechanisms can be applied in combination with in chemico and in silico approaches to hazard and risk assessment.

Keywords Skin sensitization · QSARs · Expert systems · In chemico · REACH

1 Introduction

1.1 Biological Mechanisms

Skin sensitization (also called delayed contact hypersensitivity, contact hypersensitivity, contact allergy, or allergic contact dermatitis) is a T cell mediated immunological response specific for the substance. Research dating back more than 7 decades

G. Patlewicz (✉)
DuPont Haskell Global Centers for Health and Environmental Sciences,
1090 Elkton Road, Newark, DE 19711, USA
e-mail: patlewig@hotmail.com

J. Devillers (ed.), *Ecotoxicology Modeling*, Emerging Topics in Ecotoxicology: Principles, Approaches and Perspectives 2, DOI 10.1007/978-1-4419-0197-2_3,

has established a very strong, although still incomplete, mechanistic understanding of the chemical and biological basis of skin sensitization [1–6]. Skin sensitization involves two phases: an induction and elicitation phase.

During induction, the sensitizing chemical penetrates the *stratum corneum* (SC) to the viable epidermis and binds to skin proteins/peptides to create an immunogenic complex. This complex is then recognized and processed by Langerhans cells (LCs) in the epidermis. Upon exposure to the immunogenic complex, the LCs begin a maturation process in which the LCs internalize and process the immunogenic complex to a form that will be recognized by T cells. These cells then migrate from the epidermis to the lymph nodes where they present the modified immunogenic complex to naïve T cells with receptors that are able to specifically recognize the immunogenic moiety and are stimulated to proliferate and circulate throughout the body. Sensitization has now been induced, i.e., the subject is now sensitized. These events are collectively referred to as the induction phase. A schematic of the main molecular initiating events thought to be involved in sensitization induction is shown in Fig. 1.

Upon subsequent exposure to the same sensitizer, protein binding and processing of the immunogenic complex by the LCs occurs after which the immunogenic complex is recognized by circulating T cells triggering a cascade of biochemical and cellular processes, which produces the clinical sensitization response, i.e., elicitation [2, 7, 8].

This sequence of events can be considered as a set of hurdles that a chemical must negotiate to induce sensitization. A sensitizing chemical must penetrate through the SC, form a stable association with carrier protein, deliver dermal trauma sufficient to induce and upregulate epidermal cytokines that are necessary for the mobilization, migration, and maturation of LCs, and be inherently immunogenic such that

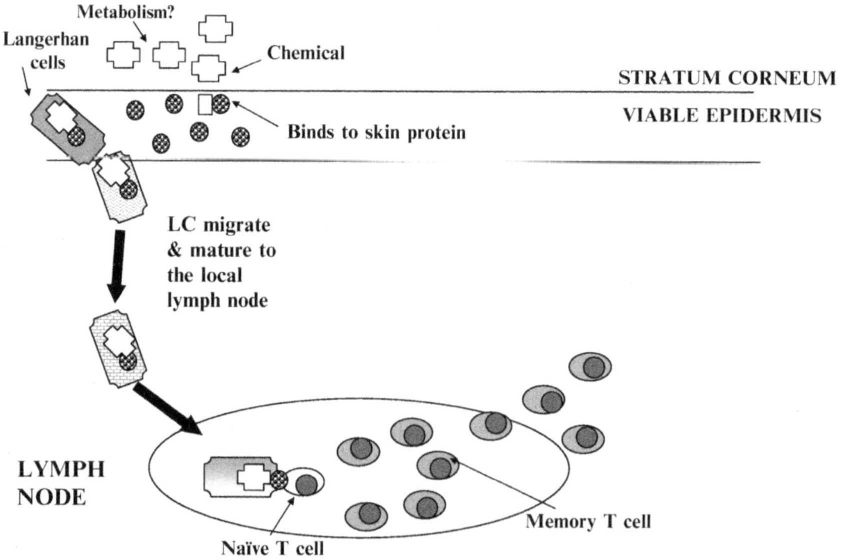

Fig. 1 The main molecular initiating events thought to occur during skin sensitization induction

a T lymphocyte response of sufficient magnitude is stimulated [8, 9]. Clearly not all of these hurdles can be and are equally important. The key is identifying which one is the rate determining step, i.e., which hurdle is dependent on the sensitizing chemical itself, as this should provide the roadmap on how to estimate the skin sensitization potential of a given chemical. To put into another context, the hurdles described above can be likened to a molecular horse race, where a horse represents one molecule of the test concentration dose that is applied in an actual sensitization test such as the local lymph node assay (LLNA). Each hurdle then is one that a molecular horse must negotiate to reach the finishing post, namely sensitization induction. Some of the horses will fail at different hurdles (e.g., the skin penetration hurdle) but the overall ratio of molecular horses that finish the race compared with those that start is effectively proportional to the sensitization potential.

We will provide a brief overview outlining the relative importance of each of these hurdles in the context of how well or how badly chemicals are able to negotiate them.

1.1.1 Role of Penetration

It seems obvious that if a chemical cannot penetrate the SC, then it should not be able to sensitize, and that other things being equal a chemical with a greater ability to penetrate than another will be the stronger sensitizer of the two. However, there are several pieces of evidence against SC penetration being a determining factor for sensitization potential, and very little evidence in favor. More details can be found in a recent paper by Roberts and Aptula, [10]. Here we will merely outline the three main pieces of evidence against penetration being a determining factor.

Vehicle Effects

One would expect large effects on sensitization potential with differing vehicles but in reality there appears no case of a compound being categorized with a different sensitizing potency, e.g., strong vs. weak when tested in two different vehicles. Small vehicle effects have been observed [11, 12], but these could be easily attributable to a limited variation in bulk properties of the test solutions – e.g., density, viscosity.

Mechanistic QSARs for Skin Sensitization

Many quantitative models correlating sensitization potential with chemical parameters have been reported, which do not require SC penetration to be modeled [13–22]. Most are based on a combination of a reactivity parameter with a hydrophobicity parameter. It is feasible that the hydrophobicity parameter in these correlations could be serving a dual purpose, but given that penetration is a function of more than just hydrophobicity alone and that including the other factors relevant to penetration such as molecular weight (MW) and melting point only serve to worsen these existing correlations; then for these cases, it seems reasonable to conclude that penetration is not normally a rate determining factor for sensitization.

Penetration Optimum

It has been reported that the optimal penetration occurs for chemicals with log P values between 2 and 3 [23]. However, there are a number of examples of known strong sensitizers, which have log P values well outside of this range [24]. Examples include poison ivy and poison oak urushiols, which have log P values > 7, or N-methyl-N-nitroso urea which has a log P of -1.29. All of these are strong sensitizers. These cases are by no means unique. There are very few cases of nonsensitization attributable to nonpenetration, and for these alternative explanations, such as depletion by reaction with hard nucleophilic groups in the SC, are possible [5, 10].

Overall we can conclude that SC penetration is not a significant hurdle in the skin sensitization process.

1.1.2 Migration of Langerhans Cells

For sensitization to occur, LC, which have acquired and processed sensitizer-modified carrier protein, have to migrate from the epidermis to the draining lymph node. Much progress has been made in recent years in understanding the rather complex details of this process, and has been well described by Corsini and Galli [25], Kimber et al. [26], and Cumberbatch et al. [27]. The LC have to detach themselves from the matrix of surrounding keratinocytes and travel along the basement membrane between the epidermis and the dermis, eventually penetrating through this membrane to reach the lymph node. In the course of this journey, they develop into mature dendritic cells (DC), losing the ability to process sensitizer-modified carrier protein, but acquire the ability to present the corresponding antigen to T cells. These events are stimulated and controlled by several cytokines (these are glycoproteins released in the epidermis, which act on the LC via binding to their receptors). An important stimulus for the production, or increased production, of such cytokines is dermal trauma [28, 29] such as irritation. Many sensitizers, for example, 2,4-dinitrochlorobenzene (DNCB), are to some extent irritant and may thereby be able to stimulate production of the cytokines necessary for sensitization. This dermal trauma is thought to be the danger signal or trigger for stimulating cytokine production.

However, there are a number of strong sensitizers, such as long chain alkene sultones and alkyl alkanesulphonates that are to all intents and purposes nonirritant, and can even be tested at 100%. Furthermore, there appear to be no known cases where a compound expected to sensitize fails to do so because of its inability to generate a danger signal. The implication then is that a danger signal can be taken for granted even for nonirritants, and thus does not constitute a rate determining hurdle in the sensitization process.

1.1.3 Antigen Recognition

The nature of the antigen, which is presented by the matured LCs to naïve T cells in the lymph node, clearly depends on the nature of the sensitizing compound: this is the basis of the specificity of the sensitization. It seems reasonable to consider whether the ability of the antigen to stimulate proliferation of T cells depends on the nature of the antigen.

However, exploring a plot of stimulation index values, taken from the LLNA, for a set of S_N2 (substitution nucleophilic bimolecular) electrophiles against their relative alkylation index (RAI) values, as calculated from the test concentrations (dose term), rate constants for reaction with butylamine, and log P values [5] shows a good fit to the curve i.e., there is no need to model differences in degree of antigen recognition in order to rationalize the data.

This then together with other evidence does suggest that differences in degree of antigen recognition play no part in determining sensitization potential.

Penetration of the SC, stimulation of migration and maturation of LCs, and antigen recognition are all important events in the induction of sensitization, but they can be taken for granted. They are neither important factors in determining whether a compound will be a sensitizer or not, nor are they important factors in determining how potent one sensitizer will be relative to another.

1.2 Chemical Mechanisms

This leaves then the formation of a stable association with carrier protein as the key factor which determines sensitization potential. The association is believed to be a covalent one whereby the chemical behaves as an electrophile and the protein as the nucleophile. The hypothesis was first articulated in 1936 by Landsteiner and Jacobs [30] followed up by others including Godfrey and Baer [31] and Depuis and Benezra [24]. For effective sensitization, a chemical must either be inherently protein reactive or be converted (chemically or metabolically) to a protein reactive species. Chemicals that are unable to associate effectively with proteins will fail to stimulate an immune response. Efforts to predict skin sensitizers have hence been focused on identifying the electrophilic features in chemicals and relating these back to skin sensitization potential.

1.2.1 Local Models: The Relative Alkylation Index Approach

The first approach in quantitative modeling of sensitization was developed by Roberts and Williams [13], who derived a mathematical model of the in vivo alkylation process (i.e., covalent binding to carrier protein) known as the RAI model.

This quantifies the degree of carrier alkylation and correlates it with sensitization potential. The model can be summarized as follows:

- The carrier protein is in a lipid environment (e.g., cell membrane), which is washed by a polar fluid (e.g., lymph or blood)
- Alkylation kinetics can be treated as pseudo first order

On the basis of these assumptions, rate equations were set up for alkylation and disappearance of the test compound from the system (by alkylation and partition into the polar fluid) and from these the RAI (1) was derived.

$$\mathrm{RAI} = \log\left(kD/P + P^2\right), \tag{1}$$

where k is the alkylation rate constant measured against a standard nucleophile, D is the molar dose, and P is the partition coefficient measured between a standard polar/nonpolar solvent. (Note that this is the inverse of the more commonly used nonpolar/polar partition coefficients.) Subsequently, the RAI model was revised so as to be based on the octanol/water coefficient, expressed as its logarithm, $\log P$. In its most general form, the RAI, an index of the relative degree of covalent binding to carrier protein, is expressed as in (2):

$$\mathrm{RAI} = \log D + a \log k + b \log P. \tag{2}$$

Thus the degree of covalent binding to carrier protein increases with increasing dose D of sensitizer, with increasing reactivity (as quantified by the rate constant or relative rate constant k for the reaction of the sensitizer with a model nucleophile) and with increasing hydrophobicity.

For specific use with LLNA data, a RAI-based potency relationship can be described by (3):

$$p\mathrm{EC3} = a \log k + b \log P + c, \tag{3}$$

where p denotes –log EC3, EC3 for this purpose being the weight percentage of chemical sensitizer required to elicit a threefold increase in lymph node cell proliferation, divided by the MW.

The RAI model has been used to evaluate a wide range of different datasets of skin sensitizing chemicals. It has in particular served well for the development of QSARs for small sets of structurally similar chemicals – i.e., local models, which are chemical class specific or to substantiate formulated hypothesis. Examples of RAI models include primary alkyl bromides [19], sultones [13], acrylates [15], sulfonate esters [32], urushiol analogues [33], aldehydes and diketones [34–36] amongst others.

As examples to illustrate the form of RAI models, two QSAR models developed in Patlewicz et al. [37] and their respective regression equations (4 and 5) are shown below:

$$p\mathrm{EC3} = 0.536 + 0.168 \log P + 0.488\, R\sigma^* + 1.313\, R'\sigma^*, \tag{4}$$

$$n = 9,\ r^2 = 0.741,\ s = 0.184,\ F = 4.77.$$

$$pEC3 = 0.245 + 0.278 \log P + 0.862\ R\sigma^*, \tag{5}$$
$$n = 12,\ r^2 = 0.825,\ s = 0.712,\ F = 21.26.$$

The σ^* values, taken from Perrin et al. [38], model the electronic effects of substituents on the rate constant, and therefore serve as surrogates for log k in the RAI expression. Equation (4) is for a set of Michael acceptor aldehydes, where $R'\sigma^*$ and $R\sigma^*$ reflect the Taft constants for the substituents on the alpha and beta carbons, respectively. Log P models the hydrophobicity. Equation (5) is for a set of Schiff base aldehydes.

As can be seen in (4) and (5), the RAI takes the form of a regression equation, which is transparent and easy to apply and interpret. However, as evidenced by the examples cited, the RAI has traditionally been used to evaluate only small sets of structurally related chemicals suggesting that the RAI model was only applicable for closely related datasets.

1.2.2 Statistical Global QSARs

Since the inception of the RAI approach and the explosion of QSAR modeling including a wide array of new statistical approaches and computing power to handle much larger volumes of data, there was an increased drive to derive models that were sufficiently general to predict the toxicity of larger sets of wide ranging chemicals.

In the area of skin sensitization, a large number of general models have been developed since the mid nineties. These statistical QSAR models are developed empirically by application of various statistical methods to sets of biological data and structural, topological, and/or geometrical information descriptors. Often known as global QSARs, they are purported to be able to make predictions for a wide range of chemicals, covering a wide range of different mechanisms of action. Probably the first global models for sensitization were those developed by Magee and Hostynek [39], who investigated the feasibility of deriving models for fragrance allergens using classification and ranking approaches. The most recent examples apply 4D molecular fingerprint similarity analysis as developed by Li et al. [40].

Although some have high predictive rates, they often lack a sound mechanistic basis. Many of these global QSARs have used selections from the same source material namely, a set of published LLNA data [41] suggesting an over emphasis on applying new statistical techniques or descriptors rather than trying to rationalize the underlying skin sensitization mechanism. In some cases, they are poorly characterized and are typically limited to providing a binary output of yes/no sensitizing activity rather than being able to rationalize potency. They have been reviewed in more detail in Patlewicz et al. [42,43]. A selection of global models have been scrutinized in depth in Roberts et al. [44] and Patlewicz et al. [45].

1.2.3 Expert Systems

There are several expert systems available for the prediction of skin sensitization including knowledge-based systems (e.g., Derek for Windows (DfW)), statistical systems (e.g., TOPKAT, MCASE) and hybrid of the two (e.g., Tissue Metabolism Simulator (TIMES)). Here we provide a summary of their main functionality and their current state of development.

DfW is a knowledge-based expert system created with knowledge of structure–toxicity relationships and an emphasis on the need to understand mechanisms of action and metabolism [46]. The skin sensitization knowledge base in DfW was initially developed in collaboration with Unilever in 1993 using its historical database of guinea pig maximization test (GPMT) data for 294 chemicals [47,48] and resulting in the extraction of 40 structural alerts. The DfW knowledge base (Version 9) contained 64 alerts for sensitization and is continuously refined as new chemical insights and data become available. Various exercises have been undertaken to suggest modifications to existing rules or hypothesize new rules. Examples include work by Barratt and Langowski [49], Zinke et al. [50], Gerner et al. [51], and Langton et al. [52].

Toxicity Prediction by Komputer Assisted Technology (TOPKAT) (http://www.accelrys.com/products/topkat/) marketed by Accelrys Inc (San Diego, USA) comprises two sets of sensitization models that were developed originally by Enslein et al. [53]. Guinea pig maximization data for 315 chemicals were assembled from various published collections in Contact Dermatitis and from the dataset published by Cronin and Basketter [54]. These data were then scaled according to classes defined by Barratt et al. [48] and two sets of models were developed using discriminant analysis. The resulting output is a computed probability discriminating for activity (sensitizing or nonsensitizing) or resolving the potency; weak/moderate vs. strong. A domain check is also included to qualify whether the predictions made are within the scope of the model's capability. The present version of TOPKAT (Version 6.2) contains data for an additional 20 chemicals [55].

CASE methodology and all its variants (e.g., MCASE, CASETOX) were developed by Klopman and Rosenkranz [56–58]. There are more than 180 modules covering various areas of toxicology and pharmacology endpoints including skin sensitization currently marketed by MultiCASE Inc. (Cleveland, Ohio, US). The CASE approach uses a probability assessment to determine whether a structural fragment is associated with toxicity. To achieve this, molecules are split into structural fragments up to a certain path length. Probability assessments determine whether fragments significantly promote or inhibit toxicity. To create models, structural fragments are incorporated into a regression analysis. The MCASE modules available for skin sensitization are described further in primary articles [59–61]. The (Q)SAR estimates for the MCASE skin sensitization model are also included in the Danish Environmental Protection Agency (EPA)'s (Q)SAR Database, which is hosted on the European Chemicals Bureau (ECB) Web site, see http://ecb.jrc.ec.europa.eu/qsar/.

The Times Metabolism Simulator platform used for predicting skin sensitization (TIMES-SS) is a hybrid expert system that was developed by the Laboratory of Mathematical Chemistry (University of Bourgas, Bulgaria), using funding and data from a Consortium comprising Industry (ExxonMobil, Procter and Gamble, Unilever) and a Regulatory Agency (Danish Environmental Protection Agency).

TIMES-SS aims to encode structure toxicity and structure metabolism relationships through a number of transformations simulating skin metabolism and interaction of the generated reactive metabolites with skin proteins. The skin metabolism simulator mimics metabolism using 2D structural information. Metabolic pathways are generated based on a set of 236 hierarchically ordered principal transformations including spontaneous reactions and enzyme-catalyzed reactions (phase I and II). The covalent reactions with proteins are described by 47 alerting groups. The associated mechanisms are in accordance with the existing knowledge on electrophilic interaction mechanisms of various structural functionalities. The integral model is essentially based on a set of submodels, associated with each of the reactive groups. Some of these reactions are additionally underpinned by mechanistically based 3D-QSARs [62, 63]. Recently an external validation activity was undertaken whereby data were generated for 40 new chemicals in the LLNA and then compared with predictions made by TIMES-SS. The results were promising with an overall good concordance (83%) between experimental and predicted values [64, 65]. The alerting groups underpinning TIMES-SS have also been scrutinized with respect to the dataset published in Gerberick et al. [41] and suggestions for improvements and refinements were proposed [66].

Overall there are reasonable prospects for the development and improvement of expert systems such as DfW, TIMES-SS but further refinement of their underlying rules is still required [52, 64, 66]. Training datasets are not always accessible (DfW being a case in point), applicability domains are not always well established, and the mechanistic interpretability is not always clear (e.g., TOPKAT). Nonetheless, these expert systems can play a valuable role in evaluating skin sensitization.

2 REACH: A Catalyst for Change

The advent of the new REACH legislation provided an impetus for the use of (Q)SARs in a regulatory context without the need for confirmatory testing [67]. Since the publication of the REACH White paper in 2001, several activities were initiated to increase acceptance of (Q)SARs. The first of these included a Workshop organised by European Chemical Industry Council (CEFIC) / The International Council of Chemical Associations (ICCA) in Setubal, Portugal in 2002 [68], which identified a number of principles for evaluating the validity of (Q)SARs. These were then evaluated by the Organisation for Economic Co-operation and Development (OECD) (as part of the Ad hoc group for (Q)SARs) and are now referred to as the "OECD principles," which read as follows:

"To facilitate the consideration of a (Q)SAR model for regulatory purposes, it should be associated with the following information:

- A defined endpoint
- An unambiguous algorithm
- A defined domain of applicability
- Appropriate measures of goodness-of-fit, robustness, and predictivity
- A mechanistic interpretation, if possible".

The principles have been summarized briefly in an OECD publication [69]. A preliminary guidance document on the characterization of (Q)SARs was then written [70], the basis of which was used in the development of the OECD Guidance document on (Q)SAR Validation [71].

REACH states that the results from valid (Q)SARs may be used in place of animal testing for the purposes of identifying the absence or presence of certain dangerous properties so long as a number of conditions are met. These include (1) the (Q)SAR is scientifically valid, (2) the substance falls within the applicability domain of the (Q)SAR model, (3) the QSAR result is suitable for classification and labeling and/or risk assessment purposes, (4) adequate reliable documentation is provided. The REACH text refers to the need to demonstrate the validity of the (Q)SAR used [1]. Validity is likely to make reference to the internationally agreed OECD principles for (Q)SAR validation already described [69, 71]. Information generated by valid (Q)SARs may be potentially used in place of experimental data, provided a number of conditions are met. The availability of a (Q)SAR for an endpoint of interest is necessary but to replace or partially replace an experimental result, the (Q)SAR data will need to be assessed for its reliability (i.e., inherent quality of the model within its applicability domain), relevance (scientific relevance and regulatory relevance), and adequacy (assessment of completeness of information in making a regulatory decision). Thus for a (Q)SAR to be adequate for a given regulatory purpose, the following conditions should ideally be fulfilled: (1) estimate generated by a valid model; (2) model applicable to the chemical of interest with sufficient reliability; and (3) model endpoint should be relevant for the regulatory purpose [72].

The principles for (Q)SAR validation identify the types of information that are considered useful for the assessment of (Q)SARs for regulatory purposes and to an extent represent best practice in their development. They do not provide criteria for the regulatory acceptance of (Q)SARs. Since no formal adoption procedure is likely under REACH, the concordance with the OECD principles is thought to be best evidenced by documenting the characteristics of both the (Q)SAR model and its prediction using specific reporting formats, so-called QSAR Model Reporting Format (QMRF) and QSAR Prediction Reporting Format (QPRF) (see http://ecb.jrc.ec.europa.eu/qsar/qsar-tools/index.php?c=QRF).

The guidance developed to date describes how (Q)SAR models may be evaluated in accordance with the OECD principles and, to an extent, what best practice should be when developing new (Q)SAR models. Skin sensitization potential needs to be assessed for chemicals above the 1 ton threshold according to Annex VII [67]. Since no in vitro replacement is currently available, nor expected to be ready in the near future [9], it is pertinent to identify robust and relevant (Q)SARs to aid in the evaluation of sensitizing potential.

2.1 *Extension of the RAI Approach*

This prompted a revisit of the available skin sensitization QSARs and what approaches might be most helpful for the purposes of hazard identification and risk assessment. As has been described earlier, the apparent array of global QSARs has been shown to be of limited performance and while expert systems have many strengths, even they have issues surrounding transparency and documentation. In light of this, one avenue that was explored by these authors was to reconsider the RAI approach and establish whether it was still relevant and whether it could be extended to have wider applicability. Studies on aldehydes and ketones in Patlewicz et al. [73] demonstrated that it was possible to derive RAI models on the basis of mechanistic applicability domain rather than just focusing on models within restricted structural classes. In Aptula et al. [74], a dataset of 41 LLNA EC3 values for a diverse range of compounds, which had been selected for internal consistency [75] were analyzed and the compounds were reclassified into reaction mechanistic applicability domains, i.e., grouping according to the reaction mechanisms whereby the compounds could react with nucleophiles. Each domain contained a diversity of structures, related by their common reaction chemistry rather than by common structural features. In addition it was found possible to rank the compounds in order of reactivity, applying established mechanistic organic chemistry principles, and in this way find clear trends, within each domain, of sensitization potential increasing with increasing reactivity. The major reaction mechanistic applicability domains identified were as follows: Michael-type addition domain, S_N2 domain, S_NAr domain, acylation domain, Schiff-base domain. Included in the Michael-type domain were pro-Michael acceptors, these being compounds which are not themselves Michael-reactive but are easily converted (e.g., by in vitro or in vivo oxidation to Michael acceptors). There was also an "unreactive" domain; compounds in this domain are expected to be nonsensitizers. Although these are not the only mechanistic applicability domains, they did provide a foundation to underpin skin sensitization with sound reaction chemistry principles, which were subsequently documented more fully in Aptula and Roberts, [76] as a set of rules for characterizing each domain (shown in Fig. 2). The rules were subsequently used to evaluate other datasets as evidenced in Roberts et al. [5, 6]. A possible future addition might be to include a domain of S_N1 based on some examples identified, where S_N1 reactions appear to offer a plausible explanation for behavior. Examples include tertiary allylic peroxides, for which protein binding by free radical reactions has also been suggested (Karlberg et al. [77] and references therein).

During the course of characterizing the domains and evaluating them with respect to different datasets, several areas where further work was needed were identified. In particular, there exists a need to gain greater understanding of the chemical basis of sensitization for certain structural classes of compounds such as aromatic compounds containing more than one hydroxyl and/or amino group, hydroperoxides, and compounds, which can readily give rise to them by autoxidation, epoxides and their autoxidation precursors. Recent advances in this area have come from Lepoittevin's group [78] and Karlberg's group (reviewed in [77]).

Work was also undertaken by Roberts et al. [22] to reanalyze the data from Patlewicz et al. [73] with the aim of developing a new mechanistic QSAR for Schiff base electrophiles (i.e., carbonyl compounds, which can bind to proteins via Schiff base formation). A QSAR was derived relating reactivity and hydrophobicity to

Mechanistic domain	Protein binding reaction	Modified protein

Michael acceptors

Identification characteristics. Double or triple bond with electron-withdrawing substituent X, such as -CHO, -COR, $-CO_2R$, -CN, $-SO_2R$, $-NO_2$...Includes para quinones and ortho quinones, often formed by oxidation of para and ortho dihydroxy aromatics acting as pro-Michael acceptors. X can also be a heterocyclic group such as 2-pyridino or 4-pyridino.

S_NAr electrophiles

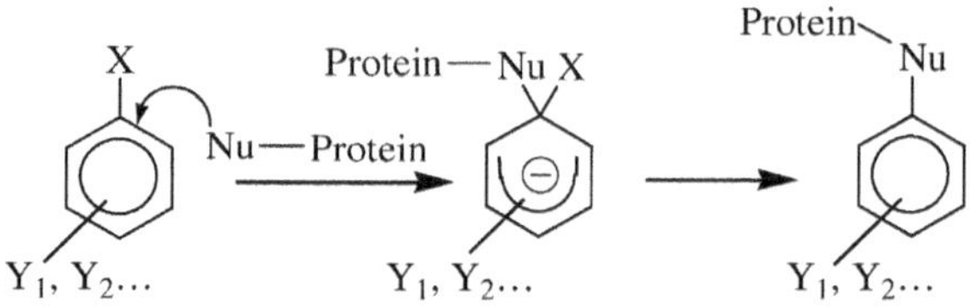

Identification characteristics. X = halogen or pseudohalogen, Y's are electron withdrawing groups (at least two) such as $-NO_2$, -CN, -CHO, $-CF_3$, -SOMe, $-SO_2Me$, ring fused nitrogen...One halogen is too weak to act as an X, but several halogens together can activate.

S_N2 electrophiles

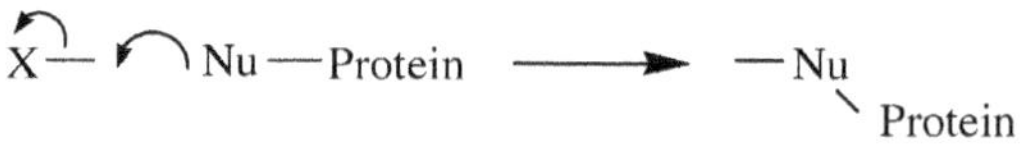

Identification characteristics. X = halogen or other leaving group, e.g. OSO_2(R or Ar), OSO_2O(R or Ar) bonded to primary alkyl, benzylic, or allylic carbon. OR and NHR or NR_2 do not usually act as leaving groups, but can do so if part of a strained 3-membered ring (e.g. epoxides, ethylenimine and substituted derivatives).

Schiff base formers

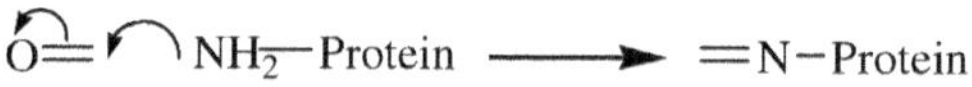

Identification characteristics. Reactive carbonyl compounds such as aliphatic aldehydes, some α,β- and α,γ-diketones, α-ketoesters. Not simple monoketones and aromatic aldehydes. Other hetero-unsaturated systems can behave analogously, e.g. C-nitroso compounds, thiocarbonyl compounds (C=S), cyanates and isocyanates, thiocyanates and isothiocyanates.

Acylating agents

Identification characteristics. X = halogen, or other group (e.g. $-OC_6H_5$) such that XH is acidic enough for X^- to act as a good leaving group. Includes anhydrides, cyclic or non-cyclic. X = -Oalkyl does not qualify, except when part of a strained lactone ring, e.g. β-propiolactone (but not γ-butyrolactone). Analogous reactions can occur with attack at sulfonyl S, phosforyl P and thioacyl C.

Fig. 2 Reaction mechanistic applicability domains

the −log of the molar EC3 (pEC3), where reactivity was modeled by $\Sigma\sigma^*$, the sum of the Taft σ^* substituent constants for the two groups attached to the carbonyl group.

A final QSAR based on 16 compounds (11 aliphatic aldehydes, 1 α-ketoester, and 4 α, β-diketones) was developed as shown by (6).

$$pEC3 = 1.12\,(\pm 0.07) \sum \sigma^* + 0.42\,(\pm 0.04)\ \log P - 0.62\,(\pm 0.13), \quad (6)$$

$$n = 16,\ r^2 = 0.952,\ r^2\text{adj} = 0.945,\ s = 0.12,\ F = 129.6.$$

The predictive performance of this QSAR was then explored across the wider Schiff Base mechanistic applicability domain, using further LLNA data for 1,3-dicarbonyl compounds taken from Gerberick et al. [41].

Two of the 1,3-dicarbonyl compounds of the further test set were classed as nonsensitizers (EC3 not reached at 40%). One of these was calculated to have EC3 = 59%; the other was calculated to have an EC3 = 39%. For the remaining seven 1,3-dicarbonyl compounds, the agreement between calculated and observed pEC3 values was good [22].

The success of this mechanistic QSAR for one of these reaction mechanistic applicability domains demonstrates how the RAI approach is more widely applicable than previously thought. A new term was also introduced for such mechanistic RAI QSARs, namely QMM Quantitative Mechanistic Models to emphasize the strong mechanistic basis. In Roberts et al. [22], this QMM for Schiff base electrophiles was also characterized with respect to the OECD principles and shown to demonstrate good concordance, suggesting appropriate validity for the purposes of the REACH requirements. An additional advantage with this QMM is that the endpoint modeled is actually the same as the regulatory endpoint, i.e., the EC3 in the LLNA which means that no translation or additional interpretation is required in using this QMM.

Thus the RAI QMM approach provides a practical and robust means of assessing skin sensitization without recourse to animal testing. In its simplest form, it relates sensitization potential to a function of hydrophobicity and reactivity.

$$\text{i.e., Sensitization} = \text{Function (Hydrophobicity} + \text{Reactivity)}.$$

Hydrophobicity as we have seen in the example above can be readily modeled using the log of the octanol-water partition coefficient. Reactivity in this case has been modeled by Taft σ^* constants. Other approaches to encode reactivity include using heats of reaction [79], or an activation index [80], or quantum chemical descriptors such as ELUMO (energy of the lowest unoccupied molecular orbital).

However, such reactivity indices are not always applicable: they may be difficult to calculate reliably for the compounds of interest, or for the reaction mechanism under consideration the appropriate parameters to model reactivity may not be obvious. This situation can be addressed by using experimental measures of reactivity of chemicals with model nucleophiles, the so-called in chemico approach. An in-depth discussion of experimental approaches for reactivity indices is given by Roberts et al. [81]. Landsteiner and Jacobs [30] pioneered this approach, using aniline in ethanol at ca. 100°C to discriminate between reactive (by what is now recognized as the S_NAr mechanism) and unreactive aromatic halo and nitro compounds, enabling them to successfully discriminate between sensitizing and nonsensitizing compounds. Butylamine kinetics in various organic solvents have been used to obtain RAI-based (combining log k_{rel} and log P) correlations with sensitization data for sultones [13], *p*-nitrobenzylhalides [14], α-(X-substituted methyl)- γ, γ-dimethyl-γ-butyrolactones [20].

Other efforts in realizing the in chemico approach came from investigating the correlation between thiol reactivity and LLNA for a set of Michael acceptors [82]. The thiol reactivity index was based on glutathione (GSH), pEC(50) thiol (EC(50) being defined as the concentration of the test substance which gave a 50% depletion of free thiol under standard conditions) in combination with a measure of cytotoxicity (pIGC(50)) to *Tetrahymena pyriformis* (TETRATOX). Thiol reactivity was found to discriminate sensitizers from nonsensitizers according to the rule: pEC(50) thiol >-0.55 indicating that the compound would be a skin sensitizer. However, because of metabolic activation a pEC(50) thiol <-0.55 does not necessarily mean that the compound will be a nonsensitizer. Excess toxicity to *T. pyriformis* (i.e., the extent of toxic potency over that expected by nonpolar narcosis) was determined to assess biological reactivity. The best discrimination based on excess toxicity in the TETRATOX assay was given by the "rule": excess toxicity >0.50 indicating that the compound would be a skin sensitizer. These approaches became more powerful (23 of the 24 compounds were predicted correctly) when used in combination. The approach was promising but not the definitive answer since thiol reactivity is only one measure of the ability of chemicals to form adducts with proteins. Other similar efforts have been undertaken by Gerberick et al. [83], who have been investigating approaches using a glutathione tripeptide or three synthetic peptides containing cysteine, lysine, or histidine residues. In Gerberick et al. [83], the reactivity of 38 chemicals with varying skin sensitizing potencies were investigated. The results revealed a correlation between skin sensitizing potency and depletion of glutathione and binding with the lysine and cysteine synthetic peptides. Further work by Gerberick et al. [84] and others has since been undertaken [85]. GSH has several advantages as a model for soft nucleophilic groups of carrier protein – it is readily available, its concentration can be analyzed by readily available methods, and, unlike simple thiols, it is odorless and nonhazardous to work with. However, it is not a chromophore so does not directly lend itself to methods such as HPLC-UV without multistep methods involving conjugation to carrier chromophores (e.g., as GSH-dinitrophenyl adducts), to detect its presence [86]. Its major limitation is that although soluble in water, it has limited solubility in organic solvents. For determining reactivity of water soluble compounds, this is not a problem, but many sensitizers are quite hydrophobic and have very limited solubility in water. Given that the nature and location of the carrier protein is not known, there is no reason to suppose that any particular model nucleophile in any particular solvent is a more realistic model than any other for the *in cutaneo* reaction. Successful correlations have been obtained with data from incubating electrophiles with simple model nucleophiles in simple solvents suggesting that the reductionist approach of using the simplest and most convenient models to gain data on relative reactivities will be sufficient for modeling the potency determining steps in the skin sensitization process.

For mechanistic domains where QMMs have not been developed, prediction is still possible, using mechanistic read across. We illustrate this below for the Michael acceptor domain, where good progress has been made toward developing a reactivity database using RC50 values for reaction with GSH [87]. The basic principle of "mechanistic read across" is that if two compounds in the same mechanistic domain are similar in their toxicity-determining parameters, they should be similar in their toxicity, irrespective of whether or not they are similar in structure. This necessitates access to a large database of skin sensitization data, one such resource is the publication by Gerberick et al. [41] where high quality LLNA data for over 200 compounds were collected. A large database of experimental kinetic reactivity data and hydrophobicity (log P) for as many of the known sensitizers as possible is also a requirement. Progress has been made toward establishing such a database [82, 87] but more resources need to be devoted to generate new reactivity data and populate this database further.

At present the reactivity database of known skin sensitizers is still quite limited, but even in its present form can still be used to make estimates by linear interpolation.

Suppose we wish to estimate the sensitization potency of 2,4-hexadienal. Its GSH RC50 is 1.533 mM [Terry Schultz, personal communication]. We can calculate its log P by the Leo and Hansch method [88]: 1.15. Using the published classification rules [76], we can classify this compound in the Michael Acceptor mechanistic domain.

In the database, we search for two compounds in the Michael acceptor domain whose RC50 and log P values are either side of the 2,4-hexadienal, and whose LLNA EC3 values are known.

We find:

- diethyl maleate, RC50 $= 3.28$; log $P = 1.84$; EC3 $= 5.8\%$
- methyl 2-nonynoate, RC50 $= 0.264$; log $P = 3.25$, EC3 $= 2.5\%$

The smaller the RC50, the more reactive it is.

Although these are not an ideal pair of compounds, in that both have log P values higher than that of the compound to be predicted, read across is still possible and relevant.

The reactivity of 2,4-hexadienal lies between those of diethyl maleate and methyl 2-nonynoate. Its hydrophobicity is similar to that of diethyl maleate. From this information, we can predict that 2,4-hexadienal's sensitization potency will be between 5.8 and 2.5. The process of determining an estimate of skin sensitization as follows: Since RC50 is inversely proportional to the rate constant, log (1/RC50) can be used to represent log k. By analogy with many other structure–sensitization correlations published for various types of compounds (e.g., sultones, lactones, alkyl alkanesulphonates, Schiff base electrophiles, etc.), relative sensitization potency (expressed as a mol% log (1/EC3) value) can be assumed to be influenced about twice as much by reactivity as by log P. Accordingly a relative sensitization parameter (RSP) can be defined as RSP = log (1/RC50) + 0.5 log P (5), and, based on the RAI model, we can write (7) as:

$$\log (1/\mathrm{EC3}_{\mathrm{mol\%}}) = a\ \mathrm{RSP} + b. \tag{7}$$

Using the RC50 and log P values, a RSP can then be calculated for each of the 3 compounds. Log $(1/\mathrm{EC3}_{\mathrm{mol\%}})$ values can be calculated from the EC3 (wt%) values and MW values for diethyl maleate and methyl 2-nonynoate:

- diethyl maleate, RSP = 0.404; $\log (1/\mathrm{EC3}_{\mathrm{mol\%}}) = 1.47$
- methyl 2-nonynoate, RSP = 2.20; $\log (1/\mathrm{EC3}_{\mathrm{mol\%}}) = 1.83$
- 2,4-hexadienal, RSP = 0.389; $\log (1/\mathrm{EC3}_{\mathrm{mol\%}}) =?$

By linear interpolation, a $\log (1/\mathrm{EC3}_{\mathrm{mol\%}})$ value of 1.47 is derived for 2,4-hexadienal. This corresponds to an estimated EC3 value of 3.25% (since the MW is 96.13). Considering the nature of these calculations is very approximate, we would quote the estimate as between 2.5 and 4% (assuming plus or minus 20% of the calculated value, and rounding to nearest 0.5). The actual observed EC3 for 2,4-hexadienal is 3.5% [41].

Read-across estimates such as the one illustrated here are very dependent on the variability of the EC3 values of the two known compounds either side of the linear interpolation, as well as the accuracy of the reactivity and log P parameters for all three compounds. Hence, interpretation of such models should be done with care and with knowledge of the inherent biological variability of the bioassay data being used to make such benchmark predictions.

3 Conclusions

There is clearly a very strong mechanistic understanding of skin sensitization, which we have tried to summarize early on. Most published skin sensitization QSARs have fallen into one of two main categories: either they are mechanistic model based RAI QSARs, typically of high statistical quality but until recently only applicable to a narrow range of closely-related structures or they are "statistical" QSARs, which aim to be global in their applicability, are variable in their successes, and lacking in real mechanistic insight [44,45].

The covalent hypothesis continues to be the most promising way of developing mechanistically-based robust QSARs. The reaction chemistry concepts recently outlined in Aptula et al. [74] and Schultz et al. [89] have changed the perspective regarding RAI QSARs. These QSARs are actually more widely applicable than originally thought. Clear chemistry-activity trends can be seen within mechanistic applicability domains leading to robust mechanistic-based QSAR models (or as defined earlier QMMs e.g. Roberts et al. [22]) as well as new insights for inclusion into expert systems such as TIMES-SS and DfW. These QMMs use reactivity and hydrophobicity as the key parameters in mathematically modeling skin sensitization.

Although hydrophobicity can be conveniently modelled using log P, the octanol-water partition coefficient, reactivity is not always readily determined from chemical structure. Initiatives are in progress to generate reactivity data for reactions relevant to skin sensitization but more resources are required to realize a comprehensive set of reactivity data [82–85, 89]. This type of data would also facilitate the derivation of in silico reactivity indices.

Our ultimate vision for skin sensitization prediction is that the animal testing laboratory should be replaced by the physical organic chemistry laboratory. Particularly bearing in mind that many compounds are easily predictable without experimentation, the experimental studies to generate the chemical data required should be no more costly or time consuming than the animal tests that have hitherto been used.

Presented with a new compound:

1. The first step is to classify it into its reaction mechanistic domain. One domain is the "unreactive" domain, populated by predicted nonsensitizers. For several mechanistic domains, there are corresponding proelectrophilic subdomains. For example, many sensitizers, such as hydroquinone and 3-alkyl/alkenyl catechols (active components of poison ivy), are thought to act as pro-Michael acceptors. Domain classification may often be possible by inspection of structure, but inevitably in some cases a confident prediction may not be possible. In such situations, experimental work will be needed to determine the reaction chemistry, in particular to determine whether the compound is electrophilic or proelectrophilic and the nature of the reactions.
2. Having assigned the compound to its reaction mechanistic applicability domain, the next step is to quantify its reactivity/hydrophobicity relative to known sensitizers in the same mechanistic applicability domain. These properties may sometimes be confidently predictable from structure, using physical organic chemistry approaches such as linear free energy relationships based on substituent constants or on molecular orbital parameters. In other cases, it will be necessary to perform physical organic chemistry measurements, such as determination of reaction kinetics and measurement of partition coefficients.
3. Having assigned the compound to its reaction mechanistic applicability domain and quantified its reactivity/hydrophobicity relative to known sensitizers in the same domain, QMM or mechanistic read-across can be used to predict the sensitization potential.

To make these types of mechanism-based read across assessments viable and practical for routine use in decision making requires the population of a database with empirical chemical kinetics reactivity data for as many known sensitizers as possible. We envisage the development and population of this reactivity database for skin sensitization to be part of a wider-ranging database being developed for reactive toxicity in general in the longer term. Reactivity of electrophiles is known to play a part in human health effects other than sensitization (e.g., liver toxicity) and in environmental toxicity [87, 90–93].

More work needs to be done on developing new methods to assess whether a new unreactive compound could be a proelectrophile, and hence likely to sensitize via abiotic or metabolic activation. Ideally an experimental method for identifying potentially sensitizing proelectrophiles would include appropriate metabolic and abiotic activating capability reflective of that in skin. Current understanding of proelectrophilic mechanisms in skin sensitization has recently been well reviewed by Karlberg et al. [77] and Gibbs et al. [94]. There are still areas of uncertainty, such that finding a suitable metabolizing system could present a significant challenge in assay development.

Currently the most pragmatic way forward is to assume that all the major proelectrophilic classes of compounds have been identified and that if a compound cannot be assigned to one of these classes and is not directly electrophilic, it is unlikely to be a sensitizer.

Chemical reactivity indices based on kinetic measurements, and application of them in mechanism-based read-across are vital areas for further development and show great promise in being able to contribute toward novel nonanimal approaches for the prediction of skin sensitization in the future.

References

1. Lepoittevin J-P, Basketter DA, Goossens A, Karlberg A-T (1997) Allergic Contact Dermatitis. The Molecular Basis. Springer: Berlin
2. Smith Pease CK (2003) From xenobiotic chemistry and metabolism to better prediction and risk assessment of skin allergy. Toxicology 192: 1–22
3. Ryan CA, Gerberick GF, Gildea LA, Huletter BC, Betts CJ, Cumberbatch M, Dearman RJ, Kimber I (2005) Interactions of contact allergens with dendritic cells: Opportunities and challenges for the development of novel approaches to hazard assessment. Toxicol Sci 88: 4–11
4. Rustenmeyer T, van Hoogstraten IMW, von Blomberg BME, Scheper R (2006) Mechanisms in allergic contact dermatitis. In: Frosch PJ, Menne T, Lepoittevin J-P (eds) Contact Dermatitis, 4th edn. Springer: Berlin, pp. 11–44
5. Roberts DW, Aptula AO, Patlewicz G (2007) Electrophilic chemistry related to skin sensitization. Reaction mechanistic applicability domain classification for a published dataset of 106 chemicals tested in the mouse local lymph node assay. Chem Res Toxicol 20: 44–60
6. Roberts DW, Patlewicz GY, Kern PS, Gerberick GF, Kimber I, Dearman RJ, Ryan CA, Basketter DA, Aptula AO (2007) Mechanistic applicability domain classification of a local lymph node assay dataset for skin sensitization. Chem Res Toxicol 20: 1019–1030
7. Basketter D, Dooms-Goossens A, Karlberg AT, Lepoittevin JP (1995) The chemistry of contact allergy: why is a molecule allergenic? Contact Dermatitis 32: 65–73
8. Kimber I, Dearman RJ (2003) What makes a chemical an allergen? Ann Allergy Asth Immunol 90(Suppl): 28–31
9. Jowsey IR, Basketter DA, Westmoreland C, Kimber I (2006) A future approach to measuring relative skin sensitising potency: A proposal. J App Toxicol 26: 341–350
10. Roberts DW, Aptula AO (2008) Determinants of skin sensitisation potential. J Appl Toxicol 28: 377–387
11. Basketter DA, Gerberick GF, Kimber I (2001) Skin sensitization, vehicle effects and the local lymph node assay. Food Chem Toxicol 39: 621–627
12. Wright ZM, Basketter, DA, Blaikie L, Cooper KJ, Warbrick EV, Dearman RJ, Kimber I (2001) Vehicle effects on skin sensitizing potency of four chemicals: Assessment using the local lymph node assay. Int J Cosmet Sci 23: 75–83

13. Roberts DW, Williams DL (1982) The derivation of quantitative correlations between skin sensitisation and physico–chemical parameters for alkylating agents and their application to experimental data for sultones. J Theor Biol 99: 807–825
14. Roberts DW, Goodwin BFJ, Williams DL, Jones K, Johnson AW, Alderson CJE (1983) Correlations between skin sensitisation potential and chemical reactivity for *p*-nitrobenzyl compounds. Food Chem Toxicol 21: 811–813
15. Roberts DW (1987) Structure–activity relationships for skin sensitisation potential of diacrylates and dimethacrylates. Contact Dermatitis 17: 281–289
16. Roberts DW (1995) Linear free energy relationships for reactions of electrophilic halo- and pseudohalobenzenes, and their application in prediction of skin sensitisation potential for S_NAr electrophiles. Chem Res Toxicol 8: 545–551
17. Roberts DW, Basketter DA (1990) A quantitative structure–activity/dose relationship for contact allergenic potential of alkyl group transfer agents. Contact Dermatitis 23: 331–335
18. Roberts DW, Basketter DA (1997) Further evaluation of the quantitative structure-activity relationship for skin-sensitizing alkyl transfer agents. Contact Dermatitis 37: 107–112
19. Basketter DA, Roberts DW, Cronin M, Scholes EW (1992) The value of the local lymph node assay in quantitative structure–activity investigations. Contact Dermatitis 27: 137–142
20. Franot C, Roberts DW, Basketter DA, Benezra C, Lepoittevin J-P (1994) Structure-activity relationships for contact allergenic potential of γγ-dimethyl-γ-butyrolactone derivatives. 2. Quantititative structure-skin sensitisation relationships for α-substituted-α-methyl-γγ-dimethyl-γ-butyrolactones. Chem Res Toxicol 7: 307–312
21. Mekenyan O, Roberts DW, Karcher W (1997) Molecular orbital parameters as predictors of skin sensitization potential of halo- and pseudohalobenzenes acting as S_NAr electrophiles. Chem Res Toxicol 10: 994–1000
22. Roberts DW, Aptula AO, Patlewicz G (2006) Mechanistic applicability domains for non-animal based prediction of toxicological endpoints. QSAR analysis of the schiff base applicability domain for skin sensitization. Chem Res Toxicol 19: 1228–1233
23. Howes D, Guy R, Hadgraft J, Heylings J, Hoeck U, Kemper F, Maibach H, Marty J-P, Merk H, Parra J, Rekkas D, Rondelli I, Schaefer H, Täuber U, Verbiese N (1996) Methods for assessing percutaneous absorption, ECVAM Workshop Report No. 13. ATLA 24: 81–106
24. Dupuis G and Benezra C (1982) Allergic contact dermatitis to simple chemicals: A molecular approach. Dekker: New York
25. Corsini E, Galli CL (2000) Epidermal cytokines in experimental contact dermatitis. Toxicology 142: 203–211
26. Kimber I, Cumberbatch M, Dearman RJ, Bhushan M, Griffiths CEM (2000) Cytokines and chemokines in the initiation and regulation of epidermal Langerhans cell mobilization. Brit J Dermatol 142: 401–412
27. Cumberbatch M, Clelland K, Dearman RJ, Kimber I (2005) Impact of cutaneous IL-10 on resident epidermal Langerhans' cells and the development of polarized immune responses. J Immunol 33: 47–62
28. Kimber I, Cumberbatch M (1992) Dendritic cells and cutaneous immune responses to chemical allergens. Toxicol Appl Pharmacol 117: 137–146
29. Kimber I, Pichowski JS, Betts CJ, Cumberbatch M, Basketter DA, Dearman RJ (2001) Alternative approaches to the identification and characterization of chemical allergens. Toxicol In Vitro 15: 307–312
30. Landsteiner K, Jacobs JL (1936) Studies on the sensitisation of animals with simple chemicals III. J Exp Med 64: 625–639
31. Godfrey H P. Baer K (1971) The effect of physical and chemical properties of the sensitizing substance on the induction and elicitation of delayed contact hypersensitivity. J Immunol 106: 431–441
32. Roberts DW, Basketter DA (2000) Quantitative structure-activity relationships: Sulfonate esters in the local lymph node assay. Contact Dermatitis 42: 154–161
33. Roberts DW, Benezra C (1993) Quantitative structure-activity relationships for skin sensitization potential of urushiol analogues. Contact Dermatitis 29: 78–83

34. Roberts DW, York M, Basketter DA (1999) Structure-activity relationships in the murine local lymph node assay for skin sensitization: Alpha, beta-diketones. Contact Dermatitis 41: 14–17
35. Patlewicz G, Basketter DA, Smith CK, Hotchkiss SA, Roberts DW (2001) Skin-sensitization structure-activity relationships for aldehydes. Contact Dermatitis 44: 331–336
36. Roberts DW, Patlewicz G (2002) Mechanism based structure-activity relationships for skin sensitisation–the carbonyl group domain. SAR QSAR Environ Res 13: 145–152
37. Patlewicz GY, Wright ZM, Basketter DA, Pease CK, Lepoittevin JP, Gimenez Arnau E (2002) Structure-activity relationships for selected fragrance allergens. Contact Dermatitis 47: 219–226
38. Perrin DD, Dempsey B, Serjeant EP (1981) pKa Prediction for Organic Acids and Bases, Chapman and Hall: London, pp. 109–126
39. Magee PS, Hostynek JJ, Maibach MI (1994) A classification model for allergic contact dermatitis. Quant Struct-Act Relat 13: 22–33
40. Li Y, Tseng YJ, Pan D, Liu J, Kern PS, Gerberick GF, Hopfinger AJ (2007) 4D-fingerprint categorical QSAR Models for skin sensitisation based on the classification of local lymph node assay measures. Chem Res Toxicol 20: 114–128
41. Gerberick GF, Ryan CA, Kern PS, Schlatter H, Dearman RJ, Kimber I, Patlewicz GY, Basketter DA (2005) Compilation of historical local lymph node data for evaluation of skin sensitisation alternative methods. Dermatitis 16: 157–202
42. Patlewicz G, Aptula AO, Roberts DW, Uriate E (2007) Skin sensitisation (Q)SARs/expert systems: From past, present to future. European Commission report EUR 21866 EN
43. Patlewicz G, Aptula AO, Roberts DW, Uriarte E (2008) A mini review of available (Q)SARs and expert systems for skin sensitisation. QSAR Comb Sci 27: 60–76
44. Roberts DW, Aptula AO, Cronin MTD, Hulzebos E, Patlewicz G (2007) Global (Q)SARs for skin sensitization – assessment against OECD principles. SAR QSAR Environ Res 18: 343–365
45. Patlewicz G, Aptula AO, Uriate E, Roberts DW, Kern PS, Gerberick GF, Ryan CA, Kimber I, Dearman R, Basketter DA (2007) An evaluation of selected global (Q)SARs/Expert systems for the prediction of skin sensitisation potential. SAR QSAR Environ Res 18: 515–541
46. Sanderson DM, Earnshaw CG (1991) Computer prediction of possible toxic action from chemical structure; the DEREK system. Human Exp Toxicol 10: 261–273
47. Barratt MD, Basketter DA, Chamberlain M, Admans GD, Langowski JJ (1994) Development of an expert system rulebase for identifying contact allergens. Toxicol In Vitro 8: 837–839
48. Barratt MD, Basketter DA, Chamberlain M, Admans GD, Langowski JJ (1994) An expert system rulebase for identifying contact allergens. Toxicol In Vitro 8: 1053–1060
49. Barratt MD, Langowski JJ (1999) Validation and subsequent development of the Derek Skin Sensitisation Rulebase by analysis of the BgVV list of contact allergens. J Chem Inf Comput Sci 39: 294–298
50. Zinke S, Gerner I, Schlede E (2002) Evaluation of a rule base for identifying contact allergens by using a regulatory database: Comparison of data on chemicals notified in the European Union with "structural alerts" used in the DEREK expert system. ATLA 30: 285–298
51. Gerner I, Barratt MD, Zinke S, Schlegel K, Schlede E (2004) Development and prevalidation of a list of structure-activity relationship rules to be used in expert systems for prediction of the skin-sensitising properties of chemicals. ATLA 32: 487–509
52. Langton K, Patlewicz GY, Long A, Marchant CA, Basketter DA (2006) Structure-activity relationships for skin sensitization: recent improvements to Derek for Windows. Contact Dermatitis 55: 342–347
53. Enslein K, Gombar VK, Blake BW, Maibach HI, Hostynek JJ, Sigman CC, Bagheri D (1997) A quantitative structure-toxicity relationships model for the dermal sensitization guinea pig maximization assay. Food Chem Toxicol 35: 1091–1098
54. Cronin MT, Basketter DA (1994) Multivariate QSAR analysis of a skin sensitisation database. SAR QSAR Environ Res 2: 159–179
55. Accelrys Inc (2004) TopKat User Guide Version 6.02, Accelrys Inc: San Diego

56. Klopman G (1984) Artificial intelligence approach to structure-activity studies. Computer automated structure evaluation of biological activity of organic molecules. J Am Chem Soc 106: 7315–7321
57. Klopman G (1992) MULTICASE 1. A hierarchical computer automated structure evaluated program. Quant Struct-Act Relat 11: 176–184
58. Rosenkranz HS, Klopman G (1995) The application of structural concepts to the prediction of the carcinogenicity of therapeutical agents. In: Wolff ME (ed) Burger's medicinal chemistry and drug discovery, 5th Edition, Volume 1, Wiley: New York, pp. 223–249
59. Graham C, Gealy R, Macina OT, Karl MH, Rosenkranz HS (1996) QSAR for allergic contact dermatitis. QSAR Comb Sci 15: 224–229
60. Gealy R, Graham C, Sussman NB, Macina OT, Rosenkranz HS, Karol MH (1996) Evaluating clinical case report data for SAR modelling of allergic contact dermatitis. Human Exp Toxicol 15: 489–493
61. Johnson R, Macina OT, Graham C, Rosenkranz HS, Cass GR, Karol MH (1997) Prioritising testing of organic compounds detected as gas phase air pollutants: structure-activity study for human contact allergens. Environ Health Perspect 105: 986–992
62. Dimitrov SD, Low LK, Patlewicz GY, Kern PS, Dimitrova GD, Comber MHI, Phillips RD, Niemela J, Bailey PT, Mekenyan OG (2005) Skin sensitization: Modeling based on skin metabolism simulation and formation of protein conjugates. Int J Toxicol 24: 189–204
63. Dimitrov S, Dimitrova G, Pavlov T, Dimitrova N, Patlewicz G, Niemela J, Mekenyan O (2005) A stepwise approach for defining the applicability domain of SAR and QSAR models. J Chem Inf Model 45: 839–849
64. Patlewicz G, Dimitrov SD, Low LK, Kern PS, Dimitrova GD, Comber MIH, Aptula AO, Phillips RD, Niemela J, Madsen C, Wedebye EB, Roberts DW, Bailey PT, Mekenyan OG (2007) TIMES-SS – A promising tool for the assessment of skin sensitization hazard. A characterisation with respect to the OECD Validation Principles for (Q)SARs. Reg Toxicol Pharmacol 48: 225–239
65. Roberts DW, Patlewicz G, Dimitrov S, Low LK, Aptula AO, Kern PS, Dimitrova GD, Comber MI, Phillips RD, Niemelä J, Madsen C, Wedebye EB, Bailey PT, Mekenyan OG (2007) TIMES-SS – A mechanistic evaluation of an external validation study using reaction chemistry principles. Chem Res Toxicol 20: 1321–1330
66. Patlewicz G, Roberts DW, Uriarte E (2008) A comparison of reactivity schemes for the prediction skin sensitisation potential. Chem Res Toxicol 21: 521–541
67. Commission of the European Communities. Regulation (EC) No 1907/2006 of the European Parliament and of the Council of 18 December 2006 concerning the Registration, Evaluation, Authorisation and Restriction of Chemicals (REACH), establishing a European Chemicals Agency, amending Directive 1999/45/EC and repealing Council Regulation (EEC) No 793/93 and Commission Regulation (EC) No 1488/94 as well as Council Directive 76/769/EEC and Commission Directives 91/155/EEC, 93/67/EEC, 93/105/EC and 2000/21/EC. Off J Eur Union L 396/1 of 30.12.2006
68. Jaworska JS, Comber M, Auer C, van Leeuwen CJ (2003) Summary of a workshop on regulatory acceptance of (Q)SARs for human health and environmental endpoints. Environ Health Perspect 22: 1358–1360
69. OECD (2004) ENV/JM/MONO/(2004)24 http://appli1.oecd.org/olis/2004doc.nsf/linkto/env-jm-mono(2004)24
70. Worth AP, Bassan A, Gallegos A, Netzeva TI, Patlewicz G, Pavan M, Tsakovska I, Vracko M (2005) The Characterisation of (quantitative) structure-activity relationships: Preliminary guidance. EUR 21866 EN
71. OECD (2007) Guidance document on the validation of (quantitative)structure-activity relationships [(Q)SAR] models. ENV/JM/MONO(2007)2. Organisation for Economic Cooperation and Development, Paris. http://www.oecd.org/dataoecd/55/35/38130292.pdf
72. Worth AP, Bassan A, de Bruijn J, Gallegos Saliner A, Netzeva T, Patlewicz G, Pavan M, Tsakovska I, Eisenreich S (2007) The role of the European Chemicals Bureau in promoting the regulatory use of (Q)SAR methods. SAR QSAR Environ Res 18: 111–125

73. Patlewicz GY, Basketter DA, Pease CK, Wilson K, Wright ZM, Roberts DW, Bernard G, Arnau EG, Lepoittevin JP (2004) Further evaluation of quantitative structure–activity relationship models for the prediction of the skin sensitization potency of selected fragrance allergens. Contact Dermatitis 50: 91–97
74. Aptula AO, Patlewicz G, Roberts DW (2005) Skin sensitization: Reaction mechanistic applicability domains for structure-activity relationships. Chem Res Toxicol 18: 1420–1426
75. Gerberick GF, Ryan CA, Kern PS, Dearman RJ, Kimber I, Patlewicz GY, Basketter DA (2004) A chemical dataset for evaluation of alternative approaches to skin-sensitization testing. Contact Dermatitis 50: 274–288
76. Aptula AO, Roberts DW (2006) Mechanistic applicability domains for nonanimal-based prediction of toxicological end points: General principles and application to reactive toxicity. Chem Res Toxicol 19: 1097–1105
77. Karlberg AT, Bergström MA, Börje A, Luthman K, Nilsson JL (2008) Allergic contact dermatitis – Formation, structural requirements, and reactivity of skin sensitizers. Chem Res Toxicol 21: 53–69
78. Eilstein J, Giménez-Arnau E, Rousset F, Lepoittevin J-P (2006) Synthesis and reactivity toward nucleophilic amino acids of 2,5-[13C]-dimethyl-benzoquinonediimine. Chem Res Toxicol 19: 1248–1256
79. Magee PS (2000) Exploring the potential for allergic contact dermatitis via computed heats of reaction of haptens with protein end-groups heats of reaction of haptens with protein end-groups by computation. Quant Struct–Act Relat 19: 356–365
80. Aptula AO, Roberts DW, Cronin MT (2005) From experiment to theory: Molecular orbital parameters to interpret the skin sensitization potential of 5-chloro-2-methylisothiazol-3-one and 2-methylisothiazol-3-one. Chem Res Toxicol 18: 324–329
81. Roberts DW, Aptula AO, Patlewicz G, Pease C (2008) Chemical reactivity indices and mechanism-based read-across for non-animal based assessment of skin sensitisation potential. J Appl Toxicol 28: 443–454
82. Aptula AO, Patlewicz G, Roberts DW, Schultz TW (2006) Non-enzymatic glutathione reactivity and *in vitro* toxicity: a non-animal approach to skin sensitization. Toxicol In Vitro 20: 239–247
83. Gerberick GF, Vassallo JD, Bailey RE, Chaney JG, Morrall SW, Lepoittevin JP (2004) Development of a peptide reactivity assay for screening contact allergens. Toxicol Sci 81: 332–343
84. Gerberick GF, Vassallo JD, Foertsch LM, Price BB, Chaney JG, Lepoittevin J-P (2007) Quantification of chemical peptide reactivity for screening contact allergens: A classification tree model approach. Toxicol Sci 97: 417–427
85. Natsch A, Gfeller H, Rothaupt M, Ellis G (2007) Utility and limitations of a peptide reactivity assay to predict fragrance allergens in vitro. Toxicol In Vitro 21: 1220–1226
86. Farriss MW, Reed DJ (1987) High-performance liquid chromatography of thiols and disulfides: Dinitrophenol derivatives. Methods Enzymol 143: 101–109
87. Yarborough JW, Schultz TW (2007) Abiotic sulfhydryl reactivity: A predictor of aquatic toxicity for carbonyl-containing α, β-unsaturated compounds. Chem Res Toxicol 20: 558–562
88. Hansch C, Leo AJ (1979) Substituent constants for correlation analysis in chemistry and biology. Wiley: New York
89. Schultz TW, Carlson RE, Cronin MTD, Hermens JLM, Johnson R, O'Brien PJ, Roberts DW, Siraki A, Wallace KB, Veith GD (2006) A conceptual framework for predicting the toxicity of reactive chemicals" modelling soft electrophilicity. SAR QSAR Environ Res 17: 413–428
90. Lipnick RL, Pritzker CS, Bentley DL (1987) Application of QSAR to model the toxicology of industrial organic chemicals to aquatic organisms and mammals. In: Hadzi D and Jerman-Blazic (eds) Progress in QSAR, Proceedings of the 6th European Symposium on Quantitative Structure-Activity Relationships, Portorose, Yugoslavia, September 22–26, 1986, Elsevier: Amsterdam, pp. 301–306
91. Lipnick RL (1988) Toxicity assessment and structure-activity relationships. In: Richardson ML (ed) Risk assessment of chemicals in the environment, Royal Society of Chemistry: London, pp. 379–397

92. Lipnick RL (1991) Outliers: their origin and use in the classification of molecular mechanisms of toxicity. Sci Total Environ 109/110: 131–153
93. Schultz TW, Ralston KE, Roberts DW, Veith GD, Aptula AO (2007) Structure–activity relationships for abiotic thiol reactivity and aquatic toxicity of halo-substituted carbonyl compounds. SAR QSAR Environ Res 18: 21–29
94. Gibbs S, van de Sandt JJ, Merk HF, Lockley DJ, Pendlington RU, Pease CK (2007) Xenobiotic metabolism in human skin and 3D human skin reconstructs: A review. Curr Drug Metab 8: 758–772

Interspecies Correlations for Predicting the Acute Toxicity of Xenobiotics

James Devillers, Pascal Pandard, Eric Thybaud, and Anne Merle

Abstract LD50 tests on rat and mouse are commonly used to express the relative hazard associated with the acute toxicity of new and existing substances. These tests are expensive, time consuming, and actively fought by Animal Rightists. Consequently, there is a need to find alternative methods. If the design of QSAR models can be used as surrogate, the search for interspecies correlations also represents a valuable alternative to the classical mammalian laboratory tests. In this chapter, the different toxicity = f(ecotoxicity) models available in the literature were first critically analyzed. In a second step, a strong bibliographical investigation was performed to collect oral, intraperitoneal, and intravenous rat and mouse LD50 data for a large collection of structurally diverse chemicals. In the meantime, EC50 data on *Vibrio fischeri* (Microtox^{TM} test) and *Daphnia magna* were also retrieved from literature. Numerous oral, intraperitoneal, and intravenous rat and mouse toxicity models were derived using *Vibrio fischeri* and *Daphnia magna* as independent variables alone or together through a stepwise regression analysis. Most of the models on *Daphnia magna* were totally new and some of them presented acceptable quality. They outperformed the Microtox^{TM} models. The usefulness of the 1-octanol/water partition coefficient (log P) as additional independent variable was also tested. The interest of nonlinear statistical tools for deriving toxicity = f (ecotoxicity) models was also experienced.

Keywords Interspecies correlation · Mammalian toxicity · *Vibrio fischeri* · *Daphnia magna* · Regression analysis · Nonlinear methods

1 Introduction

The debate surrounding the use of animals for research and testing can be traced back to the eighteenth century when the English utilitarian philosopher and jurist, Jeremy Bentham (1748–1832), who is at the origin of the word "deontology,"

J. Devillers (✉)
CTIS, 3 Chemin de la Gravière, 69140 Rillieux La Pape, France
e-mail: j.devillers@ctis.fr

J. Devillers (ed.), *Ecotoxicology Modeling*, Emerging Topics in Ecotoxicology: Principles, Approaches and Perspectives 2, DOI 10.1007/978-1-4419-0197-2_4,

focused attention on animal rights and on their capacity to suffer. In 1831, the well-known English neurophysiologist Marshall Hall (1790–1857) clearly formulated five basic principles to better control the use of animals in experiments and to take into account their suffering [1]. These principles are at the origin of the UK regulation concerning animal experiments. In fact, surprisingly, before 1986, legislation on the protection of animals used in research and testing existed in only a limited number of European countries. Indeed, in 1986, for the first time, the European Parliament adopted legislation aiming at the protection of laboratory animals. Thus, after many years of discussions, the Council of Europe approved regulations on the protection of vertebrate animals used for experimental and other scientific purposes (Convention ETS 123) [2]. In 1986, Directive 86/609/EEC [3], based on the Convention ETS 123 but more concise and restrictive, was also adopted. This Directive contains provisions on the housing and care of laboratory animals, the education and training of persons manipulating animals, the use of non-wild animals, and more generally the promotion of alternative methods to reduce the number of animals, especially the vertebrates used in the laboratories [4]. While the EU Member States are bound to implement the provisions of Directive 86/609/EEC via their national legislation, Convention ETS 123 takes effect only when ratified by a Member State [4,5]. It is noteworthy that due to the development of sciences, especially in the field of biomedicine and also the evolution of the mentalities, in 2002, the European Commission was called by the European Parliament to prepare a proposal for a revision of Directive 86/609/EEC [4]. This work is currently in its final stage. Broadly speaking the revised Directive will reinforce the well known three R's (reduction, replacement, refinement) pioneered by Russell and Burch [6].

Despite regulatory guidelines aiming at reducing the use of animals in experiments, the total number of animals used for experimental and other scientific purposes in 2005 in the 25 Member States was about 12 million. Rodents together with rabbits represented almost 78% of the total and mice were by far the most commonly used species covering 53% of the total use, followed by rats with 19% [7]. More than 60% of animals were used in research and development for human and veterinary medicine, dentistry, and in fundamental biology studies. Production and quality control of products and devices in human medicine, veterinary medicine, and dentistry required the use of 15.3% of the total number of animals reported in 2005. Toxicological and other safety evaluations represented 8% of the total number of animals used for experimental purposes. This represents a significant decrease with 2002 where this percentage was equal to 9.9 [7]. Undoubtedly, this tendency will continue in the future with Registration, Evaluation, Authorization and Restriction of Chemicals (REACH) [8], the new European Community Regulation on chemicals and their safe use entered into force on June 1, 2007. The Article 13, entitled "General requirements for generation of information on intrinsic properties of substances" stresses that regarding "human toxicity, information shall be generated whenever possible by means other than vertebrate animal tests, through the use of alternative methods, for example, in vitro methods or qualitative or quantitative structure–activity relationship models or from information from structurally related substances (grouping or read-across)".

This evolution of the regulation of chemicals prompted us to evaluate the interest of the interspecies correlations for estimating the acute toxicity of chemicals to rats and mice, which are widely used for testing the acute toxicity of new and existing substances. In a first step, literature was investigated to retrieve equations in the general form $LD50_{tox} = f(LC50_{ecotox}$ or $EC50_{ecotox})$ where LC50 refers to the concentration inducing 50% of mortality among the tested population and the EC50 stands for the effective concentration required to induce a 50% effect in the tested organisms, in both cases, in comparison with a control. These different models were critically analyzed. In a second step, attempts were made to derive new equations allowing the prediction of the acute toxicity of chemicals to rat and mouse from LC50 and EC50 data obtained on invertebrates.

2 Bibliographical Survey

2.1 *Methodological Framework*

Bibliographical investigations were made in journals, books, and reports as well as in bibliographical and factual databases. In that case, a Boolean search was made from the following keywords connected by the logical operators AND, OR, and NOT:

- Correlation, relationship, comparison, intercomparison
- Model, predictive, prediction, in vivo
- Species, interspecies, toxicity, ecotoxicity
- Invertebrate, Daphnid, alga, earthworm, nematode, Microtox^{TM}, *Vibrio fischeri*, *Daphnia*, *Tetrahymena pyriformis*, *Eisenia fetida*
- Vertebrate, rat, mammal, mouse, mice, human

Only the original publications including regression equations with their statistical parameters were selected. In addition, only the in vivo/in vivo correlations were considered, the in vitro/in vivo and in vitro/in vitro correlations being voluntarily eliminated from the present study. Here, the term in vitro only refers to animal and human cell lines.

2.2 *Correlations of Ecotoxicity Test Data with Rat or Mouse LD50 Data*

2.2.1 Correlations of Bacteria Test Data with Rat or Mouse LD50 Data

The standard Microtox^{TM} test involving the bioluminescent bacterium *Vibrio fischeri*, formerly known as *Photobacterium phosphoreum*, is a commonly used ecotoxicological bioassay whose EC50 values (inhibition of bioluminescence), generally recorded after 5, 15 or 30 min, have been correlated to EC50 and LC50 values of numerous nonmammalian species [9]. Conversely, the number of papers aiming at correlating Microtox^{TM} data with rat or mouse acute toxicity data is very limited.

In 1992, Fort [10] proposed two regression equations allowing the prediction of *Vibrio fischeri* (*V.f.*) EC50 values from oral and intravenous (i.v.) mouse LD50 values, (1) and (2). The EC50 and LD 50 values were expressed in mg/l and mg/kg, respectively.

$$\log(\text{EC50 } V.f.) = 0.55 \log(\text{LD50 Mouse oral}) - 0.13, \tag{1}$$
$$n = 123, r = 0.29, p = 0.0012.$$
$$\log(\text{EC50 } V.f.) = 1.6 \log(\text{LD50 Mouse i.v.}) - 1.8, \tag{2}$$
$$n = 51, r = 0.73, p < 0.0001.$$

Although a weak correlation was obtained with the oral LD50 data, a more interesting relationship was recorded with the intravenous LD50 data but the size of the training set was about twice less important.

Kaiser and coworkers [11] tried to extend these results from larger datasets and by considering the oral, i.v., and intraperitoneal (i.p.) routes of exposure for rat and mouse. The EC50 and LD50 values were expressed in mmol/l and mmol/kg, respectively. This yielded the design of six equations (3)–(8).

$$\log(1/\text{LD50 Rat oral}) = 0.20 \log(1/\text{EC50 } V.f.) - 0.96, \tag{3}$$
$$n = 471,\ r = 0.35,\ \text{se} = 0.74.$$
$$\log(1/\text{LD50 Mouse oral}) = 0.20 \log(1/\text{EC50 } V.f.) - 0.86, \tag{4}$$
$$n = 344,\ r = 0.35,\ \text{se} = 0.72.$$
$$\log(1/\text{LD50 Rat i.p.}) = 0.29 \log(1/\text{EC50 } V.f.) - 0.48, \tag{5}$$
$$n = 195,\ r = 0.48,\ \text{se} = 0.82.$$
$$\log(1/\text{LD50 Mouse i.p.}) = 0.25 \log(1/\text{EC50 } V.f.) - 0.49, \tag{6}$$
$$n = 378,\ r = 0.43,\ \text{se} = 0.70.$$
$$\log(1/\text{LD50 Rat i.v.}) = 0.40 \log(1/\text{EC50 } V.f.) - 0.25, \tag{7}$$
$$n = 54,\ r = 0.73,\ \text{se} = 0.79.$$
$$\log(1/\text{LD50 Mouse i.v.}) = 0.35 \log(1/\text{EC50 } V.f.) - 0.30, \tag{8}$$
$$n = 165,\ r = 0.68,\ \text{se} = 0.61.$$

Before outlier removal, the oral route entry regressions for rat (3) and mouse (4) present the same slope and correlation coefficient. Moreover the intercepts and standard error of estimates (se) are very close. Inspection of (5)–(8) shows that for the i.p. and i.v. routes of exposure, a rather good similarity also exists between the regressions for rat and mouse. This similarity between the two mammalian species for the same exposure route prompted the authors to extend the datasets by means of (9)–(11).

$$\log(1/\text{LD50 Rat oral}) = 0.97 \log(1/\text{LD50 Mouse oral}) - 0.04, \quad (9)$$
$$n = 330,\ r = 0.94,\ r^2 = 0.88,\ \text{se} = 0.30.$$
$$\log(1/\text{LD50 Rat i.p.}) = 1.02 \log(1/\text{LD50 Mouse i.p.}) - 0.02, \quad (10)$$
$$n = 162,\ r = 0.96,\ r^2 = 0.92,\ \text{se} = 0.28.$$
$$\log(1/\text{LD50 Rat i.v.}) = 0.99 \log(1/\text{LD50 Mouse i.v.}) - 0.10, \quad (11)$$
$$n = 41,\ r = 0.97,\ r^2 = 0.94,\ \text{se} = 0.29.$$

Using the extended rat datasets for each of the oral, i.p., and i.v. exposure routes, linear regressions were then determined vs. the corresponding Microtox^{TM} data yielding (12)–(14). Deletion of outliers allowed the increase of the statistical parameters of the models, (15)–(17).

$$\log(1/\text{LD50 Rat oral}) = 0.19 \log(1/\text{EC50 } V.f.) - 0.95, \quad (12)$$
$$n = 531,\ r = 0.33,\ \text{se} = 0.72,\ F = 63.4.$$
$$\log(1/\text{LD50 Rat i.p.}) = 0.25 \log(1/\text{EC50 } V.f.) - 0.50, \quad (13)$$
$$n = 427,\ r = 0.43,\ \text{se} = 0.70,\ F = 95.7.$$
$$\log(1/\text{LD50 Rat i.v.}) = 0.35 \log(1/\text{EC50 } V.f.) - 0.20, \quad (14)$$
$$n = 180,\ r = 0.66,\ \text{se} = 0.65,\ F = 139.7.$$
$$\log(1/\text{LD50 Rat oral}) = 0.20 \log(1/\text{EC50 } V.f.) - 1.03, \quad (15)$$
$$n = 506,\ r = 0.41,\ \text{se} = 0.59,\ F = 102.2.$$
$$\log(1/\text{LD50 Rat i.p.}) = 0.26 \log(1/\text{EC50 } V.f.) - 0.57, \quad (16)$$
$$n = 406,\ r = 0.51,\ \text{se} = 0.59,\ F = 141.5.$$
$$\log(1/\text{LD50 Rat i.v.}) = 0.36 \log(1/\text{EC50 } V.f.) - 0.26, \quad (17)$$
$$n = 171,\ r = 0.75,\ \text{se} = 0.52,\ F = 219.3.$$

Inspection of (12)–(17) shows that the data ranges and slopes of the regressions are unaffected by the outlier removal but that the correlation coefficients and more important the F tests and standard errors of the estimates are much improved.

2.2.2 Correlations of Protozoan Test Data with Rat or Mouse LD50 Data

Collections of chemicals have been tested on the freshwater ciliate protozoan *Tetrahymena pyriformis* (*T.p.*) but surprisingly, their use in the design of toxicity = f (ecotoxicity) models is very limited. Thus, Sauvant et al. [12] evaluated the effects of $BaCl_2$ salt, $CdCl_2$, $CoCl_2$, $CrCl_3$, $CuCl_2$, $FeCl_3$, GeO_2, $HgCl_2$, $MnCl_2$, $NbCl_5$, $Pb(NO_3)_2$, $SbCl_3$, $SnCl_4$, $TiCl_4$, $VOSO_4$, and $ZnCl_2$ on the growth rate of *T. pyriformis*. IC50s (inhibitory concentration 50%) were expressed in mmol/l and correlated with corresponding rat oral LD50 values (mmol/kg) retrieved from literature, yielding (18).

$$\log(\text{LD50 Rat oral}) = 0.571 \log(\text{IC50 } T.p.) + 0.389, \quad n = 16,\ r = 0.463,\ p = 0.07. \tag{18}$$

2.2.3 Correlations of Rotifer Test Data with Rat or Mouse LD50 Data

The 24-h LC50 values of the first ten chemicals of the multicentre evaluation of in vitro cytotoxicity (MEIC) program were tested against the estuarine rotifer *Brachionus plicatilis* (*B.p.*) and the freshwater rotifer *Brachionus calyciflorus* (*B.c.*) [13]. These chemicals were the following: paracetamol (CAS RN 103-90-2), acetylsalicylic acid (CAS RN 50-78-2), ferrous sulfate heptahydrate (CAS RN 7782-63-0), amitriptyline HCl (CAS RN 549-18-8), isopropanol (CAS RN 67-63-0), ethanol (CAS RN 64-17-5), methanol (CAS RN 67-56-1), ethylene glycol (CAS RN 107-21-1), diazepam (CAS RN 439-14-5), and digoxin (CAS RN 20830-75-5). The acute toxicity data, expressed in μmol/l, were compared by regression analysis with oral LD50 (μmol/kg) in rat, mouse, and man (HLD = human oral lethal dose). Diazepam and digoxin were excluded from the regressions (19)–(24) because their 24-h LC50 values on the two rotifers were only determined as >35,100 μmol/l and >12,800 μmol/l, respectively. Even if the r^2 values of (19)–(24) are high, the interest of these models is very limited due the nature and limited number of chemicals.

$$\log(\text{LD50 Mouse oral}) = 0.48 \log(\text{LC50 } B.p.) + 2.08, \quad n = 8,\ r^2 = 0.86. \tag{19}$$

$$\log(\text{LD50 Rat oral}) = 0.48 \log(\text{LC50 } B.p.) + 2.12, \quad n = 8,\ r^2 = 0.92. \tag{20}$$

$$\log(\text{LD50 HLD oral}) = 0.44 \log(\text{LC50 } B.p.) + 1.94, \quad n = 8,\ r^2 = 0.83. \tag{21}$$

$$\log(\text{LD50 Mouse oral}) = 0.43 \log(\text{LC50 } B.c.) + 2.40, \quad n = 8,\ r^2 = 0.81. \tag{22}$$

$$\log(\text{LD50 Rat oral}) = 0.42 \log(\text{LC50 } B.c.) + 2.44, \quad n = 8,\ r^2 = 0.88. \tag{23}$$

$$\log(\text{LD50 HLD oral}) = 0.40 \log(\text{LC50 } B.c.) + 2.22, \quad n = 8,\ r^2 = 0.81. \tag{24}$$

2.2.4 Correlations of Crustacean Test Data with Rat or Mouse LD50 Data

The water flea *Daphnia magna* (*D.m.*) is one of the most widely used invertebrates in freshwater aquatic toxicology. The criterion of acute toxicity determined with this organism is the effective concentration yielding the complete immobilization of 50% of the population of *Daphnia* after 24 or 48 h of exposure (24-h or 48-h EC50). To be considered as immobilized, the animals have to be unable to swim after a gentle agitation of the test vessel. Different authors have tried to correlate

LD50 values recorded in rat and/or mouse to EC50 values obtained on *D.m.* Thus, Khangarot and Ray [14] tested various organic and inorganic chemicals on young *D.m.* and the obtained 48-h EC50 values (mg/l) were correlated to oral rat and mouse LD50 values (mg/kg) retrieved from literature, (25) and (26). Even if the correlation coefficients of (25) and (26) are very high, the interest of these two models is reduced due to limited size of their training set.

$$\text{LD50 (Rat oral)} = 2.056\ \text{EC50}\ (D.m.) + 776.2, \quad (25)$$
$$n = 13,\ r = 0.992.$$
$$\text{LD50 (Mouse oral)} = 1.020\ \text{EC50}\ (D.m.) + 312.94, \quad (26)$$
$$n = 10,\ r = 0.991.$$

Interestingly, Enslein and coworkers [15] tried to increase the performances of a simple LD50 vs. EC50 model (27) by introducing molecular descriptors in the equation from a stepwise regression analysis. This yielded a new model (28) showing better statistics. In both equations, LC50 and LD50 values were expressed in mmol/l and mmol/kg, respectively.

$$\log\text{(LD50 Rat)} = f(\log\ \text{LC50}\ D.m.), \quad (27)$$
$$n = 147,\ r^2 = 0.53,\ s = 0.53.$$
$$\log\text{(LD50 Rat)} = 0.287\ D.m. - 0.520\ \text{aryl nitro} + 0.362\ \text{DIFPAT5}$$
$$+ 0.328\ \text{Nb electron releasing groups on a benzene ring} - 0.496\ \text{Ring perimeter}$$
$$- 0.608\ NH_2,\ \text{NH or 3-branched aliphatic amine}$$
$$- 0.408\ \text{aryl alcohol} - 0.619\ \text{methylene diphenyl linkage}$$
$$- 0.826\ \text{aliphatic ether} + 0.337\ \text{primary aliphatic hydroxyl}$$
$$- 0.568\ \text{any carbamate} - 0.487\ \text{pentane fragment}$$
$$- 0.279\ \text{propane/propene fragment} + 3.415, \quad (28)$$
$$n = 147,\ r^2 = 0.75,\ s = 0.40.$$

Inverse relationships (e.g., (29) and (30)) were also proposed by these authors [16] but only the model specifically designed for cholinesterase-inhibiting compounds was interesting (30). In (30), MW is the molecular weight and ${}^2\chi^{\text{v}}$ is the valence path molecular connectivity index of second order.

$$\log\ (1/\text{EC50}\ D.m.) = f\log\ (1/\text{LD50 Rat oral}), \quad (29)$$
$$n = 182,\ r^2 = 0.452,\ s = 1.116.$$
$$\log\ (1/\text{EC50}\ D.m.) = 0.738\ \log\ (1/\text{LD50 Rat oral})$$
$$+ 6.399\ \text{MW} - 0.147\ {}^2\chi^{\text{v}} - 9.29, \quad (30)$$
$$n = 12,\ r^2 = 0.80,\ s = 0.432,\ F = 10.66.$$

Calleja and Persoone [13] also tested the first ten chemicals of the MEIC program against the halophytic anostracan *Artemia salina* (*A.s.*) and the freshwater anostracan *Streptocephalus proboscideus* (*S.p.*). The LC50s recorded after 24 h of exposure and expressed in μmol/l were compared by regression analysis to oral LD50 (μmol/kg) in rat, mouse, and man (HLD). Equations at eight chemicals (i.e., (31), (32), (35), (37), (38), (41)) were established without digoxin and diazepam, while for the equations at nine chemicals (i.e., (33), (34), (36), (39), (40), (42)) only the former compound was excluded. As previously indicated, the interest of these equations is rather limited.

$$\log(\text{LD50 Mouse oral}) = 0.54 \log(\text{LC50 } A.s.) + 1.86, \quad n = 8,\ r^2 = 0.89. \tag{31}$$

$$\log(\text{LD50 Rat oral}) = 0.51 \log(\text{LC50 } A.s.) + 2.01, \quad n = 8,\ r^2 = 0.87. \tag{32}$$

$$\log(\text{LD50 Mouse oral}) = 0.56 \log(\text{LC50 } A.s.) + 1.76, \quad n = 9,\ r^2 = 0.90. \tag{33}$$

$$\log(\text{LD50 Rat oral}) = 0.49 \log(\text{LC50 } A.s.) + 2.09, \quad n = 9,\ r^2 = 0.87. \tag{34}$$

$$\log(\text{LD50 HLD oral}) = 0.45 \log(\text{LC50 } A.s.) + 1.82, \quad n = 8,\ r^2 = 0.80. \tag{35}$$

$$\log(\text{HLD oral}) = 0.54 \log(\text{LC50 } A.s.) + 1.44, \quad n = 9,\ r^2 = 0.80. \tag{36}$$

$$\log(\text{LD50 Mouse oral}) = 0.49 \log(\text{LC50 } S.p.) + 2.30, \quad n = 8,\ r^2 = 0.98. \tag{37}$$

$$\log(\text{LD50 Rat oral}) = 0.41 \log(\text{LC50 } S.p.) + 2.63, \quad n = 8,\ r^2 = 0.75. \tag{38}$$

$$\log(\text{LD50 Mouse oral}) = 0.51 \log(\text{LC50 } S.p.) + 2.12, \quad n = 9,\ r^2 = 0.94. \tag{39}$$

$$\log(\text{LD50 Rat oral}) = 0.42 \log(\text{LC50 } S.p.) + 2.56, \quad n = 9,\ r^2 = 0.77. \tag{40}$$

$$\log(\text{LD50 HLD}) = 0.44 \log(\text{LC50 } S.p.) + 2.15, \quad n = 8,\ r^2 = 0.94. \tag{41}$$

$$\log(\text{HLD oral}) = 0.50 \log(\text{LC50 } S.p.) + 1.81, \quad n = 9,\ r^2 = 0.82. \tag{42}$$

2.2.5 Correlations of Fish Test Data with Rat or Mouse LD50 Data

The relative vulnerability of most of the fish species to pollutants and the ecological importance of these organisms in the functioning of the ecosystems have contributed to their selection as surrogates to assess the aquatic ecotoxicity of chemicals yielding the production of collections of acute toxicity data for all the kinds of compounds susceptible to contaminate the environment. Moreover, due to the taxonomical position of these organisms, numerous equations have been proposed for predicting the acute toxicity of chemicals to rat or mouse from fish LC50s.

Relationships between rat oral LD50 (mmol/kg) and *Lepomis macrochirus* (*L.m.*) and *Pimephales promelas* (*P.p.*) 96-h LC50 (μmol/l) values obtained from the US water quality criteria documents for 47 priority pollutants, including nine organochlorine pesticides were examined by Janardan et al. [17]. Interestingly, these authors also tried to derive regression equations on data obtained from uniform protocol studies for fish [18] and male and female rats [19]. It is noteworthy that this second set only included chlorinated, organophosphorus, and carbamate pesticides. The inverse correlations were also considered because the regression analysis used in this study, which considered separate error terms for the x and y variables, provided different statistics for them. Thus, 20 different models were proposed by Janardan et al. [17] (43)–(62).

$$\log(\text{LD50 Rat}) = 0.43 \log(\text{LC50 } L.m.) - 0.056, \quad (43)$$
$$n = 44,\ r = 0.74 \text{ (priority pollutants).}$$
$$\log(\text{LD50 Rat male}) = 0.47 \log(\text{LC50 } L.m.) - 0.272, \quad (44)$$
$$n = 48,\ r = 0.73 \text{ (priority pollutants + pesticides).}$$
$$\log(\text{LD50 Rat female}) = 0.49 \log(\text{LC50 } L.m.) - 0.313, \quad (45)$$
$$n = 45,\ r = 0.75 \text{ (priority pollutants + pesticides).}$$
$$\log(\text{LD50 Rat male}) = 0.46 \log(\text{LC50 } L.m.) + 0.125, \quad (46)$$
$$n = 12,\ r = 0.76 \text{ (chlorinated pesticides).}$$
$$\log(\text{LD50 Rat female}) = 0.66 \log(\text{LC50 } L.m.) + 0.345, \quad (47)$$
$$n = 11,\ r = 0.92 \text{ (chlorinated pesticides).}$$
$$\log(\text{LC50 } L.m.) = 1.21 \log(\text{LD50 Rat}) + 0.539, \quad (48)$$
$$n = 44,\ r = 0.71 \text{ (priority pollutants).}$$
$$\log(\text{LC50 } L.m.) = 1.04 \log(\text{LD50 Rat male}) + 0.428, \quad (49)$$
$$n = 48,\ r = 0.66 \text{ (priority pollutants + pesticides).}$$
$$\log(\text{LC50 } L.m.) = 1.04 \log(\text{LD50 Rat female}) + 0.492, \quad (50)$$
$$n = 45,\ r = 0.68 \text{ (priority pollutants + pesticides).}$$
$$\log(\text{LC50 } L.m.) = 1.45 \log(\text{LD50 Rat male}) - 0.639, \quad (51)$$
$$n = 12,\ r = 0.88 \text{ (chlorinated pesticides).}$$
$$\log(\text{LC50 } L.m.) = 1.51 \log(\text{LD50 Rat female}) - 0.521, \quad (52)$$
$$n = 11,\ r = 0.999 \text{ (chlorinated pesticides).}$$
$$\log(\text{LD50 Rat}) = 0.35 \log(\text{LC50 } P.p.) - 0.161, \quad (53)$$
$$n = 38,\ r = 0.63 \text{ (priority pollutants).}$$
$$\log(\text{LD50 Rat male}) = 0.33 \log(\text{LC50 } P.p.) - 0.34, \quad (54)$$
$$n = 28,\ r = 0.58 \text{ (priority pollutants + pesticides).}$$

$$\log(\text{LD50 Rat female}) = 0.36 \log(\text{LC50 } P.p.) - 0.259, \quad (55)$$
$n = 25,\ r = 0.67$ (priority pollutants + pesticides).

$$\log(\text{LD50 Rat male}) = 0.59 \log(\text{LC50 } P.p.) + 0.192, \quad (56)$$
$n = 9,\ r = 0.999$ (chlorinated pesticides).

$$\log(\text{LD50 Rat female}) = 0.28 \log(\text{LC50 } P.p.) + 0.380, \quad (57)$$
$n = 8,\ r = 0.999$ (chlorinated pesticides).

$$\log(\text{LC50 } P.p.) = 1.37 \log(\text{LD50 Rat}) + 0.799, \quad (58)$$
$n = 38,\ r = 0.77$ (priority pollutants).

$$\log(\text{LC50 } P.p.) = 1.15 \log(\text{LD50 Rat male}) + 0.820, \quad (59)$$
$n = 28,\ r = 0.65$ (priority pollutants + pesticides).

$$\log(\text{LC50 } P.p.) = 1.53 \log(\text{LD50 Rat female}) + 0.689, \quad (60)$$
$n = 25,\ r = 0.83$ (priority pollutants + pesticides).

$$\log(\text{LC50 } P.p.) = 1.70 \log(\text{LD50 Rat male}) - 0.326, \quad (61)$$
$n = 9,\ r = 0.98$ (chlorinated pesticides).

$$\log(\text{LC50 } P.p.) = 1.29 \log(\text{LD50 Rat female}) - 0.490, \quad (62)$$
$n = 8,\ r = 0.96$ (chlorinated pesticides).

Significant relationships between species were obtained for the priority pollutants, priority pollutants plus pesticides, and chlorinated pesticides. Conversely, the authors did not find an acceptable correlation when only the organophosphate and carbamate pesticide toxicities were compared between rat and fish. In the same way, no significant relationships were obtained between the rat and fishes over all classes of pesticides (equations missing). From the scatter in plots for the priority pollutants, Janardan et al. [17] deducted that fish were relatively more sensitive (LC50/LD50 < 1) than rats for substances with an LD50 < 1 mmol/kg (rat) and less sensitive (LC50/LD50 > 1) for substances with an LD50 > 1 mmol/kg. From the regression models, they showed that the two fish species presented about the same sensitivity to the priority pollutants but bluegill (*L.m.*) was less sensitive than fathead minnow (*P.p.*) to pesticides.

Kaiser and coworker [20] proposed a rat vs. fathead minnow model (*P.p.*) with a larger domain of application (63). The performance of the model increased with a three parameter equation also including *V.f.* EC50 and 1-octanol/water partition coefficient (log P) data (64).

$$\log(1/\text{LD50 Rat}) = 0.36 \log(1/\text{LC50 } P.p.) - 1.16, \quad (63)$$
$n = 91,\ r^2 = 0.34.$

$$\log(1/\text{LD50 Rat}) = f(\log 1/\text{EC50 } V.f.,\ \log 1/\text{LC50 } P.p.,\ \log \text{P}), \quad (64)$$
$n = 91,\ r^2 = 0.41.$

Hodson [21] compared the toxicity of industrial chemicals to *Oncorhynchus mykiss* (formerly *Salmo gairdneri*), as shown by i.p. injections (i.p. LD50), oral dosing (oral LD50), and aqueous exposure (LC50), with published values for i.p. LD50s and oral LD50s of mice and rats. Prior correlation analysis, the toxicity data were expressed on a millimolar basis. Twenty equations were obtained (65)–(84). When mouse and rat oral LD50s are compared with fish i.p. LD50s, the correlation coefficients are equal to 0.807 and 0.897, respectively (65) and (66) but when the comparison is made between i.p. LD50s, (67) and (68), r is improved to 0.936 and 0.933, respectively. Despite small sample sizes, there is a strong relationship between fish oral LD50s and rat ($r = 0.827$) and mouse ($r = 0.914$) i.p. LD50s, (69) and (70). Conversely, the rat and mouse oral LD50s are not strongly related to fish oral LD50s, (71) and (72).

$$\log(\text{LD50 Mouse oral}) = 0.8046 \log(\text{LD50 } O.m. \text{ i.p.}) + 0.4267, \quad (65)$$
$$n = 13,\ r = 0.807,\ p < 0.05.$$
$$\log(\text{LD50 Rat oral}) = 0.9429 \log(\text{LD50 } O.m. \text{ i.p.}) + 0.1495, \quad (66)$$
$$n = 25,\ r = 0.8971,\ p < 0.05.$$
$$\log(\text{LD50 Mouse i.p.}) = 0.8288 \log(\text{LD50 } O.m. \text{ i.p.}) - 0.1831, \quad (67)$$
$$n = 12,\ r = 0.936,\ p < 0.05.$$
$$\log(\text{LD50 Rat i.p.}) = 1.0051 \log(\text{LD50 } O.m. \text{ i.p.}) - 0.1693, \quad (68)$$
$$n = 16,\ r = 0.933,\ p < 0.05.$$
$$\log(\text{LD50 Rat i.p.}) = 1.4080 \log(\text{LD50 } O.m. \text{ oral}) - 0.5262, \quad (69)$$
$$n = 6,\ r = 0.827,\ p < 0.05.$$
$$\log(\text{LD50 Mouse i.p.}) = 0.8264 \log(\text{LD50 } O.m. \text{ oral}) - 0.2479, \quad (70)$$
$$n = 7,\ r = 0.914,\ p < 0.05.$$
$$\log(\text{LD50 Mouse oral}) = 0.6831 \log(\text{LD50 } O.m. \text{ oral}) + 0.3048, \quad (71)$$
$$n = 7,\ r = 0.657.$$
$$\log(\text{LD50 Rat oral}) = 0.8255 \log(\text{LD50 } O.m. \text{ oral}) + 0.1606, \quad (72)$$
$$n = 9,\ r = 0.588.$$

There is considerably more variation in comparisons of i.p. LD50s to fish LC50s (73)–(75). The best relationship between i.p. LD50s and LC50s is for rat (74) while the poorest is for mouse, (73).

$$\log(\text{LD50 Mouse i.p.}) = 0.7709 \log(\text{LC50 } O.m.) + 1.1008, \quad (73)$$
$$n = 8,\ r = 0.48.$$
$$\log(\text{LD50 Rat i.p.}) = 0.5891 \log(\text{LC50 } O.m.) + 1.2136, \quad (74)$$
$$n = 11,\ r = 0.83,\ p < 0.05.$$
$$\log(\text{LD50 } O.m. \text{ i.p.}) = 0.8883 \log(\text{LC50 } O.m.) + 1.4608, \quad (75)$$
$$n = 13,\ r = 0.60,\ p < 0.05.$$

$$\log(\text{LD50 Mouse oral}) = 0.7576 \log(\text{LC50 } O.m.) + 1.6616, \quad n = 10,\ r = 0.19. \tag{76}$$

$$\log(\text{LD50 Rat oral}) = 0.7086 \log(\text{LC50 } O.m.) + 1.6553, \quad n = 15,\ r = 0.62,\ p < 0.05. \tag{77}$$

$$\log(\text{LD50 } O.m. \text{ oral}) = 0.797 \log(\text{LC50 } O.m.) + 1.5216, \quad n = 9,\ r = 0.58. \tag{78}$$

An attempt was also made by Hodson [21] to relate fish and mammal oral and i.p. LD50s to fish LC50s amended by the octanol/water partition coefficient (P). The assumption was that P could correct differences in toxicity due to the effect of partitioning of chemicals on uptake and toxicity during aqueous exposure. Six new regression equations were produced (79)–(84).

$$\log(\text{LD50 } O.m. \text{ i.p.}) = 2.464 \log(\text{LC50 } O.m. \times P) - 0.499, \quad n = 11,\ r = 0.32. \tag{79}$$

$$\log(\text{LD50 Rat i.p.}) = -2.128 \log(\text{LC50 } O.m. \times P) + 0.619, \quad n = 9,\ r = 0.01. \tag{80}$$

$$\log(\text{LD50 Mouse i.p.}) = 1.601 \log(\text{LC50 } O.m. \times P) - 0.172, \quad n = 7,\ r = 0.21. \tag{81}$$

$$\log(\text{LD50 } O.m. \text{ oral}) = 1.8661 \log(\text{LC50 } O.m. \times P) + 0.129, \quad n = 7,\ r = 0.45. \tag{82}$$

$$\log(\text{LD50 Rat oral}) = 1.9138 \log(\text{LC50 } O.m. \times P) + 0.089, \quad n = 13,\ r = 0.53. \tag{83}$$

$$\log(\text{LD50 Mouse oral}) = 1.7397 \log(\text{LC50 } O.m. \times P) + 0.193, \quad n = 8,\ r = 0.62. \tag{84}$$

Inspection of (73)–(84) shows that an improvement was only noted for the equations dealing with mouse oral LD50s (i.e., (76) vs. (84)).

Delistraty et al. [22] examined acute toxicity relationships over several exposure routes in rainbow trout (*O.m.*) and rats. An initial database of 217 chemicals (126 pesticides and 91 nonpesticides) was constituted. 1-octanol/water partition coefficient (log P) values for the organic molecules were also retrieved from literature. LC50 and LD50 values were expressed in mmol/l and mmol/kg, respectively. The authors showed that the stratification of the data into pesticides and nonpesticides did not particularly improve predictions of trout LC50s from rat oral LD50s (85)–(87). Addition of log P in the model (88) increased the r and r^2 values but it is noteworthy that the number of chemicals used to derive the model was lower (i.e., 213 vs. 145).

$$\log(\text{LC50 } O.m.) = 0.722 \log(\text{LD50 Rat oral}) - 2.16, \quad n = 213,\ r = 0.512,\ r^2 = 0.262. \tag{85}$$

$$\log(\text{LC50 } O.m.) = 0.476 \log(\text{LD50 Rat oral}) - 2.42, \quad (86)$$
$$n = 125,\ r = 0.380,\ r^2 = 0.144 \text{ (pesticides)}.$$
$$\log(\text{LC50 } O.m.) = 0.925 \log(\text{LD50 Rat oral}) - 1.98, \quad (87)$$
$$n = 88,\ r = 0.540,\ r^2 = 0.292 \text{ (nonpesticides)}.$$
$$\log(\text{LC50 } O.m.) = 0.644 \log(\text{LD50 Rat oral}) - 0.463 \log P - 0.953, \quad (88)$$
$$n = 145,\ r = 0.729,\ r^2 = 0.531.$$

Trout LC50 values were also predicted from rat LD50 data with regressions matched on exposure route. Statistically significant models were obtained for the three routes of exposure (89)–(91). Addition of log P in the models (92)–(94) did not improve the models, except for the i.p. route, (94).

$$\log(\text{LC50 } O.m. \text{ oral}) = 0.918 \log(\text{LD50 Rat oral}) + 0.153, \quad (89)$$
$$n = 27,\ r = 0.907,\ r^2 = 0.823.$$
$$\log(\text{LC50 } O.m. \text{ dermal}) = 0.794 \log(\text{LD50 Rat dermal}) + 0.384, \quad (90)$$
$$n = 11,\ r = 0.914,\ r^2 = 0.835.$$
$$\log(\text{LC50 } O.m. \text{ i.p.}) = 0.852 \log(\text{LD50 Rat i.p.}) + 0.355, \quad (91)$$
$$n = 13,\ r = 0.761,\ r^2 = 0.579.$$
$$\log(\text{LC50 } O.m. \text{ oral}) = 0.970 \log(\text{LD50 Rat oral}) + 0.050 \log P + 0.035, \quad (92)$$
$$n = 25,\ r = 0.904,\ r^2 = 0.817.$$
$$\log(\text{LC50 } O.m. \text{ dermal}) = 0.890 \log(\text{LD50 Rat dermal}) + 0.079 \log P + 0.094, \quad (93)$$
$$n = 8,\ r = 0.755,\ r^2 = 0.570.$$
$$\log(\text{LC50 } O.m. \text{ i.p.}) = 0.463 \log(\text{LD50 Rat i.p.}) - 0.324 \log P + 1.08, \quad (94)$$
$$n = 13,\ r = 0.912,\ r^2 = 0.832.$$

Models for predicting trout LC50s from rat inhalation (inh) LD50s were also designed by Delistraty [23] (95)–(98). Toxicity data were expressed in mmol/l, ppmw (parts per million by weight), or ppmv (parts per million by volume). Addition of molecular descriptors only slightly increased the performances of the best one parameter equation (i.e., (96) vs. (99)).

$$\log(\text{LC50 } O.m. \text{ mmol/l}) = 0.953 \log(\text{LD50 Rat inh mmol/l}) + 0.235, \quad (95)$$
$$n = 60,\ r = 0.678,\ r^2 = 0.459.$$
$$\log(\text{LC50 } O.m. \text{ mmol/l}) = 0.955 \log(\text{LCT50 Rat inh mmol-h/l}) - 0.126, \quad (96)$$
$$n = 46,\ r = 0.745,\ r^2 = 0.556.$$
$$\log(\text{LC50 } O.m. \text{ ppmw}) = 0.899 \log(\text{LD50 Rat inh ppmv}) - 1.46, \quad (97)$$
$$n = 15,\ r = 0.592,\ r^2 = 0.350.$$

$$\log(\text{LC50 } O.m.\ \text{ppmw}) = 1.16 \log(\text{LCT50 Rat inh ppmv-h}) - 3.22, \quad (98)$$
$$n = 11,\ r = 0.747,\ r^2 = 0.558.$$
$$\log(\text{LC50 } O.m.\ \text{mmol/l}) = 0.725 \log(\text{LCT50 Rat inh mmol-h/l})$$
$$-3.19 \log \text{MW} - 0.266 \log \text{VP} + 0.263 \log S\ (\text{mmol/l}) + 5.72, \quad (99)$$
$$n = 38,\ r = 0.873,\ r^2 = 0.763.$$

2.3 Main Characteristics of the Published Models

This literature survey clearly reveals a limited number of toxicity = f (ecotoxicity) models. The available correlations between toxicological and ecotoxicological endpoints only deal with a limited number of species as well as reduced sets of chemicals. The interspecies correlations are established considering rodents (rat and mouse) and aquatic species (mainly fish and bacteria but also crustaceans and rotifers).

Generally, models are designed for predicting mammalian toxicity from aquatic toxicity data but the converse is also found. The 99 correlation equations collected from literature were only derived from acute toxicity data on pure chemicals (LD50s, EC50s, LC50s). No interspecies relationships were investigated using chronic or sublethal effects, probably due to the lack of such data. The toxicity data are always retrieved from literature, while the ecotoxicity data can be obtained from experiments [12–14].

Several exposure routes were considered for mammalian species (i.e., oral, dermal, intraperitoneal, intravenous, inhalation) leading to the development of specific predictive models presenting different qualities. Thus, for example, Kaiser et al. [11] found that the interspecies relationships between *Vibrio fischeri*, rat, and mouse increased significantly from oral, to intraperitoneal, and to intravenous data. Regarding the rats and mice, generally no distinction is made between males and females in the modeling processes.

Most of the models for predicting mammalian toxicity from aquatic toxicity data were designed from simple linear regression analysis. However, it is noteworthy that some authors successfully included molecular descriptors in their equation, especially the 1-octanol/water partition coefficient (log P). Interestingly, Kaiser and Esterby [20] established a predictive model for rat toxicity using the results of tests performed on *Vibrio fischeri, Pimephales promelas*, and from log P.

Despite some significant correlations, it appears that the toxicity = f (ecotoxicity) models found in the literature cannot be used in practice. Indeed, most of them were established from a limited number of chemicals. Thus, more than 70% of the models found in the literature were derived from less than 50 chemicals. Moreover, it is important to note that chemicals are very often eliminated before or during the regression processes without clear justifications.

This prompted us to develop new models focused on the prediction of rat and mouse LD50s from invertebrate EC50s or LC50s. The selected species were *Daphnia magna* and *Vibrio fischeri* because these organisms are widely used for assessing the hazard of chemicals and hence, collections of EC50s for these organisms are available in the literature.

3 Design of New Toxicity = *f* (Ecotoxicity) Models

3.1 Data Sources, Notations, and Treatments

LD50s for rats and mice were retrieved from CD-ROMs (e.g., Merck Index, ECDIN, IUCLID) and data banks such as MSDS (http://physchem.ox.ac.uk/MSDS/) or Extonet (http://extoxnet.orst.edu/pips/ghindex.html) but also directly from scientific articles, books, and reports. Scripts in Python were written to navigate into this wealth of toxicological information and to structure and gather the most interesting one. CAS RNs were also retrieved for all the collected chemicals to eliminate the problem of compounds indexed twice or more with different names. This allowed us to eliminate a little bit less than 2,000 LD50s. At the end of the refining process, the toxicological database included about 23,000 rat and mouse oral, i.p., and i.v. LD50s for more than 7,000 organic and inorganic chemicals.

The same strategy was adopted for collecting EC50 values for *Daphnia magna*. Regarding *Vibrio fischeri*, a different approach was used. All the data included in the book of Kaiser and Devillers [24] were first gathered. Thus 1,800 EC50s for 1,290 organic and inorganic molecules and their corresponding CAS RN were collected and structured via Python scripts. This database was then completed from online bibliographic searching in ScienceDirect, Medline, and Google. This allowed us to retrieve 150 additional EC50s corresponding to 110 new molecules. After retrieval of the missing CAS RNs and the removal of duplicates, 82 molecules with their EC50s on *Vibrio fischeri* were added to the initial Microtox^{TM} database.

It is important to note that water solubility data were also collected when available to validate the ecotoxicological data.

For both types of data, the results were not averaged when different values were gathered for the same endpoint and chemical. In that case, the most reliable data were selected. Reliability was mainly based on the existence of test protocols but also on peer review exercises made by experts.

Furthermore, for modeling purposes, it was decided to convert all the (eco)toxicity data into log ($1/C$, C in mmol/kg or mmol/l).

Because sometimes the literature survey showed that the 1-octanol/water partition coefficient (log P) yielded interesting results when introduced as additional variable in the toxicity = f (ecotoxicity) models, it was also decided to consider this parameter in the modeling process. All the log P values were calculated from the KowWin v. 1.67 program [25].

3.2 Linear Regressions of Rat and Mouse LD50s vs. Microtox™ 5-, 15-, and 30-min EC50s

Because Kaiser et al. [11] obtained rather significant correlations between Microtox™ EC50 data and oral, i.p., and i.v. rat and mouse LD50 data, in a first step, an attempt was made to at least confirm their results. However, while Kaiser et al. [11] did not differentiate the time of exposure for *Vibrio fischeri* in their modeling strategy, it was decided to derive different Microtox™ models for the data recorded after 5, 15, and/or 30 min of exposure and the oral, i.p., and i.v. rat and mouse LD50 data. The Rv. 2.3.1 program written in *R* and freely available from the CRAN library, was used for deriving the different toxicity = f (ecotoxicity) models.

Interspecies regressions between mouse and rat LD50 values were first derived for the three routes of exposure yielding (100)–(105).

$$\log(1/\text{LD50 Rat oral}) = 1.01 \log(1/\text{LD50 Mouse oral}), \quad (100)$$
$$n = 633,\ r^2 = 0.89,\ s = 0.29,\ F = 5{,}288.$$
$$\log(1/\text{LD50 Mouse oral}) = 0.88 \log(1/\text{LD50 Rat oral}) - 0.07, \quad (101)$$
$$n = 633,\ r^2 = 0.89,\ s = 0.27,\ F = 5{,}288.$$
$$\log(1/\text{LD50 Rat i.p.}) = 0.95 \log(1/\text{LD50 Mouse i.p.}) - 0.01, \quad (102)$$
$$n = 306,\ r^2 = 0.91,\ s = 0.28,\ F = 3{,}183.$$
$$\log(1/\text{LD50 Mouse i.p.}) = 0.96 \log(1/\text{LD50 Rat i.p.}) - 0.01, \quad (103)$$
$$n = 306,\ r^2 = 0.91,\ s = 0.28,\ F = 3{,}183.$$
$$\log(1/\text{LD50 Rat i.v.}) = 0.99 \log(1/\text{LD50 Mouse i.v.}) + 0.04, \quad (104)$$
$$n = 145,\ r^2 = 0.95,\ s = 0.27,\ F = 2{,}593.$$
$$\log(1/\text{LD50 Mouse i.v.}) = 0.96 \log(1/\text{LD50 Rat i.v.}) - 0.02, \quad (105)$$
$$n = 145,\ r^2 = 0.95,\ s = 0.26,\ F = 2{,}593.$$

In Kaiser et al. [11], mouse toxicity data were only used as independent variables. Nevertheless, comparison of (9)–(11) with (100), (102), and (104) clearly shows that the latter group of models outperforms the former having better statistics and presenting a much larger domain of application.

Rat and mouse oral LD50s were correlated to *V.f.* EC50 values recorded after 5, 15, and 30 min of exposure. The corresponding equations are given below.

$$\log(1/\text{LD50 Rat oral}) = 0.24 \log(1/\text{EC50 } V.f. - 5\text{ min}) - 0.97, \quad (106)$$
$$n = 339,\ r^2 = 0.15,\ s = 0.75,\ F = 60.1.$$
$$\log(1/\text{LD50 Rat oral}) = 0.22 \log(1/\text{EC50 } V.f. - 15\text{ min}) - 1.01, \quad (107)$$
$$n = 297,\ r^2 = 0.14,\ s = 0.71,\ F = 48.8.$$
$$\log(1/\text{LD50 Rat oral}) = 0.24 \log(1/\text{EC50 } V.f. - 30\text{ min}) - 1.01, \quad (108)$$
$$n = 272,\ r^2 = 0.17,\ s = 0.67,\ F = 53.5.$$

$$\log(1/\text{LD50 Mouse oral}) = 0.20 \log(1/\text{EC50 } V.f. - 5\text{ min}) - 0.93, \quad (109)$$
$$n = 251,\ r^2 = 0.12,\ s = 0.74,\ F = 34.0.$$
$$\log(1/\text{LD50 Mouse oral}) = 0.21 \log(1/\text{EC50 } V.f. - 15\text{ min}) - 1.03, \quad (110)$$
$$n = 222,\ r^2 = 0.15,\ s = 0.67,\ F = 39.7$$
$$\log(1/\text{LD50 Mouse oral}) = 0.19 \log(1/\text{EC50 } V.f. - 30\text{ min}) - 0.93, \quad (111)$$
$$n = 209,\ r^2 = 0.11,\ s = 0.69,\ F = 26.3.$$

Equations (106)–(111) show rather poor statistical parameter values but it is important to note that no outlier removal was performed to follow the modeling strategy adopted by Kaiser et al. [11]. Otherwise, the intercepts and slopes of these equations do not differ significantly of those obtained by Kaiser et al. [11] for the oral route entry models for the rat and mouse, (3) and (4).

In the same way, rat and mouse i.p. LD50s were correlated to *V.f.* EC50 values recorded after 5, 15, and 30 min of exposure yielding (112)–(117).

$$\log(1/\text{LD50 Rat i.p.}) = 0.31 \log(1/\text{EC50 } V.f. - 5\text{ min}) - 0.65, \quad (112)$$
$$n = 142,\ r^2 = 0.26,\ s = 0.77,\ F = 48.7.$$
$$\log(1/\text{LD50 Rat i.p.}) = 0.31 \log(1/\text{EC50 } V.f. - 15\text{ min}) - 0.63, \quad (113)$$
$$n = 122,\ r^2 = 0.26,\ s = 0.77,\ F = 41.4.$$
$$\log(1/\text{LD50 Rat i.p.}) = 0.35 \log(1/\text{EC50 } V.f. - 30\text{ min}) - 0.70, \quad (114)$$
$$n = 126,\ r^2 = 0.31,\ s = 0.74,\ F = 55.7.$$
$$\log(1/\text{LD50 Mouse i.p.}) = 0.29 \log(1/\text{EC50 } V.f. - 5\text{ min}) - 0.54, \quad (115)$$
$$n = 216,\ r^2 = 0.24,\ s = 0.71,\ F = 66.1.$$
$$\log(1/\text{LD50 Mouse i.p.}) = 0.30 \log(1/\text{EC50 } V.f. - 15\text{ min}) - 0.60, \quad (116)$$
$$n = 187,\ r^2 = 0.25,\ s = 0.69,\ F = 63.2.$$
$$\log(1/\text{LD50 Mouse i.p.}) = 0.30 \log(1/\text{EC50 } V.f. - 30\text{ min}) - 0.59, \quad (117)$$
$$n = 169,\ r^2 = 0.26,\ s = 0.66,\ F = 57.3.$$

Equations (112)–(117) present better statistical parameter values than (106)–(111) but they were derived from fewer learning sets. The same tendency was observed when rat and mouse i.v. LD50s were correlated to *V.f.* EC50 values recorded after 5, 15, and 30 minutes (118)–(123).

$$\log(1/\text{LD50 Rat i.v.}) = 0.43 \log(1/\text{EC50 } V.f. - 5\text{ min}) - 0.24, \quad (118)$$
$$n = 44,\ r^2 = 0.55,\ s = 0.74,\ F = 51.0.$$
$$\log(1/\text{LD50 Rat i.v.}) = 0.41 \log(1/\text{EC50 } V.f. - 15\text{ min}) - 0.44, \quad (119)$$
$$n = 30,\ r^2 = 0.65,\ s = 0.63,\ F = 51.7.$$
$$\log(1/\text{LD50 Rat i.v.}) = 0.44 \log(1/\text{EC50 } V.f. - 30\text{ min}) - 0.38, \quad (120)$$
$$n = 29,\ r^2 = 0.70,\ s = 0.62,\ F = 62.4.$$

$$\log(1/\text{LD50 Mouse i.v.}) = 0.41 \log(1/\text{EC50 } V.f. - 5\text{ min}) - 0.36, \quad (121)$$
$$n = 79,\ r^2 = 0.50,\ s = 0.64,\ F = 78.4.$$
$$\log(1/\text{LD50 Mouse i.v.}) = 0.41 \log(1/\text{EC50 } V.f. - 15\text{ min}) - 0.46, \quad (122)$$
$$n = 70,\ r^2 = 0.56,\ s = 0.60,\ F = 86.8.$$
$$\log(1/\text{LD50 Mouse i.v.}) = 0.39 \log(1/\text{EC50 } V.f. - 30\text{ min}) - 0.39, \quad (123)$$
$$n = 57,\ r^2 = 0.66,\ s = 0.51,\ F = 106.$$

The slopes and intercepts of (106)–(123) appear rather similar for the same route of exposure. This similarity increases when, within each route of exposure, the equations are matched according to the time of exposure (i.e., 5, 15, 30 min). There is a significant increase in the regression slopes in the order oral < intraperitoneal < intravenous exposure. As stressed by Kaiser et al. [11], the change in the slopes results from the corresponding decrease in metabolic degradation with a decreasing requirement for cell membrane diffusion and resulting higher efficacy of the intravenous route relative to the oral exposure. Correlations with rats always outperform those with mice except when the rat and mouse oral LD50s are correlated to the Microtox^{TM} 15-min EC50s (i.e., (107) vs. (110)). This is surprising due to the high level of correlation that exists between the two mammalian species (100)–(105).

While Kaiser et al. [11] did not distinguish the time of exposure with *Vibrio fischeri*, in the present study, different equations were produced with the Microtox^{TM} data recorded after 5, 15, and 30 min of exposure. This difference of strategy with Kaiser et al. [11] cannot explain alone the difference of size of the learning sets between the two studies, especially if we consider that our databases were larger than those of Kaiser et al. [11]. This claim can be easily verified when we compare the size of the learning sets used in both studies for deriving the correlations between rat and mouse LD50 values. For the oral, i.p., and i.v. routes of exposure the learning sets used by Kaiser et al. [11] included 330, 162, and 41 chemicals, respectively (9)–(11). In the present study, the same sets included 633 (100), 306 (102), and 145 molecules (104), respectively.

This difference in the number of chemicals used in the linear regressions of rat and mouse LD50s vs. Microtox^{TM} EC50s should be also explained by the fact that Kaiser et al. [11] did not take into account the hydrosolubility values of the chemicals in the selection of their ecotoxicity data. Thus, for example, inspection of Table 4 (page 1,604) of their paper shows that they selected a value of 1.41 (in log $1/C$ mmol/l) for the EC50 of p, p'-DDT against *Vibrio fischeri*. This corresponds to an EC50 value equal to 13.39 mg/l while the hydrosolubility of this chemical at 15 °C is only 0.017 mg/l [26].

In addition, while in the present study, the EC50 values only determined with superior or inferior limits were discarded from the database, this rule was not adopted by Kaiser et al. [11]. Thus, for example, an intensive bibliographical search on the Microtox^{TM} toxicity of mitomycin C only provided one reference showing that the EC50 values of this chemical against *Vibrio fischeri* after 5, 10, 15, and 20 min of exposure were < 16, <16.1, <15.2, and <13.7 mg/l, respectively [27]. Surprisingly, Kaiser et al. [11] selected a value of 1.39 (in log $1/C$ mmol/l), which is equivalent to 13.7 mg/l (see Table 4, page 1,604).

Consequently, even if the toxicity $= f$ (ecotoxicity) models of the present study were derived from fewer training sets than those of Kaiser et al. [11], they present better foundations.

Inspection of our Microtox™ database showed that while some chemicals were characterized by 5-, 15-, and 30-min EC50s, for others, the EC50 values were only available for one or two times of exposure. This prompted us to first derive equations allowing the prediction of 30-min EC50s from 5 and 15-min EC50s and then to use the observed and calculated Microtox™ 30-min EC50s for computing new toxicity $= f$ (ecotoxicity) models from larger training sets. This work is presented in the next section. It is noteworthy that all the calculations were made with Statistica ver. 6 (StatSoft, Paris).

3.3 *Linear Regressions of Rat and Mouse LD50s vs. Microtox™ 30-min* EC50s*

The two models allowing the prediction of Microtox™ 30-min EC50 values from 5-min EC50s or 15-min ECs are given below, (124) and (125). They are highly statistically significant. Inspection of these models let to suppose that no difference exist between the EC50 data recorded after 5, 15, or 30 min of exposure. Although it is true for chemicals, it is totally wrong for others [24, 28]. Undoubtedly, the best strategy would consist in the design of specific models for encoding these different particularities but for the sake of simplicity we decided not to do so.

$$\log 1/\text{EC50} - 30\text{ min} = \log 1/\text{EC50} - 5\text{ min} + 0.03, \quad (124)$$
$$n = 951,\ r = 0.98,\ s = 0.22,\ F = 22{,}785,\ p < 10^{-5}.$$
$$\log 1/\text{EC50} - 30\text{ min} = \log 1/\text{EC50} - 15\text{ min} - 0.01, \quad (125)$$
$$n = 903,\ r = 0.996,\ s = 0.1,\ F = 108{,}400,\ p < 10^{-5}.$$

In the models, the *V.f.* variable being constituted of observed and approximated Microtox™ 30-min EC50 values, it is spotted by an asterisk to avoid confusions with the previous models.

$$\log(1/\text{LD50 Rat oral}) = 0.25 \log(1/\text{EC50 } V.f. - 30\text{ min}^*) - 1.00, \quad (126)$$
$$n = 407,\ r = 0.46,\ s = 0.66,\ F = 106,\ p < 10^{-5}.$$
$$\log(1/\text{LD50 Mouse oral}) = 0.23 \log(1/\text{EC50 } V.f. - 30\text{ min}^*) - 0.98, \quad (127)$$
$$n = 297,\ r = 0.43,\ s = 0.69,\ F = 67.3,\ p < 10^{-5}.$$
$$\log(1/\text{LD50 Rat i.p.}) = 0.33 \log(1/\text{EC50 } V.f. - 30\text{ min}^*) - 0.77, \quad (128)$$
$$n = 159,\ r = 0.65,\ s = 0.61,\ F = 117,\ p < 10^{-5}.$$
$$\log(1/\text{LD50 Mouse i.p.}) = 0.32 \log(1/\text{EC50 } V.f. - 30\text{ min}^*) - 0.60, \quad (129)$$
$$n = 239,\ r = 0.60,\ s = 0.61,\ F = 132,\ p < 10^{-5}.$$

$$\log(1/\text{LD50 Rat i.v.}) = 0.42\,(\log 1/\text{EC50 } V.f. - 30\,\text{min}^*) - 0.29, \quad (130)$$
$$n = 49,\ r = 0.79,\ s = 0.71,\ F = 75.2,\ p < 10^{-5}.$$
$$\log(1/\text{LD50 Mouse i.v.}) = 0.42 \log(1/\text{EC50 } V.f. - 30\,\text{min}^*) - 0.44, \quad (131)$$
$$n = 92,\ r = 0.84,\ s = 0.50,\ F = 207,\ p < 10^{-5}.$$

Equations (126)–(131) significantly outperform (106)–(123). This is mainly due to the removal of some outliers. In the first step of this study, the goal was mainly to confirm or infirm the results obtained by Kaiser et al. [11], and hence it was necessary to follow at best their methodology, which first consisted in considering the whole datasets without outlier removal. In a second step, also in agreement with the strategy used by Kaiser et al. [11], an attempt was made to optimize a little bit the equations. Most of the eliminated outliers were inorganic chemicals for which it was difficult to know whether the Microtox^{TM} EC50 values were reported to the element, the salt, etc. Inspection of (126)–(131) shows that the slopes of the models are similar for the same route of exposition in rats and mice.

3.4 Linear Regressions of Rat and Mouse LD50s vs. *Daphnia magna* 48-h* EC50s

The database on *Daphnia magna* including EC50 values recorded after 24 and 48 h of exposure, a regression equation was first computed to convert the 24-h EC50s into 48-h EC50s, (132).

$$\log 1/\text{EC50} - 48\,\text{h} = 0.99 \log(1/\text{EC50} - 24\,\text{h}) + 0.29, \quad (132)$$
$$n = 258,\ r = 0.97,\ s = 0.40,\ F = 4{,}769.$$

Equation (132) presents a high predictive power as well as a large domain of application, which is clearly shown in Fig. 1.

The models allowing the prediction of LD50s in rats and mice after oral, i.p., and i.v. absorption are presented below. Because the models include observed and calculated *Daphnia magna* 48-h EC50 values from (132), an asterisk is used to characterize the independent variable.

$$\log(1/\text{LD50 Rat oral}) = 0.30 \log(1/\text{EC50 } D.m. - 48\,\text{h}^*) - 1.13, \quad (133)$$
$$n = 588,\ r = 0.66,\ s = 0.63,\ F = 448,\ p < 10^{-5}.$$
$$\log(1/\text{LD50 Mouse oral}) = 0.28 \log(1/\text{EC50 } D.m. - 48\,\text{h}^*) - 1.06, \quad (134)$$
$$n = 374,\ r = 0.64,\ s = 0.64,\ F = 254,\ p < 10^{-5}.$$
$$\log(1/\text{LD50 Rat i.p.}) = 0.35 \log(1/\text{EC50 } D.m. - 48\,\text{h}^*) - 0.72, \quad (135)$$
$$n = 191,\ r = 0.75,\ s = 0.62,\ F = 237,\ p < 10^{-5}.$$
$$\log(1/\text{LD50 Mouse i.p.}) = 0.33 \log(1/\text{EC50 } D.m. - 48\,\text{h}^*) - 0.62, \quad (136)$$
$$n = 261,\ r = 0.69,\ s = 0.62,\ F = 241,\ p < 10^{-5}.$$

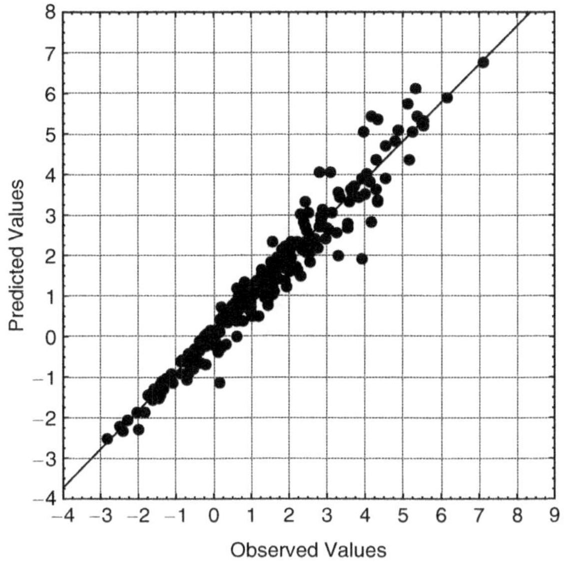

Fig. 1 Observed vs. calculated *Daphnia* 48-h EC50 values (− log (mmol/l)) from model (132)

$$\log(1/\text{LD50 Rat i.v.}) = 0.43 \log(1/\text{EC50 } D.m. - 48\,\text{h}^*) - 0.28, \quad (137)$$
$$n = 61,\ r = 0.87,\ s = 0.62,\ F = 181,\ p < 10^{-5}.$$
$$\log(1/\text{LD50 Mouse i.v.}) = 0.38 \log(1/\text{EC50 D.m.} - 48\,\text{h}^*) - 0.39, \quad (138)$$
$$n = 108,\ r = 0.79,\ s = 0.61,\ F = 175,\ p < 10^{-5}.$$

The observed vs. calculated LD50s from (133) to (138) are displayed in Figs. 2–7.

Inspection of (126)–(138) shows that it is preferable to predict rat and mouse oral and i.p. LD50s and rat i.v. LD50s from EC50s obtained from *Daphnia magna* instead of *Vibrio fischeri*, while it is the converse regarding the intravenous route of exposure in mouse.

From these results, it was interesting to test whether the use of *Vibrio fischeri* and *Daphnia magna* as independent variables in the rat and mouse regression equations improved their predictive power. The obtained results are presented in the next section.

3.5 Linear Regressions of Rat and Mouse LD50s vs. Microtox™ 30-min EC50s + Daphnia magna 48-h* EC50s*

The confrontation of the oral LD50s on rat and mouse with the EC50 values for *Vibrio fischeri* and *Daphnia magna* did not yield statistically significant two parameter equations. Conversely, statistically valid models were obtained with the i.p. LD50s, (139) and (140).

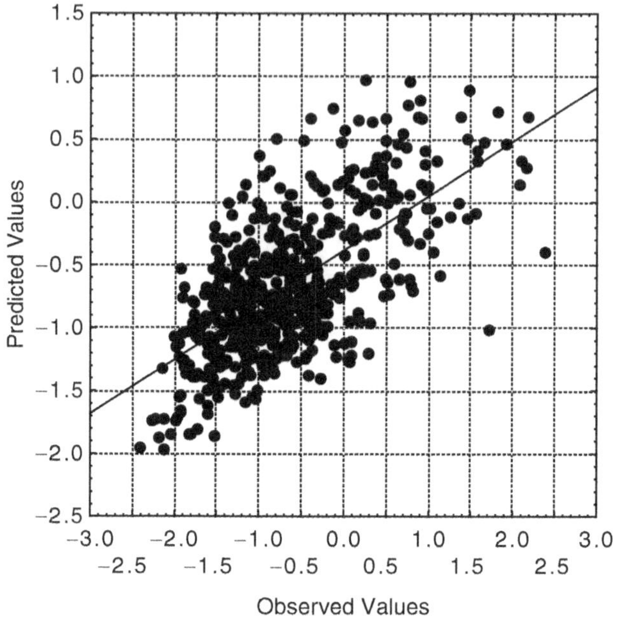

Fig. 2 Observed vs. calculated rat oral LD50 values (− log (mmol/kg)) from model (133)

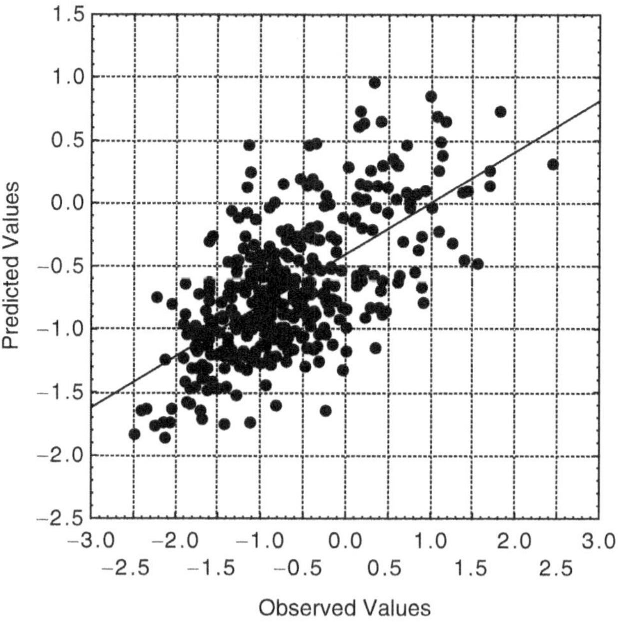

Fig. 3 Observed vs. calculated mouse oral LD50 values (− log (mmol/kg)) from model (134)

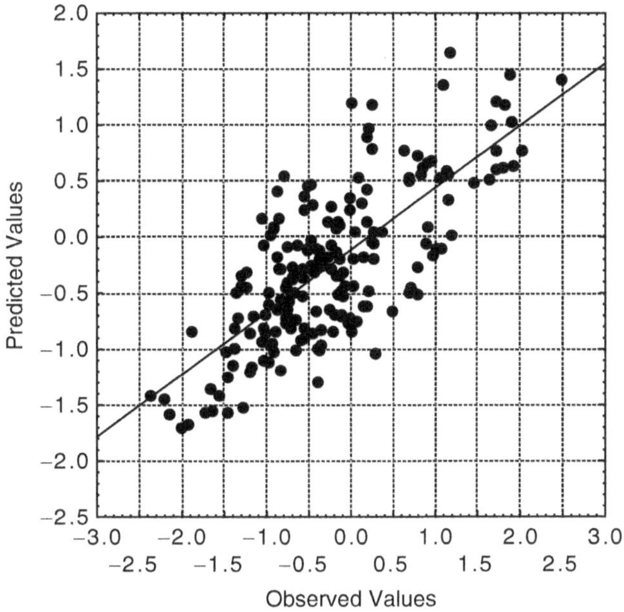

Fig. 4 Observed vs. calculated rat intraperitoneal LD50 values (− log (mmol/kg)) from model (135)

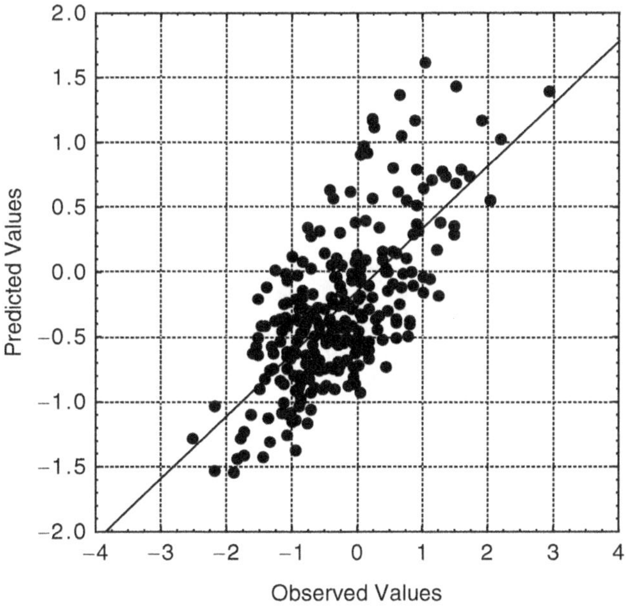

Fig. 5 Observed vs. calculated mouse intraperitoneal LD50 values (− log (mmol/kg)) from model (136)

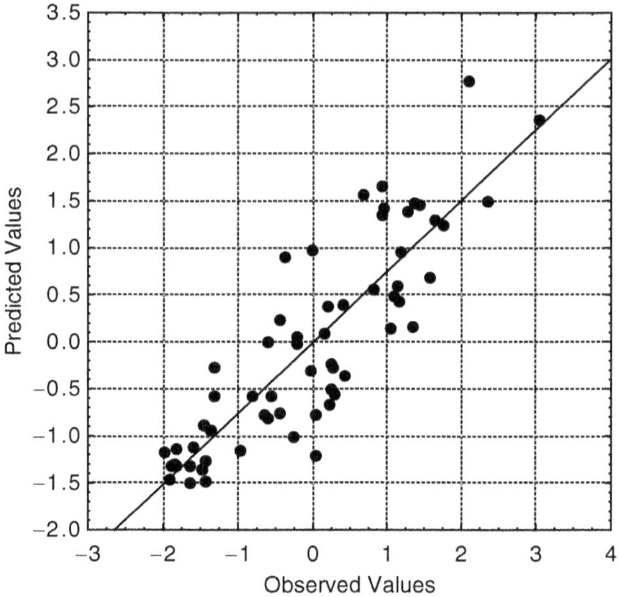

Fig. 6 Observed vs. calculated rat intravenous LD50 values (− log (mmol/kg)) from model (137)

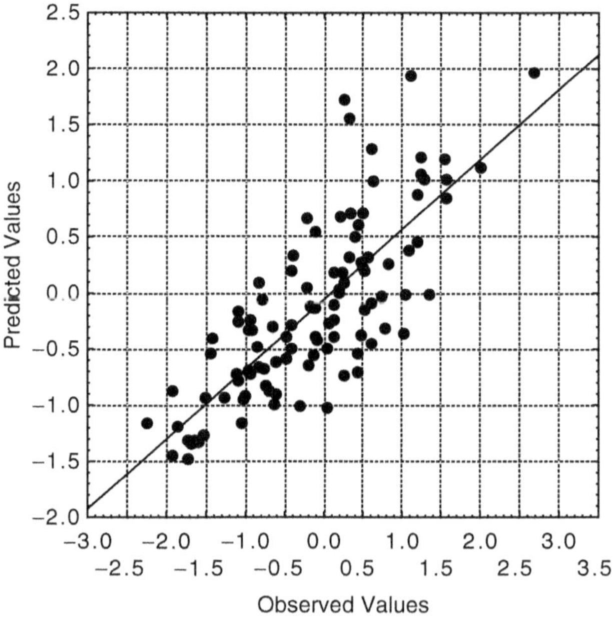

Fig. 7 Observed vs. calculated mouse intravenous LD50 values (− log (mmol/kg)) from model (138)

$$\log(1/\text{LD50 Rat i.p.}) = 0.13 \log(1/\text{EC50 } D.m. - 48\,\text{h}^*) + 0.25 \log(1/\text{EC50 } V.f. - 30\,\text{min}^*) - 0.87, \quad (139)$$
$$n = 99,\ r = 0.75,\ s = 0.55,\ F = 61.0,\ p < 10^{-5}.$$
$$\log(1/\text{LD50 Mouse i.p.}) = 0.23 \log(1/\text{EC50 } D.m. - 48\,\text{h}^*) + 0.14 \log(1/\text{EC50 } V.f. - 30\,\text{min}^*) - 0.75, \quad (140)$$
$$n = 128,\ r = 0.73,\ s = 0.55,\ F = 71.4,\ p < 10^{-5}.$$

Equation (139) presents a better correlation coefficient and standard error than (128), which only includes *Vibrio fischeri* as independent variable but the former model was obtained from 99 chemicals while the latter was derived from 159 compounds. In the same way, while (139) shows slightly better statistics than (135) with only *Daphnia magna* as independent variable, the size of its training set is about twice less important (i.e., 99 vs. 191).

Equation (140) outperforms the corresponding univariate regression equations (i.e. (129) and (136)) but again there are important differences in the size of the training sets (i.e. 128 vs. 239 and 261).

Although the confrontation of the rat intravenous LD50s with the EC50 values for *Vibrio fischeri* and *Daphnia magna* did not yield a statistically significant two parameter equation, an interesting model was obtained with the mouse data (141).

$$\log(1/\text{LD50 Mouse i.v.}) = 0.14 \log(1/\text{EC50 } D.m. - 48\,\text{h}^*) + 0.33 \log(1/\text{EC50 } V.f. - 30\,\text{min}^*) - 0.58, \quad (141)$$
$$n = 51,\ r = 0.89,\ s = 0.47,\ F = 92.3,\ p < 10^{-5}.$$

Again, (141) outperforms (131) and (138) but having a lower training set, its domain of application is also less important.

It is interesting to note that in (139)–(141), *Daphnia magna* and *Vibrio fischeri* contribute positively for predicting rat and mouse LD50s.

Because the 1-octanol/water partition coefficient (log P) seemed yield interesting results when introduced as additional variable in the toxicity = f (ecotoxicity) models [20, 22], it was also decided to consider this important physicochemical parameter as additional independent variable in the models. The results obtained with this descriptor of the hydrophobicity of chemicals are presented in the next section.

3.6 Introduction of log P in the Regressions of Rat and Mouse LD50s vs. Vibrio and Daphnia EC50s

A stepwise regression analysis was first used to correlate rat and mouse LD50 data with Microtox^{TM} 30-min EC50 or daphnid 48-h EC50 data, and log P values calculated from the KowWin v. 1.67 program [25].

Regarding the oral and intravenous LD50 data, only a two parameter equation was obtained for the oral toxicity on rat (142) the others being not statistically significant.

$$\log(1/\text{LD50 Rat oral}) = 0.32 \log(1/\text{EC50 } D.m. - 48\,\text{h}^*) - 0.03 \log P - 1.16, \quad (142)$$
$$n = 478,\ r = 0.71,\ s = 0.54,\ F = 247,\ p < 10^{-5}.$$

Even if the contribution of the 1-octanol/water partition coefficient (log P) in (142) is low, the introduction of this hydrophobic parameter in the model increases its quality. This is clearly shown when the statistical parameters of (133) and (142) are compared as well as the scatterplots of the LD50s obtained from both models (Figs. 2 and 8).

Four equations (143)–(146) were successfully computed for predicting the i.p. toxicity of chemicals to rat and mouse from *Vibrio fischeri* or *Daphnia magna* EC50 data and log P.

$$\log(1/\text{LD50 Rat i.p.}) = 0.36 \log(1/\text{EC50 } V.f. - 30\,\text{min}^*) - 0.1 \log P - 0.67, \quad (143)$$
$$n = 149,\ r = 0.64,\ s = 0.58,\ F = 51.0,\ p < 10^{-5}.$$
$$\log(1/\text{LD50 Rat i.p.}) = 0.40 \log(1/\text{EC50 } D.m. - 48\,\text{h}^*) - 0.05 \log P - 0.75, \quad (144)$$
$$n = 150,\ r = 0.81,\ s = 0.55,\ F = 138,\ p < 10^{-5}.$$

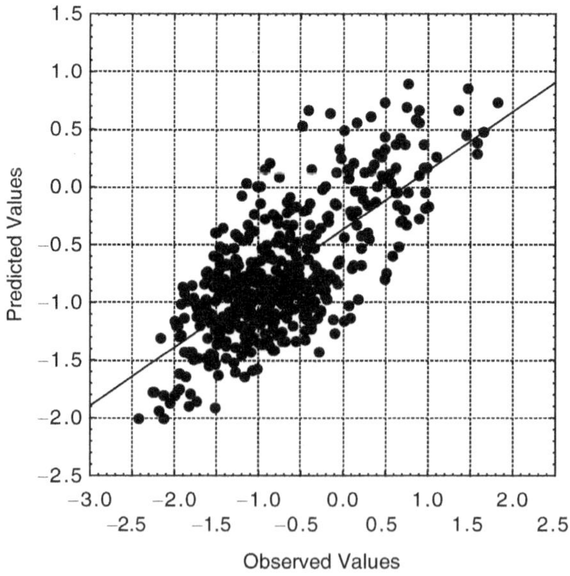

Fig. 8 Observed vs. calculated rat oral LD50 values (− log (mmol/kg)) from model (142)

$$\begin{aligned}\log(1/\text{LD50 Mouse i.p.}) &= 0.34 \log(1/\text{EC50 } V.f. - 30\,\text{min}^*) \\ &\quad - 0.07 \log P - 0.55, \qquad (145) \\ n = 227,\ r = 0.59,\ s &= 0.58,\ F = 59.5,\ p < 10^{-5}.\end{aligned}$$

$$\begin{aligned}\log(1/\text{LD50 Mouse i.p.}) &= 0.37 \log(1/\text{EC50 } D.m. - 48\,\text{h}^*) \\ &\quad - 0.07 \log P - 0.66, \qquad (146) \\ n = 202,\ r = 0.72,\ s &= 0.59,\ F = 108,\ p < 10^{-5}.\end{aligned}$$

The influence of log P in (143)–(146) is very limited. Furthermore, while the introduction of log P in the models with *Daphnia magna* slightly increases their quality, this is the converse regarding *Vibrio fischeri*.

Last, a stepwise regression analysis used to correlate LD50 data with Microtox^{TM} 30-min EC50 data, daphnid 48-h EC50 data and log P values only yielded satisfying results for the rat and mouse intraperitoneous toxicity data, the former model (147) outperforming the latter, (148).

$$\begin{aligned}\log(1/\text{LD50 Rat i.p.}) &= 0.21 \log(1/\text{EC50 } D.m. - 48\,\text{h}^*) \\ &\quad + 0.24 \log(1/\text{EC50 } V.f. - 30\,\text{min}^*) \\ &\quad - 0.15 \log P - 0.72, \qquad (147) \\ n = 91,\ r = 0.80,\ s &= 0.46,\ F = 49.9,\ p < 10^{-5}.\end{aligned}$$

$$\begin{aligned}\log(1/\text{LD50 Mouse i.p.}) &= 0.25 \log(1/\text{EC50 } D.m. - 48\,\text{h}^*) \\ &\quad + 0.16 \log(1/\text{EC50 } V.f. - 30\,\text{min}^*) \\ &\quad - 0.13 \log P - 0.63, \qquad (148) \\ n = 120,\ r = 0.74,\ s &= 0.48,\ F = 47.8,\ p < 10^{-5}.\end{aligned}$$

Equations (147) and (148) outperform (139) and (140) without log P. Inspection of these equations shows the *Daphnia magna* and *Vibrio fischeri* contribute positively to the toxicity, the values of their coefficients in the regression equations being only slightly different. Conversely, in both equations, log P contributes negatively to the toxicity. The observed and calculated i.p. LD50s with these two models are displayed in Figs. 9 and 10.

3.7 Linear vs. Nonlinear Interspecies Toxicity Modeling

Because numerous QSAR studies have shown that the artificial neural networks (ANNs) outperformed the classical linear methods to find complex relationships between the structure of the molecules and their biological activity ([28, 29], see also Chap. 1), an attempt was made to test the usefulness of these nonlinear tools for designing toxicity $= f$ (ecotoxicity) models. The ANN the most suited for this kind of study was a three layer perceptron (TLP) [30]. A TLP requiring at least three neurons in the input layer and two neurons in the hidden layer to be used reasonably, only (147) and (148) were concerned by the comparison exercise.

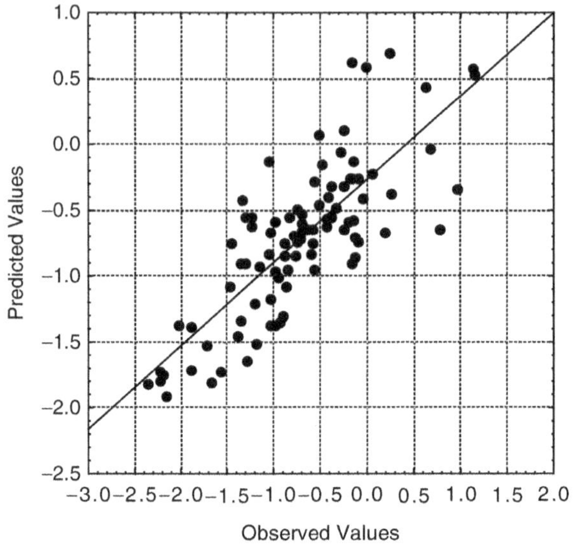

Fig. 9 Observed vs. calculated rat intraperitoneal LD50 values (− log (mmol/kg)) from model (147)

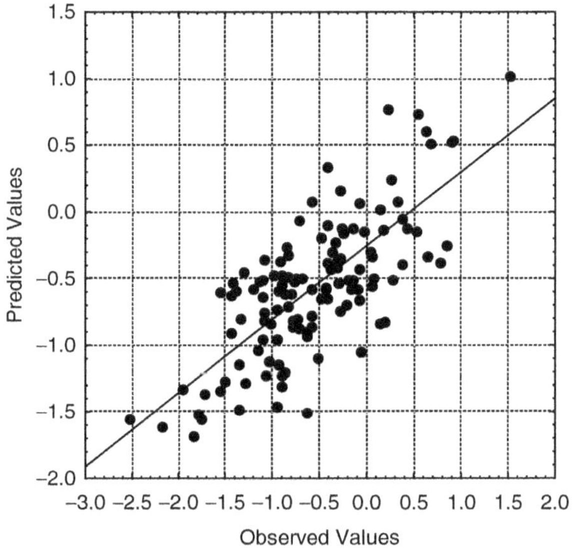

Fig. 10 Observed vs. calculated mouse intraperitoneal LD50 values (− log (mmol/kg)) from model (148)

Thus, a TLP was used for predicting i.p. rat and mouse LD50 data from Microtox^{TM} 30-min EC50, daphnid 48-h EC50, and log P data. Ten percent of the data used for deriving (147) and (148) were randomly selected for constituting the external testing sets, the remaining data being used as learning sets to train a 3/2/1

TLP. A classical min/max transformation was used to scale the data. All the calculations were performed with the Statistica ANN software (Statsoft, Paris). Different learning algorithms (back-propagation, conjugate gradient descent, Levenberg-Marquardt, quick-prop), activation and transfer functions were tested. Despite more than 7,500 runs for each endpoint by randomly varying the composition of the training and testing sets as well as the architecture of the ANN, it was impossible to obtain significantly better results for the testing sets than those obtained with a regression analysis performed from the same chemical data sets. The results were the same or at least slightly better than those recorded with the stepwise regression analysis. Obviously, the results obtained with the learning sets were always better than those recorded with regression analysis but it is not surprising, the TLPs being powerful learning devices.

It is noteworthy that no attempts were made to increase the number of neurons in the hidden layer.

Because the support vector machines (SVMs) [31] have shown their interest for classification problems and more recently to correlate data of various origins, an attempt was also made to test their usefulness on the different learning and testing sets previously used with the ANNs. The e1071 program, written in R and freely available from the CRAN library, was employed for deriving the different toxicity $= f$ (ecotoxicity) models. Unfortunately, due to different constraints, only a limited number of runs was performed. They provided better results on the external testing sets than the TLP but additional investigations are absolutely necessary to correctly estimate the performances and interest of SVM for predicting LD50s from ecotoxicity data.

Last, different nonlinear regression analyses were tested by using Statistica v. 6 as well as CurveExpert v. 1.3. The most interesting regression analysis presented the following form: $y = a(1 - e^{-bx})$. However, it was impossible to obtain significantly better results for the testing sets than those obtained with a regression analysis performed under the same conditions. Consequently, it was not justified to select these nonlinear equations as final models.

4 Conclusion

Surprisingly, there is a limited number of models available in the literature for predicting the acute toxicity of chemicals to rats and mice from ecotoxicity test data. Furthermore, when available, most of the models were derived from limited datasets. Thus about 70% of the models found in the literature were obtained from less than 50 chemicals. Moreover, it is important to note that chemicals are very often eliminated before or during the regression processes without clear justifications. Consequently, despite some significant correlations, these models cannot be used in practice. This prompted us to develop new models focused on the prediction of rat and mouse LD50s from invertebrate EC50s or LC50s. The selected species were *Daphnia magna* and *Vibrio fischeri* because these organisms are widely used for assessing the hazard of chemicals and hence, collections of EC50s for these organisms are available in the literature.

Consequently, in a first step, a strong bibliographical investigation was performed to collect oral, i.p., and i.v. rat and mouse LD50 data. In the meantime, EC50 data for *Vibrio fischeri* (Microtox^{TM} test) and *Daphnia magna* were also retrieved from literature. Python scripts were used to structure and format the toxicity and ecotoxicity data to facilitate their statistical manipulation.

A collection of oral, i.p., and i.v. rat and mouse toxicity models was derived using *Vibrio fischeri* and *Daphnia magna* as independent variables alone or together through a stepwise regression analysis. Most of the models on *Daphnia magna* are totally new. They outperform those obtained with *Vibrio fischeri*. The usefulness of the 1-octanol/water partition coefficient (log P) as additional independent variable was also tested. When included in the models, its contribution is always negative and generally marginal, except in the case of the three parametric equations including *Daphnia magna* and *Vibrio fischeri* for the prediction of rat and mouse intraperitoneal LD50s.

The interest of nonlinear statistical tools for deriving toxicity $= f$(ecotoxicity) models was also experienced. The results obtained with a three-layer perceptron and different nonlinear regressions were disappointing. The SVMs seemed to yield more interesting results but more investigations should be necessary to see whether they are more suited than classical regression analysis for deriving toxicity $= f$ (ecotoxicity) models.

Acknowledgment The financial support from the French Environment and Energy Management Agency (ADEME) is gratefully acknowledged (Contract 04 94 C 0023).

References

1. Gauthier C, Griffin G (2005) Using animals in research, testing and teaching. Rev Sci Tech Off Int Epiz 24: 735–745
2. Council of Europe (CoE) (1986) European Convention for the protection of vertebrate animals used for experimental and other scientific purposes. Strasbourg, 18.III.1986. CoE, Strasbourg
3. European Community (1986) Council Directive 86/609/EEC of 24 November 1986 on the approximation of laws, regulations and administrative provisions of the Member States regarding the protection of animals used for experimental and other scientific purposes. Off J Eur Communities L 358: 1–28
4. van Zutphen B (2007) Legislation of animal use – Developments in Europe. AATEX 14: 805–809
5. Bayvel ACD (2005) The use of animals in agriculture and science: Historical context, international considerations and future direction. Rev Sci Tech Off Int Epiz 24: 791–797
6. Russell WMS, Burch RL (1959) The principles of humane experimental technique. Methuen & Co. Ltd, London (altweb.jhsph.edu/publications/humane_exp/het-toc.htm (accessed on 11 January 2008)
7. European Community (2007) Fifth report on the statistics on the number of animals used for experimental and other scientific purposes in the member states of the European Union. Brussels, 5.11.2007, COMM(2007) 675 final
8. Regulation (EC) No 1907/2006 of the European Parliament and of the Council of 18 December 2006 concerning the Registration, Evaluation, Authorisation and Restriction of Chemicals

(REACH), establishing a European Chemicals Agency, amending Directive 1999/45/EC and repealing Council Regulation (EEC) No 793/93 and Commission Regulation (EC) No 1488/94 as well as Council Directive 76/769/EEC and Commission Directives 91/155/EEC, 93/67/EEC, 93/105/EC and 2000/21/EC
9. Kaiser KLE (1998) Correlations of *Vibrio fischeri* bacteria test data with bioassay data for other organisms. Environ Health Perspect 106(Suppl 2): 583–591
10. Fort FL (1992) Correlation of Microtox EC_{50} with mouse LD_{50}. In Vitro Toxicol 5: 73–82
11. Kaiser KLE, McKinnon MB, Fort FL (1994) Interspecies toxicity correlations of rat, mouse and *Photobacterium phosphoreum*. Environ Toxicol Chem 13: 1599–1606
12. Sauvant MP, Pepin D, Bohatier J, Groliere CA, Guillot J (1997) Toxicity assessment of 16 inorganic environmental pollutants by six bioassays. Ecotoxicol Environ Safety 37: 131–140
13. Calleja MC, Persoone G (1992) Cyst-based toxicity tests. IV. The Potential of ecotoxicological tests for the prediction of acute toxicity in man as evaluated on the first ten chemicals of the MEIC programme. ATLA 20: 396–405
14. Khangarot BS, Ray PK (1988) The confirmation of a mammalian poison classification using a water flea (*Daphnia magna*) screening method. Arch Hydrobiol 113: 447–455
15. Enslein K, Tuzzeo TM, Borgstedt HH, Blake BW, Hart JB (1987) Prediction of rat oral LD50 from *Daphnia magna* LC50 and chemical structure. In Kaiser KLE (ed) QSAR in Environmental Toxicology – II, Reidel Publishing Company, Dordrecht
16. Enslein K, Tuzzeo TM, Blake BW, Hart JB, Landis WG (1989) Prediction of *Daphnia magna* EC50 values from rat oral LD50 and structural parameters. In Suter GW, Lewis MA (eds) Aquatic Toxicology and Environmental Fate, Vol. 11, ASTM STP 1007, Philadelphia
17. Janardan SK, Olson CK, Schaeffer DJ (1984) Quantitative comparisons of acute toxicity of organic chemicals to rat and fish. Ecotoxicol Environ Safety 8: 531–539
18. Johnson WW, Finley MT (1980) Handbook of acute toxicity of chemicals to fish and aquatic invertebrates. Resource Publication 135. US Department of the Interior, Fish and Wildlife Service, Washington, DC
19. Gaines TB (1969) Acute toxicity of pesticides. Toxicol Appl Pharmacol 14: 515–534
20. Kaiser KLE, Esterby SR (1991) Regression and cluster analysis of the acute toxicity of 267 chemicals to six species of biota and the octanol/water partition coefficient. Sci Total Environ 109/110: 499–514
21. Hodson PV (1985) A comparison of the acute toxicity of chemicals to fish, rats and mice. J Appl Toxicol 5: 220–226
22. Delistraty D, Taylor B, Anderson R (1998) Comparisons of acute toxicity of selected chemicals to rainbow trout and rats. Ecotoxicol Environ Safety 39: 195–200
23. Delistraty D (2000) Acute toxicity to rats and trout with a focus on inhalation and aquatic exposures. Ecotoxicol Environ Safety 46: 225–233
24. Kaiser KLE, Devillers J (1994) Ecotoxicity of chemicals to *Photobacterium phosphoreum*. Gordon and Breach Science Publishers, Reading
25. Meylan WM, Howard PH (1995) Atom/fragment contribution method for estimating octanol–water partition coefficients. J Parm Sci 84: 83–92
26. Biggar JW, Riggs RL (1974) Apparent solubility of organochlorine insecticides in water at various temperatures. Hilgardia 42: 383–391
27. Yates IE (1985) Differential sensitivity to mutagens by *Photobacterium phosphoreum*. J Microbiol Meth 3: 171–180
28. Devillers J, Domine D (1999) A noncongeneric model for predicting toxicity of organic molecules to *Vibrio fischeri*. SAR QSAR Environ Re. 10: 61–70
29. Devillers J (1996) Neural networks in QSAR and drug design. Academic Press, London
30. Eberhart RC, Dobbins RW (1990) Neural network PC tools. A practical guide, Academic Press, San Diego
31. Vapnik V (1995) The nature of statistical learning theory. Springer-Verlag, Berlin

Use of Multicriteria Analysis for Selecting Ecotoxicity Tests

James Devillers, Pascal Pandard, Anne-Marie Charissou, and Antonio Bispo

Abstract It is now well admitted that a battery of ecotoxicity tests should be designed by accounting for the requirements of a specific scenario such as classification of wastes or remediation efficiency of contaminated soils. The development of a single battery of tests for all applications is thereafter recognized not to be relevant. The selection of tests for constituting a battery may be established according to expert judgments, decision criteria such as cost, ecological relevance, sensitivity of selected organisms, standardization of the methods, implementation of the test protocols or after statistical analysis of test results obtained on a large series of bioassays. In this chapter, a methodological framework, based on the combination of an original multicriteria method called SIRIS and multivariate analyses, is presented for selecting ecotoxicity tests for assessing the level of contamination of soils. The interest of this approach that simultaneously accounts for ecological, technical, and economical constraints is discussed.

Keywords Multicriteria analysis · SIRIS method · Test battery · a priori selection · Soil contamination

1 Introduction

The preservation of the structure and functioning of the ecosystems as well as the protection of human health rely on the hazard assessment of man-made chemicals that is mainly based on the use of (eco)toxicity tests. The purpose of (eco)toxicity testing is to generate quantitative or qualitative information on the unwanted effects of xenobiotics for their regulation as requested by numerous regulatory authorities worldwide, for estimating acceptable concentrations of pollutants in water and food, for setting permissible exposure limits for workers, and for protecting the biota [1–3].

In this context, numerous acute and chronic ecotoxicity tests have been developed over the last 30 years for the environmental hazard assessment of chemicals. Because a single species or endpoint cannot adequately reflect contaminant effects

J. Devillers (✉)
CTIS, 3 Chemin de la Gravière, 69140 Rillieux La Pape, France
e-mail: j.devillers@ctis.fr

J. Devillers (ed.), *Ecotoxicology Modeling*, Emerging Topics in Ecotoxicology: Principles, Approaches and Perspectives 2, DOI 10.1007/978-1-4419-0197-2_5,

to all biota in the ecosystem under study, it is common to use several test species generally representing different trophic levels to try to reflect the environmental situation as realistically as possible (see e.g., [4–10]). However, although the literature documents the use of ecotoxicity test batteries, there are only a limited number of studies dealing with the selection of the test battery components. Broadly speaking, two main strategies can be used to reach this goal. The selection can be made a posteriori from the use of multivariate analyses such as principal components analysis [11–14], hierarchical cluster analysis [14, 15], correspondence factor analysis [15], multidimensional scaling [16], and nonlinear mapping [14]. However, recently we stressed [14] that very often the potentialities of multivariate analyses were not enough exploited and at best a kind of typology of the different tests was provided without the proposal of an optimal battery including a minimum number of tests. Conversely, with the a priori methods, the selection is made independently of the results, being based on various criteria such as standardization of the method, ecological relevance of test organisms, and cost [17–19]. Although the pros and cons of the different scientific and technical criteria have been deeply analyzed [18, 19], to our knowledge, no attempt has been made to propose a statistical method allowing to organize and aggregate this heterogeneous information to facilitate the selection of test batteries. This is the main objective of this chapter. Because it is a complex problem accounting for variables of different nature and weights that can yield different solutions depending on initial decision criteria, multicriteria analysis (MCA) appeared particularly suited for the a priori selection of tests to constitute an optimal battery. Indeed, MCA is particularly applicable to cases where a simple cost-benefit analysis fails. It is an aid to decision-making that helps stakeholders to organize the available information, explore with transparency the effects of their own decisions, and minimize the possibility for postdecision disappointments [20]. Consequently, MCA has found applications in numerous domains such as air quality control [21, 22], climate change evaluation [23, 24], river and water resource management [25, 26], flood risk perception and management [27, 28], forest management [29–32], contaminated site assessment [33–36], waste and sewage sludge management [37, 38], disease control program evaluation [39, 40], hazard and risk assessment of chemicals [41–43] and nanomaterials [44], phylogeny analysis [45], and financial and socio-economic analyses [46–51].

In this chapter, MCA and multivariate analyses were used for selecting ecotoxicity tests for assessing the level of contamination of soils on the basis of ecological, technical, and economical constraints.

2 Materials and Methods

2.1 Test Selection and Description

An extensive bibliographical investigation was made from the following two syntagms:

- Syntagm A: Comparison or choice or selection or classification or discrimina! or (multivariate and analysis) ranking or model,
  or
  PCA or (Principal component analysis) or PLS or regression or (neural network) or classification or fuzzy or tree or SIMCA or co-inertia or (factorial and analysis),
  and
  battery or bioassay or (toxicity and testing) or biomarker or biotest or ecotoxicity or acute or chronic or genotox! or teratogen! or (sensitivity and species and distribution).
- Syntagm B: Battery or bioassay or (toxicity and testing) or biomarker or biotest or ecotoxicity or acute or chronic or genotox! or teratogen! or (sensitivity and species and distribution),
  and
  regulation or (quality and criteria) or (threshold and value) or (limit and value) or (decision and scheme) or (cut-off and value) or (guide and value),
  and
  soil or sediment or sludge or waste or (dredged and material) or effluent or wastewaters or lixiviate or leachate or eluate or percolate or (water and extract) or compost or amendment or ((polluted or contaminated) and site).

Syntagm A allowed us to retrieve publications dealing with methods for selecting batteries of bioassays without focusing on specific scenarios. Conversely, this was the main goal of syntagm B to secure the retrieval of publications dealing with the use of batteries of bioassays to assess the toxicity of complex environmental matrices for regulation purposes.

The combined use of the two syntagms yielded the retrieval of a first list of 177 references, which was reduced to 134 after the elimination of 43 false positives. After the reading of these 134 documents, only 53 were found to fully satisfy our criteria of quality and the goal of the study [4, 11–13, 16, 52–99]. They allowed us to extract a first list of assays, which was completed by the new standardized tests dealing with the quality of soils and still not cited in the literature. A total of 115 tests was obtained. They were representative of terrestrial, aquatic, and sediment media and included different endpoints measured on organisms occupying different trophic levels in the environment as well as populations. The 115 tests are listed in Table 1 with information on their standardization status (Yes/No), the environmental matrix used to perform the tests, which can be solid (Soil) or/and liquid (Liq.), and the type of toxicity, which was encoded according to the following categories: acute (Ac), chronic (Ch), genotoxic/teratogen (GT), behavior (Bh), biosensor (Bs), biomarker (Bm), functioning of the ecosystems (Fe), and structure of the ecosystems (Se).

2.2 Multicriteria Analysis

Because the system of integration of risk with interaction of scores (SIRIS) method has clearly shown its interest and potentialities in environmental and occupational

Table 1 List of selected tests

Test number	Test description	Standardized	Matrix	Toxicity type
1	Inhibition of nitrification of soil microorganisms	Y[a]	Soil[a]	Fe[a]
2	Inhibition of respiration of soil microorganisms	Y	Soil	Fe
3	Bait-lamina test (28 days)	N	Soil	Fe
4	Mycorrhizal fungi (*Glomus mosseae*) – spore germination test (14 days)	Y	Soil	Ac
5	Inhibition of growth of higher plants (monocotyledonous species commercially available, 14 days)	Y	Soil	Ac
6	Inhibition of growth of higher plants (dicotyledonous species commercially available, 14 days)	Y	Soil	Ac
7	Inhibition of growth of higher plants (wild monocotyledonous species, 14 days)	Y	Soil	Ac
8	Inhibition of growth of higher plants (wild dicotyledonous species, 14 days)	Y	Soil	Ac
9	Inhibition of root elongation of higher plants (4–5 days)	N	Liq.	Ac
10	Inhibition of root elongation of higher plants (4 days)	Y	Soil	Ac
11	Inhibition of seed emergence of higher plants	Y	Soil	Ac
12	Inhibition of growth of *Lactuca sativa* (28 days)	N	Soil	Ch
13	*Allium cepa* micronucleus assay (24 h)	N	Liq.	GT
14	*Tradescantia* micronucleus assay	Y	Soil	GT
15	*Vicia faba* micronucleus assay (48 h)	Y	Liq.	GT
16	*Nicotiana tabacum* mutation test (45 days)	N	Soil	GT
17	Abundance of soil nematodes	N	Soil	Se
18	Mortality of enchytraeids (*Enchytraeus crypticus* or *E. albidus*)	N	Soil	Ac
19	Inhibition of reproduction of enchytraeids (*E. crypticus* or *E. albidus*)	Y	Soil	Ch
20	Mortality of earthworms (*Eisenia fetida* or *E. andrei*, 14 days)	Y	Soil	Ac
21	Inhibition of reproduction of earthworms (*E. fetida*, 56 days)	Y	Soil	Ch
22	Earthworm (*E. fetida*) avoidance test	Y	Soil	Bh

(continued)

Table 1 (continued)

Test number	Test description	Standardized	Matrix	Toxicity type
23	Mortality of earthworms (*Lumbricus terrestris*, 14 days)	Y	Soil	Ac
24	Mortality of springtails (*Folsomia candida*, 4 days)	N	Soil	Ac
25	Inhibition of reproduction of springtails (*F. candida*, 28 days)	Y	Soil	Ch
26	Mortality of *Cetoniinae* (*Oxythyrea funesta*, 10 days)	Y	Soil	Ac
27	Inhibition of growth of *Cetoniinae* (*O. funesta*)	N	Soil	Ch
28	Inhibition of growth of garden snails (*Helix aspersa*, 28 days)	Y	Soil	Ch
29	Inhibition of growth of *Bacillus sp.* (24 h)	N	Liq.	Ac
30	Umu-test (*Salmonella typhimurium*)	Y	Liq.	GT
31	Ames test (*S. typhimurium*)	Y	Liq.	GT
32	SOS-DNA repair – Vitotox® (*S. typhimurium*)	N	Liq.	GT
33	L-arabinose resistance test	N	Liq.	GT
34	Inhibition of root colonization of *Medicago trunculata* (28 days)	Y	Soil	Ch
35	Chronic toxicity in higher plants (*Avena sativa*; 49–56 days)	Y	Soil	Ch
36	Chronic toxicity in higher plants (*Brassica napus*; 35–42 days)	Y	Soil	Ch
37	*Escherichia coli* toxichromopad	N	Liq.	Ac
38	*E. coli* toxichromotest	N	Liq.	Ac
39	*E. coli* Metplate	N	Liq.	Ac
40	*E. coli* SOS chromotest	N	Liq.	GT
41	Inhibition of respiration of *Pseudomonas putida*	N	Liq.	Ac
42	Inhibition of growth of *P. putida*	Y	Liq.	Ch
43	Inhibition of growth of *P. fluorescens* (7 days)	N	Liq.	Ch
44	Biomet assay *(Alcaligenes eutrophus)*	N	Liq.	Bs
45	Inhibition of growth of *Microcystis aeruginosa* (4 days)	Y	Liq.	Ch
46	Inhibition of respiration of activated sludge	Y	Liq.	Ac
47	Inhibition of nitrification of activated sludge	Y	Liq.	Ac
48	Activated sludge – Polytox	N	Liq.	Ch
49	Activated sludge – ATP luminescence test	N	Liq.	Ac
50	Activated sludge – inhibition of L-alanine-aminopeptidase	N	Liq.	Ac

(continued)

Table 1 (continued)

Test number	Test description	Standardized	Matrix	Toxicity type
51	Luminotox (fluorescence of stabilized photosynthetic enzyme complexes, 15 min)	N	Liq.	Ac
52	Inhibition of growth of *Pseudokirchneriella subcapitata* (previously *Selenastrum capricornutum*; 3 or 4 days)	Y	Liq.	Ch
53	*Pseudokirchneriella subcapitata* – inhibition of esterase (15 min)	N	Liq.	Ac
54	*P. subcapitata* Hsp 70		Liq.	Bm
55	Inhibition of growth of *Scenedesmus subspicatus* (3 or 4 days)	Y	Liq.	Ch
56	Inhibition of growth of *S. quadricauda* (3 days)	Y	Liq.	Ch
57	Inhibition of growth of *S. pannonicus* (3 days)	Y	Liq.	Ch
58	*Chlamydomonas reinhardii* (photosynthesis: EPR, 24 h)	N	Liq.	Ch
59	Inhibition of growth of *C. reinhardii* (10 days, continuous system)	N	Liq.	Ch
60	*Nitellopsis obtusa* – cell membrane depolarization (45 min)	N	Liq.	Ac
61	Inhibition of growth of *Lemna minor* (4 days)	Y	Liq.	Ch
62	Inhibition of growth of *L. minor* (7 days)	Y	Liq.	Ch
63	Mortality of *Spirostomum ambiguum* (Spirotox; 24 h)	N	Liq.	Ac
64	*Tetrahymena thermophila* – behavioral toxicity test (24 h)	N	Liq.	Bh
65	Inhibition of population growth of *T. pyriformis* (40–48 h)	N	Liq.	Ch
66	Inhibition of growth of *Colpidium campylum*	N	Liq.	Ch
67	Inhibition of reproduction of *Brachionus calyciflorus* (48 h)	Y	Liq.	Ch
68	Mortality of *B. calyciflorus* (24 h)	N	Liq.	Ac
69	Inhibition of growth of *Panagrellus redivivus* (4 days)	N	Liq.	Ch
70	Inhibition of mobility of *Plectus acuminatus* (4 days)	N	Liq.	Ac
71	Inhibition of growth of *Hydra oligactis* (21 days)	N	Liq.	Ch
72	Inhibition of mobility of *Daphnia magna* (24 or 48 h)	Y	Liq.	Ac
73	Inhibition of reproduction of *D. magna* (21 days)	Y	Liq.	Ch

(continued)

Table 1 (continued)

Test number	Test description	Standardized	Matrix	Toxicity type
74	*D. magna* – beta-galactosidase activity (1–2 h)	N	Liq.	Ac
75	Inhibition of reproduction of *Ceriodaphnia dubia* (7 days)	Y	Liq.	Ch
76	Mortality of *C. dubia* (24 or 48 h)	Y	Liq.	Ac
77	Mortality of *Thamnocephalus platyurus* (24 h)	N	Liq.	Ac
78	*Dreissena polymorpha* – Comet assay	N	Liq.	GT
79	*D. polymorpha* – phagocytic activity	N	Liq.	Bm
80	Inhibition of reproduction of *Lymnaea stagnalis* (40 days)	N	Liq.	Ch
81	*Xenopus laevis* – Fetax assay	Y	Liq.	GT
82	*X. laevis* – micronucleus assay (12 days)	Y	Liq.	GT
83	*X. laevis* – DNA adducts (12 days)	N	Liq.	GT
84	Mortality of *X. laevis* (12 days)	Y	Liq.	Ac
85	*X. laevis* – growth and development	N	Liq.	Ch
86	Mortality of *Cyprinus carpio* (4 days)	Y	Liq.	Ac
87	Mortality of *Danio rerio* (4 days)	Y	Liq.	Ac
88	*D. rerio* – Early Life Stage toxicity test (11 days)	Y	Liq.	Ch
89	Mortality of *Poecilia reticulata* (48 h)	Y	Liq.	Ac
90	*P. reticulata* – prolonged toxicity test (28 days)	Y	Liq.	Ch
91	Mortality of *Oncorhynchus mykiss* (4 days)	Y	Liq.	Ac
92	Inhibition of growth of *O. mykiss* (7 days)	Y	Sed.	Ch
93	*Oryzias latipes* – growth and behavior (40 days)	N	Liq.	Ch
94	*Pimephales promelas* – mortality and growth of larvae (7 days)	Y	Liq.	Ch
95	*Chironomus riparius* – mortality and development (28 days)	Y	Sed.	Ch
96	*C. riparius* – mortality and development (10 days)	Y	Sed.	Ac
97	Mortality of *Hyalella azteca* (10 days)	Y	Sed.	Ac
98	Inhibition of light emission of *V. fischeri* (previously *Photobacterium phosphoreum*; 5, 15, or 30 min)	Y	Liq.	Ac
99	*V. fischeri* – chronic test	N	Liq.	Ch
100	*V. fischeri* – solid-phase test	N	Soil/Sed.	Ac

(continued)

Table 1 (continued)

Test number	Test description	Standardized	Matrix	Toxicity type
101	*V. fischeri* – flash test (30 s)	N	Sed.	Ac
102	*V. fischeri* – Mutatox test (24 h)	N	Liq.	GT
103	Inhibition of growth of *Euplotes vanuus* (48 h)	N	Liq.	Ch
104	Inhibition of growth of *Skeletonema costatum* (4 days)	Y	Liq.	Ch
105	Inhibition of growth of *Dunaliella tertiolecta* (4 days)	N	Liq.	Ch
106	Mortality of *Echinocardium cordatum* (14 days)	N	Liq.	Ac
107	*Paracentrotus lividus* – embryo-larval test (48 h)	N	Liq.	GT
108	*Ciona intestinalis* – embryo larval test (24 h)	N	Liq.	GT
109	Mortality of *Artemia salina* (24 h)	N	Liq.	Ac
110	Mortality of *Acartia tonsa* (48 h)	Y	Liq.	Ac
111	Mortality of *Corophium volutator* (10 days)	N	Liq.	Ac
112	Mortality of *Eohaustorius estuarius* (10 days)	Y	Sed.	Ac
113	*Palaemon serratus* – larval mortality (3 days)	N	Liq.	Ac
114	*Mytilus galloprovincialis* – embryo-larval test (48 h)	Y	Liq.	GT
115	Macroinvertebrate communities – biotic indexes	N	Sed.	Se

[a] See text for significance

toxicology [41–43], it was used in this study. Briefly, the method needs first to select the number of criteria (variables) necessary to correctly describe the studied phenomenon. Obviously, this number depends on the complexity of the problem. The more complex the problem, the larger the number of variables it is necessary to consider, at least in a first step. Indeed, in all the modeling processes, it is preferable to optimize the number of variables (parsimony principle). The selected variables, which can be qualitative or quantitative, are then transformed into modalities coded as favorable (f) or unfavorable (d) or as favorable (f), moderately favorable (m), or unfavorable (d). If need be, it is possible to define more modalities. The threshold limits are generally determined from expert judgments. The variables are ranked according to their relative importance. Indeed, it is obvious that in a multicriteria decision system, all the selected variables do not have the same importance with respect to the final decision. In other words, they do not have the same weight. When facing this kind of situation, it is of common use to introduce coefficients in the calculation procedure to modify the final weight of some variables. This strategy suffers from a lack of transparency and is too rigid. In SIRIS, the variables are ranked by decreasing order of importance, and this order will have an impact on the results. It is noteworthy that two variables can have the same importance. Last

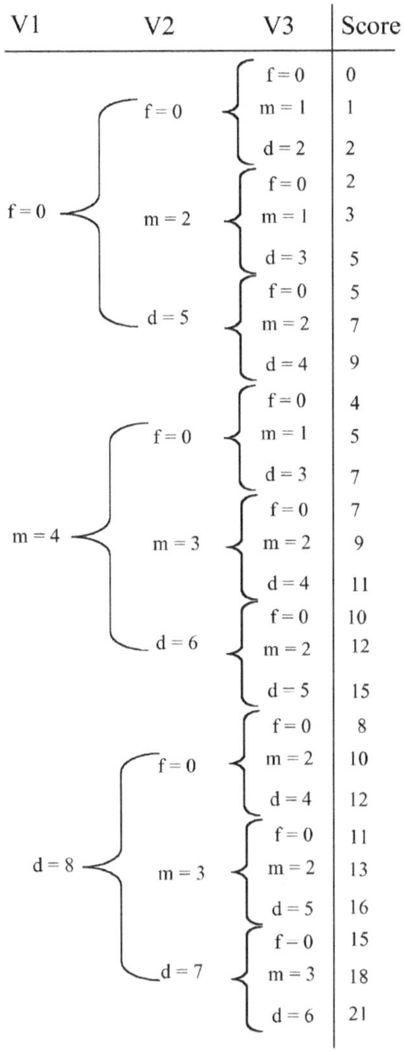

Fig. 1 Scale of SIRIS scores calculated for a system at three variables (V1, V2, V3) presenting three modalities and with V1 > V2 > V3

a min/max scale of scores is calculated according to specific incremental rules [41]. An illustrative example of calculation is given in Fig. 1.

The 115 selected tests (Table 1) were described by 17 technical and scientific criteria (variables), each being defined by two or three characteristics (modalities). They are listed below with their code, the modalities favorable, moderately favorable, and unfavorable being encoded as f, m, and d, respectively.

Technical criteria:

- Supply of test organisms (STO): negligible time (f), < ½ day by week (m), > ½ day by week (d).
- Test volume or quantity (TVQ): <10 mL or 10 g (f), 10–100 mL or g (m), ≥100 mL or g (d).

- Test duration (TDU): $\leq$ 4 days (f), 5–20 days (m), $\geq$ 21 days (d).
- Test room (TRO): usual (f), specialized (m), very specialized (e.g. T °C, moisture) (d).
- Material (MAT): usual (f), specialized (e.g. microscope) (m), very specialized (e.g. radioactive meter) (d).
- Cost (COS): < 1-day technician (f), 1–4 day technician (m), > 4 day technician (d).
- Competences (COM): technician (f), higher (d).
- Scope (SCO): several environmental matrices (f), one matrix: solid or liquid substrate (d).
- Test starting (TST): immediate (f), 1–3 days (m), > 3 days (d).
- Perception by a nonspecialized public (PNP): easy (e.g., mortality, growth) (f), moderate (e.g., luminescence) (m), difficult (e.g. DNA adducts) (d).

Scientific criteria:

- Ecological relevance (ERL): high (f), intermediate (m), low (d).
- Contact duration (COD): > 70% of the life cycle of the organism (f), 10–70% of the life cycle (m), < 10% of the life cycle (acute test) (d).
- Level of standardization (LOS): standardized (f), under preparation (m), not standardized (d).
- Comparison of laboratory and field testing (CLF): toxicological effects that can be measured in the field (f), toxicological effects correlated with toxicological effects that can be measured in the field (m), toxicological effects non relevant in field testing (d).
- Mechanism of action (MOA): function (e.g., enzymatic reaction) (f), organ (e.g., roots) (m), whole organism (e.g., mortality) (d).
- Genetic stability (GST): asexual reproduction (e.g., parthenogenesis) (f), sexual reproduction (d).
- Test limitations and constraints (TLC): low (f), intermediate (m), high (d).

All the calculations were made with the SIRIS-2D software [100].

3 Results and Discussion

The SIRIS method allows us to simultaneously consider variables encoding very different information. However, our practical experience shows that it is generally preferable to construct several scales of SIRIS scores (one per category of variables) instead of a unique scale. This was also verified in the present study. Thus, the technical and scientific criteria were considered separately yielding the construction of two scales of SIRIS scores. The hierarchy of the variables in the SIRIS method being problem-dependent, the variables were ranked by accounting for the main goal of the study, which was the selection of tests for assessing the level of contamination of soils. Thus, the hierarchies were C1: STO > TDU > TVQ > TST > PNP > SCO > COS > COM > TRO > MAT and S1: COD > ERL > LOS > CLF > GST > TLC > MOA for the technical and scientific criteria, respectively. The typology of the 115 tests characterized by their two SIRIS scores

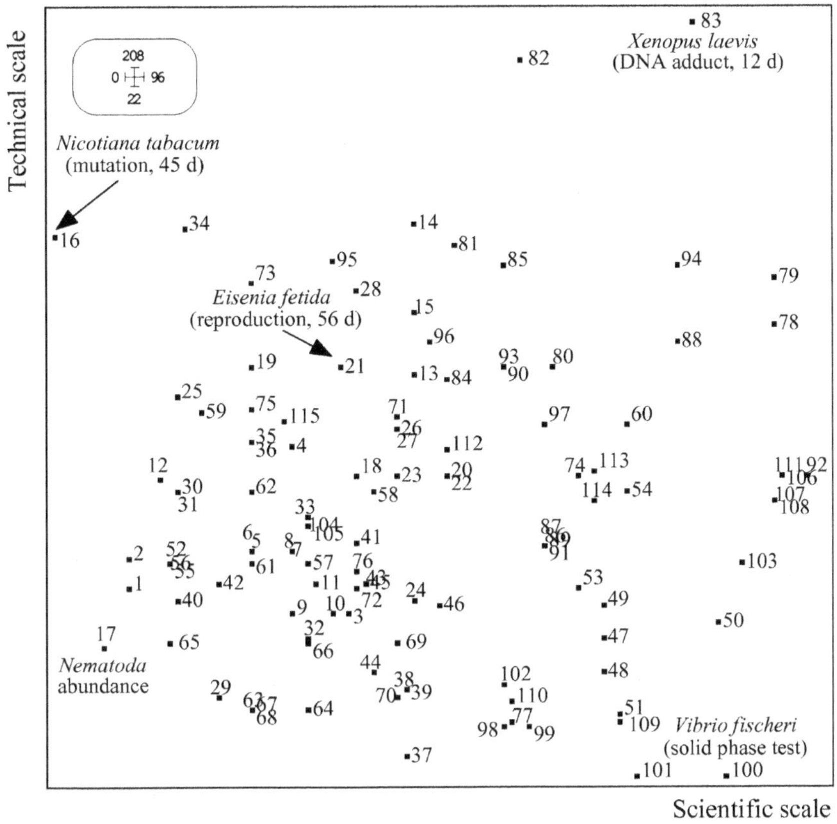

Fig. 2 SIRIS map of the 115 ecotoxicity tests. See Table 1 for the correspondence between the numbers and the test names

is displayed in Fig. 2. The tests located in the left bottom part of Fig. 2 are the most interesting because they present low score values on the scientific and technical scales. They can be considered as good candidates for the constitution of a relevant test battery. At the opposite, the tests located in the top right part of Fig. 2 (e.g., #82 and #83) are not suited for the selected scenario due to the high values of their SIRIS scores on the technical and scientific scales.

It was interesting to compare the results obtained with the SIRIS method to those produced by another statistical approach. In the SIRIS method, the initial variables being transformed into modalities, it seemed legitimate to select a multiple correspondence factor analysis (MCFA) [101] after transformation of the matrix of test/modalities into a complete disjunctive data matrix. The calculations were made with ADE-4 [102], a powerful statistical software program specifically designed for the analysis of environmental data.

The distribution of the 115 tests (Table 1) on the first factorial plan (i.e., F1F2), which accounts for 27.76% of the total inertia of the system, shows a Guttman effect also called horseshoe shape (Fig. 3). The interpretation of this figure has to be made

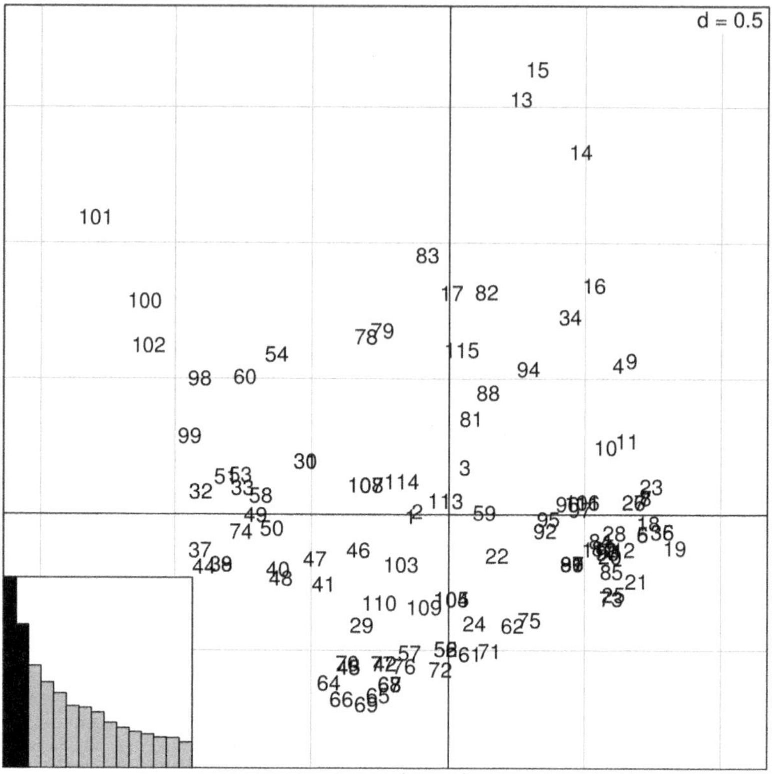

Fig. 3 F1F2 factorial plan (F1, horizontal axis) of the MCFA showing the display of the 115 tests (see Table 1)

from the corresponding map dealing with the variables at two or three modalities. To interpret the relative location of the tests in Fig. 3, it should be necessary to simultaneously consider the 17 doublets or triplets of points belonging to the space of variables. The task is highly difficult, even impossible if all the variables are represented on the same graph. To overcome this problem, a collection of 17 maps, one per variable, was drawn (Fig. 4). In addition, the doublets and triplets of modalities were represented from their symbols f, m, and d for favorable, moderately favorable, and unfavorable, respectively. Last, the different symbols were linked by arrows from f to m or d.

Thus, for example, the tests #100 (*Vibrio fischeri*, solid-phase test) and #101 (*V. fischeri*, flash test, 30 s), which are located in the top left part of Fig. 3, show, for example, favorable modalities for variables STO (supply of test organisms), TVQ (test volume or quantity) and TDU (test duration) and unfavorable modalities for the variables ERL (ecological relevance) and COD (contact duration). However, it is important to note that the COD variable is poorly expressed on FIF2 (Fig. 4). The STO, TDU, and TVQ variables occupying the top of the hierarchy of the technical criteria and the COD and ERL variables occupying the top of the hierarchy of the

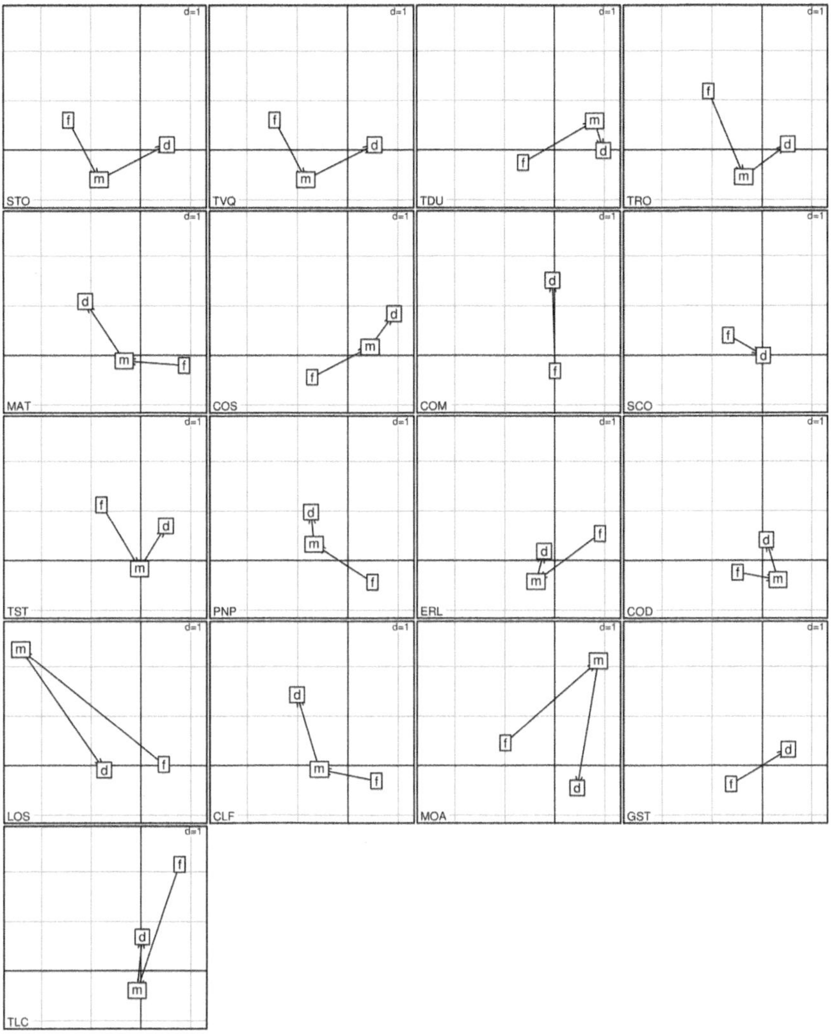

Fig. 4 F1F2 factorial plan (F1, horizontal axis) of the MCFA showing the display of the 17 variables, each of them being displayed separately (see text)

scientific criteria, it is obvious that the space occupied by these tests in Fig. 3 corresponds to the right bottom part of Fig. 2. The same type of reasoning can be applied to the points (i.e., tests) located in the left bottom part of Fig. 3, which can be interpreted from Fig. 4 and which have also to be compared with the points located in the left bottom part of Fig. 2. However, it is noteworthy that the technical scale of the SIRIS map (Fig. 2) is more difficult to find in Fig. 3. It is partially expressed in the second part of the horseshoe shape (Fig. 3), actually the top right part and as previously indicated, the location of these points can also be explained from inspection of the variables displayed in Fig. 4.

Because the F1F2 factorial plan only accounts for 27.76% of the variance of the system, it is obvious that some tests are badly represented in Fig. 3. Consequently, to correctly perform a typology of the 115 tests, it is necessary to consider the other factorial plans. This was done and as illustration only the F4F5 factorial plan, which accounts for 13.3% of the variance of the system, is given here (Figs. 5 and 6). The space of objects (Fig. 4), under the strong dependence of the test #101 that is the unique test presenting a modality m regarding the LOS variable, allows us to find the technical axis of the SIRIS map (Fig. 2).

Consequently, the SIRIS method and a classical multivariate method, such as MCFA, yield the same typology of the 115 tests.

Unfortunately, the typology obtained from the double scale of SIRIS scores (Fig. 2) is not satisfying for all the 115 studied tests. The test #20 dealing with the mortality of earthworms (*Eisenia fetida* or *E. andrei*) and which is located in about the middle part of Fig. 2 presents SIRIS scores equal to those of the earthworm (*E. fetida*) avoidance test #22 while for the former test the time of exposure is 14 days and for the latter the time is only of 48 h. Even if the experience on this rather new test is limited, the avoidance test seems to be more sensitive than the acute toxicity test on *E. fetida* (#20). Its location on the map could penalize this type

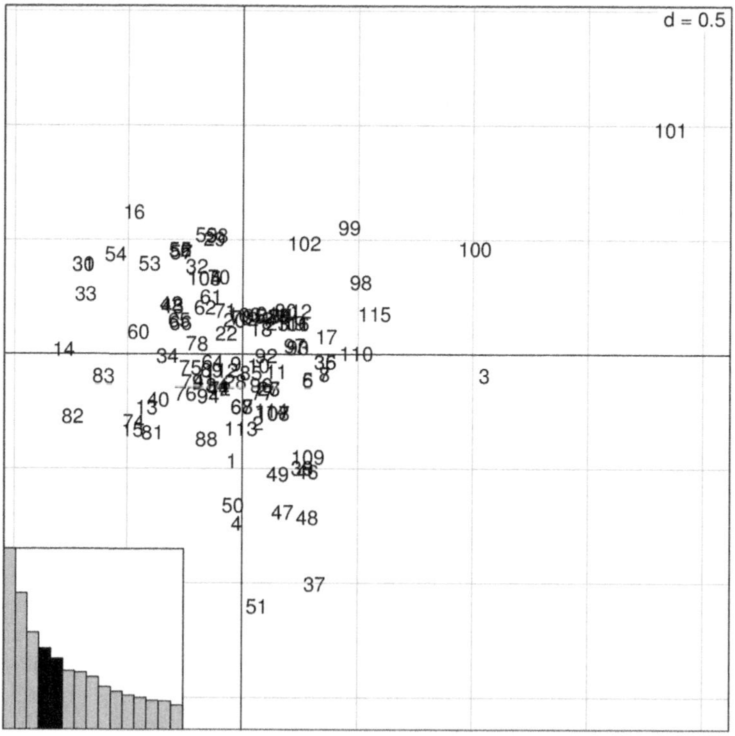

Fig. 5 F4F5 factorial plan (F4, horizontal axis) of the MCFA showing the display of the 115 tests (see Table 1)

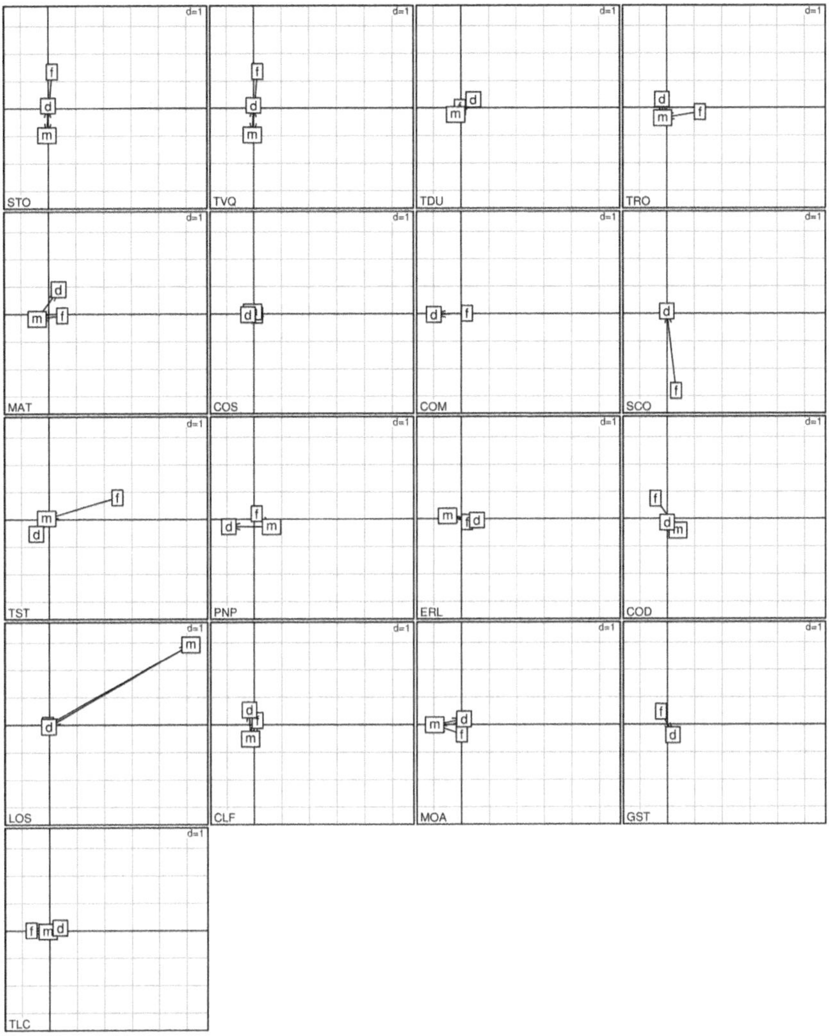

Fig. 6 F4F5 factorial plan (F4, horizontal axis) of the MCFA showing the display of the 17 variables, each of them being displayed separately (see text)

of biological response in comparison to the classical toxicity endpoints recorded with this species and dealing with mortality and reproduction. In the same way, the superimposition of the tests #67 (inhibition of reproduction of *Brachionus calyciflorus* in 48 h) and #68 (mortality of *B. calyciflorus* in 24 h) in Fig. 2 is not logical.

It is obvious that the choice of a hierarchy for the criteria (i.e., variables) highly influences the values of the SIRIS scores (Fig. 1). Consequently, with the SIRIS method, it is always preferable to test the effects of different hierarchies on the results. In this study, this was particularly needed for the technical criteria. Indeed, while for the scientific criteria, a full consensus existed between the experts for ranking the variables, it was more difficult to reach this goal with the technical criteria. Consequently, three other hierarchies of the technical criteria were tested (i.e., C2, C3, C4). The three new technical scales of SIRIS scores were used with the previous scientific scale of SIRIS scores to draw three new SIRIS maps (not shown). These maps were compared with Fig. 2. Although the main trends of the typology of the 115 tests did not change, it was difficult to precisely evaluate the influence of the changes in the hierarchies on the relative position of the tests on the SIRIS maps. To overcome this problem, the SIRIS scores calculated from the different hierarchies of technical criteria were compared by means of triangular representations [102].

Briefly, a triangular representation is a graphical method allowing to display all the information regarding the variability of a distribution of frequencies at three categories. Each side of the triangle corresponds to one category and is graduated from 0 to 1. Each element is divided by the total sum per row. If one element presents the same value for the three variables, the result will be always 1/3, 1/3, 1/3. Consequently, the goal of this method is not to perform global comparisons but to detect how the different elements behave each other. Figure 7 shows how to find the coordinates of a point in the triangle. The triangular representation is particularly suited to reveal the variations in the classification of the tests obtained from the different hierarchies of technical criteria. It allows to underline those that are the most sensitive to changes and the way the criteria act within each hierarchy. Thus, for example, Fig. 8 compares the SIRIS codes obtained for the technical criteria by using the initial hierarchy (C1, Fig. 2) and two new hierarchies called C2 and C3. Obviously, the triangles C1-C2-C4 and C1-C3-C4 were also considered (figures not shown). In Fig. 8, the large top left triangle is graduated from 0 to 1. Conversely, the large top right triangle is a zoom of the central part of the previous triangle. The scales of the zooms are always represented by small grey triangles located in the left top part of the graduated triangles. In the bottom square, the zooming of the cloud of points

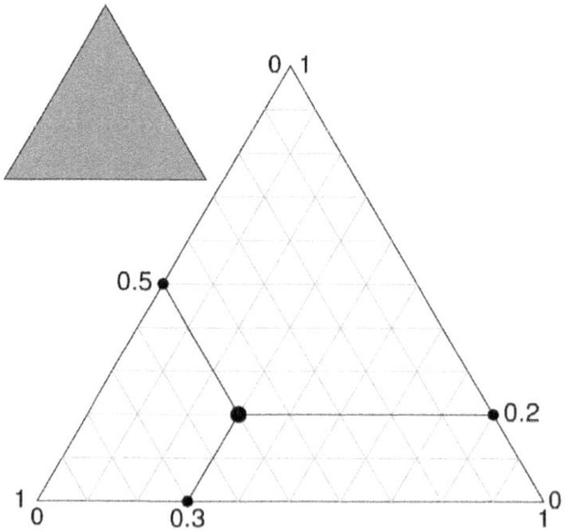

Fig. 7 Theoretical example showing how the relative position of a point has to be interpreted in the triangular representation

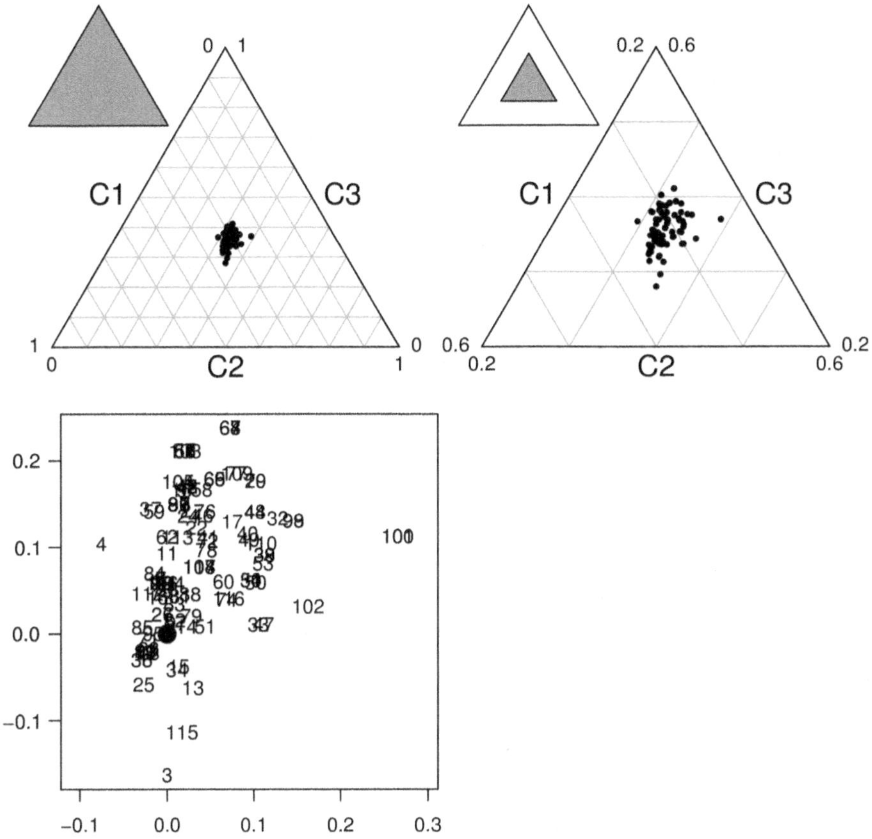

Fig. 8 Triangular representation of SIRIS scores calculated from three hierarchies of technical criteria (C1, C2, C3)

is maximal. The large black point symbolizes the center of the triangle. If a test is located on this point, this means that the SIRIS scores for the three hierarchies are equal. The interpretation of the figure is rather straightforward. Thus, for example, the test #3 (bait-lamina test in 28 days) is located in the bottom part of the square because its score SIRIS are equal for C1 and C2 (i.e., 63) and its score SIRIS for C3 equals 49. The test #115 (macroinvertebrate communities, biotic indexes), located just above, shows SIRIS scores of 109, 112, and 93 for C1, C2, and C3, respectively. The test #4 (mycorrhizal fungi (*Glomus mosseae*), spore germination test in 14 days) in the left part of the figure shows SIRIS scores of 103, 90, and 112 for C1, C2, and C3, respectively. At the opposite, the tests #100 (*V. fischeri*, solid-phase test) and #101 (*V. fischeri*, flash test in 30 s), which are superimposed, present SIRIS scores of 24, 39, and 37 for C1, C2, and C3, respectively.

Comparison of the different triangular representations clearly reveals that the more the changes occur at the top of the hierarchy of the technical criteria, the more the typology of the 115 tests is changed. Moreover, the improvements in the location of tests in the SIRIS maps are always made detrimental to the position of others. Consequently, it was decided to keep the initial hierarchy of the technical criteria used to construct Fig. 2 (i.e., C1).

Figure 2 is particularly suited to estimate the relevance of batteries of tests and to optimize their selection. To discuss these points, three different batteries in relation with the assessment of contaminations in soils and one with wastes were considered. Thus, the ISO/FDIS 17616 project for soil quality assessment [103] proposes to use a battery including tests of inhibition of nitrification (#1 in Table 1) and respiration (#2) of soil microorganisms, inhibition of growth of monocotyledonous (#5) and dicotyledonous (#6) plants, mortality of earthworms (#20), inhibition of mobility of *Daphnia magna* (#72) and light emission of *V. fischeri* (#98), and the umu-test on *Salmonella typhimurium* (#30). A battery including the tests #5, #6, #20, and #98, the spore germination test on *G. mosseae* (#4), and the test on the inhibition of growth of *Pseudokirchneriella subcapitata* (#52) or *Scenedesmus subspicatus* (#55) was proposed in the VADETOX program [104]. A battery including the tests #5, #6, #20, #52, #72, #98, and the test on the inhibition of reproduction of *Ceriodaphnia*

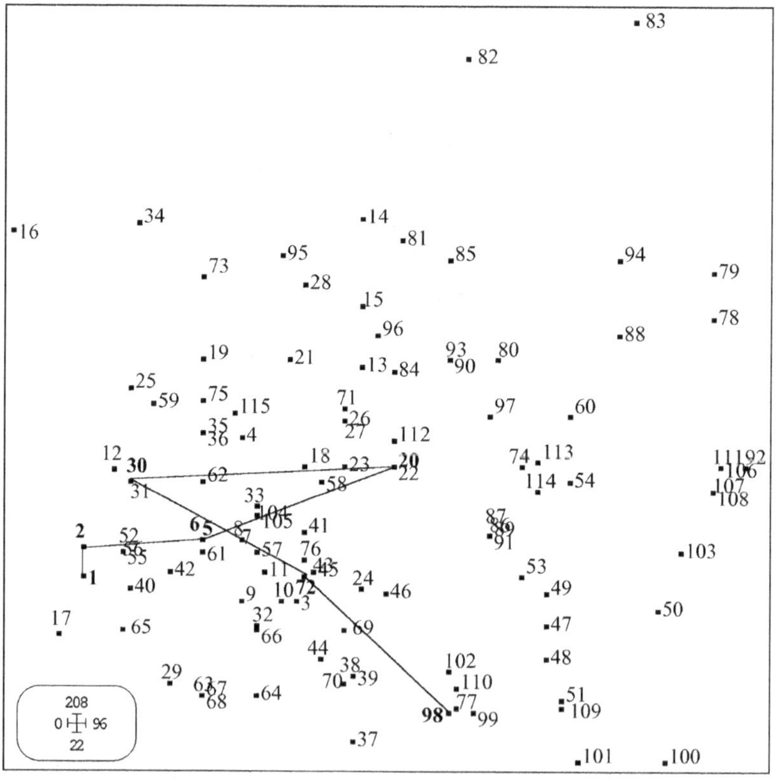

Fig. 9 Location in the SIRIS map of the battery of tests recommended by the ISO 17616 standardization project

dubia (#75) was proposed to assess the ecotoxicity of wastes [105]. Last, to assess the ecological risk and bioremediation efficiency of oil-polluted soils, van Gestel et al. [74] proposed a battery including the tests #52, #72, the bait-lamina test (#3), the tests of inhibition of root elongation (#9) and seed emergence (#11) of higher plants, the test of inhibition of reproduction of earthworms (#21), the tests of mortality (#24) and inhibition of reproduction (#25) of springtails, the test of inhibition of growth of *Bacillus sp.* (#29), and the test of inhibition of mobility of *Plectus acuminatus* (#70). These batteries are represented in Figs. 9–12 to estimate their scientific and technical reliability.

It is interesting to note that all these batteries combine direct tests and assays performed on water extract of the selected matrices. However, they deal with a limited number of tests in comparison with the list of tests retrieved from literature (Table 1). Most of the tests of these batteries, except those regarding soil functional activities, were developed initially for chemicals and are commonly used to assess

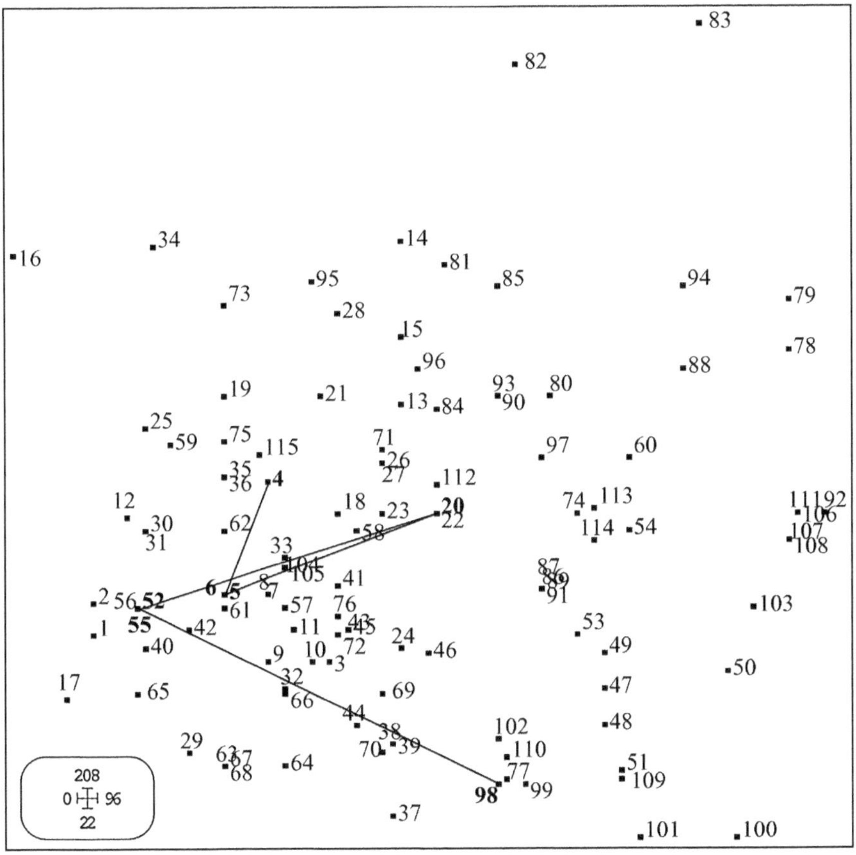

Fig. 10 Location in the SIRIS map of the battery of tests recommended by the ADEME VADETOX study

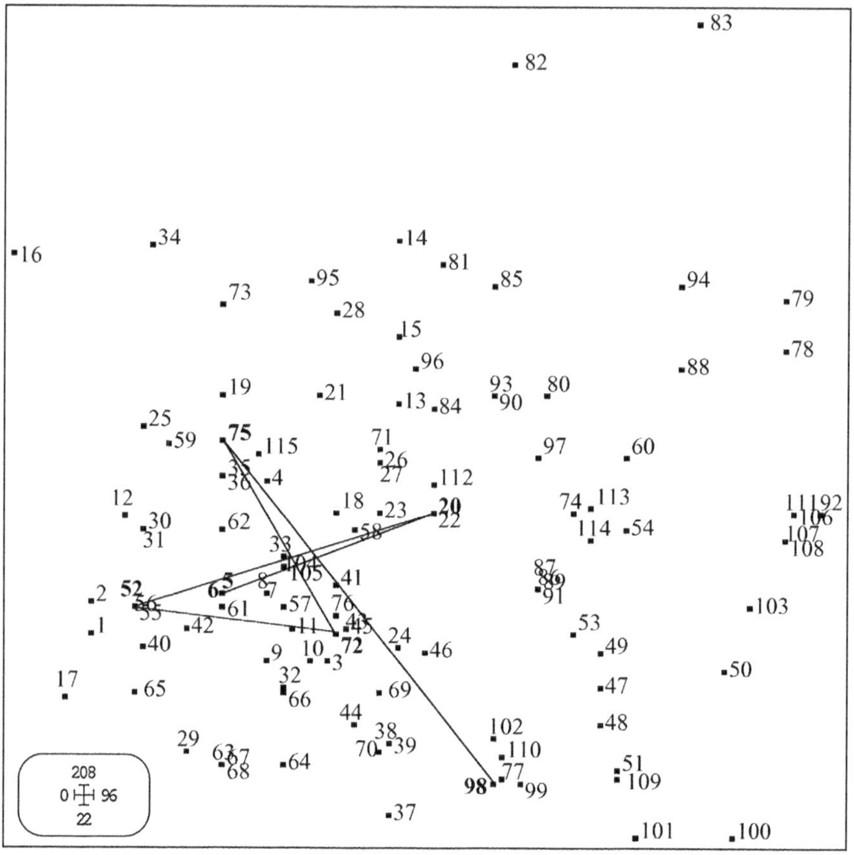

Fig. 11 Location in the SIRIS map of the battery of tests recommended to assess waste ecotoxicity

water or soil quality. These selected tests are generally located in the left inferior quadrant of the SIRIS map (Figs. 9–12) revealing that they are scientifically relevant for assessing the level of contamination of soils, and they do not suffer from important technical problems. Only one battery includes a test of genotoxicity on bacteria (test #30 in Fig. 9), which is performed only on water extract.

The location of the tests #20 (mortality of earthworms) and #98 (inhibition of light emission of *V. fischeri*) on the SIRIS map (Figs. 9–11) is in favor of their exclusion because they do not fully satisfy the criteria of selection for the constitution of test batteries. However, their integration in most of the batteries can be justified for the former test that it is absolutely necessary to estimate the toxicity of pollutants on the soil invertebrates. The presence of the latter test in the batteries can be also easily explained because it allows to rapidly obtain a response and in addition, it is very cheap, even if its ecological relevance is low regarding the studied scenarios.

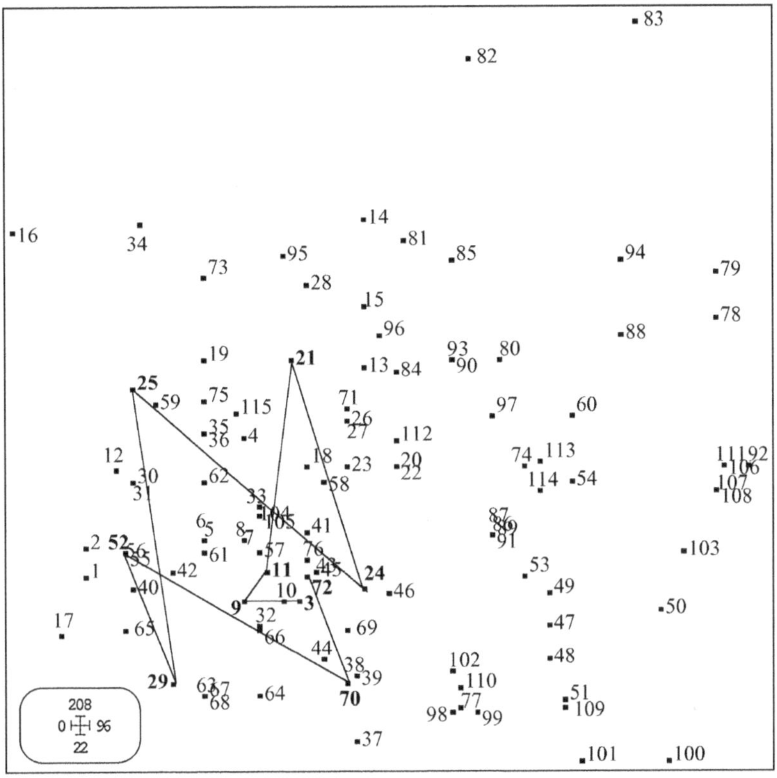

Fig. 12 Location in the SIRIS map of the battery of tests recommended by van Gestel et al. [74]

4 Conclusion

In this chapter, an a priori method for the selection of a battery of tests is presented and exemplified in a case study dealing with the selection of assays for assessing the level of contamination of soils. The selection, which is based on the use of the SIRIS multicriteria method, accounts for scientific and technical criteria. The two categories of criteria were ranked according to their order of importance in relation with the studied scenario. The SIRIS method was used to calculate a scientific and a technical score for 115 tests retrieved from literature. A typology of the 115 tests was performed from the SIRIS scores and confirmed from MCFA. Four different batteries of tests, recommended for characterizing the ecotoxicity of soils and wastes, were represented on the SIRIS map of 115 tests and briefly discussed for estimating their reliability.

The advantages of our methodological approach are numerous. The method is highly flexible and allows to account for any kind of criteria. Indeed, it is obvious that the selection of a test cannot be only based on scientific criteria. Other criteria such as its cost, technical difficulty, and duration also have to be considered.

With the SIRIS method, the relevant criteria can be considered together or split into different categories. For this kind of ecotoxicological study, our experience shows that the latter strategy, which yields different scales of scores, is the most interesting. In all cases, the criteria have to be ranked from expert judgments.

Regulatory agencies are interested in the scoring methods because they are well suited for classification problems. In this context, a SIRIS map derived from two scales of scores appears particularly interesting to perform typologies. It can also be used to optimize selections and propose improvements. Thus, from the SIRIS map of tests, it is not only possible to select the most interesting battery of tests in term of scientific and technical criteria but it is also possible to pinpoint tests that are scientifically relevant for the studied scenario but which need additional work to improve some of their technical criteria, which currently penalize their inclusion in a battery.

Acknowledgment The financial support from the French Environment and Energy Management Agency (ADEME) is gratefully acknowledged (Contract 0475C0081).

References

1. Anonymous (2006) Toxicity testing for assessment of environmental agents: Interim report. Committee on Toxicity Testing and Assessment of Environmental Agents, National Research Council, National Academy of Sciences, USA
2. Anonymous (2007) Toxicity testing in the 21st century: A vision and a strategy. Committee on Toxicity Testing and Assessment of Environmental Agents, National Research Council, National Academy of Sciences, USA
3. Junghans M, Schaefer M, Drost W, Hassold E, Stock F, Dünne M, Juffernholz T, Meyer W, Ranke J (2008) Reconsidering environmental effects assessment of chemicals: Proposal for a dynamic testing strategy. Basic Appl Ecol 9: 356–364
4. Lambolez L, Vasseur P, Ferard JF, Gisbert T (1994) The environmental risks of industrial waste disposal: an experimental approach including acute and chronic toxicity studies. Ecotoxicol Environ Safety 28: 317–328
5. Repetto G, Jos A, Hazen MJ, Molero ML, del Peso A, Salguero M, Castillo PD, Rodriguez-Vicente MC, Repetto M (2001) A test battery for the ecotoxicological evaluation of pentachlorophenol. Toxicol In Vitro 15: 503–509
6. Ward ML, Bitton G, Townsend T, Booth M (2002) Determining toxicity of leachates from Florida municipal solid waste landfills using a battery-of-tests approach. Environ Toxicol 17: 258–266
7. Nalecz-Jawecki G, Grabinska-Sota E, Narkiewicz P (2003) The toxicity of cationic surfactants in four bioassays. Ecotoxicol Environ Safety 54: 87–91
8. Wilke BM, Riepert F, Koch C, Kühne T (2008) Ecotoxicological characterization of hazardous wastes. Ecotoxicol Environ Safety 70: 283–293
9. Alvarenga P, Palma P, Gonçalves AP, Fernandes RM, de Varennes A, Vallini G, Duarte E, Cunha-Queda AC (2008) Evaluation of tests to assess the quality of mine-contaminated soils. Environ Geochem Health 30: 95–99
10. Macken A, Giltrap M, Foley B, McGovern E, McHugh B, Davoren M (2008) A model compound study: The ecotoxicological evaluation of five organic contaminants employing a battery of marine bioassays. Environ Pollut 153: 627–637

11. Clément B, Persoone G, Janssen C, Le Dû-Delepierre A (1996) Estimation of the hazard of landfills through toxicity testing of leachates: I. Determination of leachate toxicity with a battery of acute tests. Chemosphere 33: 2303–2320
12. Rojickova-Padrtova R, Marsálek B, Holoubek I (1998) Evaluation of alternative and standard toxicity assays for screening of environmental samples: selection of an optimal test battery. Chemosphere 37: 495–507
13. Manusadzianas L, Balkelyte L, Sadauskas K, Blinova I, Pollumaa L, Kahru A (2003) Ecotoxicological study of Lithuanian and Estonian wastewaters: selection of the biotests, and correspondence between toxicity and chemical-based indices. Aquat Toxicol 63: 27–41
14. Pandard P, Devillers J, Charissou AM, Poulsen V, Jourdain MJ, Férard JF, Grand C, Bispo A (2006) Selecting a battery of bioassays for ecotoxicological characterization of wastes. Sci Total Environ 363: 114–125
15. Devillers J, Elmouaffek A, Zakarya D, Chastrette M (1988) Comparison of ecotoxicological data by means of an approach combining cluster and correspondence factor analyses. Chemosphere 17: 633–646
16. Ren S, Frymier PD (2003) Use of multidimensional scaling in the selection of wastewater toxicity test battery components. Water Res 37: 1655–1661
17. Keddy CJ, Greene JC, Bonnell MA (1995) Review of whole-organism bioassays: Soil, freshwater sediment, and freshwater assessment in Canada. Ecotoxicol Environ Safety 30: 221–251
18. van Gestel CAM, Leon CD, Van Straalen NM (1997) Evaluation of soil fauna ecotoxicity tests regarding their use in risk assessment. In: Tarradellas J, Bitton G, Rossel D (eds) Soil ecotoxicology, CRC Press, Boca Raton, FL
19. Breitholtz M, Rudén C, Ove Hansson S, Bengtsson BE (2006) Ten challenges for improved ecotoxicological testing in environmental risk assessment. Ecotoxicol Environ Safety 63: 324–335
20. Belton V, Stewart TJ (2002) Multiple criteria decision analysis: An integrated approach. Kluwer Academic, Dordrecht
21. Teng JY, Tzeng GH (1994) Multicriteria evaluation for strategies of improving and controlling air quality in the super city: A case study of Taipei city. J Environ Manag 40: 213–229
22. Ayoko GA, Morawska L, Kokot S, Gilbert D (2004) Application of multicriteria decision making methods to air quality in the microenvironments of residential houses in Brisbane, Australia. Environ Sci Technol 38: 2609–2616
23. Konidari P, Mavrakis D (2006) Multi-criteria evaluation of climate policy interactions. J Multi-Crit Decis Anal 14: 35–53
24. Konidari P, Mavrakis D (2007) A multi-criteria evaluation method for climate change mitigation policy instruments. Energy Policy 35: 6235–6257
25. Hermans C, Erickson J, Noordewier T, Sheldon A, Kline M (2007) Collaborative environmental planning in river management: An application of multicriteria decision analysis in the White River Watershed in Vermont. J Environ Manage 84: 534–546
26. Joubert A, Stewart TJ, Eberhard R (2003) Evaluation of water supply augmentation and water demand management options for the city of Cape Town. J Multi-Crit Decis Anal 12: 17–25
27. Raaijmakers R, Krywkow J, van der Veen A (2008) Flood risk perceptions and spatial multi-criteria analysis: An exploratory research for hazard mitigation. Nat Hazards 46: 307–322
28. Kenyon W (2007) Evaluating flood risk management options in Scotland: A participant-led multi-criteria approach. Ecol Econ 64: 70–81
29. Espelta JM, Retana J, Habrouk A (2003) An economic and ecological multi-criteria evaluation of reforestation methods to recover burned *Pinus nigra* forests in NE Spain. Forest Ecol Manage 180: 185–198
30. Henig MI, Weintraub A (2006) A Dynamic objective-subjective structure for forest management focusing on environmental issues. J Multi-Crit Decis Anal 14: 55–65
31. Briceno-Elizondo E, Jäger D, Lexer MJ, Garcia-Gonzalo J, Peltola H, Kellomäki S (2008) Multi-criteria evaluation of multi-purpose stand treatment programmes for Finnish boreal forests under changing climate. Ecol Ind 8: 26–45

32. Gomontean B, Gajaseni J, Edwards-Jones G, Gajaseni N (2008) The development of appropriate ecological criteria and indicators for community forest conservation using participatory methods: A case study in northeastern Thailand. Ecol Ind 8: 614–624
33. Critto A, Torresan S, Semenzin E, Giove S, Mesman M, Schouten AJ, Rutgers M, Marcomini A (2007) Development of a site-specific ecological risk assessment for contaminated sites: Part I. A multi-criteria based system for the selection of ecotoxicological tests and ecological observations. Sci Total Environ 379: 16–33
34. Semenzin E, Critto A, Carlon C, Rutgers M, Marcomini A (2007) Development of a site-specific ecological risk assessment for contaminated sites: Part II. A multi-criteria based system for the selection of bioavailability assessment tools. Sci Total Environ 379: 34–45
35. Semenzin E, Critto A, Rutgers M, Marcomini A (2008) Integration of bioavailability, ecology and ecotoxicology by three lines of evidence into ecological risk indexes for contaminated soil assessment. Sci Total Environ 389: 71–86
36. Promentilla MAB, Furuichi T, Ishii K, Tanikawa N (2008) A fuzzy analytic network process for multi-criteria evaluation of contaminated site remedial countermeasures. J Environ Manage 88: 479–495
37. Rousis K, Moustakas K, Malamis S, Papadopoulos A, Loizidou M (2008) Multi-criteria analysis for the determination of the best WEEE management scenario in Cyprus. Waste Manage 28: 1941–1954
38. Bellehumeur C, Vasseur L, Ansseau C, Marcos B (1997) Implementation of a multicriteria sewage sludge management model in the southern Québec municipality of Lac-Mégantic, Canada. J Environ Manage 50: 51–66
39. Baltussen R, ten Asbroek AHA, Koolman X, Shrestha N, Bhattarai P, Niessen LW (2007) Priority setting using multiple criteria: should a lung health programme be implemented in Nepal? Health Policy Plan 22: 178–185
40. Rakotomanana F, Randremanana RV, Rabarijaona LP, Duchemin JB, Ratovonjato J, Ariey F, Rudant JP, Jeanne I (2007) Determining areas that require indoor insecticide spraying using multi criteria evaluation, a decision-support tool for malaria vector control programmes in the central highlands of Madagascar. Int J Health Geog 6: 1–11
41. Vaillant M, Jouany JM, Devillers J (1995) A multicriteria estimation of the environmental risk of chemicals with the SIRIS method. Toxicol Model 1: 57–72
42. Guerbet M, Jouany JM (2002) Value of the SIRIS method for the classification of a series of 90 chemicals according to risk for the aquatic environment. Environ Impact Assess Rev 22: 377–391
43. Vincent R, Bonthoux F, Lamoise C (2000) Evaluation du risque chimique. Hiérarchisation des risques potentiels. Cahiers Notes Doc Hyg Sec Travail 178: 29–34
44. Linkov I, Satterstrom FK, Steevens J, Ferguson E, Pleus RC (2007) Multi-criteria decision analysis and environmental risk assessment for nanomaterials. J Nanoparticle Res 9: 543–554
45. Rota-Stabelli O, Telford MJ (2008) A multi criterion approach for the selection of optimal outgroups in phylogeny: Recovering some support for Mandibulata over Myriochelata using mitogenomics. Mol Phylogenet Evol 48: 103–111
46. Phillips L, Bana e Costa CA (2007) Transparent prioritisation, budgeting and resource allocation with multi-criteria decision analysis and decision conferencing. Ann Oper Res 154: 51–68
47. Locatelli B, Rojas V, Salinas Z (2008) Impacts of payments for environmental services on local development in northern Costa Rica: A fuzzy multi-criteria analysis. Forest Policy Econ 10: 275–285
48. Bottani E, Rizzi A (2008) An adapted multi-criteria approach to suppliers and products selection-An application oriented to lead-time reduction. Int J Product Econ 111: 763–781
49. Esteves AM (2008) Mining and social development: Refocusing community investment using multi-criteria decision analysis. Res Policy 33: 39–47
50. Zucca A, Sharifi AM, Fabbri AG (2008) Application of spatial multi-criteria analysis to site selection for a local park: A case study in the Bergamo Province, Italy. J Environ Manage 88: 752–769

51. Chou TY, Hsu CL, Chen MC (2008) A fuzzy multi-criteria decision model for international tourist hotels location selection. Int J Hosp Manage 27: 293–301
52. Schaefer M (2004) Assessing 2,4,6-trinitrotoluene (TNT)-contaminated soil using three different earthworm test methods. Ecotoxicol Environ Safety 57: 74–80
53. Aït-Aïssa S, Pandard P, Magaud H, Arrigo A, Thybaud E, Porcher JM (2003) Evaluation of an in vitro hsp70 induction test for toxicity assessment of complex mixtures: comparison with chemical analyses and ecotoxicity tests. Ecotoxicol Environ Safety 54: 92–104
54. Beiras R, Fernández N, Bellas J, Besada V, González-Quijano A, Nunes T (2003) Integrative assessment of marine pollution in Galician estuaries using sediment chemistry, mussel bioaccumulation, and embryo-larval toxicity bioassays. Chemosphere 52: 1209–1224
55. Chenon P, Gauthier L, Loubieres P, Severac A, Delpoux M (2003) Evaluation of the genotoxic and teratogenic potential of a municipal sludge and sludge-amended soil using the amphibian *Xenopus laevis* and the tobacco: *Nicotiana tabacum* L. var. xanthi Dulieu. Sci Total Environ 301: 139–50
56. Chevre N, Gagne F, Blaise C (2003) Development of a biomarker-based index for assessing the ecotoxic potential of aquatic sites. Biomarkers 8: 287–98
57. Chevre N, Gagne F, Gagnon P, Blaise C (2003) Application of rough sets analysis to identify polluted aquatic sites based on a battery of biomarkers: a comparison with classical methods. Chemosphere 51: 13–23
58. Farré M, Barcelo D (2003) Toxicity testing of wastewater and sewage sludge by biosensors, bioassays and chemical analysis. Trends Anal Chem 22: 299–310
59. Isidori M, Lavorgna M, Nardelli A, Parrella A (2003) Toxicity identification evaluation of leachates from municipal solid waste landfills: A multispecies approach. Chemosphere 52: 85–94
60. Mueller DC, Bonner JS, McDonald SJ, Autenrieth RL, Donnelly KC, Lee K, Doe K, Anderson J (2003) The use of toxicity bioassays to monitor the recovery of oiled wetland sediments. Environ Toxicol Chem 22: 1945–1955
61. Sheehan P, Dewhurst RE, James S, Callaghan A, Connon R, Crane M (2003) Is there a relationship between soil and groundwater toxicity? Environ Geochem Health 25: 9–16
62. Stronkhorst J, Schipper C, Brils J, Dubbeldam M, Postma J, van de Hoeven N (2003) Using marine bioassays to classify the toxicity of Dutch harbor sediments. Environ Toxicol Chem 22: 1535–1547
63. Tsui MTK, Chu LM (2003) Aquatic toxicity of glyphosate-based formulations: comparison between different organisms and the effects of environmental factors, Chemosphere 52: 1189–1197
64. Vigano L, Arillo A, Buffagni A, Camusso M, Ciannarella R, Crosa G, Falugi C, Galassi S, Guzzella L, Lopez A, Mingazzini M, Pagnotta R, Patrolecco L, Tartari G, Valsecchi S (2003) Quality assessment of bed sediments of the Po River (Italy). Water Res 37: 501–518
65. Bekaert C, Ferrier V, Marty J, Pfohl-Leszkowicz A, Bispo A, Jourdain MJ, Jauzein M, Lambolez-Michel L, Billard H (2002) Evaluation of toxic and genotoxic potential of stabilized industrial waste and contaminated soils. Waste Manag 22: 241–247
66. Dalzell DJB, Alte S, Aspichueta E, de la Sota A, Etxebarria J, Gutierrez M, Hoffmann CC, Sales D, Obst U, Christofi N (2002) A comparison of five rapid direct toxicity assessment methods to determine toxicity of pollutants to activated sludge, Chemosphere 47: 535–545
67. De Boer WJ, Besten PJ, Ter Braak CF (2002) Statistical analysis of sediment toxicity by additive monotone regression splines. Ecotoxicology 11: 435–450
68. Gagne F, Blaise C, Aoyama I, Luo R, Gagnon C, Couillard Y, Campbell P, Salazar M (2002) Biomarker study of a municipal effluent dispersion plume in two species of freshwater mussels. Environ Toxicol 17: 149–159
69. Lapa N, Barbosa R, Morais J, Mendes B, Méhu J, Santos Oliveira JF (2002) Ecotoxicological assessment of leachates from MSWI bottom ashes, Waste Manag 22: 583–593
70. Schultz E, Vaajasaari K, Joutti A, Ahtiainen J (2002) Toxicity of industrial wastes and waste leaching test eluates containing organic compounds. Ecotoxicol Environ Safety 52: 248–255

71. Cordova Rosa EV, Simionatto EL, de Souza Sierra MM, Bertoli SL, Radetski CM (2001) Toxicity-based criteria for the evaluation of textile wastewater treatment efficiency. Environ Toxicol Chem 20: 839–845
72. Machala M, Dusek L, Hilscherova K, Kubinova R, Jurajda P, Neca J, Ulrich R, Gelnar M, Studnickova Z, Holoubek I (2001) Determination and multivariate statistical analysis of biochemical responses to environmental contaminants in feral freshwater fish *Leuciscus cephalus*, L. Environ Toxicol Chem 20: 1141–1148
73. Renoux AY, Tyagi RD, Samson R (2001) Assessment of toxicity reduction after metal removal in bioleached sewage sludge. Water Res 35: 1415–1424
74. van Gestel CA, van der Waarde JJ, Derksen JG, van der Hoek EE, Veul MF, Bouwens S, Rusch B, Kronenburg R, Stokman GN (2001) The use of acute and chronic bioassays to determine the ecological risk and bioremediation efficiency of oil-polluted soils. Environ Toxicol Chem 20: 1438–1449
75. Juvonen R, Martikainen E, Schultz E, Joutti A, Ahtiainen J, Lehtokari M (2000) A battery of toxicity tests as indicators of decontamination in composting oily waste. Ecotoxicol Environ Safety 47: 156–166
76. Radix P, Léonard M, Papantoniou C, Roman G, Saouter E, Gallotti-Schmitt S, Thiébaud H, Vasseur P (2000) Comparison of four chronic toxicity tests using algae, bacteria, and invertebrates assessed with sixteen chemicals. Ecotoxicol Environ Safety 47: 186–194
77. Ziehl TA, Schmitt A (2000) Sediment quality assessment of flowing waters in South-West Germany using acute and chronic bioassays. Aquat Ecosys Health Manag 3: 347–357
78. Bierkens J, Klein G, Corbisier P, Van Den Heuvel R, Verschaeve L, Weltens R, Schoeters G (1998) Comparative sensitivity of 20 bioassays for soil quality. Chemosphere 37: 2935–2947
79. Robidoux PY, Lopez-Gastey J, Choucri A, Sunahara GI (1998) Procedure to screen illicit discharge of toxic substances in septic sludge received at a wastewater treatment plant. Ecotoxicol Environ Safety 39: 31–40
80. Tarkpea M, Andrén C, Eklund B, Gravenfors E, Kukulska Z (1998) A biological and chemical characterization strategy for small and medium-sized industries connected to municipal sewage treatment plants. Environ Toxicol Chem 17: 234–250
81. Clément B, Colin JR, Le Dû-Delepierre A (1997) Estimation of the hazard of landfills through toxicity testing of leachates: 2. Comparison of physico-chemical characteristics of landfill leachates with their toxicity determined with a battery of tests. Chemosphere 35: 2783–2796
82. Chang LW, Meier JR, Smith MK (1997) Application of plant and earthworm bioassays to evaluate remediation of a lead-contaminated soil. Arch Environ Contam Toxicol 32: 166–171
83. Cheung YH, Neller A, Chu KH, Tam NFY, Wong CK, Wong YS, Wong MH (1997) Assessment of sediment toxicity using different trophic organisms. Arch Environ Contam Toxicol 32: 260–267
84. Maxon CL, Barnett AM, Diener DR (1997) Sediment contaminants and biological effects in southern California: Use of multivariate statistical approach to assess biological impact. Environ Toxicol Chem 16: 775–784
85. Sherry J, Scott B, Dutka B (1997) Use of various acute, sublethal, and early life-stage tests to evaluate the toxicity of refinery effluents. Environ Toxicol Chem 16: 2249–2257
86. Sweet LI, Travers DF, Meier PG (1997) Chronic toxicity evaluation of wastewater treatment plant effluents with bioluminescent bacteria: A comparison with invertebrates and fish. Environ Toxicol Chem 16: 2187–2189
87. Vahl HH, Karbe L, Westendorf J (1997) Genotoxicity assessment of suspended particulate matter in the Elbe river: Comparison of *Salmonella* microsome test, arabinose resistance test, and umu-test. Mutat Res 394: 81–93
88. Naudin S, Garric J, Vindimian E, Bray M, Migeon B, Vollat B, Lenon G (1995) Influence of the sample preservation mode to assess the chronic toxicity of an industrial effluent. Ecotoxicol Environ Safety 30: 54–62
89. Hund K, Traunspurger W (1994) Ecotox-evaluation strategy for soil bioremediation exemplified for a PAH-contaminated site. Chemosphere 29: 371–390

90. Schafer H, Hettler H, Fritsche U, Pitzen G, Roderer G, Wenzel A (1994) Biotests using unicellular algae and ciliates for predicting long-term effects of toxicants. Ecotoxicol Environ Safety 27: 64–81
91. Agences de l'eau (2001) Bioessais sur sédiments – Méthodologies et application à la mesure de la toxicité de sédiments naturels. Les Etudes des Agences de l'Eau n° 76
92. Bastien C, Martel L (1995) Sélection des tests de toxicité en milieu aquatique applicables aux sédiments contaminés - Projet de coopération Franco-Québécoise. Direction des Laboratoires, Ministère de l'Environnement
93. Ingersoll CG, Ankley GT, Benoit DA, Brunson EL, Burton GA, Dwyer FJ, Hoke RA, Landrum PF, Norberg-King TJ, Winger PV (1995) Toxicity and bioaccumulation of sediment-associated contaminants using freshwater invertebrates: A review of methods and applications. Environ Toxicol Chem 14: 1885–1894
94. Jauzein M, Jourdain MJ, Bispo A (1997) Mise au point de méthodes de caractérisation de l'écotoxicité des sols et des déchets – Development of methods to assess soil and waste ecotoxicity - Programme Ecotoxicité des Sols et des Déchets – Document de synthèse – Rapport 3
95. Slooff W, Canton JH (1983) Comparison of the susceptibility of 11 freshwater species to 8 chemical compounds. II. (semi)chronic toxicity tests. Aquat Toxicol 4: 271–281
96. Vindimian E, Garric J, Flammarion P, Thybaud E, Babut M (1999) An index of effluent aquatic toxicity designed by partial least squares regression, using acute and chronic tests and expert judgments. Environ Toxicol Chem 18: 2386–2391
97. Sekkat N, Guerbet M, Jouany JM (2001) Etude comparative de huit bioessais à court terme pour l'évaluation de la toxicité de lixiviats de déchets urbains et industriels. Rev Sci Eau 14: 63–72
98. Fochtman P, Raszka A, Nierzedska E (2000) The use of conventional bioassays, microbiotests, and some "rapid" methods in the selection of an optimal test battery for the assessment of pesticides toxicity. Environ Toxicol 15: 376–384
99. Vaajasaari K, Joutti A, Schultz E, Selonen S, Westerholm H (2002) Comparisons of terrestrial and aquatic bioassays for oil-contaminated soil toxicity. J Soils Sediments 2: 194–202
100. SIRIS-2D (version 2.04). CTIS, France
101. Escofier B, Pagès J (1991) Presentation of correspondence analysis and multiple correspondence analysis with the help of examples. In: Devillers J, Karcher W (eds) Applied multivariate analysis in SAR and environmental studies, Kluwer Academic, Dordrecht
102. Thioulouse J, Chessel D, Dolédec S, Olivier JM (1997) ADE-4: A multivariate analysis and graphical display software. Stat Comput 7: 75–83
103. ISO/FDIS 17616 (2007) "Soil Quality – Guidance for the choice and evaluation of bioassays for ecotoxicological characterization of soils and soil material Geneva: International Standard Organization: 17pp
104. Bispo A, Feix I, Schwartz C, Morel JL, Charissou AM, Jourdain MJ, Gauthier L, Cluzeau D, Ablain F (2005) Ecotoxicological evaluation of the wastes and their products on agrosystem: Towards an integrated process for risk prevention for ecosystems. Synthèse du programme ADEME-VADETOX, contrat n° 0375C0010
105. MATE (1998) Critères et méthodes d'évaluation de l'écotoxicité des déchets. Paris: Ministère de l'Aménagement du Territoire et de l'Environnement: 19 pp

Physiologically Based Toxicokinetic (PBTK) Modeling in Ecotoxicology

Kannan Krishnan and Thomas Peyret

Abstract Physiologically based toxicokinetic [(PBTK), or alternatively referred to as physiologically based pharmacokinetic (PBPK)] models are quantitative descriptions of absorption, distribution, metabolism, and excretion of chemicals in biota. PBTK models are increasingly being used as an effective tool for designing toxicology experiments and for conducting extrapolations essential for risk assessments. This chapter describes the basic concepts, equations, parameters, and software essential for developing PBTK models. QSAR methods for estimating input parameters as well as data sources containing relevant parameters for model development in rats, mice, cattle, birds, and fish are summarized. Model templates for creating PBTK models in fish and terrestrial species are presented. Several examples of model simulations are presented along with a brief discussion of how PBTK models can be applied to make significant advances in ecotoxicology and ecotoxicological risk assessments.

Keywords Toxicokinetics · Pharmacokinetics · PBPK model · PBTK model · Physiologically based models

1 Introduction

Upon exposure to chemicals and their degradation products in air, water, soil, or food, the target and nontarget organisms in the environment may absorb them leading to metabolism and distribution in various tissues, including the target tissue(s) where toxicity may result. Knowledge of the "dose to target tissue" provides a better scientific basis (than the exposure dose) for understanding the dose–response relationships and conducting risk assessments [1]. The dose to target tissue(s) is the net result of the rate and magnitude of the processes of absorption, distribution, metabolism, and excretion (ADME) in biota. Since it is not always feasible or

K. Krishnan (✉)
Département de santé environnementale et santé au travail, 2375 chemin de la Cote Ste Catherine, Pavillon Marguerite d'Youville, Université de Montréal, Montreal, PQ, Canada
e-mail: kannan.krishnan@umontreal.ca

J. Devillers (ed.), *Ecotoxicology Modeling*, Emerging Topics in Ecotoxicology: Principles, Approaches and Perspectives 2, DOI 10.1007/978-1-4419-0197-2_6,

possible to measure target tissue concentration of the toxic moiety associated with various exposure routes, doses, scenarios, and species, toxicokinetic models are increasingly being sought as valuable tools in ecotoxicology and human health risk assessment [2].

Toxicokinetic (or pharmacokinetic) models are mathematical descriptions of ADME in biota. These models facilitate quantitative descriptions of the temporal change in the concentrations of chemicals and/or their metabolites in biological matrices (e.g., blood, tissue, urine, alveolar air) of the exposed organism. Toxicokinetic models often describe the organism as a set of compartments that are characterized physiologically or empirically [3]. The classical or empirical toxicokinetic models are developed on the basis of fitting to experimental data. In this case, the number of compartments (usually one or two), their volumes as well as rates of processes are estimated by fitting the model to experimental data on the time course of chemical concentration in the biological matrix of interest (e.g., blood, urine) [3]. As such, these models can be used confidently for interpolation but not for extrapolation of kinetics and target tissue dose associated with various scenarios, species, and routes of exposure. In this regard, physiologically based toxicokinetic [(PBTK), or alternatively referred to as physiologically based pharmacokinetic (PBPK)] models are particularly useful [4]. The PBTK models are quantitative descriptions of ADME of chemicals based on interrelationships among critical biological, physicochemical, and biochemical determinants [5]. This chapter describes the development and computational implementation of PBTK models, using chemicals and animal species of interest to ecotoxicologists and risk assessors.

2 PBTK Modeling: The Process

Typically, the process of PBTK modeling starts with the definition of the problem and identification of a strategy for the model simulations to resolve the issue at hand (e.g., interspecies differences in toxic response, intraspecies variability in target tissue dose, route-to-route extrapolation of NOAEC). Then, available data on toxicity, toxicokinetics, metabolism, as well as physicochemical properties of a chemical are evaluated to facilitate the identification of a conceptual model and an initial set of parameters. Subsequently, equations representing ADME are written and solved using software that facilitates the solution of ordinary differential equations [4–6]. The model simulations then are compared with experimental data in order to accept the model or to refine the model following uncertainty/sensitivity analyses. The following sections outline briefly the key principles and current practices for developing PBTK models.

2.1 Conceptual Model

The conceptual PBTK model corresponds to the diagrammatic representation, in the form of boxes and arrows, of the key elements of the organism that determine

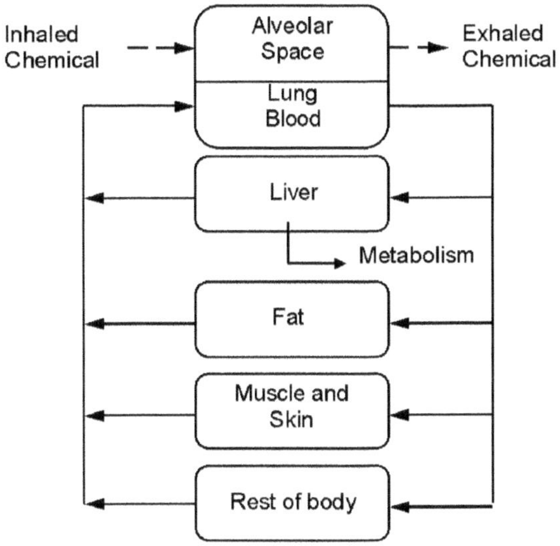

Fig. 1 Conceptual representation of a PBTK model

the exposure as well as ADME of a chemical. Figure 1 represents the conceptual PBTK model for simulating inhalation exposure to a volatile organic chemical. The following elements are considered in developing a conceptual PBTK model for a given chemical [5]:

- Target organ
- Portals of entry
- Metabolism
- Lipophilicity
- Mass balance

In PBTK models intended for application in ecotoxicology and risk assessment, the inclusion of the target organ as a separate compartment should be given consideration. However, if the metabolic and accumulation characteristics of a target organ are not significantly different from other organs (e.g., brain vs. other richly perfused tissues), there is no need to represent the target organ as a separate compartment. In some cases, the chemical concentration in systemic circulation (e.g., blood) might be sufficient for use as a surrogate of target organ exposure.

Next, the portals of entry should be considered for representation as separate compartments in PBTK models. In this context, gills or lungs for describing exposure via inspired water/air, GI tract in the case of oral exposure, and skin for simulating dermal contact are included as separate compartments. In some cases, the portal of entry is not characterized or represented explicitly but the rate of the amount absorbed (i.e., input to the system) is computed [7].

The metabolizing tissues are often characterized as separate compartments since (1) they significantly influence the overall kinetic behavior of a chemical and (2) the time course of chemicals in these tissues would be different from other tissues. Accordingly, liver is often represented as a separate compartment. When extrahepatic metabolism is significant and intertissue difference in the rate of metabolism is significant, various tissues are represented individually (e.g., brain, testes, kidney, etc.) [8].

In PBTK models for lipophilic or superlipophilic chemicals, adipose tissue is represented as a separate compartment. This compartment essentially represents the total volume of fat depots found in the various regions of the body (e.g., abdominal fat, omental fat, subcutaneous fat). For hydrophilic chemicals such as methanol or metals, it would be irrelevant to consider representing the fat depot as a separate compartment since it does not contribute to their storage or kinetic behavior.

Finally, the remaining tissues receiving blood supply should be represented as one compartment ("rest of body") or more ["richly perfused tissues" and "poorly perfused tissues" (PPT)], even if they do not qualify under any of the aforementioned considerations (i.e., target organ, portals of entry, metabolism and storage). To maintain mass balance of chemical in the system, it is essential to track its fate in all tissue compartments that receive it through blood flow. This aspect is illustrated by the inclusion of a compartment referred to as "rest of body" in Fig. 1.

2.2 *Quantitative Descriptions of ADME*

The quantitative descriptions of ADME included in PBTK models are based on the consideration of following key questions:

- *Absorption*: How does the chemical enter systemic circulation (i.e., blood)?
- *Distribution*: How is the chemical distributed to the various organs?
- *Metabolism*: What are the sites and pathways of biotransformation?
- *Excretion*: What are the routes and processes of elimination from the body?

2.2.1 Absorption

Absorption is the process by which a chemical in the microenvironment crosses the biological barrier to enter systemic blood circulation [9]. The biological barriers relate to stratum corneum in the case of dermal exposures, alveoli in the case of pulmonary exposure, or membranes in the GI tract for oral exposure [10]. The intravenous (i.v.), subcutaneous, and intraperitoneal (i.p.) routes may also be of relevance for interpreting experimental toxicology studies, even though they are not directly relevant to environment-related exposures. The process of absorption as described in PBTK models facilitates the computation of the rate of change in the amount of chemical at the portal of entry or chemical concentration in arterial blood (C_a) as a function of exposure concentration. Accordingly, for inhalation exposures, C_a in rodents is obtained as follows [11]:

$$C_a = \frac{Q_p C_{inh} + Q_c C_v}{Q_c + \left(\frac{Q_p}{P_b}\right)}, \quad (1)$$

where C_{inh} is the concentration in inhaled air, C_a is the arterial blood concentration, C_v is the concentration in mixed venous blood, P_b is the blood:air partition coefficient, Q_c is the cardiac output, and Q_p is the alveolar ventilation rate.

For dermal exposures, the concentration following absorption is calculated by solving the following mass-balance differential equation [12]:

$$\frac{\mathrm{d}C_{\text{sk}}}{\mathrm{d}t} = \frac{\left(Q_{\text{sk}}(C_a - C_{\text{vsk}}) + K_p \times A[C_{\text{water}} - (C_{\text{sk}}/P_{\text{s:e}})]\right)}{V_{\text{sk}}}, \tag{2}$$

where A is the area of skin exposed, C_{sk} is the concentration in skin, C_{vsk} is the concentration in venous blood leaving the skin, C_{water} is the chemical concentration in the microenvironment contacting the skin, $\mathrm{d}C_{\text{sk}}/\mathrm{d}t$ is the rate of change in the concentration of chemical in the skin, K_p is the permeability coefficient, Q_{sk} is the rate of blood flow to skin, $P_{\text{s:e}}$ is the skin:environment partition coefficient, and V_{sk} is the volume of skin.

Oral uptake of chemicals is often described as a first-order process as follows [7]:

$$\frac{\mathrm{d}A_o}{\mathrm{d}t} = K_o A_{\text{stom}}, \tag{3}$$

where $\mathrm{d}A_o/\mathrm{d}t$ refers to the rate of amount of chemical absorbed orally, K_o is the oral absorption constant, and A_{stom} is the amount of chemical remaining in the stomach.

2.2.2 Distribution

Distribution, in the present context, refers to the uptake of a chemical from systemic circulation by the metabolizing, storage, and excretory organs [9]. The volume of blood, volume of tissues as well as the extent of protein binding together determine the extent of dilution or distribution of an absorbed chemical. The volume of distribution then reflects the "apparent" volume of blood in which a chemical is distributed and as such it does not reflect a true, measurable physiological volume. V_d actually corresponds to the volume of blood plus the sum of the product of tissue volumes and tissue:blood partition coefficients [3].

The distribution of chemical in each tissue compartment may be limited either by the membrane (diffusion-limited uptake) or by the blood flow (perfusion-limited uptake) [13]. In the case of perfusion-limited uptake, the flux of a chemical $\left(V_t \times \frac{\mathrm{d}C_t}{\mathrm{d}t}\right)$ is proportional to its concentration gradient (ΔC) as defined by Fick's law of simple diffusion:

$$V_t \times \frac{\mathrm{d}C_t}{\mathrm{d}t} = k\Delta C, \tag{4}$$

where V_t is the tissue volume and k is the proportionality constant.

Since the blood flow to tissues is essentially equal to the proportionality constant, $V_t \times \frac{\mathrm{d}C_t}{\mathrm{d}t}$ in PBTK model compartments is calculated as follows:

$$V_t \times \frac{\mathrm{d}C_t}{\mathrm{d}t} = \text{input (mg/h)} - \text{output (mg/h)}, \tag{5}$$

where input is the blood flow rate × concentration entering the tissue, and output is the blood flow rate × concentration leaving the chemical.

Notationally, the earlier equation is written as follows:

$$V_t \frac{dC_t}{dt} = Q_t - (C_a - C_{vt}), \quad (6)$$

where Q_t is the tissue blood flow, C_a is the arterial blood concentration (entering), and C_{vt} is the venous blood concentration (leaving).

For high molecular weight compounds (dioxins, PCBs, hexachlorobenzene), diffusion is often the rate-limiting process such that their uptake through the tissue subcompartments must be considered [13]. This requires that the mass-balance differential equations be developed both for the tissue blood and cellular matrix subcompartments of each tissue.

2.2.3 Metabolism

Metabolism refers to the biotransformation and conjugation of xenobiotics (commonly termed as phase I and phase II processes) in the biota [9,10]. Phase I reactions are often mediated by enzymes with finite binding sites (cytochrome P-450, flavin monooxygenases, xanthine oxidase, amine oxidase, monoamine oxidase, carbonyl reductases, etc.) whereas phase II reactions are frequently limited by the availability of cofactors (sulfate, glutathione, glucuronide, etc.). The rate of metabolism $\left(\frac{dA_{met}}{dt}\right)$ in PBTK models is described as a first order, second order, or saturable process as follows:

$$\text{First order}: \frac{dA_{met}}{dt} = K_f C_{vt} V_t, \quad (7)$$

$$\text{Second order}: \frac{dA_{met}}{dt} = K_s C_{vt} V_t C_{cf}, \quad (8)$$

$$\text{Saturable}. \frac{dA_{met}}{dt} = \frac{V_{max} C_{vt}}{K_m + C_{vt}}, \quad (9)$$

where K_f is the first-order metabolism constant (e.g., h^{-1}), C_{cf} is the concentration of cofactor in tissue, "t," V_t is the volume of the tissue, K_s is the second-order metabolism constant (e.g., L/mg/h), V_{max} is the maximum velocity of enzymatic reaction (e.g., mg/h), and K_m is the Michaelis–Menten affinity constant (e.g., mg/L).

Conjugation reactions have often been described as a second order process with respect to the concentration of the cofactor and the chemical [8, 14], whereas the phase I reactions are described as a first-order process at very low exposure concentrations or as a saturable process according to Michaelis–Menten equation. Alternative formulations based on intrinsic clearance and hepatic clearance can also be used in this regard [5, 15].

It is important to note that the metabolism descriptions captured in (7)–(9) are based on venous blood concentration (i.e., free concentration) of chemical and they conform to the "venous equilibration model" descriptions. Since these equations are solved along with (6), the blood flow limitation (i.e., perfusion limitation) of metabolism is automatically accounted for. Alternative models of hepatic metabolism (e.g., parallel tube model, distributed sinusoidal perfusion model) have been described in the literature but they appear to yield essentially the same results regarding the whole-body clearance of chemicals [5].

2.2.4 Excretion

Excretion refers to the removal of the chemical and/or its metabolite from systemic circulation [9]. The principal routes of excretion include urine, bile, feces, and exhaled air. Exhalation of volatile chemicals from systemic circulation is described in PBTK models on the basis of the rate of respiration, cardiac output, and the blood:air partition coefficient. Biliary and fecal excretion rates depend upon the rate of bile flow, the rate of transfer into and reabsorption from the bile as well as the molecular weight of the chemical or its conjugated metabolite [10]. Urinary excretion is modeled as a function of the rates of filtration, reabsorption, and secretion. The amount of chemical filtered ($\mathrm{d}F/\mathrm{d}t$) equals the glomerular filtration rate (GFR) and the blood concentration of unbound chemical (Cu):

$$\mathrm{d}F/\mathrm{d}t = \mathrm{GFR} \times \mathrm{Cu}. \tag{10}$$

The rate of change in the concentration of chemical or its metabolite in the urine ($\mathrm{d}U/\mathrm{d}t$) equals the following:

$$\mathrm{d}U/\mathrm{d}t = U_o \times \mathrm{Cu}, \tag{11}$$

where U_o is the urinary output (mL/min) and Cu is the chemical concentration in urine (mg/mL).

The amount of chemical secreted (or reabsorbed) can, in turn, be calculated on the basis of the difference between the amount in urine and the amount filtered [16].

2.3 *Estimation of Parameters*

To solve the equations constituting the PBTK model, the numerical values of input parameters – physiological parameters, partition coefficients, and biochemical rate constants – should be known. The numerical values of physiological parameters such as breathing rate, skin surface area, cardiac output, tissue blood flow rates, and tissue volumes can either be obtained from published literature or determined experimentally. Compilations of physiological parameters for several terrestrial animal species (rat, mouse, swine, cow, goat, etc.) [17–19] as well as for aquatic life,

Table 1 Reference values of tissue volumes and blood flow rates in mouse and rat [18]

Physiological parameters	Mouse	Rat
Body weight (kg)	0.025	0.25
Tissue volume (L)		
Liver	0.0014	0.01
Fat	0.0025	0.0175
Organs	0.0013	0.0125
Muscle	0.0175	0.1875
Cardiac output (L/min)	0.017	0.083
Blood flow to organs (L/min)		
Liver	0.0043	0.0208
Fat	0.0015	0.0075
Organs	0.0087	0.0423
Muscle	0.0026	0.0125
Minute volume (L/min)	0.037	0.174
Alveolar ventilation (L/min)	0.025	0.117

particularly fish [20, 21], have appeared in the literature. Table 1 presents reference values for PBTK modeling in rats and mice whereas Table 2 summarizes the physiological parameter values for fish.

The blood:microenvironment (i.e., blood:air, blood:water) as well as tissue:blood partition coefficients required for PBTK modeling may be obtained (1) from steady-state toxicokinetic data obtained following repeated dosing [22], (2) with in vitro systems using ultrafiltration, equilibrium dialysis, or headspace equilibrium technique [23–25], or (3) by in silico approaches based on molecular and biological determinants [26–32]. The latter approaches have either been developed on the basis of mechanistic determinants (i.e., *n*-ctanol:water partition coefficient, *n*-octanol:air partition coefficient, blood and tissue content of neutral lipids, phospholipids, and water) [28, 29] or on the basis of fitting to experimental data obtained with limited number of chemicals [30–32]. Figure 2 presents examples of partition coefficients for chloroethanes, obtained in vitro using fish tissues.

Metabolism parameters (V_{max}, K_m, K_f) are obtained by analyzing in vivo toxicokinetic data or using subcellular preparations (e.g., postmitochondrial preparations, microsomes), hepatocytes, or tissue slices in vitro. Reviews of the various in vitro systems for estimating metabolism rate constants and the issues related to the scaling of V_{max}, K_m, or their ratio (i.e., intrinsic clearance) in fish and mammals have recently been published [33–36]. It is reasonable to use the K_m obtained in in vitro studies directly in the PBTK models, since the same isozyme is involved in the metabolism of a given chemical both in vitro and in vivo; however, the V_{max} obtained in vitro should be scaled to the whole liver (or the appropriate metabolizing organ) based on the difference in the enzyme content between in vitro and in vivo situations, as follows [37]:

$$V_{max\,(in\ vivo)} = V_{max\,(in\ vitro)} \times \text{protein concentration} \times \text{tissue volume}. \qquad (12)$$

Table 2 Physiological parameters for rainbow trout, fathead minnow, and channel catfish [20, 21]

Parameter	Rainbow trout	Channel catfish	Fathead minnow Parameter	Value
Body weight (kg)	1	1	Body weight (g)	0.147
Cardiac output (L/h)	2.06	2.5	Cardiac output (mL/h)	1.646
Effective ventilation (L/h)	7.20	9.00	Mean oxygen consumption rate (mg/h)	0.124
Skin surface area (cm^2)	1025	1,025		–
Gill surface area (cm^2)	2,000	1,500		–
Fat blood flow (L/h)	0.18	0.28	Fat blood flow (fraction of cardiac output)	0.01
Liver blood flow (L/h)	0.06	0.09		–
Poorly perfused tissue blood flow (L/h)	1.08	1.38	Carcass blood flow (fraction of cardiac output)	0.4
Richly perfused tissue blood flow (L/h)	0.47	0.40	Viscera blood flow (fraction of cardiac output)	0.54
Kidney blood flow (L/h)	0.12	0.16		–
Skin blood flow (L/h)	0.15	0.19	Skin blood flow (fraction of cardiac output)	0.05
Fat volume(L)	0.098	0.066	Fat volume (fraction of body weight)	0.022
Liver volume(L)	0.012	0.015		–
Poorly perfused tissue volume(L)	0.718	0.765	Carcass volume (fraction of body weight)	0.888
Richly perfused tissue volume(L)	0.063	0.044	Viscera volume (fraction of body weight)	0.09
Kidney volume(L)	0.009	0.010		–
Skin volume(L)	0.100	0.100		–

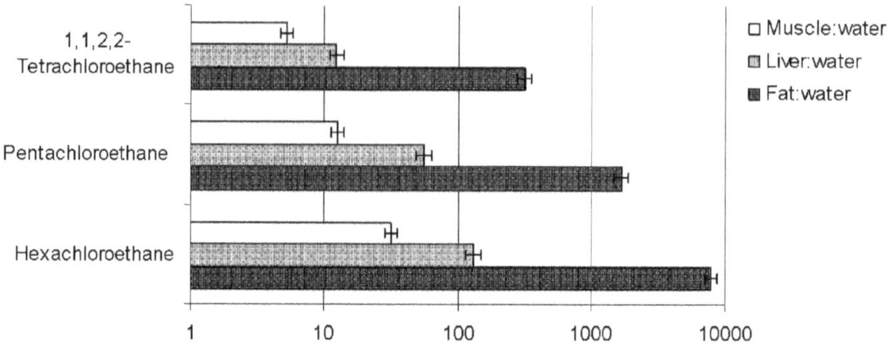

Fig. 2 In vitro tissue:water partition coefficients of 1,1,2,2-tetrachloroethane, pentachloroethane, hexachloroethane. Data (mean ± SE) from Bertelsen et al. [32]

Other model parameters associated with the processes of absorption, macromolecular binding, and excretion may be determined by conducting time-course analysis in vivo or in vitro. A reasonable strategy for accurate estimation of specific biochemical parameters in vivo is to conduct experiments under conditions where toxicokinetic behavior of a chemical is related to one or two dominant factors and thereby derive estimates of these parameters [5].

2.4 QSAR Methods for Parameter Estimation

The development of QSARs for estimating chemical-specific parameters of PBTK models is an evolving area of research. Thus, when the partition coefficients and metabolism constants for a particular chemical are not available, they may be predicted using QSARs for purposes of PBTK modeling. There is some success with in silico prediction of partition coefficients for PBTK models [24–30]. Fragment constant method has been used to predict partition coefficients for PBPK models of volatile organic chemicals, as follows [26, 38, 39]:

$$\log P_{\mathrm{t}} = \sum_{i=1}^{n} f_i C_i, \tag{13}$$

where f_i is the frequency of occurrence of the molecular fragment i (e.g., CH_3, Br, F, aromatic cycle), C_i is the contribution of the fragment i, and P_{t} is the partition coefficient required for PBTK modeling (e.g., blood:air, tissue:blood).

The QSARs of this type have only been applied to low molecular weight volatile organic chemicals that are lipophilic; these approaches are yet to be developed for other classes of chemicals, particularly for molecules of larger size (e.g., pesticides, dioxins, etc.). The challenge with such substances is related to the need for QSARs for tissue diffusion coefficients, macromolecular binding association constants, absorption rates (i.e., dermal permeability coefficient, oral absorption rate constant), and metabolism rates.

Despite the limited past effort in developing QSARs, the prediction of metabolic constants (V_{max}, K_m) continues to be the bottleneck for implementing QSAR-based PBTK models [40, 41]. At the present time, there does not exist a methodology for predicting a priori the values of the metabolic constants in various species of concern for ecotoxicological risk assessment. In the absence of a validated QSAR for metabolic constants, an useful modeling strategy involves setting the hepatic extraction to 0 or 100% in the PBTK models. Using this approach, one can predict the plausible range of the blood concentration–time profiles prior to in vivo experimentation [5].

2.5 Model Validation

Once the model structure, equations, and parameters are assembled and the code written in the simulation/programming language, comparison of simulations with experimental data is undertaken to validate/refine the model. The primary objective of model validation/evaluation process is then to determine whether all major TK determinants/processes have been adequately identified and characterized. The choice of method(s) for evaluating and validating the model (visual inspection, discrepancy indices, statistical tests including residual analysis) would depend upon the purpose for which the model is used. Even though quantitative tests of goodness of fit are useful, it is equally important to consider the ability of the model to provide an accurate prediction of the general trend of the time-course data (i.e., bumps, valleys).

The lack of precise knowledge about the parameter values may contribute to uncertain predictions of dose metrics whereas the variability of parameter values in a population would lead to difference in toxicokinetics and tissue dose of the chemical being modeled. In this regard, uncertainty analysis (i.e., evaluation of the impact of the lack of precise knowledge about the numerical value of the input parameter or model structure on the model output) as well as variability analysis (i.e., evaluation of the impact of the range of parameter values expected in a population on model output) are relevant. These analyses, often conducted along with a sensitivity analysis, help refine the confidence in PBTK models intended for use in risk assessment [2]. When conducting these analyses, it is important to ensure that the resulting model and parameters are within plausible range or representative of the reality. Particularly [4],

- The numerical values of physiological parameters are within known, plausible limits.
- The sum of tissue volumes is lower than or equal to the body weight.
- The sum of tissue blood flows is equal to cardiac output.
- The mass balance is respected (chemical absorbed = chemical in body + chemical eliminated).

- The covariant nature of the parameters is appropriately respected (e.g., the animal with the lowest breathing rate cannot be the one receiving the highest cardiac output).

3 PBTK Models for the Rat

In this chapter, the rat is used to illustrate the development and implementation of PBTK models for terrestrial species. The initial modeling example relates to the exposure of adult male rats to toluene (a volatile organic chemical) in the inhaled air. The rat PBTK model for toluene consists of four compartments representing the adipose tissue, the PPT, richly perfused tissues (RPT), and the metabolizing tissue (liver) interconnected by systemic circulation as well as a gas-exchange lung compartment. In this model, the PPT compartment comprises muscle and skin whereas the RPT compartment represents brain and the viscera, including kidney, thyroid, bone marrow, heart, testes, and hepatoportal system. For solving the set of equations constituting the inhalation PBTK model for toluene, a modeling software that contains integration algorithms for solving differential equations is required. Alternatively, the various parameters and equations can be entered within an EXCEL® spreadsheet and solutions obtained by integrating according to Euler algorithm [6]. Accordingly, the first step would be to enter the numerical values of the PBPK model parameters and identify them appropriately so that these can be used anywhere in the spreadsheet. For example, the numerical value contained in cell D5 is referred to as Q_c (Table 3). Since the cardiac output is referred to as Q_c in this example, whenever Q_c is typed in any other part of the spreadsheet, the numerical value found in cell D5 will be imported automatically. Table 4 lists all the model equations and the way in which they are entered into the EXCEL® spreadsheets. In effect, two equations per compartment are written to facilitate the tracking of the temporal evolution of (1) the rate of change in concentration (dC/dt) (i.e., a differential equation) as well as (2) the concentration (C) (i.e., integral of the differential equation – based on Euler algorithm).

Accordingly, as shown in Fig. 3, calculations for the four tissue compartments occupy eight columns (columns E–L) and the calculation/representation of the simulation time, exposure concentration, arterial concentration, and venous blood concentration occupies additionally one column each (columns B, C, D, and M, respectively). In this example, each line in the EXCEL® spreadsheet represents the state of the system (in terms of chemical concentrations) at every integration/communication interval (i.e., 0.005 h). Every time the numerical values in cells corresponding to input parameters are changed, the solution to the set of this inhalation PBTK model equations for toluene in the rat is generated anew (Fig. 3).

The earlier PBTK model can be expanded to include the uptake of chemicals by various exposure routes (e.g., oral, inhalation, dermal) (Fig. 4). Appendix 1 presents the model code for constructing such a PBTK model to simulate the toxicokinetics of perchloroethylene in rats exposed by the oral, dermal, and inhalation routes [43]. The simulations obtained with this multiroute PBTK model are depicted in Fig. 5.

Table 3 List of parameters for the four-compartmental rat PBTK model, their numerical values, and location in EXCEL® spreadsheet

Parameters	Abbreviation[a]	Numeric values[b]	Cell location[c]
Cardiac output	Q_c	5.25 L/h	D5
Alveolar ventilation rate	Q_p	5.25 L/h	D6
Fat blood flow	Q_f	0.47 L/h	D7
Hepatic blood flow	Q_l	1.31 L/h	D8
Poorly perfused tissues (PPT)	Q_s	0.79 L/h	D9
Richly perfused tissues (RPT)	Q_r	2.68 L/h	D10
Fat volume	V_f	0.022 L	E7
Liver volume	V_l	0.012 L	E8
PPT volume	V_s	0.174 L	E9
RPT volume	V_r	0.012 L	E10
Fat:blood partition coefficient	P_f	56.72	F7
Liver:blood partition coefficient	P_l	4.64	F8
Poorly perfused tissue:blood partition coefficient	P_s	1.54	F9
Richly perfused tissue:blood partition coefficient	P_r	4.64	F10
Blood:air partition coefficient	P_b	18	G7
Maximal velocity of metabolism	V_{max}	1.66 mg/h	H8
Michaelis–Menten affinity constant	K_m	0.55 mg/L	I8

[a] The various model parameters are referred to, using these abbreviations in the spreadsheet
[b] All parameter values were based on Tardif et al. [42]
[c] The cell locations provided here correspond to the column and row coordinates respectively, that is, the alphabetical letters denote the columns and the Arabic numerals correspond to the rows of the spreadsheet

4 PBTK Models for Cattle

For PBTK modeling of chemical contaminants in cows, the conceptual model structure and mass-balance differential equations are essentially the same as shown for the rat in the preceding section. Even though there are important physiological differences (e.g., digestive system), the level of detail to be included in the PBTK model would depend upon the purpose for which the model is being developed. For simulating the transfer of lipophilic contaminants to milk in lactating cows, a simple model structure shown in Fig. 6 is sufficient.

The rate of change of the amount of chemical in blood ($\mathrm{d}A_b/\mathrm{d}t$) can be calculated as follows [44]:

$$\frac{\mathrm{d}A_b}{\mathrm{d}t} = \sum Q_t \times \left(C_b - \frac{C_t}{P_t}\right) - C_{\mathrm{Lmilk}} \times P_{\mathrm{milk}} \times C_b, \tag{14}$$

where Q_t refers to the tissue blood flow, C_t is the concentration of chemical in the tissue, P_t is the tissue:blood partition coefficient, C_b is the blood concentration, C_{Lmilk} is the milk clearance, and P_{milk} is the milk:blood partition coefficient.

Table 4 Equations used in the calculation of tissue, arterial, and venous blood concentrations of toluene, and their expression in EXCEL® spreadsheet

Compartment	Equations[a]	Expression in EXCEL®[b]
Arterial blood	$C_{a,n} = \frac{Q_c \times C_{v,n-1} + Q_p \times C_{inh,n}}{Q_c + \frac{Q_p}{P_b}}$	D36 = $[(Q_c \times U35) + (Q_p \times C36)/(Q_c + (Q_p/P_b))]$
Liver	$dC_l/dt_n = Q_{l/Vl} \times (C_{a,n} - C_{l,n-1/Pl}) - \frac{V_{max/Vl} \times C_{l,n-1}/P_l}{K_m + C_{l,n-1}/P_l}$	E36 = QL/VL × (D36 − F35/PL) − ((V_{max}/VL × F35/PL)/(K_m + F35/PL))
	$C_l = \text{INTEG}(dC_l/dt)$	F36 = E36 × t + F35
Fat	$dC_f/dt_n = Q_f/V_f \times (C_{a,n} - C_{f,n-1}/P_f)$	G36 = QF/VF × (D36 − H35/PF)
	$C_f = \text{INTEG}(dC_f/dt)$	H36 = G36/Pl
Poorly perfused tissues	$dC_s/dt_n = Q_s/V_s \times (C_{a,n} - C_{s,n-1}/P_s)$	I36 = QS/VS × (D36 − J35/PS)
	$C_s = \text{INTEG}(dC_s/dt)$	J36 = I36 × t + J35
Richly perfused tissues	$dC_r/dt_n = Q_r/V_r \times (C_{a,n} - C_{r,n-1}/P_r)$	K36 = QR/VR × (D36 − L35/PR)
	$C_r = \text{INTEG}(dC_r/dt)$	L36 = K36 × t + L35
Venous blood	$C_{V,n} = \frac{Q_l \times C_{l,n}/P_l + Q_f \times C_{f,n}/P_f + Q_s \times C_{s,n}/P_s + Q_r \times C_{r,n}/P_r}{Q_c}$	M36 = (QL × F36/PL) + (QF × H36/PF) + (QS × J36/PS) + (QR × (L36/PR))/QC

[a] Subscripts n and $n-1$ refer to the current and previous simulation times
[b] Abbreviation and values of model parameters are given in Table 3

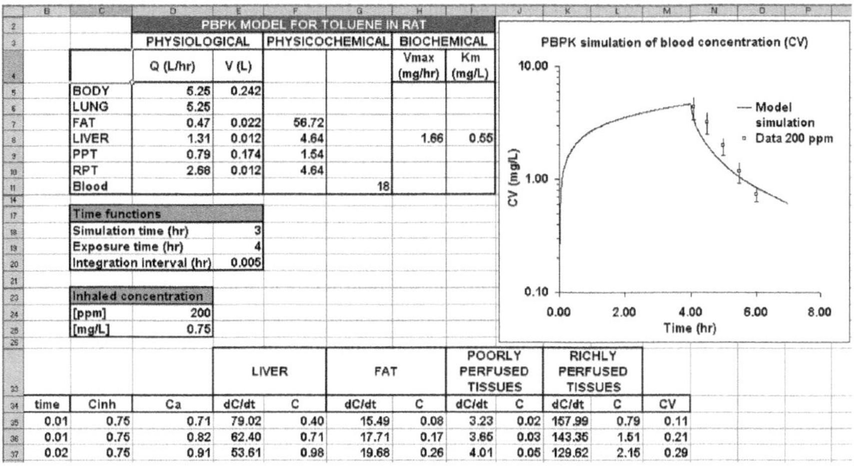

PBPK MODEL FOR TOLUENE IN RAT

	PHYSIOLOGICAL		PHYSICOCHEMICAL		BIOCHEMICAL	
	Q (L/hr)	V (L)			Vmax (mg/hr)	Km (mg/L)
BODY	5.25	0.242				
LUNG	5.25					
FAT	0.47	0.022	56.72			
LIVER	1.31	0.012	4.64		1.66	0.55
PPT	0.79	0.174	1.54			
RPT	2.68	0.012	4.64			
Blood				18		

Time functions	
Simulation time (hr)	3
Exposure time (hr)	4
Integration interval (hr)	0.005

Inhaled concentration	
[ppm]	200
[mg/L]	0.75

			LIVER		FAT		POORLY PERFUSED TISSUES		RICHLY PERFUSED TISSUES		
time	Cinh	Ca	dC/dt	C	dC/dt	C	dC/dt	C	dC/dt	C	CV
0.01	0.75	0.71	79.02	0.40	15.49	0.08	3.23	0.02	157.99	0.79	0.11
0.01	0.75	0.82	62.40	0.71	17.71	0.17	3.65	0.03	143.35	1.51	0.21
0.02	0.75	0.91	53.61	0.98	19.68	0.26	4.01	0.05	129.62	2.15	0.29

Fig. 3 Print out of a computer screen depicting an EXCEL® spreadsheet with (1) the PBTK model simulation of venous blood concentrations of toluene during and following a 4-h exposure of rats to 200 ppm of this chemical, (2) the numerical values of the PBTK model parameters, as well as (3) a portion of the raw numbers corresponding to calculations of tissue and blood concentrations, simulated at time intervals of 0.005 h

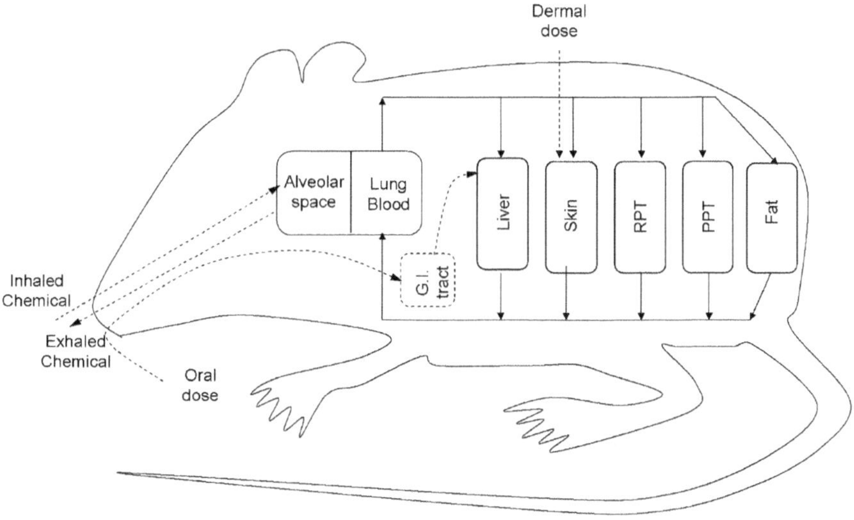

Fig. 4 Conceptual representation of a multiroute PBTK model for perchloroethylene in the rat

The PBTK model has been used to simulate the toxicokinetics and milk transfer of 2,3,7,8-TCDD [44]. Tables 5 and 6 represent parameters useful for PBTK modeling in cows, namely, the physiological parameters and tissue composition data. PBTK models, similar to those for cows, have also been constructed for sheep, goats, and swine [45–49].

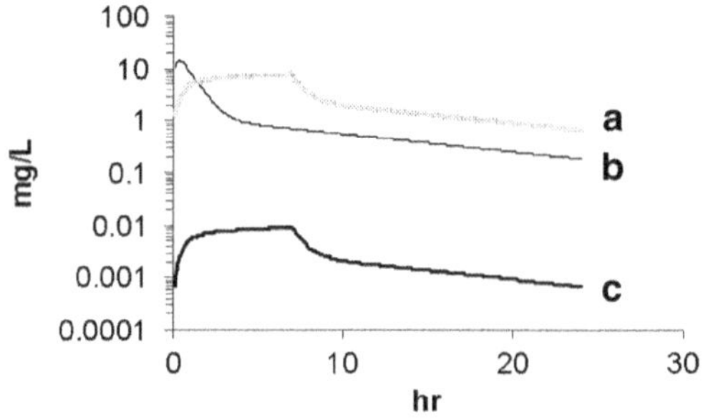

Fig. 5 PBTK model-based route to route extrapolation of the toxicokinetics of perchloroethylene in the rat: (**a**) Inhalation: 100 ppm for 4 h; (**b**) Oral dose: 10 mg/kg; (**c**) Dermal contact: 100 ppm in air for 4 h

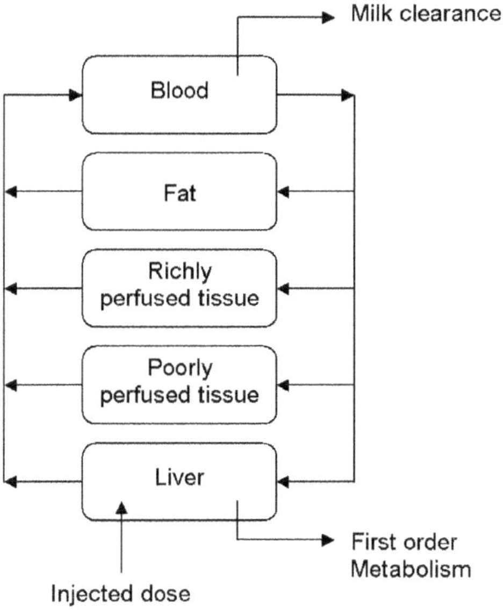

Fig. 6 Conceptual representation of the PBTK model for lipophilic compounds in dairy cattle

Table 5 Physiological parameters for the cattle PBPK model [44]

Parameter	Lactating cow	Cow
Blood volume (L)	42	42
Liver volume (L)	8.5	8.5
Fat volume (L)	61	135
Richly perfused tissue volume (L)	31	31
Poorly perfused tissue volume (L)	310	385
Cardiac output (L/day)	86,500	43,250
Liver blood flow (L/day)	39,600	19,800
Fat blood flow (L/day)	3,300	1,650
Richly perfused tissue blood flow (L/day)	26,300	13,150
Poorly perfused tissue blood flow (L/day)	17,300	8,650

Table 6 Lipid content of richly perfused tissues in dairy cattle [17]

Tissues	Total lipid (fraction of tissue weight)	Neutral lipids (fraction of total lipids)	Phospholipids (fraction of total lipids)
Brain	0.0800	0.4795	0.5205
Gall bladder	0.0150	0.9000	0.1000
Heart	0.0600	0.6250	0.3750
Kidney	0.0670	0.7725	0.2275
Pancreas	0.1610	0.2430	0.7570
Spleen	0.0506	0.2500	0.7500
Stomach	0.0285	0.1874	0.8126
Adrenal glands	0.0472	0.9885	0.0115
Parathyroid and thyroid glands	0.0140	0.6250	0.3750

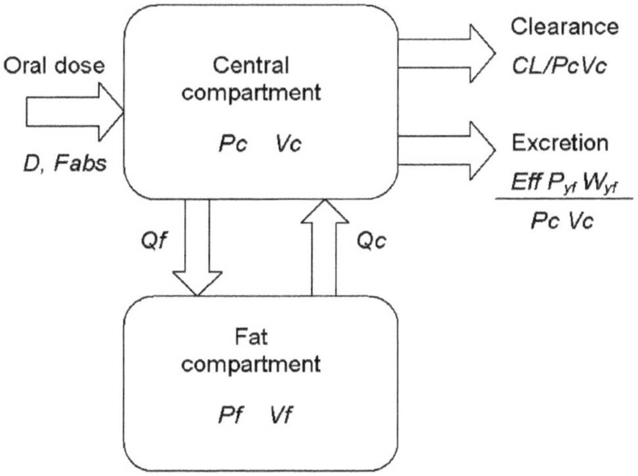

Fig. 7 Conceptual representation of the PBTK model for dioxins in chicken [50]. *c* Central compartment, *f* fat compartment, *yf* yolk fat, *P* partition coefficient, *V* compartment volume, *Q* flow, *D* dose, *Fabs* fraction absorbed, *Eff* laying efficiency

5 PBTK Models for Birds

PBTK modeling of chemicals in birds is limited to one published effort. Van Eijkeren et al. [50] simulated the transfer and toxicokinetics of dioxins and dioxin-like PCBs in chicken following exposure to contamined feed. The model is essentially simple (Fig. 7), consisting of two compartments: one central and one adipose tissue compartment. The adipose tissue compartment comprises the abdominal fat, subcutaneous fat as well as intermuscular fat. All the other tissues of the chicken are lumped within the central compartment. Compartments are characterized by their volume of distribution (characterized by the compartment volume V and partition coefficient P). Compartments are interconnected by flows (Q), as in other PBTK

Table 7 Physiological parameters for the hen [51–67]

Parameters	Value
Cardiac output (QC; mL/min)	234–430
Blood flow to richly perfused tissues (fraction of QC)	0.52
Blood flow to yolk (fraction of QC)	0.05
Blood volume (mL/kg body weight)	90
Volume of fat as fraction of body weight	0.04–0.07
Volume of liver as fraction of body weight	0.02–0.0262
Volume of richly perfused tissues as fraction of body weight	0.098
Volume of poorly perfused tissues as fraction of body weight	0.53–0.66

models. The toxicant elimination from the central compartment is described using a clearance term as well as yolk fat excretion. The yolk fat excretion, in turn, is computed on the basis of the laying efficiency *Eff*, the distribution volume of yolk fat as well as the volume of distribution of the central compartment (Volume V_c and partition P_c). Table 7 lists the parameters useful for the development of PBTK models in hen.

6 PBTK Models for Fish

The process and approach involved in the development of PBTK models for aquatic species are similar to that of terrestrial species discussed earlier. The PBTK models are instrumental in simulating the kinetics of ADME in fish exposed to chemicals in water or diet. Although the conceptual model and the mathematical descriptions of tissue distribution and metabolism are identical to that of the rats (Fig. 8), the mechanistic determinants and equations determining absorption and elimination in the fish differ from that of the rat. Principally, the calculation of the arterial blood concentration resulting from branchial flux differs from the manner in which inhalation exposure is modeled in the rat. This is due to the fact that, in the rat, the uptake and equilibrium during pulmonary exposures are driven by the ratio of chemical concentration in arterial blood and ambient air [30, 68, 69]. However, in the fish, due to countercurrent flow, the arterial blood equilibrates with inspired water (blood flow-limited exchange) whereas the venous blood equilibrates with expired water (ventilation-limited exchange). Chemical flux at fish gills can be calculated as function of the exchange coefficient and difference in chemical activities in venous blood and inspired water as follows [20]:

$$F_g = k_x \times \left(C_{insp} - C_{ven}/P_{bw}\right). \tag{15}$$

The exchange coefficient k_x, in turn, is equal to Q_g (respiratory volume) when the equilibration of inspired water with arterial blood limits chemical exchange,

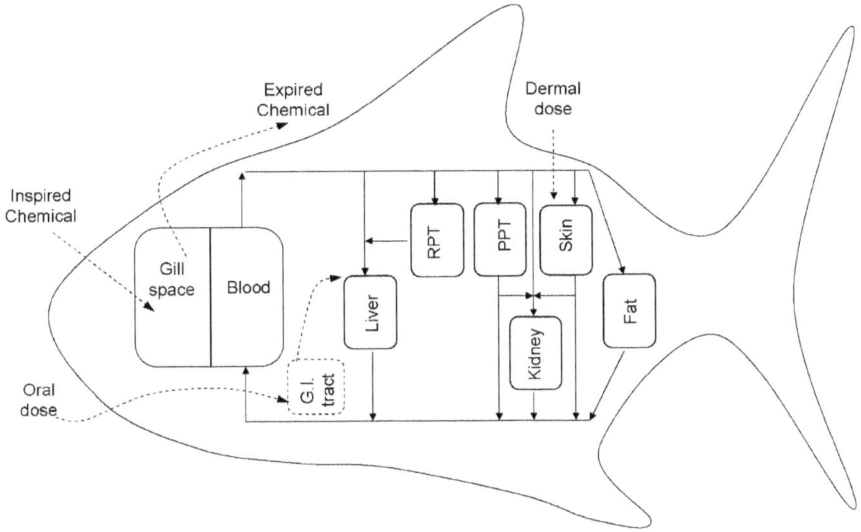

Fig. 8 Conceptual representation of a PBTK model for fish. *PPT* poorly perfused tissues, *RPT* richly perfused tissues, *GI* gastrointestinal. Based on Nichols et al. [20]

or to $Q_c P_{bw}$ (cardiac output times blood:water partition coefficient) when the equilibration of venous blood with expired water limits chemical exchange. The aforementioned general equation can then be rewritten as follows [20, 68, 69]:

$$F_g = \min\left(Q_g, Q_c \times P_{bw}\right) \times \left(C_{insp} - C_{ven}/P_{bw}\right). \tag{16}$$

Figure 9 illustrates the set of equations that need to be solved for predicting the toxicokinetics of inspired chemicals in the fish. Additional equations for dermal and dietary uptake, as required, can be included for simulating other exposure scenarios [20, 21]. The template presented in Fig. 10 is essentially the same as the one for the rat. Comparing with the rat PBTK model (Table 4), it can be seen that (1) the alveolar ventilation rate (Q_p) in the rat was replaced by the effective ventilation rate (Q_g) in the fish, and (2) the numerical values of all other parameter differ between rats and fish. Further, in contrast to the rat model, the arterial blood concentration (C_a) in the fish model was calculated as follows:

$$C_a = C_v + \frac{\mathrm{GUL} \times \left(C_{inh} - \frac{C_v}{P_b}\right)}{Q_c}. \tag{17}$$

Therefore, the equation in cell D36 of the spreadsheet for the fish becomes as follows:

$$\mathrm{D36} = \mathrm{U35} + \mathrm{GUL} \times \frac{(\mathrm{C36} - \mathrm{U35}/P_b)}{Q_c}, \tag{18}$$

where GUL is the gill uptake limitation [$= \min(Q_g,\ Q_c \times P_{bw})$] [68].

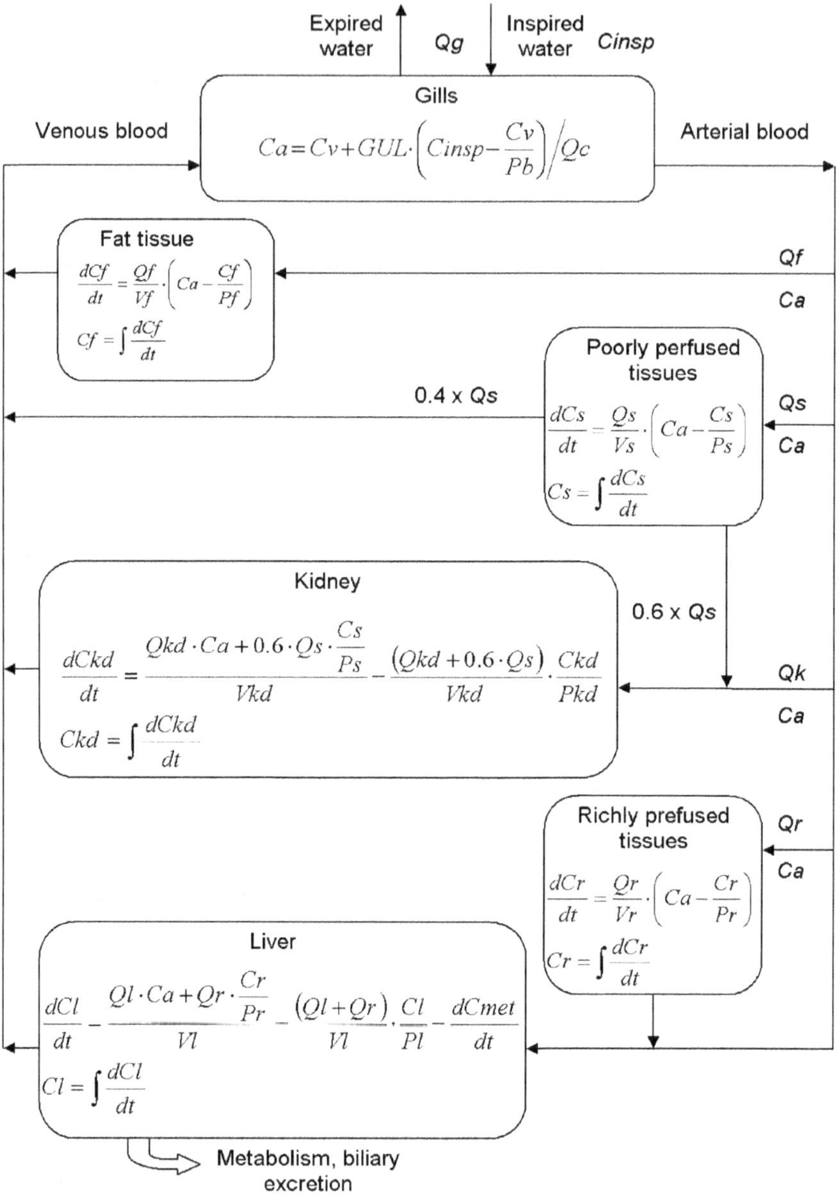

Fig. 9 Mathematical representation of a fish PBTK model. Based on Nichols et al. [68].

In Figure 10, the parameters as well as simulations of the PBTK model for 1,1,2,2-tetrachloroethane in rainbow trout are presented (exposure concentration = 1.06 mg/L, exposure duration = 4 h, cardiac output = 2.07 L/h; effective ventilation rate = 7.20 L/h = 0.18 L/h; hepatic blood flow = 0.06 L/h; blood flow to PPT = 1.24 L/h; blood flow to richly perfused tissues (RPT) = 0.48 L/h;

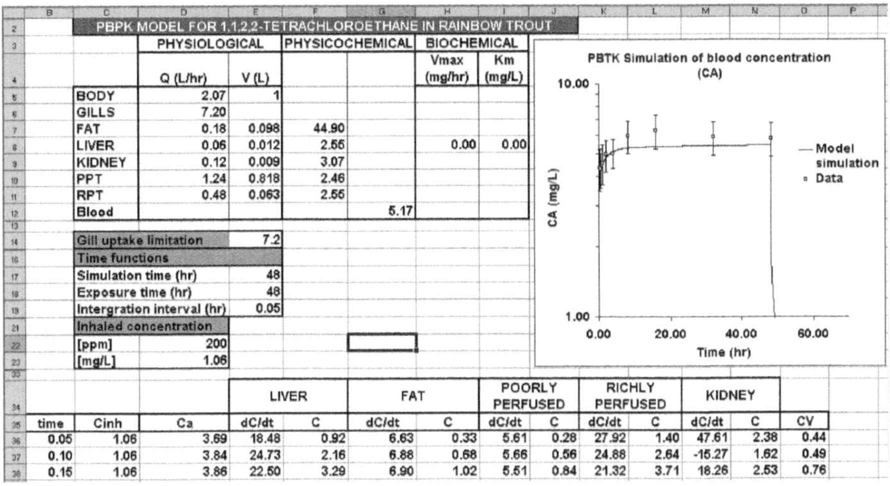

Fig. 10 Printout of a computer screen depicting an EXCEL® spreadsheet with (1) the PBTK model simulation of arterial blood concentration (CA) in rainbow trout during and following 48-h exposure to 1,1,2,2-tetrachoroethane in water (1.062 mg/L), (2) the numerical values of the PBTK model parameters, as well as (3) a portion of the raw numbers corresponding to calculations of tissue and blood concentrations, simulated at time intervals of 0.05 h

fat volume = 0.098 L; liver volume = 0.012 L; PPT volume = 0.818 L; RPT volume = 0.071 L; fat:blood partition coefficient = 44.9; liver:blood partition coefficient = 2.55; PPT:blood partition coefficient = 2.46; RPT:blood partition coefficient = 2.55; blood:water partition coefficient = 5.17; and the metabolism rate = 0) [30]. The code for this fish PBTK model, in textual form, is presented in Appendix 2.

Table 8 lists PBTK models developed so far in various fish species whereas Tables 2 and 9 provide summary of parameters useful for constructing PBTK models in channel catfish, rainbow trout as well as fathead minnows.

7 Applications in Ecotoxicology and Risk Assessment

Simulation models are effective tools for designing toxicology experiments and for conducting extrapolations essential for risk assessments. A principal advantage of such tools as the PBTK models is that they allow the evaluation of the various plausible hypotheses by computer simulation. One can ask questions of the "if... then..." nature. For example, for a given set of physiological and biochemical parameters, what is the impact of change in chemical structure (or lipophilicity) on tissue dose and toxicokinetics? Conversely, for a given chemical, what would be the impact of changes in physiology (i.e., reflective of species differences) on the target tissue dose? The model can then be used to generate quantitative predictions of the expected outcome and are tested experimentally. The PBTK models can be

Table 8 PBPK models developed for fish

Species	Chemical	Reference
Dogfish shark (*Squalus acanthias*)	Phenol red and its glucuronide	[70]
Sting rays (*Dasyatidae sabina and Dasyatidae sayi*)	Methotrexate	[71]
Rainbow trout (*Oncorhynchus mykiss*)	Pentachloroethane	[68]
	Pyrene	[72]
	1,1,2,2-Tetrachloroethane, pentachloroethane, hexachloroethane	[20, 30]
	2, 2′, 5, 5′-Tetrachlorobiphenyl	[73]
	Paraoxon	[74]
Brook trout (*Salvenus fontinalis*)	2,3,7,8-Tetrachlorodibenzo-*p*-dioxin	[75]
Lake trout (*Salvelinus namaycush*)	1,1,2,2-Tetrachloroethane, pentachloroethane, hexachloroethane	[76]
Channel catfish (*Ictalurus punctatus*)	1,1,2,2-Tetrachloroethane, pentachloroethane, hexachloroethane	[20, 77]
Fathead minnows (*Pimephales promelas*)	1,1,2,2-Tetrachloroethane, pentachloroethane, hexachloroethane	[21]
	2, 2′, 5, 5′-Tetrachlorobiphenyl	[78]
Tilapia (*Oreochromis mossambicus*)	Arsenic	[79, 80]

Table 9 Lipid content of catfish, fathead minnow, and trout tissues [32]

		Fraction of tissue weight[a]		
Tissue	Species	Nonpolar lipid	Total lipid	Water
Blood	Catfish	0.006 ± 0.001	0.013 ± 0.001	0.839 ± 0.004
	Fathead minnow	0.009 ± 0.001	0.019 ± 0.001	0.876 ± 0.014
	Trout	0.007 ± 0.001	0.014 ± 0.001	0.839 ± 0.004
Fat	Catfish	0.886 ± 0.016	0.899 ± 0.016	0.050 ± 0.015
	Fathead minnow	1.001 ± 0.030	1.010 ± 0.033	0.016 ± 0.009
	Trout	0.934 ± 0.005	0.942 ± 0.006	0.050 ± 0.015
Liver	Catfish	0.016 ± 0.002	0.039 ± 0.003	0.735 ± 0.005
	Trout	0.018 ± 0.002	0.045 ± 0.004	0.746 ± 0.003
Muscle	Catfish	0.002 ± 0.000	0.009 ± 0.001	0.791 ± 0.003
	Fathead minnow	0.017 ± 0.001	0.025 ± 0.001	0.806 ± 0.010
	Trout	0.022 ± 0.001	0.030 ± 0.003	0.769 ± 0.004

[a] Mean ± standard error

viewed as dynamic constructs that can be continually updated as significant new information and data become available.

The principal application of PBTK models in ecotoxicology and risk assessment is to predict the toxicokinetics, bioconcentration, bioaccumulation, and target tissue dose of the parent chemical or its reactive metabolite. Using the tissue dose of the putative toxic moiety of a chemical in risk assessment calculations provides a better basis of relating to the observed toxic effects than the external or exposure concentrations of the parent chemical [81–83]. Because PBTK models facilitate the prediction of target tissue dose for various exposure scenarios, routes, doses, and species [81, 83], they are instrumental in addressing the uncertainty associated with the conventional extrapolation approaches and uncertainty factors employed in risk assessments [1, 2, 81, 84].

In conclusion, the PBTK models represent a systematic approach for identification, characterization, and integration of the mechanistic determinants of uptake, metabolism, distribution, and excretion of chemicals in biota. The development of such mechanistic toxicokinetic models should contribute to refinement/reduction of animal use in toxicology and ecotoxicology studies, as well as the enhancement of the scientific basis of risk assessments.

References

1. Andersen ME, Clewell HJ III, Gargas ML (1991) Physiologically-based pharmacokinetic modeling with dichloromethane, its metabolite carbon monoxide and blood carboxyhemoglobin in rats and humans. Toxicol Appl Pharmacol 108: 14–27
2. United States Environmental Protection Agency (U.S. E.P.A.) (2006) Approaches for the application of physiologically based pharmacokinetic (PBPK) models and supporting data in risk assessment (Final Report). U.S. Environmental Protection Agency, Washington, DC
3. Gibaldi M, Perrier D (1982) Pharmacokinetics. Marcel Dekker, New York
4. Chiu WA, Barton HA, DeWoskin RS, Schlosser P, Thompson CM, Sonawane B, Lipscomb JC, Krishnan K (2007) Evaluation of physiologically based pharmacokinetic models for use in risk assessment. J Appl Toxicol 27: 218–237
5. Krishnan K, Andersen ME (2007) Physiologically based pharmacokinetic modeling in toxicology. In: Hayes AW (ed) Principles and methods of toxicology, 5th edn. Taylor & Francis, Boca Raton, FL
6. Haddad S, Pelekis M, Krishnan K (1996) A methodology for solving physiologically based pharmacokinetic models without the use of simulation softwares. Toxicol Lett 85: 113–126
7. Krishnan K, Andersen ME (2001) Physiologically based pharmacokinetic modeling in toxicology. In: Hayes AW (ed) Principles and methods of toxicology, 4th edn. Taylor & Francis, Philadelphia, PA
8. Krishnan K, Gargas ML, Fennell TR, Andersen ME (1992) A physiologically-based description of ethylene oxide dosimetry in the rat. Toxicol Ind Health 8: 121–140
9. Weber LJ (1982) Aquatic toxicology. Raven Press, New York
10. Klaassen CD (1999) Casarett and Doull's toxicology: The basic science of poisons: Companion handbook, 5th edn. McGraw-Hill, Health Professions Division, New York
11. Ramsey JC, Andersen ME (1984) A physiologically based description of the inhalation pharmacokinetics of styrene in rats and humans. Toxicol Appl Pharmacol 73: 159–175
12. McDougal JN, Jepson GW, Clewell HJ, MacNaughton MG, Andersen ME (1986) A physiological pharmacokinetic model for dermal absorption of vapors in the rat. Toxicol Appl Pharmacol 85: 286–294
13. Gerlowski LE, Jain RK (1983) Physiologically based pharmacokinetic modeling: Principles and applications. J Pharm Sci 72: 1103–1127
14. D'Souza RW, Francis WR, Andersen MW (1988) Physiological model for tissue glutathione depletion and decreased resynthesis after ethylene dichloride exposures. J Pharmacol Exp Ther 245: 563–568

15. Nichols JW, Schultz IR, Burkhard LP (2006) In vitro–in vivo extrapolation of quantitative hepatic biotransformation data for fish I. A review of methods, and strategies for incorporating intrinsic clearance estimates into chemical kinetic models. Aquat Toxicol 78: 74–90
16. Dewoskin RS, Thompson CM (2008) Renal clearance parameters for PBPK model analysis of early lifestage differences in the disposition of environmental toxicants. Regul Toxicol Pharmacol 51: 66–86
17. Mitruka BM, Rawnley HM (1977) Clinical biochemical and haematological reference values in normal and experimental animals. Masson Publishing, New York
18. Arms AD, Travis CC (1988) Reference physiological parameters in pharmacokinetic modeling. Office of Health and Environmental Assessment, US EPA, Washington, DC. NTIS PB 88-196019
19. Brown RP, Delp MD, Lindstedt SL, Rhomberg LR, Belisle RP (1997) Physiological parameter values for physiologically based pharmacokinetic models. Toxicol Ind Health 13: 407–484
20. Nichols JW, McKim JM, Lien GJ, Hoffman AD, Bertelsen SL, Elonen CM (1996) A physiologically based toxicokinetic model for dermal absorption of organic chemicals by fish. Fundam Appl Toxicol 31: 229–242
21. Lien GJ, Nichols JW, McKim JM, Gallinat CA (1994) Modeling the accumulation of three waterborne chlorinated ethanes in fathead minnows (*Pimephales promelas*): A physiologically based approach. Environ Toxicol Chem 13: 1195–1205
22. Gallo JM, Lam FC, Perrier DG (1987) Area method for the estimation of partition coefficients for physiological pharmacokinetic models. J Pharmacokinet Biopharm 15: 271–280
23. Lin JH, Sugiyama Y, Awazu S, Hanano M (1982) In vitro and in vivo evaluation of the tissue to blood partition coefficients for physiological pharmacokinetic models. J Pharmacokinet Biopharm 10: 637–647
24. Gargas ML, Burgess RJ, Voisard DE, Cason GH, Andersen ME (1989) Partition coefficients of low molecular weight volatile chemicals in various liquids and tissues. Toxicol Appl Pharmacol 98: 87–99
25. Sato A, Nakajima T (1979) Partition coefficients of some aromatic hydrocarbons and ketones in water, blood and oil. Br J Ind Med 36: 231–234
26. Béliveau M, Tardif R, Krishnan K (2003) Quantitative structure–property relationships for physiologically based pharmacokinetic modeling of volatile organic chemicals in rats. Toxicol Appl Pharmacol 189: 221–232
27. Payne MP, Kenny LC (2002) Comparison of models for the estimation of biological partition coefficients. J Toxicol Environ Health A 65: 897–931
28. Poulin P, Krishnan K (1996) A mechanistic algorithm for predicting blood:air partition coefficients of organic chemicals with the consideration of reversible binding in hemoglobin. Toxicol Appl Pharmacol 136: 131–137
29. Poulin P, Krishnan K (1996) A tissue composition-based algorithm for predicting tissue:air partition coefficients of organic chemicals. Toxicol Appl Pharmacol 136: 126–130
30. Nichols JW, McKim JM, Lien GJ, Hoffman AD, Bertelsen SL (1991) Physiologically based toxicokinetic modeling of three waterborne chloroethanes in rainbow trout (*Oncorhynchus mykiss*). Toxicol Appl Pharmacol 110: 374–389
31. DeJongh J, Verhaar HJM, Hermens JLM (1997) A quantitative property–property relationship (QPPR) approach to estimate in vitro tissue–blood partition coefficients of organic chemicals in rats and humans. Arch Toxicol 72: 17–25
32. Bertelsen S, Hoffman AD, Gallinat CA, Elonen CM, Nichols JW (1998) Evaluation of log Pow and tissue lipid content as predictors of chemical partitioning to fish tissues. Environ Toxicol Chem 17: 1447–1455
33. Nichols JW, Schultz IR, Burkhard LP (2006) In vitro–in vivo extrapolation of quantitative hepatic biotransformation data for fish. I. A review of methods, and strategies for incorporating intrinsic clearance estimates into chemical kinetic models. Aquat Toxicol 78: 74–90
34. Nichols JW, Fitzsimmons PN, Burkhard LP (2007) In vitro–in vivo extrapolation of quantitative hepatic biotransformation data for fish. II. Modeled effects on chemical bioaccumulation. Environ Toxicol Chem 26: 1304–1319

35. Barter ZE, Bayliss MK, Beaune PH, Boobis AR, Carlile DJ, Edwards RJ, Houston JB, Lake BG, Lipscomb JC, Pelkonen OR, Tucker GT, Rostami-Hodjegan A (2007) Scaling factors for the extrapolation of in vivo metabolic drug clearance from in vitro data: Reaching a consensus on values of human microsomal protein and hepatocellularity per gram of liver. Curr Drug Metab 8: 33–45
36. Lipscomb JC, Poet TS (2008) In vitro measurements of metabolism for application in pharmacokinetic modeling. Pharmacol Ther 118: 82–103
37. Krishnan K, Gargas ML, Andersen ME (1993) In vitro toxicology and risk assessment. Altern Meth Toxicol 9: 185–203
38. Béliveau M, Krishnan K (2005) A spreadsheet program for modeling quantitative structure–pharmacokinetic relationships for inhaled volatile organics in humans. SAR QSAR Environ Res 16: 63–77
39. Basak SC, Mills D, Gute BD (2006) Prediction of tissue: Air partition coefficients – Theoretical vs. experimental methods. SAR QSAR Environ Res 17: 515–32
40. Gargas ML, Seybold PG, Andersen ME (1988) Modeling the tissue solubilities and metabolic rate constant (Vmax) of halogenated methanes, ethanes, and ethylenes. Toxicol Lett 43: 235–256
41. Waller CL, Evans MV, McKinney JD (1996) Modeling the cytochrome P450-mediated metabolism of chlorinated volatile organic compounds. Drug Metab Dispos 24: 203–210
42. Tardif R, Charest-Tardif G, Brodeur J, Krishnan K (1997) Physiologically based pharmacokinetic modeling of a ternary mixture of alkyl benzenes in rats and humans. Toxicol Appl Pharmacol 144: 120–134
43. Poet TS, Weitz KK, Gies RA, Edwards JA, Thrall KD, Corley RA, Tanojo H, Hui X, Maibac HI, Wester RC (2002) PBPK modeling of the percutaneous absorption of perchloroethylene from a soil matrix in rats and humans. Toxicol Sci 67: 17–31
44. RijksInstituut voor Volksgezondheid en Milieu/National institute of public health and the environment (RIVM) (1999) Model for estimating initial burden and daily absorption of lipophylic contaminants in cattle, Report no. 643810 005. RIVM, The Netherlands
45. Craigmill AL (2003) A physiologically based pharmacokinetic model for oxytetracycline residues in sheep. J Vet Pharmacol Ther 26: 55–63
46. Buur JL, Baynes RE, Craigmill AL, Riviere JE (2005) Development of a physiologic-based pharmacokinetic model for estimating sulfamethazine concentrations in swine and application to prediction of violative residue in edible tissues. Am J Vet Res 66: 1686–1693
47. Buur JL, Baynes RE, Smith G, Riviere JE (2006) Use of probabilistic modeling within a physiologically based pharmacokinetic model to predict sulfamethazine residue withdrawal times in edible tissues in swine. Antimicrob Agents Chemother 50: 2344–2351
48. Villesen HH, Foster DJR, Upton RN, Somogyl AA, Martinez A, Grant C (2006) Cerebral kinetics of oxycodone in conscious sheep. J Pharm Sci 95: 1666–1676
49. Jensen ML, Foster D, Upton R, Grant C, Martinez A, Somogyi A (2007) Comparison of cerebral pharmacokinetics of buprenorphine and norbuprenorphine in an in vivo sheep model. Xenobiotica 37: 441–457
50. Van Eijkeren JC, Zeilmaker MJ, Kan CA, Traag WA, Hoogenboom LA (2006) A toxicokinetic model for the carry-over of dioxins and PCBs from feed and soil to eggs. Food Addit Contam 23: 509–517
51. Altman PL, Dittmer DS (1971) Respiration and circulation. Fes Amer Soc Exptl Biol, Bethesda, MD
52. Farrell AP (1991) Circulation of body fluids. In: Posser CL (ed) Environmental and metabolic animal physiology, Vol. 4, 4th edn. Wilex-Liss, New York
53. Piiper J, Drees F, Scheid P (1970) Gas exchange in the domestic fowl during spontaneous breathing and artificial ventilation. Resp Physiol 9: 234–245
54. Boelkins JN, Mueller WJ, Hall KL (1973) Cardiac output distribution in the laying hen during shell formation. Comp Biochem Physiol A 46: 735–743
55. Moynihan JB, Edwards NA (1975) Blood flow in the reproductive tract of the domestic hen. Comp Biochem Physiol A 51: 745–748

56. Sapirstein LA, Hartman FA (1959) Cardiac output and its distribution in the chicken. Am J Physiol 196: 751–752
57. Freeman BM (1971) The corpuscles and physical characteristics of blood. In: Bell DJ, Freeman BM (eds) Physiology and biochemistry of the domestic fowl, Vol. 2. Academic press, London
58. Jean-Blain M, Alquier J (1948) Les aliments d'origine animale destinés à l'homme. Vigot, Paris
59. Clackson MJ, Richards TG (1971) The liver with special reference to bile formation. In: Bell DJ, Freeman BM (eds) Physiology and biochemistry of the domestic fowl, Vol 2. Academic press, London
60. Wolfenson D, Berman A, Frei YF, Snapir N (1978) Measurement of blood flow distribution by radioactive microspheres in the laying hen (Gallus domesticus). Comp Biochem Physiol A 61: 549–555
61. Wolfenson D, Frei N, Shapir N, Berman A (1981) Heat stress effect on capillary blood flow and its redistribution in the laying hen. Pflügers Arch 390: 86–93
62. Gilbert AB (1971) The female reproductive effort. In: Bell DJ, Freeman BM (eds) Physiology and biochemistry of the domestic fowl, Vol 3. Academic Press, London
63. Wight PAL (1971) The pineal gland. In: Bell DJ, Freeman BM (eds) Physiology and biochemistry of the domestic fowl, Vol 1. Academic Press, London
64. Payne LN (1971) The lymphoid system. In: Bell DJ, Freeman BM (eds) Physiology and biochemistry of the domestic fowl, Vol 2. Academic Press, London
65. Skadhauge E (1983) Excretion. In: Freeman BM (ed) Physiology and biochemistry of the domestic fowl, Vol. 4. Academic Press, London
66. Grubb BR (1983) Allometric relations of cardiovascular function in birds. Am J Physiol 245: H592–H597
67. Hamilton PB, Garlich JD (1971) Aflatoxin as a possible cause of fatty liver syndrome in laying hens. Poultry Sci 50: 800–804
68. Nichols JW, McKim JM, Andersen ME, Gargas ML, Clewell HJ III, Erickson RJ (1990) A physiologically based toxicokinetic model for the uptake and disposition of waterborne organic chemicals in fish. Toxicol Appl Pharmacol 106: 433–447
69. Erickson RJ, McKim JM (1990) A model for exchange of organic chemicals at fish gills: Flow and diffusion limitations. Aquat Toxicol 18: 175–198
70. Bungay PM, Debrick RL, Guarino AM (1976) Pharmacokinetic modeling of the digfish shark (*Squalus acanthias*): Distribution and urinary and biliary excretion of phenol red and its glucuronide. J Pharmacokinet Biopharm 4: 377–388
71. Zaharko DS, Debrick RL, Oliviero VT (1972) Prediction of the distribution of methotrexate in the sting rays *Dasyatidae sabina* and sayi by use of a model developed in mice. Comp Biochem Physiol A 42: 183–194
72. Law FCP, Abedini S, Kennedy CJ (1991) A biologically based toxicokinetic model for pyrene in Rainbow trout. Toxicol Appl Pharmacol 110: 390–402
73. Nichols JW, Fitzsimmons PN, Whiteman FW, Dawson TD, Babeu L, Juenemann J (2004) A Physiologically based toxicokinetic model for dietary uptake of hydrophobic organic compounds by fish. I. Feeding studies with 2,2′,5,5′-tetrachlorobiphenyl. Toxicol Sci 77: 206–218
74. Abbas R, Hayton WL (1997) A physiologically based pharmacokinetic and pharmacodynamic model for paraoxon in rainbow trout. Toxicol Appl Pharmacol 145: 192–201
75. Nichols JW, Jensen KM, Tietge JE, Johnson RD (1998) Physiologically based toxicokinetic model for maternal transfer of 2,3,7,8-tetrachlorodibenzo-*p*-dioxin in brook trout (*Salvelinus fontinalis*). Environ Toxicol Chem 17: 2422–2434
76. Lien GJ, McKim JM, Hoffman AD, Jenson CT (2001) A physiologically based toxicokinetic model for lake trout (*Salvelinus namaycush*). Aquat Toxicol 51: 335–350
77. Nichols JW, McKim JM, Lien GJ, Hoffman AD, Bertelsen SL, Gallinat CA (1993) Physiologically-based toxicokinetic modeling of three waterborne chloroethanes in channel catfish, *Ictalurus punctatus*. Aquat Toxicol 27: 83–112
78. Lien GJ, McKim JM (1993) Predicting branchial and cutaneous uptake of 2,2′,5,5′-tetrachlorobiphenyl in fathead minnows (*Pimephales promelas*) and Japanese medaka (*Oryzias latipes*): Rate limiting factors. Aquat Toxicol 27: 15–32

79. Liao CM, Liang HM, Chen BC, Singh S, Tsai JW, Chou YH, Lin WT (2005) Dynamical coupling of PBPK/PD and AUC-based toxicity models for arsenic in tilapia *Oreochromis mossambicus* from blackfoot disease area in Taiwan. Environ Pollut 135: 221–233
80. Ling MP, Liao CM, Tsai JW, Chen BC (2005) A PBTK/TD modeling-based approach can assess arsenic bioaccumulation in farmed Tilapia (*Oreochromis mossambicus*) and human health risks. Integr Environ Assess Manag 1: 40–54
81. Andersen ME, MacNaughton MG, Clewell HJ III, Paustenbach DJ (1987) Adjusting exposure limits for long and short exposure period using a physiological pharmacokinetic model. Am Ind Hyg Ass J 48: 335–343
82. Benignus VA, Boyes WK, Bushnell PJ (1998) A dosimetric analysis of behavioral effects of acute toluene exposure in rats and humans. Toxicol Sci 43: 186–195
83. Clewell HJ III, Andersen ME, Barton HA (2002) A consistent approach for the application of pharmacokinetic modeling in cancer and noncancer risk assessment. Environ Health Perspect 110: 85–93
84. Lipscomb JC, Ohanian EV (2007) Toxicokinetics and risk assessment. Informa Healthcare, New York

Appendix 1: Code for the PBTK Model of Perchloroethylene in the Rat

```
! PARAMETERS
! Constants
CONSTANT BW = 0.25 ! Body weight (kg)
CONSTANT KQC = 15 ! Cardiac output (L/hr/kg)
CONSTANT KQP = 15 ! Alveolar ventilation (L/hr/kg)
CONSTANT SURF = 70 ! Skin surface area (cm²)
! Volumes (fraction of body weight)
CONSTANT KVF = 0.09 ! Fat volume
CONSTANT KVL = 0.049 ! Liver volume
CONSTANT KVS = 0.72 ! Volume of poorly perfused tissues
CONSTANT KVR = 0.05 ! Volume of richly perfused tissues
CONSTANT KVSK = 0.10 ! Skin volume
! Tissues blood flows (Fraction of QC)
CONSTANT KQF = 0.05 ! Fat blood flow(l/hr/QC)
CONSTANT KQL = 0.26 ! Liver blood flow (l/hr/QC)
CONSTANT KQR = 0.44 ! Richly perfused tissue blood flow (l/hr/QC)
CONSTANT KQS = 0.20 ! Poorly perfused tissues blood flow (l/hr/QC)
CONSTANT KQSK = 0.05 ! Skin blood flow (l/hr/QC)
! Partition coefficients
CONSTANT PB = 18.85 ! Blood:air
CONSTANT KPL = 70.3 ! Liver:air
CONSTANT KPF = 1638.0 ! Fat:air
CONSTANT KPR = 70.3 ! Richly perfused tissues:air
CONSTANT KPS = 20.0 ! Poorly perfused tissues:air
CONSTANT KPSK = 41.50 ! Skin:air
CONSTANT KPW = 0.79 ! Water:air
```

```
! Biochemical and chemical parameters
CONSTANT MW = 165.834 ! Molecular weight (g/mol)
CONSTANT KVMAX = 0.9 ! Maximum velocity (mg/hr/kg)
CONSTANT KM = 5.62 ! Michaelis constant (mg/L)
CONSTANT Kp = 0.09 ! Skin permeability constant (cm/hr)
CONSTANT Ka = 3.00 ! Oral absorption constant (hr^-1)
CONSTANT F = 1.00 ! Bioavailability
! Simulation parameters
CONSTANT TSTOP = 24. ! Simulation length (hr)
CONSTANT DUREE = 7. ! Exposition length (hr)
CONSTANT CINT = 0.1 ! Communication interval
! Exposure concentrations
CONSTANT CONC = 0.0 ! Atmospheric Concentration (ppm)
CONSTANT DOSE = 0.0 ! Oral dose (mg/kg)
! Scaled parameters
! Volumes
VR = KVR*BW ! Volume of richly perfused tissues (L)
VS = KVS*BW ! Volume of poorly perfused tissues (L)
VF = KVF*BW ! Volume of fat (L)
VL = KVL*BW ! Volume of liver (L)
VSK = KVSK*BW ! Volume of skin (L)
! Flows
QC = KQC*BW**0.7 ! Cardiac output (L/hr)
QP = KQP*BW**0.7 ! Alveolar ventilation (L/hr)
QF = KQF*QC ! Blood flow to fat (L/hr)
QL = KQL*QC ! Blood flow to liver (L/hr)
QR = KQR*QC ! Blood flow to richly perfused tissues (L/hr)
QS = KQS*QC ! Blood flow to poorly perfused tissues (L/hr)
QSK = KQSK*QC ! Blood flow to skin (L/hr)
! Metabolic constant mg/hr
VMAX = KVMAX*BW**0.74 ! Maximum velocity (mg/hr)
! Partition Coefficients
PL = KPL/PB ! Liver:blood
PF = KPF/PB ! Fat:blood
PR = KPR/PB ! Richly perfused tissues:blood
PS = KPS/PB ! Poorly perfused tissues:blood
PSK = KPSK/PB ! Skin:blood
PSKm = KPSK/1 ! Skin:air
! MODEL EQUATIONS
! Calculation of blood concentration
CA = (QC*CV+QP*CI)/(QC+(QP/PB)) ! Arterial blood concentration (mg/L)
CV = (QF*CF/PF+QL*CL/PL+QR*CR/PR+QS*CS/PS+QSK*CSK/PSK)/
QC ! Venous blood concentration (mg/l)
```

```
! Perchloroehtylene in stomach
RAG = Ka*AG ! Rate of absorption in stomach
AG = DOSE*F - INTEG(RAG, 0.0) ! Quantity in stomach
! Liver compartment
RCL = (QL*(CA-CL/PL)-RAM + RAG)/VL ! Rate of change in concentration
(mg/L/hr)
CL = INTEG(RCL,0) ! Concentration in liver (mg/L)
! Perchloroethylene metabolism
RAM = VMAX*CL/PL/(KM + CL/PL)
AM = (RAM,0.)
! Fat compartment
RCF = QF/VF*(CA-CF/PF) ! Rate of change in concentration (mg/L/hr)
CF = INTEG(RCF,0) ! Concentration in Fat (mg/L)
! Richly perfused compartment
RCR = QR/VR*(CA-CR/PR) ! Rate of change in concentration (mg/L/hr)
CR = INTEG(RCR,0) ! Concentration in Richly perfused tissues (mg/L)
! Poorly perfused tissues
RCS = QS/VS*(CA-CS/PS) ! Rate of change in concentration (mg/L/hr)
CS = INTEG(RCS,0) ! Concentration in Poorly perfused tissues (mg/L)
! Skin compartment
RCSK = QSK/VSK*(CA-CSK/PSK) + Kp*SURF/1000*(Cliq-CSK/PSKm)
! Rate of change in concentration (mg/L/hr)
CSK = INTEG(RCSK,0) ! Concentration in skin (mg/L)
```

Appendix 2: Code for the PBTK Model of 1,1,2,2-Tetrachloroethane in the Rainbow Trout

```
! PARAMETERS
! Constants
! Physiological parameters
CONSTANT BW = 1. ! Body weight (L)
CONSTANT QG = 7.2 ! Effective respiratory volume (L/hr)
CONSTANT QC = 2.07 ! Cardiac output (L/hr)
! Volumes (Fraction of body weight)
CONSTANT VFC = 0.098 ! Fat volume
CONSTANT VLC = 0.012 ! Liver volume
CONSTANT VSC = 0.818 ! Volume of poorly perfused tissues
CONSTANT VRC = 0.063 ! Volume of richly perfused tissues
CONSTANT VKC = 0.009 ! Volume of kidney
! Tissues blood flows (Fraction of cardiac output (QC))
CONSTANT QFC = 0.085 ! Blood flow to fat (L/hr/QC)
CONSTANT QLC = 0.029 ! Blood flow to liver (L/hr/QC)
CONSTANT QSC = 0.600 ! Blood flow to poorly perfused tissues (L/hr/QC)
```

```
CONSTANT QRC = 0.230 ! Blood flow to richly perfused tissues (L/hr/QC)
CONSTANT QKC = 0.056 ! Blood flow to kidney (L/hr/QC)
! Metabolic constants
CONSTANT VMAX = 0 0 ! Maximum velocity (mg/hr)
CONSTANT KM = 0.000001 ! Michaelis constant (mg/L; non-zero value)
! Partition coefficients
CONSTANT PL = 2.55 ! Liver:blood
CONSTANT PF = 44.9 ! Fat:blood
CONSTANT PR = 2.55 ! Richly perfused tissues:blood
CONSTANT PS = 2.46 ! Poorly perfused tissues:blood
CONSTANT PK = 3.07 ! Kidney:blood
CONSTANT PB = 5.17 ! Blood:water
! Simulation parameters
CONSTANT TSTOP = 70 ! Simulation length (hr)
CONSTANT DUREE = 48. ! Exposition length (hr)
CONSTANT CINT = 0.25 ! Communication interval
! Exposure parameters
CONSTANT CONC = 1.06 ! Inspired concentration (mg/L)
! Scaled parameters
! Flows
QF = QFC*QC ! Fat blood flow (L/hr)
QS = QSC*QC ! Poorly perfused tissues blood flow (L/hr)
QR = QRC*QC ! Richly perfused tissues blood flow (L/hr)
QL = QLC*QC ! Liver blood flow (L/hr)
QK = QKC*QC ! Kidney blood flow (L/hr)
! Volumes
VR = VRC*BW ! Richly perfused tissues volume (L)
VS = VSC*BW ! Poorly perfused tissues volume (L)
VF = VFC*BW ! Fat volume (L)
VL = VLC*BW ! Liver volume (L)
VK = VKC*BW ! Kidney volume (L)
! Gill uptake limitation
IF(QC*PB .GT. QR) THEN
GUL = QG !Water flow limited gill uptake
ELSE
GUL = QC*PB !Blood flow limited gill uptake
END IF
! MODEL EQUATIONS
! Calculation of blood concentration'
CA = CV + GUL*(CI-CV/PB)/QC ! Arterial blood concentration (mg/L)
CV = (QF*CF/PF + (QL + QR)*CL/PL + 0.4*QS*CS/PS + (QK + 0.6*QS)*
CK/PK)/QC ! Venous blood concentration (mg/L)
```

```
! Liver compartment
RCL = (QL*CA + QR*CR/PR)/VL-(QL + QR)/VL*CL/PL-RCM
! Rate of change in tissue concentration (mg/L/hr)
CL = INTEG(RCL,0) ! Concentration in the liver (mg/L)
! Liver Metabolism
RCM = (VMAX*CL/PL)/(KM + CL/PL)/VL ! Rate of concentration metabolized
CM = INTEG(RCM,0.) ! Concentration metabolized
! Fat compartment
RCF = QF/VF*(CA-CF/PF) ! Rate of change in tissue concentration (mg/L/hr)
CF = INTEG(RCF,0) ! Concentration in the fat (mg/L)
! Richly perfused tissues compartment
RCR = QR/VR*(CA-CR/PR) ! Rate of change in tissue concentration (mg/L/hr)
CR = INTEG(RCR,0) ! Concentration in the richly perfused tissues (mg/L)
! Poorly perfused tissues compartment
RCS = QS/VS*(CA-CS/PS) ! Rate of change in tissue concentration (mg/L/hr)
CS = INTEG(RCS,0) ! Concentration in the poorly perfused tissues (mg/L)
! Kidney compartment
RCK = (QK*CA + 0.6*QS*CS/PS)/VK-(QK + 0.6*QS)/VS*CK/PK
! Rate of change in tissue concentration (mg/L/hr)
CK = INTEG(RCK,0) ! Concentration in the Kidney
```

Earthworms and Their Use in Eco(toxico)logical Modeling

Willie J.G.M. Peijnenburg and Martina G. Vijver

Abstract A healthy terrestrial food web is essential for the sustainable use of soils. Earthworms are key species within terrestrial food webs and perform a number of essential functionalities like decomposition of organic litter, tillage and aeration of the soil, and enhancement of microbial activity. Chemicals may impact the functions of the soil by directly affecting one or more of these processes or by indirectly reducing the number and activity of soil engineers like earthworms. The scope of this chapter is on the assessment and modeling of the interactions of chemicals with earthworms and the resulting impacts. It is the aim of this contribution to provide a general review of the research that were undertaken to increase our understanding of the underlying processes.

Chemicals may induce a variety of adverse effects on ecosystems. Chemical speciation, bioavailability, bioaccumulation, toxicity, essentiality, and mixture effects are key issues in assessing the hazards of chemicals. Although it is possible to group chemicals with regard to their fate and effects, a plethora of chemical and biological processes affects actually occurring effects. These effects are usually modulated by (varying) environmental conditions. Using the basic processes underlying the uptake characteristics and the adverse effects of organic pollutants and metals on earthworms as an illustration, an overview will be given of the interactions between the chemistry and biology of pollutants, mostly at the interface of biological and environmental matrices. The impact of environmental conditions on uptake and toxicity of chemicals for soil dwelling organisms will explicitly be accounted for. The environmental chemistry of organic compounds and metals, as well as the resulting methods for assessing chemical availability are assumed as tokens and the emphasis is thus on the biological processes that affect the fate and effects of contaminants following interaction of the earthworms with the bioavailable fraction.

Keywords Oligochaetes · Physiology · Pollutants · Uptake · Accumulation modeling · Effect modeling

W.J.G.M. Peijnenburg (✉)
Laboratory for Ecological Risk Assessment, National Institute of Public Health and the Environment, P.O. Box 1, 3720 BA Bilthoven, The Netherlands
e-mail: WJGM.Peijnenburg@rivm.nl

J. Devillers (ed.), *Ecotoxicology Modeling*, Emerging Topics in Ecotoxicology: Principles, Approaches and Perspectives 2, DOI 10.1007/978-1-4419-0197-2_7,

1 Earthworms: Relevance, Preferences, and Interactions

1.1 Earthworms and Their Environmental Relevance

Soils are used for a large number of strongly varying purposes, including agriculture, forestry, gardening, and playing fields. A healthy terrestrial food web is essential for the sustainable use of soils for these and other purposes. The soil food web is the set of organisms that work underground to help sustain the essential functions of soil. There are billions of organisms that make up the soil food web. These include bacteria, fungi, protozoa, nematodes, arthropods, and earthworms. Each type of organism plays an important role in keeping the soil healthy. Earthworms take a special place in this respect as not only they eat about every other particle in the soil, but also when they eat they leave behind "castings" which are high in organic matter and plant nutrients, and are a valuable fertilizer. By actively adding earthworms to the soil, soils get in a better condition and their fertility is further improved.

Widely respected ecologists like Darwin and Righi were among the first scientists to recognize the importance of species in general and earthworms in particular. During 40 years of active research on endangered earthworms in tropical areas, Righi published about 100 papers on earthworm taxonomy, physiology, ecology, and biogeography: see for instance Fragoso et al. [1] for a review on the influence of Righi on tropical earthworm taxonomy. It was Charles Darwin [2] who considered earthworms as one of our planet's most important caretakers. "I doubt," he said, "whether there are many other animals which have played so important a part in the history of the world, as have these lowly organized creatures." Darwin was the first to describe how earthworms tilled the soil, swallowing and ejecting soil as castings, or worm manure. He estimated that an acre of garden soil could contain over 50,000 earthworms and yield 18 tons of organic castings per year (scientists now figure worms can number over one million per acre). Darwin's naturalist approach and his long-term experience in observing the behavior of different animals helped him distinguish various possible "functions" of earthworms. He briefly alluded to different functional groups of worms:

1. Deep burrowing and shallow burrowing species
2. Large-compact and small-granular casters
3. Litter and soil feeders

These characteristics are among the most important characteristics currently used in various functional classifications of the soil fauna and earthworms. The most widely used recent functional classifications are those of Bouché [3], Lee [4,5], and Lavelle [6]. These classes generally include three main groups (endogeic, anecic, and epigeic earthworms) that are defined on the predominant habitat of a species. Although these three subgroups have been proposed, some earthworm species do not seem to fit into any particular category or, rather, fit in between proposed categories (e.g., epi-endogeic and endo-anecic). Other earthworm's classifications include those of Lavelle [7] and Lavelle et al. [8], into ecosystem engineers and litter transformers, and of Blanchart et al. [9] into compacting and decompacting species. These schemes attempt to integrate knowledge on feeding habits and functional significance of earthworms in the soil. Darwin's contributions in this area deal primarily with the influence of earthworms on soil physical processes, although he also touches upon the selection and processing of particular leaf litters.

Earthworms move through the soil creating tunnels, and thus areas that can be filled by air and water. Fields that are "tilled" by earthworm tunneling can absorb water at a rate 4–10 times that of fields without earthworm tunnels. This reduces water runoff, restores groundwater, and helps store more water for dry spells. Burrowing also helps nutrients enter the subsoil at a faster rate and opens up pathways for roots to grow into. During droughts the tunnels allow plant roots to penetrate deeper, to reach the water they need to thrive. Earthworms help to keep the soil healthy by moving organic matter from the surface into the soil. By speeding up the breakdown of plant material, earthworms also speed up the rate at which nutrients are recycled back to the plants. Earthworms are thus an essential part of the soil food functioning. Without them, all the organic matter would build up on the soil surface.

The capability of changing the soil structure by preferential feeding on organic material by earthworms was the basis for vermiculturing of organic-matter-rich waste materials. Together with bacteria, earthworms are the major catalyst for decomposition in a healthy vermicomposting system, although other soil species also play a contributing role: these include insects, other worms, and fungi/molds. Vermicompost is a nutrient-rich, natural fertilizer and soil conditioner. The earthworm species (or composting worms) most often used are Red Wigglers (*Eisenia fetida*) or Red Earthworms (*Lumbricus rubellus*). These species are commonly found in organic-rich soils throughout Europe and North America and especially prefer the special conditions in rotting vegetation, compost, and manure piles. To benefit from their active stimulation of soil processes, earthworms nowadays are commercially available. Mail-order suppliers or angling (fishing) shops keep earthworms in bred and composting worms are sold for vermicomposting practices and sold as bait. Small-scale vermicomposting is well-suited to turn kitchen waste into high-quality soil, where space is limited. Thanks to the pioneering work of Dr. Clive Edwards [10] in the area of vermicomposting that this technique is now widely applicable to generate soil structure and soil quality enhancing compost. Vermicomposts can also be used in pollutant bioremediation for organic contaminants and heavy metals as the microbial degradation of the organic pollutants is accelerated dramatically and the heavy metals become irreversibly bound into the humic materials that are formed, so that they are not available to plants. Dr. Zharikov's research [11] into methods of soil purification revealed that earthworms are also capable of enhancing the cleaning of the contaminated soils by stimulating the growth of microorganisms that breakdown the contaminants.

As there is no doubt that the earthworm can be of major benefit to a healthy soil ecosystem, it is important to understand the key role of earthworms in many biogeochemical cycles and in soil development as related to the impacts of land uses. This is particularly true in relation to restoration of damaged ecosystems and to preventive maintenance to avoid damage.

1.2 Earthworms and Their Preferences

Environmental factors that provide the most dominant impacts on earthworm populations are moisture, temperature, and pH, on top of food resource quantity and quality. Soil moisture affects earthworm abundance, activity patterns, and thus geographic distribution. Earthworms tend to dig deeper or even tend to go into a diapauses during periods of prolonged drought. During rainy periods earthworm species tend to surface to escape from drowning. Soil temperature influences seasonal activity, limiting earthworms during warm and cold periods. Soil pH often is cited as a limiting factor on earthworm distributions. For instance, the best studied group (European Lumbricidae) generally does not inhabit soils with pH below 4.0. Other taxa tolerate lower pH values, including some Pacific coast native species (pH 3.1–5.0; [12]), thereby indicating that soil acidity might be less limiting for certain earthworm species than for others.

Soil climate determines the periods of earthworm activity. Within a habitat type, variations in soil climatic factors occur (because of slope, aspect, soil particle size distribution, and drainage characteristics) that result in variation in earthworm activity period and earthworm abundance. A forested habitat probably has a relatively buffered soil climate compared to the more exposed grasslands and agricultural land. Grassland temperature and moisture regimes are probably more extreme and could accentuate the effects of slope, soil properties, and other site characteristics. An agricultural cycle having long periods of bare ground could further intensify the impact of weather on earthworms.

The quantity and quality of food influences earthworm abundance. Food sources are all types of organic matter. Organic matter may render the soil strongly acidic, could be rich in digestibility reducing compounds, or could have a high carbon-to-nitrogen ratio. These qualities tend to reduce earthworm populations. Lack of organic matter is generally a significant limiting factor for earthworms. The fact that most agricultural soils are depleted of organic matter, likely accounts for lower abundance of earthworms in agricultural land or recently abandoned cropland.

1.3 Earthworms and Essential Elements

Natural and man-made chemical substances may severely interfere the natural fluctuations in earthworm populations in specific habitats. Availability or the lack of essential nutrients on the one hand shapes natural ecosystems, whereas on the other hand excess amounts of bioavailable nutrients and micropollutants reduce the natural abundance of species and affect the natural ecosystem functioning. This observation was the basis for the concept of optimal concentration of essential elements (OCEE). This concept was among others proposed to account for metal-specific aspects of essentiality and homeostasis.

A first attempt to account for the metal-specific aspects of essentiality and homeostasis was achieved by the optimal concentration of essential elements-no risk area

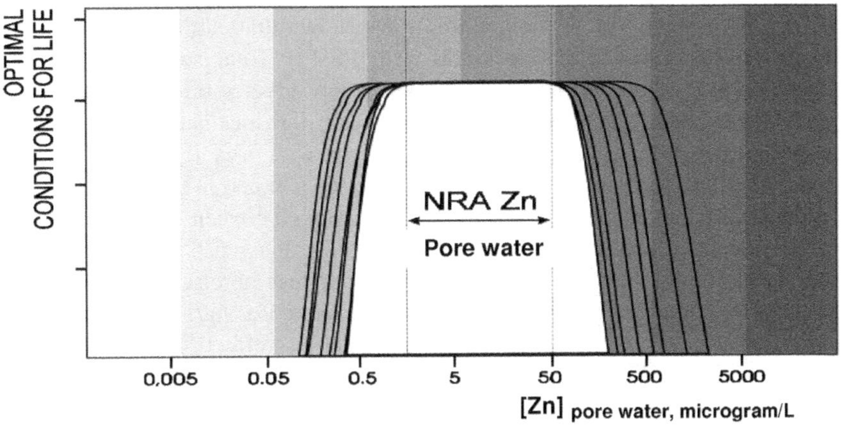

Fig. 1 Hypothetical presentation of the OCEE curves of all individual organisms in a given environment. The *inner envelope* of these *curves* represents the no risk area (NRA) for that given environment in which all organisms are protected from both toxicity and deficiency (adopted from [13])

concept (OCEE-NRA) based upon the assumption that all OCEEs for all individual organisms belonging to a certain habitat type (ecoregions) are centered on the natural essential element (metal) background concentration typical for that habitat. Figure 1 gives a schematic representation of the OCEE-NRA concept: at low nutrient levels, adverse effects are observable related to lack of nutrients; increased levels of essential elements induce toxicity. Furthermore, research results indicated that the sensitivity of the toxicity response of an organism to an essential metal is a function of the essential element concentration in which it was cultivated. Acclimatization explains the decrease in sensitivity at higher background concentrations in the culture medium. The recognition and demonstration that organisms do belong to different OCEE-NRAs underscore the relevance of this concept and have lead to the fundamentals of the metallo-region concept. The major technical difficulties for the integration of the OCEE-NRA concept into regulatory frameworks for environmental risk assessment are the spatial and temporal variability in natural background levels as well as the variability in physicochemical conditions influencing metal bioavailability and toxicity.

Apart from agriculturally oriented studies on optimal levels of essential elements, relatively little quantitative information is available on deficiency levels of most nutrients for earthworms.

1.4 Earthworms and Pollutants

As earthworms ingest large amounts of soil or specific fractions of soil (i.e., organic matter), they are continuously exposed to contaminants through their alimentary surfaces [14]. Moreover, several studies have shown that earthworm skin is a significant route of contaminant uptake as well [15–17]. Toxic substances and excess nutrients are accumulated and subsequently exert adverse effects by a variety of interactive modes of action, both with regard to the mechanisms of uptake and the mechanisms of toxicity. Whereas interactions with organic micropollutants are strongly modulated by organic carbon pools in the soil and in the fat tissue, uptake and effects of metals are modulated by interactions between the various soil and pore water constituents. Soil constituents serve in this sense as capacity controlling factors modulating the bioavailable pool whereas pore water parameters like pH, dissolved organic carbon, and macronutrients like Ca/Mg/Na serve as intensity-controlling factors as they modulate actually occurring effect.

It is the aim of this chapter to exemplify the use of earthworm as a key species in soil toxicity testing. Based on ecological considerations, the objective of this contribution is to give a general overview on the accumulation of chemicals by earthworms and the toxic effects exerted due to interactions of these animals with micropollutants. Providing an in-depth discussion of the basic phenomena underlying accumulation and adverse effects is not the primary aim. Instead, a short overview will be provided of the approaches used in testing assessing and modeling bioaccumulation and ecotoxicity.

2 Earthworms as Model Organism

2.1 Bioindicators for Chemical Stress

Bioindicators are used as representatives of parts of ecosystems or of one or more functions [18]. The basic consideration of the use of biomarkers is that living organisms provide the best reflection of the actual state of ecosystems and of changes therein. These measures can be done on either structure or functioning of ecosystems. For both type of measurements, oligochaete are generally regarded as highly suitable bioindicators. Their importance in the structure of ecosystems can be explained because they are an ecologically dominant invertebrate group. Moreover, earthworms occur in many different soils from temperate to tropical areas. Also their importance in food chains, with earthworms being a food source for many organisms such as birds and mammals, has implicated that many ecological studies have focused on studying the ecology and ecotoxicology of the earthworm. Thereupon, most oligochaetic species are easy to handle and to culture under laboratory settings [19]. Respecting this, earthworm species are often used as test organisms to determine the effect and accumulation of chemicals from soil [19–23]. Due to their behavior and morphology, earthworms are in close contact with the aqueous and solid phases of the soil. From experimental studies it could be concluded that for both inorganic [17] and organic [16] contaminants earthworms are exposed to pollutants in the soil mainly via the pore water. Most oligochaetic species are not extremely sensitive to low levels of chemicals [24, 25]. The chemical composition of their body is fairly constant, which facilitates the understanding of the mechanisms of toxicity. Their internal organization is not highly complex, and possesses strongly differentiated organs. Moreover, it is described very well in literature [18].

2.2 *Ecophysiology of Earthworms*

Oligochaete worms have a thick mucus layer that surrounds the epidermis [26], through which respiration and the excretion of waste products occur. This mechanism makes the earthworms sensitive to water loss. The digestive interior of oligochaete species is well investigated [27]. There is evidence that the uptake of food via the gut is not a heterogeneous process during the gut passage. During ingestion mucus is mixed with the food. In the first part of the digestive system of an oligochaete, calciferous glands actively release Ca^{2+} in the gut contents. The crop is used for storage of the gut content, before mechanical grinding and digestion in the gizzard. The gizzard opens up into the intestine, which forms the largest part of the alimentary canal. Gut conditions in the final part of the digestive system (the intestine) are actively regulated by excretion of NH_4^+. A typhlosole (see Fig. 2), a dorsal infolding of the gut epithelium effectively increasing the internal surface, is present along the anterior and mid intestine, thereby also increasing the secretory and absorptive surface areas. The pH along the entire digestive tract is quite constant between 6 and 7, and the digestion is driven by enzymes [28]. The gut pH is often higher than the bulk soil pH, especially in earthworms inhabiting acid soils.

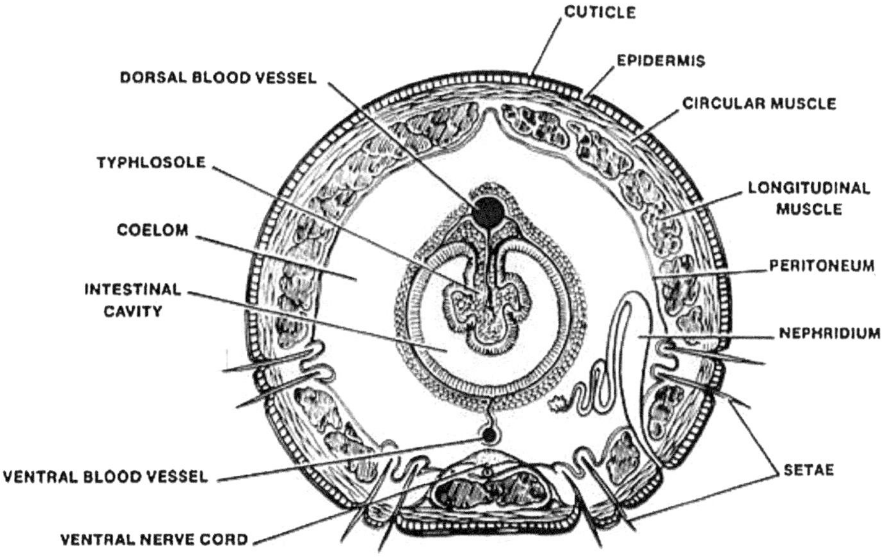

Fig. 2 Cross section of the posterior body cavity of earthworms

The largest part of the body burden is bound in the chloragogenous tissue [29] located around the digestive tract (see Fig. 3). The cells of this tissue (chloragocytes) contain many chloragosomes, including calcium granula (type A) and sulfur-rich granules (type B). All granulum types are likely to play a role in the homeostasis of essential elements but also for detoxification of chemicals that entered the body. The resorption capacity of the digestive tract is most efficient in the posterior alimentary canal.

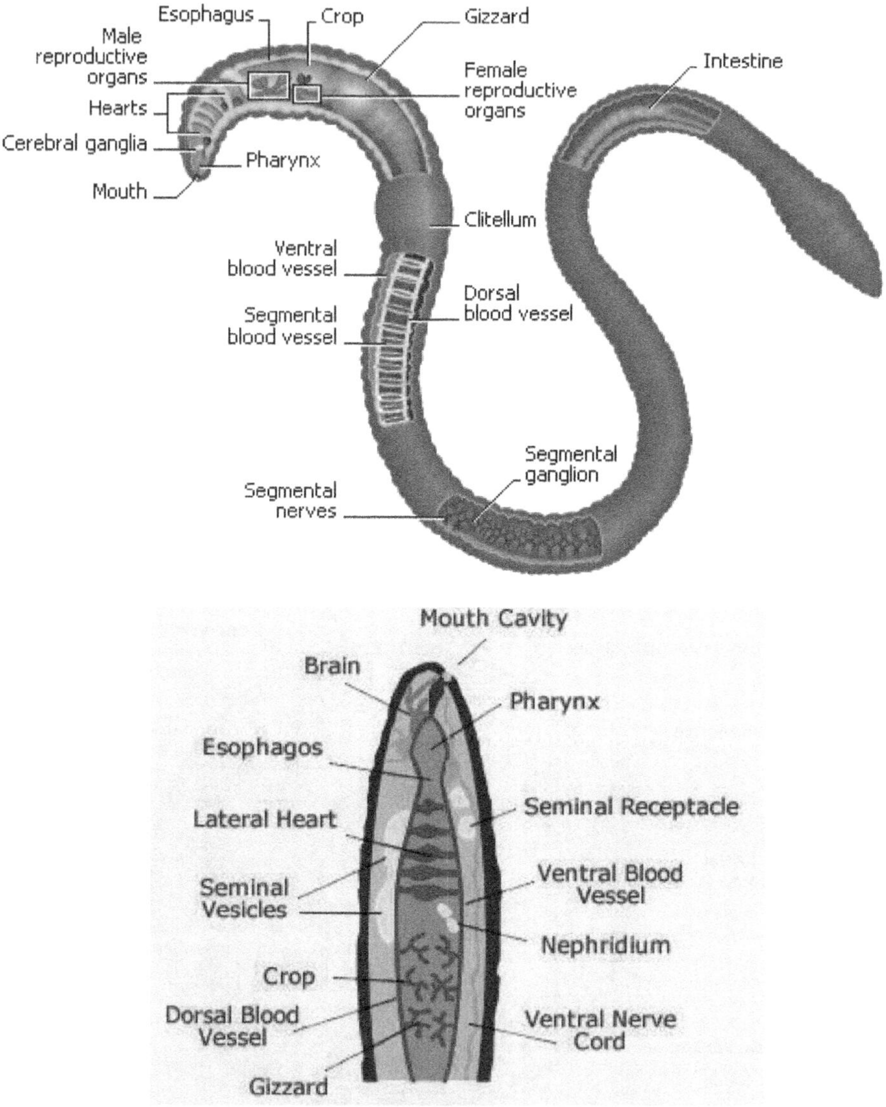

Fig. 3 Schematic of the anatomy of earthworms

3 Accumulation

Biological uptake of most synthetic (hydrophobic) organic contaminants occurs by simple passive diffusion across a cell membrane. Membrane carriers are not involved and the biological effect of organic contaminants is often (but surely not always) characterized by narcosis, implying that the extent of adverse effect of organic contaminants is proportional to the value of the octanol–water partition coefficient. In contrast, as metals generally exist in strongly hydrated species, they are unable to traverse biological membranes by simple diffusion. In general, the interaction of metals with organisms is somehow related to a liquid phase, according to the principles of the Free Ion Activity Model (FIAM) [30]. The mechanisms can be described as follows:

1. Advection or diffusion of the metal from the bulk solution to the biological surface
2. Diffusion of the metal through the outer "protective layer"
3. Sorption/surface complexation of the metal at passive binding sites within the protective layer, or at sites on the outer surface of the plasma membrane
4. Uptake of the metal (transport across the plasma membrane)

Membrane transport occurs by facilitated transport, usually passive (i.e. not against a concentration gradient), and necessarily involves either membrane carriers or channels. The chemical binds to the carrier protein and is carried through the membrane by a process that requires no cellular energy. There is some specificity to the carrier protein binding, and so the process is applicable only for selected chemicals. Transport of essential metals is for instance facilitated by carriers or pores specific to the element, although metals are also transported on carriers designed for elements of similar physicochemical characteristics.

3.1 *BCFs and BAFs*

The terms "Biota Concentration Factors" (BCFs) and "Bioaccumulation Factors" (BAFs) can be defined as similar words and are both used to quantify to which extent chemicals are transported from the exposure medium into organisms. By definition, the higher the BCF value, the more chemicals are taken up and the higher the potential risk regarding adverse effects on the organism itself and at higher trophic levels. Most studies report relationships between internal and external concentrations (BCF) where steady state is assumed [31, 32]. An extended overview of BCFs in earthworms for organic chemicals is given by Jager [33], whereas Sample et al. [34] developed and tested uptake factors and regression models for uptake factors for metals in earthworms. The bioconcentration factors found for PCBs were between 7,200 (low mol. PCB) and 126,000 (high mol. PCB). BCFs for chlorobenzenes in earthworms ranged from 12 to 4,000. Pesticides display widely varying BCF values: ranging from less than 1 (for instance Aldicarb: 0.7) to over 5,000

(Lindane). In general, BCFs increase with hydrophobic properties of organic chemicals albeit that especially biotransformation may lower apparent BCF values. BCF values are known to be species and soil dependent. As an example Kelsey et al. [35] determined the BAF in four field-weathered soils for an epigeic species *Eisenia fetida*, an anecic species *Lumbricus terrestris*, and an endogeic species *Aporrectodea caliginosa*. The epigeic species had BCFs that were approximately tenfold higher than those for the other species. With regard to contaminant-residence time, the BAF for *E. fetida* was lower in weathered soils relative to that in freshly amended soils, but age of p,p'-DDE did not significantly alter the BAF for *A. caliginosa* [35]. The biota-soil accumulation factors (BSAFs) observed for individual PAHs in field-collected earthworms (*A. caliginosa*) were up to 50-fold lower than the BSAFs predicted using equilibrium-partitioning theory [36].

An overview of BCFs in earthworms for inorganic chemicals is given by Janssen et al. [31]. Bioaccumulation factors varied between metals. The BCF of As ranged from 0.1 to 3, Cd ranged from 1 to 203, Cr ranged from 0.03 to 0.5, Cu ranged from 0.2 to 8, Ni ranged from 0.07 to 0.6, Pb ranged 0.005 to 1.3, and Zn ranged from 0.1 to 18. In general, BCFs for metals decrease with higher exposure concentrations [37]. The same inverse relationship was found in aquatic systems between bioaccumulation factors and, trophic transfer factors and exposure concentrations [38].

A general finding is that BCFs decline with increasing pollutant concentration in soil. The uptake and adverse effects of chemicals to earthworms can be modified dramatically by soil physical/chemical characteristics, yet expressing exposure as total chemical concentrations does not address this problem. Bioavailability can be incorporated into ecological risk assessment during risk analysis, primarily in the estimation of exposure. However, in order to be used in the site-specific ecological risk assessment of chemicals, effects concentrations must be developed from laboratory toxicity tests based on exposure estimates utilizing techniques that measure the bioavailable fraction of chemicals in soil, not total chemical concentrations [39]. The final and most difficult task in any assessment is to relate body residues to levels known, or suspected, to be associated with adverse biological responses. To address this, physiological knowledge on chemical distribution over the body should be combined with the knowledge on accumulation. Paracelsus stated in 1564 that "What is there that is not poison? Solely the dose determines that a thing is not a poison" [40]. We should add to this statement that also the biological significance of accumulation is of importance [41].

3.2 *More Compartments*

Earthworms are able to accumulate organics to a great extent. The ability to deal with high levels of accumulated organics can be ascribed to the manifestation of organics to bind to fatty tissues [42–45]. Bioaccumulation of organics also can be ascribed using multiple compartments, although two compartments are usually sufficient.

Earthworms are also able to accumulate metals to a great extent. The ability to deal with high levels of accumulated metals can be ascribed to the slow turnover of the tissues in which metals accumulate. Morgan et al. [46] found distinct differences in the distribution of various metals throughout the earthworms' body, whereby the sequestration on chlorogocytes played a dominant role, resulting in different patterns of tissue accumulation [47] and different tolerances [48]. Metals such as Cd and Cu are predominantly bound to metal-binding proteins [49] and with these proteins, the metal moves through the body to organs and tissues in which it is deposited in inorganic forms. Cd was retrieved in high amounts from the nephridia and to a lower extent from the body wall of earthworms [50]. Pb is found in waste nodules located in the coelomic fluid [51]. The granulas contain many essential and nonessential chemicals. For instance Cd preferentially binds to sulfur-rich granules instead of oxygen-rich granules, and hence is found in the type B granules, also called cadmosomes.

A pragmatic method to describe and quantify the internal sequestration of metals is found in Vijver et al. [41,52].

3.3 *How to Perform Experiments for Optimal Results*

Dynamic biological measures of bioavailability – thus the rate at which organisms take up contaminants from the environment – are the best and according to the latest scientific state-of- the-art on how to derive indicators of bioavailability [53]. Actual uptake and elimination fluxes are very difficult to measure. A pragmatic solution to overcome this problem is to measure body burdens as a function of time in an organism exposed to the medium tested. Parameter estimation is done by curve fitting the accumulation data. In the most simple case, the exposure concentration is constant, and as soon as the organism is exposed, internal concentrations are increasing [53,54].

By this way an accumulation curve can be fitted according to the following general equation (most simple form):

$$Q = C_0 + (a/k)\mathrm{e}^{-kt}. \tag{1}$$

In this equation, Q is the amount of chemical accumulated at equilibrium or at steady-state conditions; C_0 the initial body burden; a the uptake flux; k the elimination rate constant; and t is the time.

Exposure of organisms under fluctuating external conditions, as is the common case in reality, can also be modeled. This is done by taking into account the kinetics of the bioavailable fraction of the chemical for a specific organism. For instance, in the case of biotransformation of the contaminant being taken up by the earthworm or in case of cocoon production, (1) transforms into [55]:

$$Q = (a_1 C_0)/(k_2 - k_0) \times (\mathrm{e}^{-(k_0 t)} - \mathrm{e}^{-(k_2 t)}). \tag{2}$$

In (2) a_1 is the uptake rate constant, k_0 the rate constant for degradation of the chemical in the medium, and k_2 is the elimination rate constant.

The common experimental set up in order to measure accumulation is often to expose relatively large numbers of earthworms, divided over a number of jars, to a soil. At different time intervals, earthworms are sacrificed and measured for their body burden. It is preferred to measure more frequently over time instead of more replicates at the same exposure time. Especially within the initial stage of the exposure and thus during initial uptake of the chemicals by the earthworms, many samples with a small time interval should be taken. The sampling strategy should be according a log-scale, with fewer measurements at the end than at the beginning in order to accurately capture initial uptake kinetics.

Accumulation is the net effect of uptake and the ability of the organism to eliminate a chemical once it has entered the body. Estimation of uptake in the presence of simultaneous elimination is improved significantly if the uptake is followed by an elimination phase without uptake, because this will yield a better estimate of the elimination rate constant, and consequently also a better estimate of the uptake parameter. Therefore, experiments usually involve an uptake phase and an elimination phase, simply by transferring the organism to a clean medium after a certain period. This situation can easily be performed when artificially spiked soils are used for the accumulation testing. However, when using natural contaminated field soils, in most cases it is difficult to find an uncontaminated field soil with similar characteristics as the contaminated field soil. Subsequently an appropriate elimination phase is difficult to test. An alternative technique allowing for the quantification of uptake and turnover kinetics in biota is isotopic labeling. The main advantage of this technique is that it overcomes the problem of selecting an unpolluted reference site and that it is nondestructive for the exposed organisms. Hence the biological variation of accumulation can be studied for single species. Moreover, it overcomes detection limitations within the body burden of earthworms, and allows insight into essential metal uptake even in the presence of highly regulated body concentrations.

3.4 Alternative Measures of Bioaccumulation

Alternatives to assess bioaccumulation without the direct measurement of internal concentrations in organisms or effects on earthworms are the use of mimic techniques (see for an overview of these techniques [56]). The use of passive sampling devices (PSDs) is an example of these kinds of mimic techniques which are potentially direct chemical indicators for assessing the bioavailability of chemicals. PSDs are constructed in several forms but often consist of lipophilic material within a semipermeable membrane, mimicking biological membranes. Exposure of biota to chemicals is assessed this way, and the techniques account for aging and mobility of chemicals in the matrix [57]. The results of Awata et al. [57] showed that concentrations as determined in the PSD were in good agreement with accumulation data in the earthworms as measured after exposure in contaminated soils. Uptake rates and maximum concentrations in PSDs were observed to positively correlate with uptake rates and maximum concentrations in earthworms for both of the soil types studied (sandy loam and silt loam). These results indicate that PSDs may be used as a surrogate for earthworms and provide a chemical technique for assessing the availability of aged chemical residues in soil. Similar findings were reported by Van der Wal et al. [58], who concluded that measuring concentrations of hydrophobic chemicals using polydimethylsiloxane solid phase microextraction (which is a kind of PSD) is a simple and reliable tool to estimate bioaccumulation in biota exposed to soil. The opposite was been concluded by Bergknut et al. [59], who showed a distinct difference between evaluated PSD techniques and bioaccumulation in earthworms. Generally, there were larger proportions of carcinogenic PAHs (4–6 fused rings) in the earthworms compared to the concentrations as found with the mimic techniques. In cases that the exposure media (e.g. soils) were heterogeneous, the PSDs had no predictive capacity.

From the information provided above it may be concluded that it will be difficult to develop a single and universally applicable chemical method that is capable of mimicking biological uptake, and thus estimating the bioavailability of chemicals. In some cases, a strong numerical relationship of bioaccumulation of chemicals with biomimetic techniques is reported; in other cases no such correlation is found. This general finding is related to the fact that accumulation by living organisms like earthworms is more dynamic than can be simulated by chemical means. Only in those cases where chemical interactions overrule organism-specific ecological impacts (like feeding behavior, regulation of body concentrations by active uptake and/or elimination, and biotransformation), a strong correlation between uptake and biometry may be found.

4 Toxicity

4.1 Toxicity Testing

4.1.1 General

Earthworms are frequently used as part of batteries of indicator species to test the effects of pollutants on ecosystems. A wide array of substrates (including artificial substrates like OECD soils – a mixture of sand, kaolinite clay, peat, and $CaCO_3$ to adjust pH), test designs, and endpoints are exploited and guidelines have been designed to standardize the assessment of adverse effects on earthworms. Apart from laboratory testing, terrestrial model ecosystems (TMEs; [60]), field enclosures, and field testing [61] are employed to increasingly mimic actually occurring effects in the field. Testing data are employed to derive models capable of predicting effects at various levels of integration, varying from simple linear regression equation based on soil or pore water characteristics up till advanced concept taking account of the specific interactions of chemicals with earthworms. The species most commonly tested in a laboratory setting are the compost worms *Eisenia andrei* and *E. fetida* as more field-relevant species like *L. rubellus* and *A. caliginosa* are difficult to rear.

A general distinction that is often made when performing earthworm testing is between acute (i.e., short exposure time) and chronic testing. For some chemicals, like for copper, this difference is often artificial as the acute-to-chronic-effect ratios are close to 1. As a rule of thumb, exposure times up till 14 days are considered to represent acute testing. Exposure times in field testing may exceed various seasons and last even for several generations of animals.

4.1.2 Biomarkers of Exposure and Toxicity

Apart from the commonly studied endpoints discussed below, the use of biological responses other than reproduction, growth, and mortality to estimate either exposure or resultant effects has received increased attention [62–65]. Biomarkers are typically biochemical changes that are induced following exposure to a contaminant. Biological responses are possible at the molecular, subcellular, and cellular level. A major reason for the interest in biomarkers is the limitation of the classical approach in ecotoxicology in which the amount of chemical present in an animal or plant is related to adverse effects on the classical endpoints. Bioavailability and toxicity differ, however, in laboratory tests compared to those observed in the field, and multiple toxicants are typically present simultaneously under field conditions. Also, only a few of the conventional endpoints can be assessed in in situ experiments. Biomarkers have the potential to circumvent the limitations mentioned as they respond only to the biologically available fraction of a pollutant, independent of mitigating effects of soil characteristics.

In order for a biomarker, or a battery of biomarkers, to be useful in effective assessment of chemicals to earthworms, a number of key features apply [66]:

1. The marker must be identified in the species of interest.
2. Knowledge is required on the range of toxic compounds that elicit a biomarkers response.
3. To estimate the magnitude of the chemical stress, a dose–response relationship between the biomarker response and the bioavailable concentration of the chemical is desirable.
4. Possibilities to link biomarkers responses to higher levels of biological hierarchy are desirable. For a biomarker to be of more use than an indicator of exposure, a correlation between the observed responses and deleterious effects at the individual or populations/community level should be established. A subcellular biomarker may, for example, act as an early warning of effects at population level.
5. For a biomarker to be useful in the field, any response should have a low inherent variability with a known (preferably: a low) dependence on physiological and physiochemical conditions. Among others, the induction time and the persistence

of a biomarker response should be known in order to estimate the likelihood and significance of detecting a response in field samples.

Up till now, various biomarkers have been developed and have been applied with varying amounts of success. The most important categories include:

1. *DNA alterations induced by reactions of contaminants with genotoxic properties*. The most common reactions are adduct formation (covalent binding of the contaminant or its metabolite to DNA), strand breakage, base exchange, and increased unscheduled DNA synthesis. Limited information is available on the environmental significance of DNA alteration at higher levels, the natural variability, and the persistence in time of DNA adducts.
2. *Induction of metal-binding proteins*. Heavy metals entering earthworms at concentrations exceeding the metabolically required metal pool may be bound and detoxified by binding to metallothionein and other metal-binding proteins. Although the role of metallothionein and other metal-binding proteins is not fully understood, these proteins are thought to be involved in the intracellular regulation of essential and nonessential metal levels in tissues. Apart from limited attempts on Cd, no studies have been undertaken to establish dose–response relationships for induction of metal proteins. Links to higher levels and natural variability also require more attention before this type of biomarker is suited for quantifying exposure and/or metal toxicity.
3. *Inhibition of enzymes*. Inhibition of cholinesterases is the most common studied biomarker of exposure of earthworms to carbamate and organophosphorus pesticides. Cholinesterases are used for the transmission of nerve signals and contaminants can cause a depression in cholinesterases activity. Depression of cholinesterases activity may well depend on the metabolic compounds rather than the parent compound and just a few studies have reported on natural variability of the cholinesterases activity in earthworms. On the other hand, inhibition of cholinesterases activity in earthworms was shown to be dose dependent in both coelomic fluid and in nerve tissue [67].
4. *Lysosomal membrane integrity*. Lysosomes are a morphological heterogeneous group of membrane-bound subcellular organelles that catabolize organelles and macromolecules. A change in lysosomal membrane stability is thought to be a general measure of stress. At the subcellular level, the lysosomal system has been identified as a particular target for toxic effects of contaminants. The neutral red retention time (NRR) is used to investigate lysosomal stability and for just a few chemicals a dose–response relationship was obtained thus far. Few studies have been concerned with the natural variability of the lysosomal membrane stability and with the establishing links with higher levels like reproductive output and mortality [68]. Aquatic studies have indicated that lysosomal response can also be induced by nonchemical stressors such as osmotic shock and dietary depletion.
5. *Immunological responses building upon the fact that the immune system is the main defense of an earthworm against invasion of foreign material and biological agents*. A wide range of chemicals has been shown to be capable of affecting

the immune system, which in severe cases may quickly result in morbidity and death. Sublethal changes in special compartments of the immune system occur first and provide early indications of toxic effects. The immunological system is known for its flexibility and adaptability and it has been observed for earthworms that the immunological depression returns to normal levels quickly after removal of the earthworms from the source of exposure. Relatively few studies have dealt with the impact of chemicals on the immune system of earthworms, and dose–response relationships as well as linkage to higher levels of effects are rarely available.

Although some biomarkers provide a forewarning of adverse effects resulting from exposure of earthworms to contaminants, more work is needed to understand the limitations of the use of biomarkers. Thus, for biomarkers to be of use as early warning tools, more effort is needed in linking biomarker responses at the subcellular and cellular levels with effects at population level under natural conditions.

4.2 The Kinds of Effects Commonly Measured

4.2.1 Laboratory Testing

Laboratory tests play an important role in earthworm testing. The endpoints mortality, reproduction, and change of body weight are standardized and well described in widely accepted guidelines for testing of chemicals [69–71]. Other endpoints like behavior, morphological changes, and physiological changes are reported occasionally, but they are not evaluated in a standardized way. All tests include a validity criterion for effects in the control, like mortality not to exceed 10%.

Mortality is usually expressed by means of LC_{50}, the dose at which 50% of the animals die. Although extrapolation of laboratory-derived test results to the field is not straightforward, this endpoint is highly relevant for the field. The performance of the standardized test is usually checked by occasional testing of a reference compound like chloracetamide in case of testing of organic pesticides.

Reproduction too is of high relevance for the field. Various endpoints may be considered, including number of cocoons, hatchability of cocoons, number of juveniles, weight of juveniles, and time needed for the juveniles to reach sexual maturation. Juvenile numbers in the control and the coefficient of variation following duplication are important validity criteria. The best way to do reproduction testing is by establishing a full dose–response relationship and subsequently evaluating the no observed effect concentration (NOEC) or the effect concentration (EC_x) at which a specific percentage of reduction of reproduction is deducible.

Body weight change is less clearly defined in testing protocols and may be interpreted in different ways. Ring tests have shown that reproducibility of body weight change is sufficient, but an inverse relationship between reproduction and body weight change was found: animals that rapidly gain weight do not reproduce at the same time and the mechanisms influencing this process are not yet fully understood.

Care should be taken in the evaluation of body weight changes when mortality occurs, as mean body weight changes may be obscured by differences in sensitivity among animals of different size and weight. This problem is less relevant when body weight change is expressed as the change in overall biomass, thus including the mortality endpoint.

Independent of the endpoint and the test duration, behavior of the earthworms is a factor complicating the interpretation of the test results. Prolonged burrowing time, prolonged crawling on the soil surface, flaccidity, hardened test animals, and color changes either may directly affect the testing results or may be an indicative of more delicate effects. A test approach that is recently getting increased attention deals with the ability of earthworms to avoid contaminated soil. This ability can act as an indicator of toxic potential in a particular soil [72] and has the potential to be used as an early screening tool in site-specific risk assessment. Avoidance tests are becoming more common in soil ecotoxicology because they are ecologically relevant and have a shorter duration time compared with standardized soil toxicity tests. Soil properties like quantity and quality of soil organic matter, texture, and soil pH can, however, modify the avoidance response, and obviously the impact of soil properties needs to be properly considered when interpreting results of avoidance tests with earthworms.

4.2.2 Terrestrial Model Ecosystems and Field Enclosures

To facilitate extrapolation of laboratory-derived testing results toward the field, TMEs and field enclosures are used to more realistically simulate field conditions [61]. Experiments with model ecosystems offer several advantages compared to field studies and simple laboratory setups. Though limited in size they bear complex biotic and abiotic interactions. The parameters under investigation can be easily modulated, environmental conditions can be controlled, and in contrast to field tests it is possible to study effects of chemicals while avoiding uncontrolled distribution of residues and metabolites within the biosphere. Despite the complexity of TMEs and field enclosures, they can be sufficiently replicated in order to establish an appropriate statistical plot design. Model ecosystems described in the literature differ in many features. This concerns size, soil structure (intact soil core vs. homogeneous filling), organisms (natural community vs. selected taxa), and the exposure site (field vs. laboratory). Thus model systems differ notably in their similarity to field conditions.

The extrapolation from model ecosystem experiments to the field situation is to be more feasible than from laboratory experiments. Experiments measuring microbial activity and availability of macronutrients showed for instance that field enclosures (exposition in the field) are more reliable in resembling the field situation than indoor TMEs (exposition in the laboratory) [73]. Nonetheless, a noncritical transfer of results from model systems to the field is not acceptable and a sound validation with appropriate field studies is recommended [74]. Different experiments with various types of model systems have been conducted and published [75–77]. The objectives of most studies were (1) to analyze the fate of chemicals, (2) to study their direct effects on organisms, (3) to validate mathematical models, and (4) to measure secondary, indirect effects on the ecosystem [78].

Recently TMEs were discussed for regulatory purposes in the environmental risk assessment of industrial chemicals, biocides, and plant protection products within the European Union [79]. Annex IV of the EU-Directive 91/414/EC [80] concerning the placing of plant protection products on the market lists the conditions (thresholds) which demand a scientific verification of laboratory effect studies with soil organisms under field conditions. Referring to Annex II, Sect. 8.4, the authors [79] conclude that TMEs are considered to be an important tool for risk assessment if they resemble conditions in the field. Thus it is apparent that a yet poorly considered objective of TME studies should concern the comparability of TME and field results.

A wide array of endpoints is potentially assessed in TMEs and field enclosures. This includes the endpoints common in laboratory testing as well as feeding activity, burrowing behavior, and avoidance testing.

4.2.3 Field Testing

Conditions in the field are highly variable and may change drastically over episodes of less than 1 day (like the day/night cycle, deposition of rain and/or snow, strongly increased temperatures in the top layer during periods of sunshine, as well as longer lasting episodes of flooding and drought). Adverse effects on earthworms are typically assessed at the species and community level in terms of abundance and population densities, and maturity. Often, biomarkers are applied to identify previous exposure and body burdens are used as indicators of effective exposure. No standardized assessment methods of field effects are available, let alone validated models to extrapolate across soils.

4.3 Factors Affecting Toxicity Test Results

Standardized toxicity testing is conducted under fixed biotic and environmental conditions that allow comparison of results among testing laboratories and facilitate interpretation of the findings. However, increased standardization inherently hinders extrapolation of test results toward the field. To improve understanding of specific differences between laboratory testing and field effects, the factors affecting differences in effective exposure and actually occurring effects in the laboratory settings and in the field need intense investigation.

An obvious factor that is of relevance in comparing test results is the test species used vs. species common in the field. Compost worms are commonly used in laboratory testing, probably due to the relative ease of culturing compost worms by means of organic rich material like dung. As noted before, typical field worms like *L. rubellus*, *L. terrestris*, *A. caliginosa*, and *Aporrectodea rosea* do not reproduce easily. This raises the question of typical differences in sensitivity toward chemical across compost worms and typical soil worms. Spurgeon and Hopkins [21] observed that although *E. fetida* was less sensitive to zinc than *L. rubellus* and *A. rosea*, the difference in toxicity was no more than a factor of 2 and was within-test variability. Heimbach [81] on the other hand observed larger differences in earthworm sensitivity to earthworms, up to a factor of 10 between *E. fetida* and *L. terrestris*.

Field populations of earthworms typically consist of a mixture of adults, subadults (nonclitellate worms), juveniles, newly hatched animals, and cocoons. Particularly severe effects of contaminants on any life stage could have severe effects on populations. On the other hand, laboratory testing is typically carried out with adult worms only. Typically, juveniles are more sensitive to toxicants than adult worms; Spurgeon and Weeks [82] showed for instance a difference of a factor of 1.9 between toxicity of zinc to juvenile and adult worms.

Exposure time is an important factor in extrapolating toxicity test results. This is especially true for chemicals (most notably metals) that display slow uptake and elimination kinetics. Typical maximum exposure times in laboratory testing of about 28 days are often too low to reach equilibration of metal levels in the organisms. This is especially true for nonessential metals as internal concentrations of essential elements are usually regulated within well-defined limits. The aspect of test duration therefore requires specific attention in extrapolating test results.

Weather conditions are another factor to consider, albeit that data on the effect of temperature and humidity on earthworm sensitivity are scarce. The most common earthworm species in the field are typically least sensitive to contaminants at temperature conditions in between 10 and 15°C.

Soil properties and pretreatment conditions are probably the most dominant factors impacting the sensitivity of earthworms. In case of metals, soil pH is a dominant factor in this respect. In general, a decrease of pH will increase metal levels in the pore water and hence toxicity, albeit that hydrogen ions are protective of metal toxicity. Soil sorption sites like organic matter and clay strongly modulate toxicity. Criel et al. [83] studied for instance the effect of soil characteristics on the toxicity of copper to terrestrial invertebrates, and performed chronic toxicity tests with *E. fetida* in 19 European field soils. Toxicity values varied largely among soils with 28d EC_{50} (concentrations causing 50% effect) ranging from 72.0 to 781 mg Cu kg^{-1} dry weight. Variation in copper toxicity values was best explained by differences in the actual cation exchange capacity (CEC) at soil pH. Using the obtained regression algorithms, the observed toxicity could – in most cases – be predicted within a factor of two.

The effect of pretreatment is most significantly related to aging of the contaminants prior to testing. Longer aging times greatly decrease toxicity for both organic chemicals and metals.

4.4 How to Model Toxicity

The example given above of Criel et al. [83] of modeling metal toxicity across a series of field soils is a nice illustration of the current state of the art. As opposed to the aquatic compartment, the interplay between the biotic and abiotic factors modulating toxicity is not yet well understood. Consequently, models for predicting toxicity toward earthworms across a wide array of soils and soil types are virtually lacking. This is especially the case for metals.

4.4.1 Organic Compounds

In case of hydrophobic organic chemicals that act strictly according to the general mechanism of polar narcosis, competition for sorption of the contaminant between the soil organic matter and the organic matter of the earthworm has been the basis for establishing the critical body residue concept (CBR) and the translation of CBRs toward critical concentrations in any of the environmental compartments, assuming on the one hand that the total body concentration of a nonpolar narcotic organic contaminant is proportional to the concentration at the target or receptor of toxicity, while on the other hand assuming that (1) the fat tissue is the main storage compartment for hydrophobic organic chemicals and (2) the fat tissue behaves similarly to the abiotic organic phases present in the system.

McCarty and Mackay [84] showed that CBRs for polar narcotics are indeed fairly constant. The latter two assumptions imply that the CBR or a specific effect level (EC_x, with x being the extent of adverse effect) is proportional to the octanol–water partitioning coefficient of the chemical:

$$\log \mathrm{CBR} \text{ or } \log \mathrm{EC}_x \approx a(\log K_{\mathrm{ow}}) + b. \tag{3}$$

Karickhoff et al. [85] were one of the first authors to show the equilibrium concept of partitioning of organic compounds by reporting that K_{ow} is proportional to the compound-specific organic-carbon normalized partition coefficient (K_{oc}). Subsequently it is K_{oc} that may be used to predict not only the degree of chemical partitioning between water and the sediment or soil organic carbon, but also the baseline-toxicity of hydrophobic organic chemicals in a specific medium varying in organic carbon content:

$$K_{\mathrm{oc}} = K_{\mathrm{d}}/f_{\mathrm{oc}}, \quad \text{with } K_{\mathrm{d}} = C_{\mathrm{w}}/C_{\text{solid phase}} \tag{4}$$

and

$$\log \mathrm{CBR} \text{ or } \log \mathrm{EC}_x = a(\log K_{\mathrm{oc}})f_{\mathrm{oc}} + b. \tag{5}$$

Although the study was not carried out with earthworms, Paumen et al. [86] recently cautioned that even minute changes in the chemical structure (in this case isomers and metabolites) of a toxicant may induce unpredictable (isomer) specific toxicity, not only emphasizing the need of chronic toxicity testing to gain insight into long-term effects but also elegantly showing the limitations the CBR concept.

Van Gestel and Ma [87] combined information on exposure routes (pore water) and toxicity data of chlorinated aromatics for two earthworm species (*E. andrei* and *L. rubellus*) in four (chlorophenols and chloroanilines) and two (chlorobenzenes) soils to derive quantitative structure activity relationships (QSARs) that may be used to predict toxicity of chloroaromatics in additional soils. The QSARs are based on lipophilicity of the test compounds, expressed in terms of their log K_{ow}. It was noted by these authors that both earthworm species are not equally sensitive to chlorobenzenes and chloroanilines, *E. andrei* is more sensitive than *L. rubellus* to chlorophenols and toxicity data of chlorosubstituted anilines, phenols, and benzenes are in close agreement with data on toxicity for fish.

A similar conclusion was drawn by Miyazaki et al. [88] for acute toxicity of chlorophenols for *E. fetida*. A different exposure modality was used by these authors to derive QSARs as the worms were exposed on filter paper wetted with a solution of the individual chlorophenols.

4.4.2 Metals

For many metals, it is the free ionic form that is most responsible for toxicity. This is despite the fact that strictly speaking, metals may be taken up via various exposure pathways and in a complexed state, bound to a number of ligands of varying binding capacity and varying binding strength. The FIAM is used to explain the relationship between speciation in the external environment and bioavailability to the organisms [30]. The FIAM produces speciation profiles of a metal in an aquatic system and provides insight into the relative bioavailabilities of the different forms of metal as well as the importance of complexation. The basic assumption underlying the FIAM is that adverse effects are proportional to the activity of the free metal ion in solution, or in the case of soils – the pore water. Although it has been shown that other species might also contribute to metal uptake and metal toxicity, most evidence supports the FIAM.

There is, however, an increasing body of evidence becoming available, showing that the toxicity caused by the free metal ion is modulated by a number of chemically induced competing processes. This observation was the basis for the development of Biotic Ligand Models (BLMs). BLM theory on the one hand incorporates the impact of water chemistry (most notably pH and DOC) on metal speciation, whereas the model on the other hand quantifies the assumption of competition between the major cations like Ca^{2+}, Mg^{2+}, Na^{+}, and H^{+}, and free metal ions for binding sites at the organism–water interface may result in a decreased toxicity of the free metal ion [89]. In some cases it is taken into account that other metal species have the potential to contribute to toxicity, like complexes with OH^{-} and $CO_3{}^{2-}$ ions and organic metabolites in case of Cu. BLMs include all these aspects and are, therefore, gaining increased interest in the scientific as well as the regulatory community. In fact, the BLM concept, now developed for Cu, Ni, Ag, and Zn, is considered as

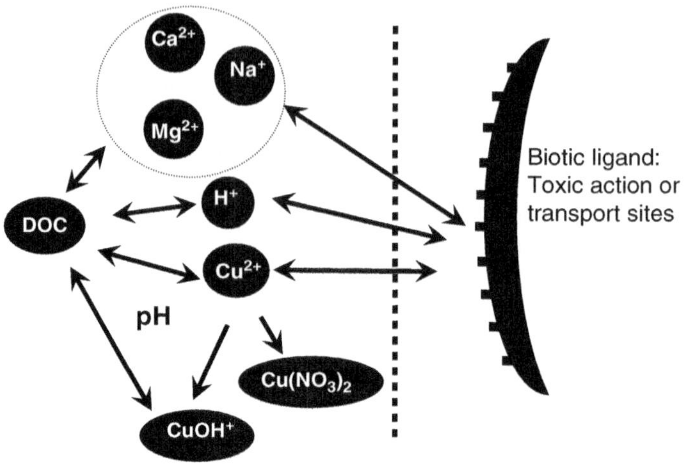

Fig. 4 Schematic overview of the processes underlying the Biotic Ligand Concept for metal toxicity, in this case copper. The *left-hand side* of the scheme depicts pore water constituents that affect copper speciation, the *right-hand side* depicts the interaction of the free copper ion with the biotic ligand of the earthworm (in this case the epidermis) as affected by competition with competing ions like $Ca^{2+}/Na^{+}/Mg^{2+}/H^{+}$

the currently most practical technique to assess the ecotoxicity of metals on a site-specific basis. Therefore, the BLM concept is now being approved in the EU. A schematic representation of the BLM concept is given in Fig. 4.

A basic assumption of the BLM is that metal toxicity occurs as the result of metal ions reacting with binding sites at the organism–water interface, represented as a metal–biotic ligand (metal–BL) complex. The concentration of this metal–BL complex is proportionally related to the magnitude of the toxic effect, independent of the physical–chemical characteristics of the test medium. Hence, the acute toxicity of a trace metal to an organism can be calculated when metal speciation, the activity of each cation in solution, and the stability constant for each cation to the BL(s) for the organism are known. BLMs have recently been developed for copper toxicity to earthworms [90]. Paquin et al. [89] provided a historical overview of the fundamentals of BLMs.

5 Conclusions

Species-specific morphological, physiological, and behavioral aspects basically determine the contribution of potential uptake pathways of nutrients and natural and anthropogenic contaminants. Intraspecies (especially including short-term weather deviations) and interspecies variances (like size and ecological preferences) will most likely modify the actual contribution of potential exposure pathways, thus modifying actually occurring adverse effects.

Earthworms are ubiquitous ecosystem engineers and litter transformers that are essential for maintaining a healthy soil ecosystem. They inhabit virtually all soil layers while they tend to move upward and downward the soil profile in response to variations in the water table. Earthworms have been studied for various decades and their intra- and interspecies variances are fairly well understood. It may be concluded that earthworms are suited organisms for ecotoxicity studies and indicator organisms for the assessment of potential risks:

1. The uptake routes of chemicals are clear, with a dominant contribution of uptake of pollutants via the pore water. For hydrophobic chemicals with log K_{OW} > approximately 6, ingestion of food and soil particles may induce additional uptake of micropollutants.
2. The magnitude of accumulation of chemicals is rather high and earthworms are therefore suited for assessing potentially bioavailable fractions and resulting adverse effects. Compartment modeling may be used to quantify accumulation as a function of time.
3. Earthworms are well suited for assessing adverse effects:
   (a) A number of toxicity endpoints (like mortality, reproductive success, growth) may relatively easily be deduced, whereas earthworms are not specifically more sensitive or less sensitive for the majority of chemicals. Van Gestel and Ma [87] found for instance that toxic effects of chlorinated aromatics are similar for earthworms and fish.
   (b) Because of their ease of cultivation and their ubiquitous nature, earthworms have frequently been the topic of study and effect and accumulation data are relatively abundant for comparative purposes and for inter- and intrasystem extrapolation.
   (c) It has been shown that it is possible to derive QSARs for predicting effects of chemicals on various earthworm species.

On the other hand it should be noted that most effect and accumulation assays haven typically been carried out in a laboratory setting. Field studies varying from TMEs [75] up till analysis of population parameters are scarce. Field studies at all levels of ecological hierarchy would be well suited for extrapolation and validation of models generated on the basis of laboratory data and would provide important tools for assessing ecosystem health.

References

1. Fragoso C, Brown G, Feijoo A (2004) The influence of Gilberto Righi on tropical earthworm taxonomy: The value of a full-time taxonomist. Pedobiologia 47: 400–404
2. Darwin C (1809–1882) The formation of vegetable mould through the action of worms, with observations on their habits. Release date 2000–10–01
3. Bouché MB (1977) Strategies lombriciennes. In: U. Lohm, T. Persson (Eds) Soil organisms as components of ecosystems. Ecol Bull (Stockholm) 25: 122–132

4. Lee KE (1959) The earthworm fauna of New Zealand. New Zeal Depart Sci Ind Res Bull 130: 486
5. Lee KE (1985) Earthworms – Their Ecology and Relationships with Soils and Land Use. Academic, New York, NY, p. 411
6. Lavelle P (1981) Stratégies de reproduction chez les vers de terre. Acta Oecol Gen 2: 117–133
7. Lavelle P (1997) Faunal activities and soil processes: Adaptive strategies that determine ecosystem function. Adv Ecol Res 27: 93–132
8. Lavelle P, Bignell D, Lepage M, Wolters V, Roger P, Ineson P, Heal OW, Dhillion S (1997) Soil function in a changing world: The role of invertebrate ecosystem engineers. Eur J Soil Biol 33: 159–193
9. Blanchart E, Lavelle P, Braudeau E, LeBissonnais Y, Valentin C (1997) Regulation of soil structure by geophagous earthworm activities in humid savannas of Côte d'Ivoire. Soil Biol Biochem 29: 431–439
10. Edwards CA (Ed.) (1998, 2004). Earthworm Ecology (1st Ed. 1998; 2nd Ed. 2004) CRC, Boca Raton FL
11. Zharikov GA, Fartukov SV, Tumansky IM, Ishchenko NV (1993) Use of the solid wastes of microbial industry by preparing worm compost. Biotechnologia 9: 21–23
12. McKey-Fender D, Fender WM, Marshall VG (1994) North American earthworms native to Vancouver Island and the Olympic Peninsula. Can J Zool 72: 1325–1339
13. Waeterschoot H, Van Assche F, Regoli L, Schoeters I, Delbeke K (2003) Metals in perspective. J Environ Monit 5: 95N–102N
14. Morgan JE, Morgan AJ (1988) Earthworms as biological monitors of Cd, Cu, Pb, and Zn in metalliferous soils. Environ Pollut 54: 123–138
15. Saxe JK, Impellitteri CA, Peijnenburg WJGM, Allen HE (2001) A novel model describing heavy metal concentrations in the earthworm *Eisenia andrei*. Environ Sci Technol 35: 4522–4529
16. Jager T, Fleuren RHLJ, Hogendoorn EA, De Korte G (2003) Elucidating the routes of exposure for organic chemicals in the earthworm, *Eisenia andrei* (Oligochaeta). Environ Sci Technol 37: 3399–3404
17. Vijver MG, Vink JPM, Miermans CJH, Van Gestel CAM (2003) Oral sealing using glue: A new method to distinguish between intestinal and dermal uptake of metals in earthworms. Soil Biol Biochem 35: 125–132
18. Didden W (2003) Oligochaetes. In: Markert BA, Breure AM, Zechmeister HG (Eds) Bioindicators and Biomonitors. Elsevier, Amsterdam
19. Løkke H, Van Gestel CAM (1998) Handbook of Soil Invertebrate Toxicity Tests. Wiley, Chichester
20. Van Gestel CAM, Dirven-van Breemen EM, Baerselman R (1993) Accumulation and elimination of cadmium, chromium and zinc and effects on growth and reproduction in *Eisenia andrei* (Oligochaeta, Annelida). Sci Total Environ Part 1: 585–597
21. Spurgeon DJ, Hopkin SP (1996) The effects of metal contamination on earthworm populations around a smelting works – quantifying species effects. Appl Soil Ecol 4: 147–160
22. Osté LA, Dolfing J, Ma W-C, Lexmond TM (2001) Cadmium uptake by earthworms as related to the availability in the soil and the intestine. Environ Toxicol Chem 20: 1785–1791
23. Lanno RP, McCarty LS (1997) Earthworm bioassays: Adopting techniques from aquatic toxicity testing. Soil Biol Biochem 29: 693–697
24. Stürzenbaum SR, Kille P, Morgan AJ (1998) Heavy metal-induced molecular responses in the earthworm, *Lumbricus rubellus* genetic fingerprinting by directed differential display. Appl Soil Ecol 9: 495–500
25. Lanno RP, Wren CD, Stephenson GL (1997) The use of toxicity curves in assessing the toxicity of soil contaminants to *Lumbricus terrestris*. Soil Biol Biochem 29: 689–692
26. Laverack MS (1963) The Physiology of Earthworms. Pergamon, Oxford
27. Wallwork JA (1983) Annelids: The First Coelomates. Studies in Biology, Earthworm Biology. Edward Arnold Publishers, London
28. Edwards CA, Lofty JR (1972) Biology of Earthworms. Chapman and hall, London

29. Ireland MP, Richards KS (1981) Metal content, after exposure to cadmium, of two earthworms of known differing calcium metabolic activity. Environ Pollut 26: 69–78
30. Campbell PGC (1995) Interactions between trace metals and aquatic organisms: A critique of the free-ion activity Model. In: Tessier A, Turner DR (Eds) Metal Speciation and Bioavailability in Aquatic Systems. Wiley, New York, NY, pp. 46–102
31. Janssen RPT, Peijnenburg WJGM, Posthuma L, Van den Hoop MAGT (1997) Equilibrium partitioning of heavy metals in Dutch field soils. I. Relationship between metal partitioning coefficients and soil characteristics. Environ Toxicol Chem 16: 2479–2488
32. Lock K, Jansen CR (2001) Zinc and cadmium body burdens in terrestrial oligochaetes: Use and significance in environmental risk assessment. Environ Toxicol Chem 20: 2067–2072
33. Jager T (1998) Mechanistic approach for estimating bioconcentration of organic chemicals in earthworms. Environ Toxicol Chem 17: 2080–2090
34. Sample BE, Suter ii GW Beauchamp JJ, Efroymson RA (1999) Literature-derived bioaccumulation models for earthworms: Development and validation. Environ Toxicol Chem 18: 2110–2120
35. Kelsey JW, White JC (2005) Multi-species interactions impact the accumulation of weathered 2,2-bis (p-chlorophenyl)-1,1-dichloroethylene (p, p′-DDE) from soil. Environ Pollut 137: 222–230
36. Kreitinger JP, Quiñones-Rivera A, Neuhauser EF, Alexander M, Hawthorne SB (2007) Supercritical carbon dioxide extraction as a predictor of polycyclic aromatic hydrocarbon bioaccumulation and toxicity by earthworms in manufactured-gas plant site soils. Environ Toxicol Chem 26: 1809–1817
37. McGeer JC, Brix KV, Skeaf JM, DeForest DK, Brigham SI, Adams WJ, Green A (2003) Inverse relationship between bioconcentration factor and exposure concentration for metals: Implications for hazard assessment of metals in the aquatic environment. Environ Toxicol Chem 22: 1017–1037
38. DeForest DK, Brix KV, Adams WJ (2007) Assessing metal bioaccumulation in aquatic environments: The inverse relationship between bioaccumulation factors, trophic transfer factors and exposure concentration. Aquat Toxicol 84: 236–246
39. Lanno R, Wells J, Conder J, Bradham K, Basta N (2004). The bioavailability of chemicals in soil for earthworms. Ecotoxicol Environ Safety 57: 39–47
40. Paracelsus (Philip T. B. von Hohenheim) (1564) Drey Bucker, The Heirs of Arnold Byrckmann, Cologne Germany
41. Vijver MG, Van Gestel CAM, Lanno RP, Van Straalen NM, Peijnenburg WJGM (2004) Internal metal sequestration and its ecotoxicological relevance – a review. Environ Sci Technol 38: 4705–4712
42. Loonen H, Muir DCG, Parsons JR, Govers HAJ (1997) Bioaccumulation of polychlorinated dibenzo-p-dioxins in sediment by oligochaetes: Influence of exposure pathway and contact time. Environ Toxicol Chem 16: 1518–1525
43. Belfroid A, Seinen W, Van Den Berg M, Hermens J, Van Gestel K (1995) Uptake, bioavailability and elimination of hydrophobic compounds in earthworms (*Eisenia andrei*) in field contaminated soil. Environ Toxicol Chem 14: 605–612
44. Belfroid A, Seinen W, Van Gestel K, Hermens J, Van Leeuwen K (1995) Modelling the accumulation of hydrophobic organic chemicals in earthworms: Application of the equilibrium partitioning theory. Environ Sci Pollut Res 2: 5–15
45. Beyer WN (1996) Accumulation of chlorinated benzenes in earthworms. Bull Environ Contam Toxicol 57: 729–736
46. Morgan JE, Morgan AJ (1990) The distribution of cadmium, copper, lead, zinc and calcium in the tissues of the earthworm *Lumbricus rubellus* sampled from one uncontaminated and four polluted soils. Oecologia 84: 559–566
47. Morgan AJ, Turner MP, Morgan JE (2002) Morphological plasticity in metal-sequestering earthworm chloragocytes: Morphometric electron microscopy provides a biomarker of exposure in field populations. Environ Toxicol Chem 21: 610–618

48. Morgan JE, Morgan AJ (1998) The distribution and intracellular compartmentation of metals in the endogeic earthworm *Aporrectodea caliginosa* sampled from an unpolluted and a metal-contaminated site. Environ Pollut 99: 167–175
49. Stürzenbaum SR, Winters C, Galay M, Morgan AJ, Kille P (2001) Metal ion trafficking in earthworms – identification of a cadmium specific metallothionein. J Biol Chem 276: 34013–34018
50. Prinsloo MW, Reinecke SA, Przybylowicz WJ, Mesjasz-Przybylowicz J, Reinecke AJ (1990) Micro-PIXE studies of Cd distribution in the nephridia of the earthworm *Eisenia fetida* (Oligochaeta). Nucl Instrum Methods Phys Res B 158: 317–322
51. Andersen C, Laursen J (1982) Distribution of heavy metals in *Lumbricus terrestris, Aporrectodea longa* and *A. rosea* measured by atomic absorption and X-ray fluorescence spectrometry. Pedobiologia 24: 347–356
52. Vijver MG, Van Gestel CAM, Van Straalen NM, Lanno RP, Peijnenburg WJGM (2006) Biological significance of metals partitioned to subcellular fractions within earthworms (*Aporrectodea caliginosa*). Environ Toxicol Chem 25: 807–814
53. Van Straalen NM, Donker MH, Vijver MG, Van Gestel CAM (2005) Bioavailability of contaminants estimated from uptake rates into soil invertebrates. Environ Pollut 136: 409–417
54. Peijnenburg W, Posthuma L, Zweers P, Baerselman R, De Groot A, Van Veen R, Jager D (1999) Relating environmental availability to bioavailability: Soil-type dependent metal accumulation in the oligochaete *Eisenia andrei*. Ecotoxicol Environ Safety 44: 294–310
55. Widianarko B, Kuntoro FX, Van Gestel CAM, Van Straalen NM (2001) Toxicokinetics and toxicity of zinc under time-varying exposure in the guppy (*Poecilia reticulata*). Environ Toxicol Chem, 20: 763–768
56. Peijnenburg WJGM, Zablotskaja M, Vijver MG (2007) Monitoring metals in terrestrial environments within a bioavailability framework and a focus on soil extraction. Ecotoxicol Environ Safety 67: 163–179
57. Awata H, Johnson KA, Anderson TA (2000) Passive sampling devices as surrogates for evaluating bioavailability of aged chemicals in soil. Toxicol Environ Chem 73: 25–42
58. Van Der Wal L, Jager T Fleuren RHLJ, Barendregt A, Sinnige TL, Van Gestel CAM, Hermens JLM (2004) Solid phase microextraction to predict bioavailability and accumulation of organic micropollutants in terrestrial organisms after exposure to a field-contaminated soil. Environ Sci Technol 38: 4842–4848
59. Bergknut M, Sehlin E, Lundstedt S, Andersson PL, Haglund P, Tysklind M (2007) Comparison of techniques for estimating PAH bioavailability: Uptake in *Eisenia fetida*, passive samplers and leaching using various solvents and additives. Environ Pollut 145: 154–160
60. Koolhaas JE, Van Gestel CAM, Römbke J, Soares AMVM, Jones SE (2004) Ring-testing and field-validation of a terrestrial model ecosystem (TME) – An instrument for testing potentially harmful substances: Effects of carbendazim on soil microarthropod communities. Ecotoxicology 13: 75–88
61. Boyle TB, Fairchild JF (1997) The role of mesocosm studies in ecological risk analysis. Ecol Appl 7: 1099–1102
62. McCarthy JF, Shugart LR (1990) Biological markers of environmental contamination. In: McCarthy JF, Shugart LR (Eds) Biomarkers of Environmental Contamination. Lewis Publishers Chelsea, MI
63. Depledge MH, Fossi MC (1994) The role of biomarkers in environmental assessment (2) invertebrates. Ecotoxicology 3: 161–172
64. Weeks JM, Svendsen C (1996) Neutral red retention by lysosomes from earthworm (*Lumbricus rubellus*) coelomocytes: A simple biomarker of exposure to soil copper. Environ Toxicol Chem 15: 1801–1805
65. Xiao NW, Song Y, Ge F, Liu XH, Yang ZY (2006) Biomarkers responses of the earthworm *Eisenia fetida* to acetochlor exposure in OECD soil. Chemosphere 65: 907–912
66. Scott-Fordsmann JJ, Weeks JM (1998) Review of selected biomarkers in earthworms. In: Sheppard S, Bembridge J, Holmstrup M, Posthuma L (Eds) Advances in Earthworm Ecotoxicology. SETAC Press, Pensacola, FL, pp. 173–189

67. Dikshith TSS, Gupta SK (1981) Carbaryl induced biochemical changes in earthworm (*Pheretima posthuma*). Indian J Biochem Biophys 18: 154
68. Maboeta MS, Reinecke SA, Reinecke AJ (2002) The relation between lysosomal biomarker and population responses in a field population of Microchaetus sp. (Oligochaeta) exposed to the fungicide copper oxychloride. Ecotoxicol Environ Safety 52: 280–288
69. International Standards Organization (1993) Soil Quality – Effects of Pollutants on Earthworms (*Eisenia fetida*). Part 1: Determination of Acute Toxicity Using Artificial Soil Substrate, Geneva, Switzerland ISO DIS 11268–1
70. International Standards Organization (1996) Soil Quality – Effects of Pollutants on Earthworms (*Eisenia fetida fetida, Eisenia fetida andrei*). Part 2: Determination of Effects on Reproduction, Geneva, Switzerland ISO DIS 11268–2
71. Organization for Economic Cooperation and Development (1984) OECD guidelines for testing of chemicals: Earthworm acute toxicity test. OECD Guideline No. 207, Paris, France
72. Natal-da-Luz T, Römbke J, Sousa JP (2008) Avoidance tests in site-specific risk assessment – influence of soil properties on the avoidance response of collembolan and earthworms. Environ Toxicol Chem 27: 1112–1117
73. Teuben A, Verhoef HA (1992) Relevance of micro- and mesocosm experiments for studying soil ecosystem processes. Soil Biol Biochem 24: 1179–1183
74. Carpenter SR (1996) Microcosm experiments have limited relevance for community and ecosystem ecology. Ecology 77: 677–680
75. Knacker T, Van Gestel CAM, Jones SE, Soares AMVM, Schallnaß HJ, Förster B, Edwards CA (2004) Ring-testing and field-validation of a terrestrial model ecosystem (TME) – An instrument for testing potentially harmful substances: Conceptual approach and study design. Ecotoxicology 13: 9–27
76. Parmelee RW, Wentsel RS, Phillips CT, Simini M, Checkai RT (1993) Soil microcosm for testing the effects of chemical pollutants on soil fauna communities and trophic structure. Environ Toxicol Chem 12: 1477–1486
77. Vink K, Van Straalen NM (1999) Effects of benomyl and diazinon on isopod mediated leaf litter decomposition in microcosms. Pedobiologia 43: 345–359
78. Boyle TP, Fairchild JF (1997) The role of mesocosm studies in ecological risk analysis. Ecol Appl 7: 1099–1102
79. Weyers A, Sokull-Klüttgen B, Knacker T, Martin S, Van Gestel CAM (2004) Use of terrestrial model ecosystem data on environmental risk assessment for industrial chemicals, biocides and plant protection products in the EU. Ecotoxicology 13: 163–176
80. European Union. Council Directive of 15 July 1991 Concerning the Placing of Plant Protection Products on the Market, 91/414/EC Brussels Belgium
81. Heimbach F (1992) Correlation between data from laboratory and field tests for investigating the toxicity of pesticides for earthworms. Soil Biol Biochem 24: 1749–1753
82. Spurgeon DJ, Weeks JM (1998) Evaluation of factors influencing results from laboratory toxicity tests with earthworms. In: Sheppard S, Bembridge J, Holmstrup M, Posthuma L (Eds) Advances in Earthworm Ecotoxicology. SETAC Press, Pensacola, FL, pp. 15–25
83. Criel P, Lock K, Van Eeckhout H, Oorts K, Smolders E, Janssen C (2008) Influence of soil properties on copper toxicity for two soil invertebrates. Environ Toxicol Chem 27: 1748–1755
84. McCarty LS, Mackay D (1993) Enhancing ecotoxicological modeling and assessment. Environ Sci Technol 27: 1719–1728
85. Karickhoff SW, Brown DS, Scott TA (1979) Sorption of hydrophobic pollutants on natural sediments. Water Res 13: 241–248
86. Paumen ML, Stol P, Ter Laak TL, Kraak MHS, Van Gestel CAM, Admiraal W (2008) Chronic exposure of the oligochaete *Lumbricus variegatus* to polycyclic aromatic compounds (PACs): Bioavailability and effects on reproduction. Environ Sci Technol 42: 3434–3440
87. Van Gestel CAM, Ma W-C (1993) Development of QSAR's in soil ecotoxicology: Earthworm toxicity and soil sorption of chlorophenols, chlorobenzenes and chloroanilines. Water Air Soil Pollut 69: 265–276

88. Miyazaki A, Amano T, Saito H, Nakano Y (2002) Acute toxicity of chlorophenols to earthworms using a simple paper contact method and comparison with toxicities to fresh water organisms. Chemosphere 47: 65–69
89. Paquin PR, Gorsuch JW, Apte S, Batley GE, Bowles KC, Campbell PGC, Delos CG, DiToro DM, Dwyer RL, Galvez F, Gensemer RW, Goss GG, Hogstrand C, Janssen CR, McGeer JC, Naddy RB, Playle RC, Santore RC, Schneider U, Stubblefield WA, Wood CM, Wu KB (2002) The biotic ligand model: A historical overview. Comp Biochem Physiol C 133: 3–35
90. Steenbergen N, Iaccino F, De Winkel M, Reijnders L, Peijnenburg W (2005) Development of a biotic ligand model and a regression model predicting acute copper toxicity to the earthworm *Aporrectodea caliginosa*. Environ Sci Technol 39: 5694–5702

The Potential for the Use of Agent-Based Models in Ecotoxicology

Christopher J. Topping, Trine Dalkvist, Valery E. Forbes, Volker Grimm, and Richard M. Sibly

Abstract This chapter introduces ABMs, their construction, and the pros and cons of their use. Although relatively new, agent-based models (ABMs) have great potential for use in ecotoxicological research – their primary advantage being the realistic simulations that can be constructed and particularly their explicit handling of space and time in simulations. Examples are provided of their use in ecotoxicology primarily exemplified by different implementations of the ALMaSS system. These examples presented demonstrate how multiple stressors, landscape structure, details regarding toxicology, animal behavior, and socioeconomic effects can and should be taken into account when constructing simulations for risk assessment. Like ecological systems, in ABMs the behavior at the system level is not simply the mean of the component responses, but the sum of the often nonlinear interactions between components in the system; hence this modeling approach opens the door to implementing and testing much more realistic and holistic ecotoxicological models than are currently used.

Keywords Population-level risk assessment · ALMaSS · Pattern-oriented modeling · ODD · Multiple stressors

1 Introduction

This chapter is intended to provide some background on agent-based models (ABMs) and the potential for their use in ecotoxicology. This is achieved by a mixture of examples and minireview of ABM issues; it is, therefore, intended as a primer for those interested in further exploring this type of modeling in ecotoxicology.

Ecotoxicology has, in common with the majority of the natural sciences, followed the basic principles of analytic thinking whereby the whole is abstractly

C.J. Topping (✉)
Department of Wildlife Ecology and Biodiversity, National Environmental Research Institute, University of Aarhus, Grenåvej 14, DK-8410 Rønde, Denmark
e-mail: cjt@dmu.dk

J. Devillers (ed.), *Ecotoxicology Modeling*, Emerging Topics in Ecotoxicology: Principles, Approaches and Perspectives 2, DOI 10.1007/978-1-4419-0197-2_8,

separated into its constituent parts in order to study the parts and their relationships. This approach to science works for physical systems such as those typically studied in physics or chemistry, but may not always be the optimal approach for biological systems with their innate complexity and interactions. For example, in the case of evaluating the impact of stressors on biological systems there is clearly a great difference between the response of animals in the laboratory, given a precisely measured and timed dose of toxicant, and the populations of the same animals moving through a real-world situation of spatiotemporal variability in toxicant concentration, interacting with each other and the biotic and abiotic components of their environment.

It is in fact rather difficult to see how the abstract laboratory test can easily be related to impacts at the population level. Following this train of thought suggests that in order to properly understand this kind of system we should perhaps embrace its complexity rather than ignore it. This means treating a system as an integrated whole whose properties arise from the relationships between the system components rather than studying the components in isolation, thus shifting from the importance of elements to the importance of organizational pattern, i.e., applying a systems approach. Luckily, the use of ABMs opens up the potential for doing just this.

1.1 What Is an ABM?

An ABM is a computational model for simulating the actions and interactions of autonomous individuals in a defined virtual world, with a view to assessing their effects on the system as a whole. This is clearly analogous to integrating the response of individuals into a population response that, when considering impact assessment in ecotoxicology, is the level at which interest and protection goals are usually aimed.

Of course, there are many models of ecological populations and many approaches, but there are a number of characteristics of ABMs that set them apart from other more traditional approaches. These characteristics can be broadly described as being their explicit consideration of spatiotemporal variability and their ability to include individual behavior, with population responses being emergent features. Thus, animal behavior such as patterns of movement can be simulated so that a dispersing animal moves in very different ways depending upon its type (e.g., bird, mouse, beetle, human). This provides a huge predictive potential compared with more aggregated approaches.

These properties have resulted in the use of ABMs in a wide and steadily increasing range of applications. In 1996, there were 31 agent-based papers published (source: ISI Web of Knowledge), but by 2006 the number had risen to 494. Some varied examples include simulations of immune system responses to perturbations [1], of ethnic diversity in economically and spatially structured neighborhoods [2], of entry and exit routes to a baseball stadium under a range of conditions including simulation of terrorist attack [3], and of urban evacuation strategies [4]. Current use of ABMs in ecotoxicology is limited, but their usage in related areas is increasing. Recent developments include models of whale watching by tour boats, including evaluation of the risks to the whale population [5], epidemiology (e.g., [6, 7]), the exploitation of limited renewable resource [8], and conservation [9–11]. ABMs help understand biological systems because, unlike physical systems, there is heterogeneity in their components, and this heterogeneity affects the overall dynamics of the system [12, 13] in short because variation in space and time matters in biological systems and ABMs deal with this very well.

In ecology, ABMs developed somewhat independently of other disciplines and are often referred to as "individual-based models" (IBMs). The distinction is, however, of little importance today, and Grimm [14] suggests not distinguishing IBMs and ABMs any longer and using both terms interchangeably. Originally the term IBM was used to emphasize the discreteness of individuals, heterogeneity among individuals, and local interactions, rather than adaptive decision making and behavior, which have been the main drivers in the development of ABMs [12, 15]. Recently however, IBMs and ABMs have merged into one big class of models [16], covering a wide range from very simple to rather complex models [17].

In this chapter, we focus on "full-fledged" ABMs, which include realistic landscapes, a high temporal and spatial resolution, individual heterogeneity, local interactions, adaptive behavior, and often also different species. This is, in terms of development time and resources needed for testing and parameterization, the most demanding type of ABMs, but also the most powerful one if it comes to the potential to validate these models and to use them for predictions of environmental scenarios that so far have not been observed. It should be kept in mind, however, that more simple ABMs also have their place in basic and applied ecology, including ecotoxicology (e.g., [18]).

1.2 Constructing ABMs

ABMs can be significantly more demanding to develop than other population models. Development starts with the creation of a conceptual model of the system of study comprising the basic simulation goals, elements of the system and their behaviors, and the endpoints of interest [16, 19]. Depending upon the goals of the model, it may utilize designed or empirically grounded agents and environments, and choices here may have significant implications for results, as we now show.

In early ABMs structural environment into which the agents are placed was created using regular geometric shapes, but it is now known that the use of unrealistic structural environments may bias results [20], and a similar argument can be made for simplification of the behaviors of agents [21]. Another problem that the ABM developer may face, which is not a problem for traditional modeling approaches, is that of concurrency. Concurrency problems occur when objects interact, especially if their interaction is controlled via some limiting resource. A good example of this is the well-known model by DeAngelis et al. [22] where wide-mouthed bass interact indirectly through their *Daphnia* food resource and directly by eating each other. By not taking account of concurrency issues the positive feedback loops emergent in the model were strengthened (see [23] for a discussion of this effect and concurrency issues in ABMs). Concurrency issues are not critical to all ABMs but in cases where they are they can increase the complexity of model design. Scheduling of the model's processes and the exact mode of updating the model's state variables are thus critical and need to be planned and communicated carefully [24, 25].

It will by now be apparent that the increase in realism made accessible by ABMs comes at a cost, both in terms of potentially huge data requirements, but also in terms of the technical ability required for model construction. However, the technical problems are eased by the emergence of software tools. Thus, models may be created using ABM "platforms," that is, libraries of predefined routines such as REPAST [26], NetLogo [27], and SWARM [28]. Models of limited complexity can be developed using these platforms, whereas more complex or computationally demanding models are usually implemented in more efficient low-level object-oriented languages such as C++ or Java. Animal, Landscape, and Man Simulation System (ALMaSS), a framework for ABMs for pesticide risk assessment [29], which is used as an example throughout this paper, was written in C++ since run times are very long, and shaving tiny fractions of seconds from loops can save many hours of simulation time with millions of agents.

While simple systems can be built by anyone of average programming ability, the effectiveness of larger scope and more realistic models depends on the ability of the programmer to code efficiently. At this level of software engineering there is a whole new skill set required by the ABM developer. For example, sorting routines are common constructs in ABMs but vary hugely in their efficiency, so choices here may dramatically affect overall runtimes. There is also the problem of code reliability. With large and complex models the scope and complexity of errors increases and code maintenance and debugging tasks can mushroom out of all proportion. This is particularly the case with highly complex multiagent communication such as between flock or family members, and it has cost many weeks of debugging in ALMaSS. Coping with such problems requires familiarity with basic computing science principles. Hence, the optimal solution is that the modeler also possesses software engineering skills, which will not only speed up the development cycle, but will also improve the model design by ensuring good code structure at an early phase. However, while there is an increase in the number of computational biologists being trained, this skill combination is still rare. Grimm and Railsback [16] therefore recommend considering close collaborations of ecological modelers and computer scientists where, however, the modeler should keep full control of the software, that is, not depend on the computer scientist to use the software and modify it.

Unfortunately no simple introduction to building ABMs currently exists. There are many good object-oriented tutorials available however, and these, combined with an understanding of the philosophy of the approach, are a good place to start. Detailed advice can be found in Grimm and Railsback [16] who provide an introduction to what they term "individual-based ecology," which encompasses the use and development of ABMs.

2 Examples Illustrating the Use of ABMs

We here present examples selected to illustrate some of the facets of using ABMs, and some of the interesting results that can emerge. The series of example applications used to illustrate the potential of ABMs in ecotoxicological research utilize a single ABM system, ALMaSS [29]. In these examples space limits a description of the manner in which conclusions were drawn, but in all cases this was by carrying out additional exploratory simulations to test the behavior of the system under different conditions, as well as detailed analysis of outputs in the light of knowledge of the model structure. In addition, we will briefly introduce two further families of ABMs, which were not developed for ecotoxicology, but which very well illustrate both the high costs for developing full-fledged ABMs and their striking predictive power, once their testing has been completed.

2.1 Introduction to ALMaSS

ALMaSS was designed as a system to evaluate the impact of human management of landscapes on key species of animals in the Danish landscape. ALMaSS was not created with a clearly focused goal in mind but to be a highly flexible system capable of simulating a wide range of interactions between landscape structure, management, and animal ecology. Thus, ALMaSS is a flexible system for implementing ABMs of selected species, with the aim of predicting the impact of changes in management of the Danish landscape.

ALMaSS can be separated into two main components: the landscape and animal models. The landscape comprises a topographical map, together with strategies of human management (primarily farming but also other management such as mowing of roadside verges), traffic and road networks, weather, submodels for calculating arthropod biomass, models for general vegetation and crop growth, and also models of the environmental fate of pesticides. These submodels and processes are updated on a daily basis during the simulation and provide the potential to model factors such as farm and crop management in great detail. The farm management modules permit the definition of different farm types each with their specific crop choices and type of management (e.g., conventional pig, arable, and dairy production, and organic variants of these).

Each farm mapped in the landscape is allotted a farm type and the farm manager, also an agent, applies management to his fields in terms of sowing crops and subsequent crop husbandry while reacting to weather and soil conditions. Crop husbandry is highly detailed (see [30]) and simulates all farming activities that would be carried out on that crop (e.g., plowing, harrowing, sowing, fertilizer applications, pesticide applications, harvest, and postharvest operations). Application of pesticides and fertilizers can be allocated specific characteristics (e.g., amount and type) and may result in changes in the vegetation growth, arthropod biomass, and provide field-specific information for animal models such as the type and amount of toxicant present.

The topographic map utilized by the landscape has a resolution of $1\,m^2$ and typically covers an area of $100\,km^2$. Combining this map with the management information, weather and vegetation growth information creates a virtual reality into which the animal models are placed. The animal models are agents designed to simulate the ecology and behavior of individual animals as closely as possible. Each agent moves around in its virtual world in much the same way that a real animal moves in the real world, picking up information from its surroundings as it goes and acting upon this in order to feed and ultimately reproduce. Changes to the agent's environment occur on a daily basis as weather changes, vegetation grows, or the farmer manages a field.

A number of animal models exist for ALMaSS. Those used as examples here are *Alauda arvensis* (skylark) [30, 31], *Microtus agrestis* (field vole) [29], *Bembidion lampros* (beetle) [32], *Erigone atra/Oedothorax fuscus* (spider) [33], and *Capreolus capreolus* (roe deer) [34]. These range from species with highly detailed behavior but low numbers (roe deer) to spiders with simple behavior but the necessity to handle over 1 million agents concurrently. However, all models conform to a basic framework, essentially a state machine, whereby:

- Each animal has an initial state that is a behavioral state.
- There is a set of possible input events.
- Transitions to new behavioral states depend on input events.
- Actions (output events) are determined by behavioral state and environmental opportunities.

Each agent will cycle through this state machine at least once per simulation day and potentially many times depending upon the inputs and outputs. For example, a vole in the state "explore" may explore his surroundings, resulting in the input that there is no food, and make a transition to the new state "dispersal"; this results in the action of dispersal that then triggers a transition to the state "explore." This cycle may repeat itself until the vole finds food, dies, or runs out of time that day (Fig. 1). Inputs may also occur as events, not under the control of the animal. For example, if our dispersing vole is run over by a car it will make an immediate transition to

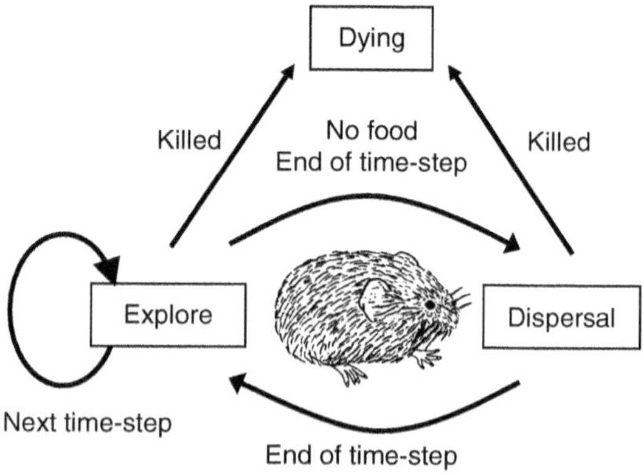

Fig. 1 A diagram of a fragment of the field vole state machine. States are denoted with *boxes*, transitions by *arrows*. See text for further explanation

the state “dying.” This event-driven interaction is also the basis for modeling topical exposure to pesticide applications, meaning that an animal may only be exposed if it is in the location where the pesticide is sprayed at the time it is sprayed.

A system such as ALMaSS has a number of potential uses in ecotoxicology. These can broadly be divided into three main categories:

- Policy scenario analysis: This utilizes the capability of the agent-based system to respond to changing inputs. For example, how will pesticide usage be affected by specific taxation measures? (see examples 1 and 4 later). Taxation is an input to the model that causes changes in farmer behavior, which result in changed patterns of pesticide use. Since the animals react to pesticides as they find them in their day-to-day activity, their behavior in turn is affected, and the sum of their behaviors results in a population response that can be evaluated.
- Risk/impact assessment and regulation: Scenarios of application of pesticides with specified properties are studied and population responses are evaluated (see examples 2 and 3). The challenge here is to define specific yet representative scenarios, since a greater range of factors is analyzed than is traditional in this area.
- Systems understanding: Perhaps the most important use of ABMs in ecotoxicology is to improve our understanding of the ecological systems and how they are affected by pesticides. ALMaSS is able to use a systems approach to investigate system properties that would be impossible or exceedingly difficult to study in real life (see examples 1–4).

2.2 Example 1: Impacts of Mechanical Weeding on Skylark Populations

Pesticide use has been an important factor in the decline of a range of European farmland bird species over the last 20 years, primarily via indirect effects on wild plants and arthropods [35, 36]. It is, therefore, desirable to use pesticides less, but policies directed toward this need to be based on good advice. With this background Odderskær et al. [37] set out to evaluate the potential impact of replacing herbicide use with mechanical weeding on inter alia skylark populations. Mechanical weeding is rarely used in conventional farming, despite its well-documented effectiveness, so there is little opportunity for observational study. The goal of the ALMaSS modeling was to assess the direct or indirect impact of mechanical weeding on birds reproducing in fields where it is applied. The problem was tackled in two stages: the first an experiment to assess the lethality of mechanical weeding to skylark nests, and the second to assess potential impacts of different management scenarios.

A range of scenarios were simulated (see [37]) but those that show the clearest results are experimental scenarios where the assumption is that all farmers in the landscape grew a single monoculture crop. Figure 2 shows the number of nests, nests with eggs (under incubation), and nests with young, which were destroyed when mechanical weeding was used in monoculture spring barley on either the 10th

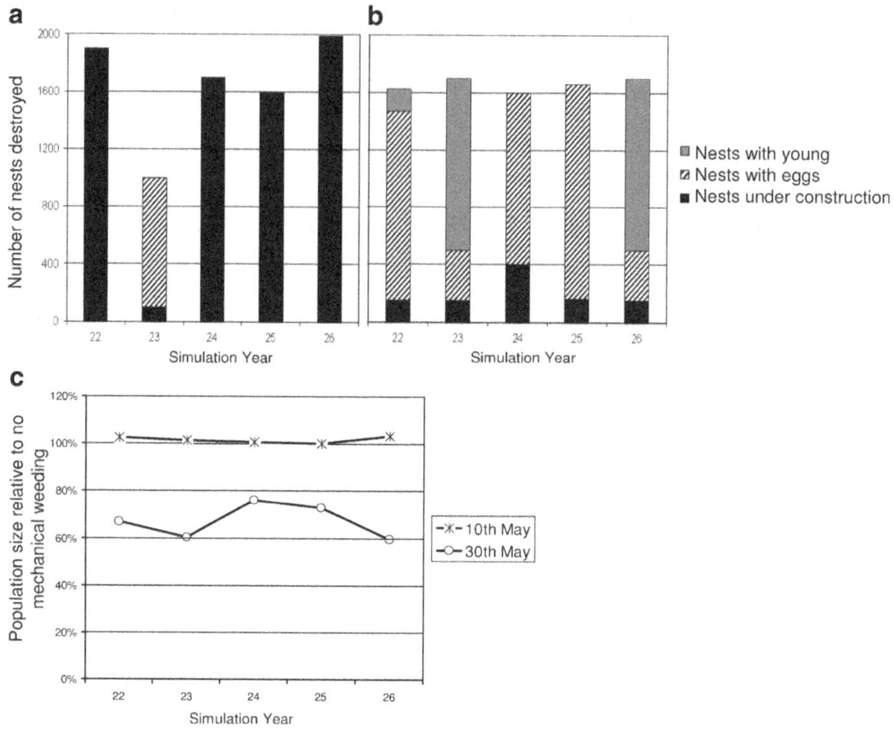

Fig. 2 Example 1: ALMaSS scenario results. **(a)** The number of nests destroyed by mechanical weeding on 10th May. **(b)** The number of nests destroyed by mechanical weeding on 30th May. **(c)** The population-level impact of mechanical weeding shown relative to a no mechanical weeding situation

or 30th May, which corresponds to mid- or late-season application. Although variable with year and therefore weather, late-season use destroyed a very large number of nests containing eggs or young, whereas the earlier application largely affected nests during nest building or egg-laying. The skylark population was consequently much reduced by late application (24–40%) whereas earlier application resulted in a slight increase of up to 3%. This increase is surprising and the model was neither specifically designed nor calibrated to make this prediction, which, therefore, can be considered an independent or secondary prediction (sensu [16]). Moreover, an ABM does not require us to just believe in the results as a black box, but allows us to try and understand why certain things happen. In this case, closer analysis of the model revealed that due to the rapid growth of the cereal crop the skylark has only a limited window of breeding opportunity between emergence and canopy closure [38–40] and is often limited to just one breeding attempt. Since the first clutch of the season is usually one egg smaller than the second clutch in this species, the early loss of a clutch was a slight benefit if the second brood could be completed before the breeding window closed. Broods lost due to weeding on 30th May (40 days from sowing) could not be replaced within the window of opportunity. These results led Odderskær et al. [37] to recommend that mechanical weeding be used up to a maximum of 30 days after sowing to avoid significant risk to skylark populations. The recommendation was not with respect to a calendar date, because it is timing with respect to the breeding window that is critical. In a subsequent independent field study [41], it was found that mechanical weeding 35 days or later after sowing caused significant reduction in skylark breeding in spring cereals. Thus, the prediction of the model was confirmed indicating that key elements of the skylark's population dynamics were captured in the model, that is, the model was structurally realistic [42].

2.3 Example 2: Risk Assessment for Beetles and Spiders Including Multiple Stressors

Regulatory authorities have strict procedures for assessing whether a pesticide presents an unacceptable risk to nontarget organisms. For example, according to EU directive 91/414 and its annexes and guidance documents, if the toxicity exposure ratio (TER) is <5, "no authorization shall be granted, unless it is clearly established through an appropriate risk assessment that under field conditions no unacceptable impact occurs after the use of the plant protection product under the proposed conditions of use" (Annex VI of EU Directive 91/414/EEC). While this criterion may seem objective and stringent it is also administratively inflexible and simplified. In this example, we demonstrate how misleading the criterion can be by evaluating pesticide impact with and without other mortality factors (multiple stressors) and by using test species with slightly differing characteristics.

ALMaSS scenarios were created using the following assumptions:

- An insecticide was applied to cereals.
- Treated cereals received from one to three applications each year in late May to July following normal farming practices for insecticides.
- No other pesticides were used anywhere in the landscape (the current regulatory standpoint).
- Exposure to the pesticide resulted in 90% mortality for all exposed beetle and spider life-stages.
- Exposure occurred when the organism was present in the field on the day of pesticide application, and all organisms present were considered to be exposed.
- Residues were not assumed to have any impact, hence only direct exposure to spray was considered toxic.
- There was no drift to off-crop areas.
- The landscape considered was a 10 km × 10 km area of Denmark near the town of Bjerringbro (56°22′N, 9°40′E) (Fig. 3).

Three factors were varied:

- The proportion of the landscape exposed was altered by assuming that insecticide was applied to 0, 25, 50, and 100% of cereal fields, and that all arable fields grew cereals.

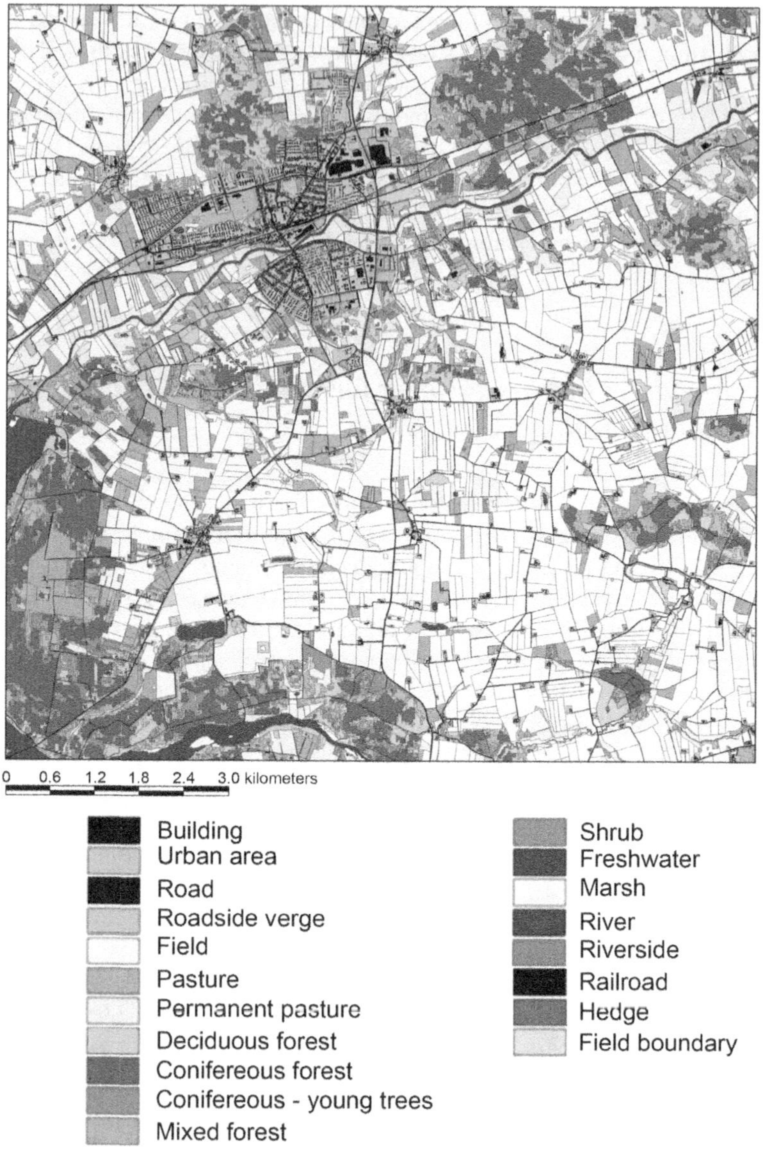

Fig. 3 A GIS representation of the Bjerringbro area in central Jutland, Denmark. This is the landscape used in all ALMaSS examples

- The implications of assumptions about other mortality factors were investigated by running four scenarios – one where the impact of soil cultivation and harvest mortalities was assessed in the absence of pesticide (scenario BM in Fig. 4b), a second scenario where only pesticide mortalities were incorporated and soil and harvest mortalities were ignored (scenario PM), and a third scenario where the

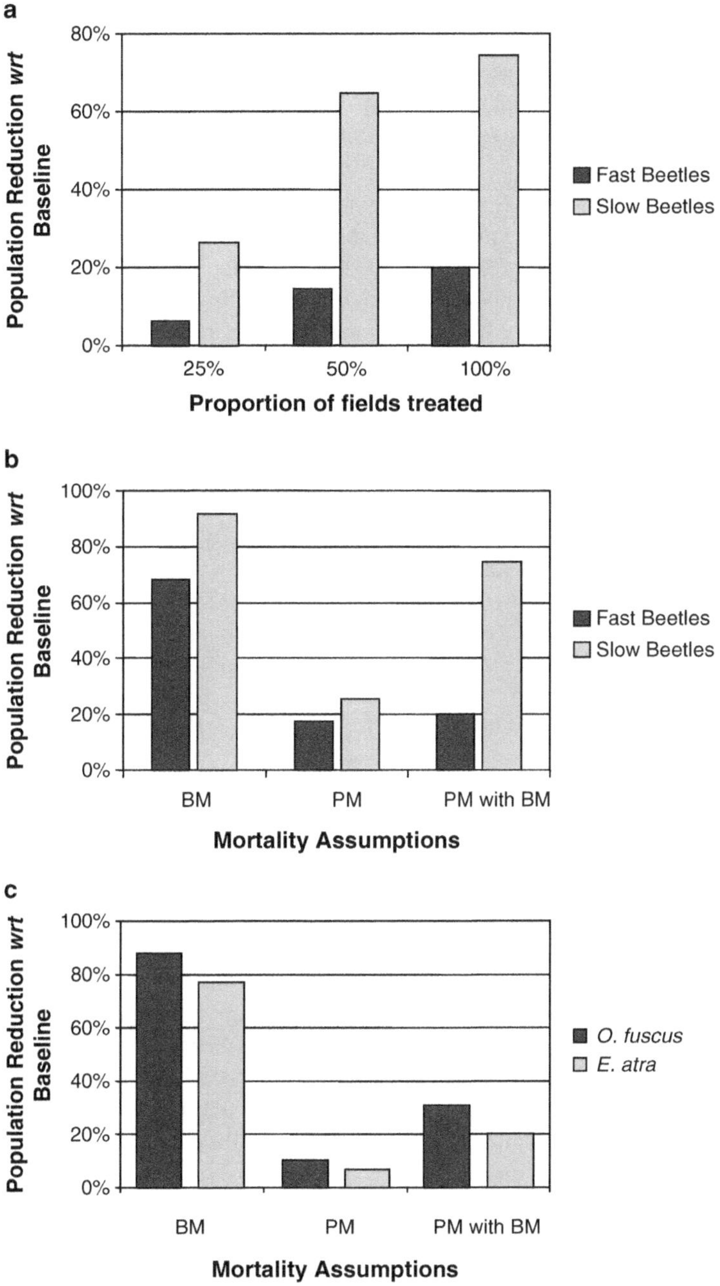

Fig. 4 Example 2: ALMaSS scenario results. Population reductions are expressed as a percentage of those in the baseline scenario (see text). **(a)** The size of population reduction in relation to the area treated with insecticide, for fast and slow moving beetles. **(b)** The size of population reduction of fast and slow beetles, BM = only agricultural operation mortalities, PM = only insecticide mortality, PM with BM = pesticide mortality assessed against a background of agricultural operation mortality. **(c)** Same as **(b)** but for two species of spider

impact of the pesticide was assessed against a background of including the soil cultivation and harvest mortalities (scenario BM with PM). Values for mortalities were available from [43], and all arable fields were assumed to grow cereals and have insecticide applications. A fourth scenario was run without pesticide or soil cultivation and harvest mortality and was used as a baseline for the results presented in Fig. 4.

- Variation in species life history was assessed in two ways. A very simple change to the beetle model was made by changing the maximum daily movement rate used by [32] to be 10 or 20 m per day (slow and fast beetles). The second assessment was made using models of two species of linyphiid spider (*Erigone atra* and *Oedothorax fuscus*), both with similar habitat requirements and both common agricultural species but differing in their breeding behavior and dispersal. *O. fuscus* has a shorter breeding season and lower dispersal ability than *E. atra*.

Twenty replicates were obtained of all scenarios with scenario runs of 55 years. The first 11 years were discarded as a burn-in period, and the results were expressed as mean population size over the last 44 years. Weather data were as used by Topping and Odderskær [30] and were a continuous loop of 11 years of weather data from a weather station near to the landscape simulated.

Results – For clarity all results are expressed as the size of the population reduction compared with a baseline scenario. Increasing the area treated with insecticide reduced beetle population size, but the effect was much more severe if the beetles moved slowly (Fig. 4a). Smaller differences were observed between fast and slow beetles in terms of their sensitivity to background and pesticide mortalities (scenarios BM, PM, and BM with PM, Fig. 4b), nor was there much difference in the responses of the two spider species (Fig. 4c). Background mortalities were generally high and much higher than those caused by the pesticide impact alone. However, if we evaluate the effects of the pesticide while controlling for background mortalities (i.e., BM vs. PM with BM) then in all cases the impact of the pesticide was greater than measured without other mortalities, and in the case of the less mobile beetle and spider it was almost four times greater.

The results demonstrate two effects. The first is that mobility clearly interacts with the pesticide application, and therefore we can get widely differing results with different life-history strategies. This effect has been shown in the real world in carabid beetles [44] and is partly due to mobile beetles and spiders being able to “leapfrog” disaster by moving from field to field and therefore having a greater probability of not being sprayed, but largely due to the faster recovery potential of mobile animals as they reinvade and breed in recently sprayed areas.

The second effect is related to the population dynamics. In cases where mortality on individuals is low the population grows and reaches a level where it becomes self-regulating through density dependence. At this point the impact of lower levels of mortality is to remove many individuals that would have died in any case, equivalent to the doomed surplus of Errington [45]; hence, impacts are lower when seen at the population level. In contrast, a population under heavy mortality, such as slow beetles under soil cultivation and harvest mortalities, is very vulnerable to a small extra mortality because this kills animals that would otherwise have contributed to population growth.

2.4 Example 3: Impacts of an Endocrine Disrupter on Vole Populations: Toxicity, Exposure, and Landscape Structure

As with example 2 with multiple stressors this analysis is derived from a risk assessment, but with the purpose of investigating the components of the assessment to gain an understanding of the field vole population response, rather than conducting a formal risk assessment. Here, we exploit the ability of ALMaSS to incorporate complex patterns of toxicity, to modify different aspects of a pesticide risk assessment, and calculate the population-level response. This flexibility allows the manipulation of all aspects of the risk assessment in an experimental way, using the model as a virtual laboratory to carry out experiments that would be impossible in the real world. Specifically we investigate how changes in toxicology, exposure, and landscape structure alter population responses, to gain insights into the properties of the system. The scenarios we present are illustrative only; for a comprehensive account, see Dalkvist et al. [46].

The toxicology investigated is unusual but closely similar to that of the fungicide vinclozolin, an endocrine disrupter where the effect is inherited epigenetically through the male germline after exposure in the uterus [47, 48]. This toxicology is challenging to model because of the epigenetic component of transmission of effects, and because expression of the toxic effects is chronic. In the model, expression of toxic effect was as either absolute sterility or a halving of the mating success of male offspring. Those with a reduced mating success passed on this genetic trait to their male offspring.

Other than the altered fertility the affected males were assumed to behave as nonaffected individuals since it was not known if the affected voles would change behavior, and the worst case was assumed. However, females mating with sterile males did not experience false pregnancies and would attempt to mate the following day if mating was unsuccessful. This is likely to be a real situation since voles are polygamous, but it is by no means certain that a female will not mate with the same infertile vole again. This depends on which male vole is closest to her at the time of mating, and it is therefore a function of the territorial behavior of the model voles. This polygamous behavior has the result that both inheritance and purging of the epigenetic effect are density dependent. This is because the probability of a nonsterile vole territory overlapping a female's territory increases with vole density. The system thus comprises complex dynamics that would be difficult to study experimentally in the real world, but is amenable to investigation in an ABM.

In all cases scenarios were constructed by modifying a single factor at a time and expressing the results as a population size relative to a baseline scenario where no pesticide was applied. The landscape used was again that shown in Fig. 3, but with some fields replaced by orchards, randomly placed until orchards occupied 10% of the total agricultural land. Landscape structure was modified in later experiments by altering the locations of patches of optimal habitat. Pesticide was applied for 30 years starting in year 31 and was followed by a 60-year recovery period again where no pesticide was applied. Thirty-five replicates of each scenario were run. For clarity the experimental scenarios were divided into two groups: one to investigate the toxicity and exposure factors and the other to evaluate landscape structural impacts.

2.4.1 Toxicity and Exposure Scenarios

Five scenarios were constructed to evaluate the impact of factors related to toxicology and exposure. These were (1) a "default" scenario with one pesticide application to all orchards on May 31. The other scenarios were constructed by varying one factor at a time of the default scenario, as follows: (2) a "clover/grass" scenario where the pesticide was sprayed on clover grass fields that replaced orchards, (3) a "two applications" scenario where the orchards had an additional pesticide treatment on 14th June, (4) a "NOEL" (no observable effect level) scenario where the effect level was altered to one quarter of the NOEL in the default scenario, and (5) a "DT_{50}" scenario where the pesticide half-life was a factor four times longer than that in the default scenario.

Toxicity and Exposure Results

The population responses differed between scenarios as shown in Fig. 5. Taking each scenario in turn:

- Clover/Grass: Spraying clover grass instead of orchards resulted in the lowest population depression of all scenarios, and the population reached full recovery within the simulation period. This might seem strange because the field vole lives in grass-vegetated areas that can function both as a continuous food supply and cover [49], and therefore exposure might be expected in a grass crop. However, clover grass fields in the modern intensive agricultural landscape are cut for silage or used for grazing livestock throughout the year, so that the voles' habitat is continually being destroyed. Consequently, these fields are not suitable breeding habitat [50–52], although they facilitate dispersal. Accordingly a small fraction of the voles were exposed to the pesticide in our simulation, resulting in a negligible population depression.

In contrast the orchards contain grass cover between the trees, which in the "default" scenario is cut once a year just before harvest, and voles living here were subject to much less disturbance. This illustrates the importance of the animals' ecology and behavior in risk assessment. It is also interesting to note that the impact at the population level in this scenario was ca. 1%, but that 4% of all male voles exhibited a toxic response (Table 1). Of these 4% only 22% carried the paternally transmitted gene, indicating that the voles that were affected were not breeding as successfully as those in other scenarios.

Two applications scenario: A second application to the orchards led to a doubling of the amount of pesticide applied in the landscape, but not a doubling of the population depression or the proportion of affected voles (Fig. 5a, b). The explanation is that the second application hits a population containing voles already affected by the first.

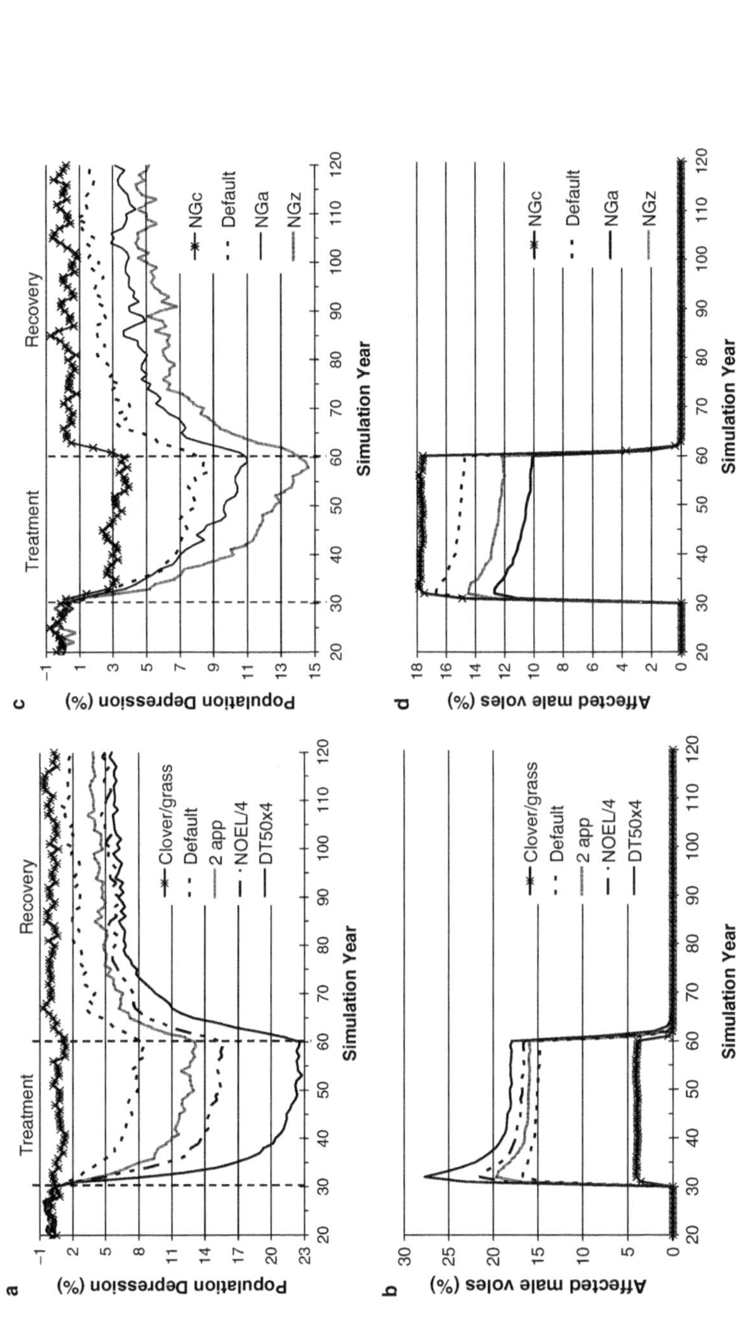

Fig. 5 Example 3: ALMaSS scenario results. **(a)** Population depression of a field vole population exposed to an endocrine disrupter with different toxicological and exposure scenarios. See text for further explanation. **(b)** The proportion of male voles impacted directly or genetically by pesticide exposure in toxicity and exposure scenarios. **(c)** Population depression of a field vole population exposed to an endocrine disrupter with different landscape structural scenarios. **(d)** The proportion of male voles impacted directly or genetically by pesticide exposure in landscape structural scenarios

Table 1 Example 3: results of ALMaSS simulations

Scenarios	Total of affected males (%)	Directly affected males as a % of total affected	Baseline population size (1,000s)
Clover/grass	4	78	58
Default	15	52	58
Two applications	16	56	58
NOEL/4	17	59	58
DT_{50}*4	18	75	58
NG around orchards	18	54	62
NG not around orchards	10	51	54
0% NG	12	51	37

The total proportion of all male voles affected by the endocrine disrupter together with the proportion of those that were directly affected by exposure in the uterus and the total mean size of the vole population in the baseline scenario for each toxicological, exposure, and landscape structural scenarios

NOEL and DT_{50} (half-life) scenarios: In the NOEL scenario toxicity increased by a factor of 4, and this resulted in a doubling of population impact than in the default scenario and a higher impact than applying the pesticide twice. However, a fourfold increase in half-life, in the DT_{50} scenario, had even more impact (Fig. 5a). The explanation can be found in the first-order kinetics of decay for the pesticide:

$$C = C_0 \, e^{-kt} \Leftrightarrow k = -(\ln(C/C_0)) \, / \, t \Leftrightarrow k = \ln 2 \, / \, DT_{50}, \tag{1}$$

where C is the concentration of the residue at time t, C_0 is the residue concentration at the start, and k is a rate constant for loss, which is dependent on DT_{50}. By halving DT_{50}, k is doubled, which increases the coefficient of the exponential curve and so reduces the period of exposure. By contrast changing NOEL is equivalent to changing the constant C in (1), which would result in a small change of the time period of exposure (t) compared with changes in k. Thus, the voles are more sensitive to alterations in half-life than to alterations in toxicity. Despite this, half-lives of pesticides receive little attention in current risk assessments.

Toxicity and Exposure Discussion

The population recovered completely by year 120 only in the Clover/Grass scenario, where a limited proportion of the voles had been affected. This result could have been related to the epigenetic effect of the pesticide, but investigation of the frequency of affected voles showed that the alteration was purged from the population after only a short period (Fig. 5b). In fact, the phenomenon was related to the spatial dynamics of the voles in this fragmented landscape. Even small perturbations of the population can mean local extinction for small subpopulations, and the time before recolonization depends on their location relative to larger source populations. If the perturbation is large then this effect is exacerbated resulting in more isolated subpopulations and consequently an elongation of the recovery period (Fig. 5a).

The unusual form of the recovery curve was a result of initial logistic population growth in core habitats, followed by delays dependent on dispersal to recolonize other areas that had been lost following pesticide application. The reverse mechanism, together with epigenetic breeding depression, explains the continual decline of the voles during the period of continuous pesticide application, as patches slowly become empty and the vole population contracts to core habitats. This spatial mechanism provides a new dimension to risk assessment since spatial dynamics are currently ignored.

2.4.2 Landscape Structural Manipulations

As shown earlier there are indications that the magnitude and effect of pesticide exposure on populations are influenced by the spatial structure of contamination in the landscape and habitat location [53–55]. Even so, the use of nonspatial approaches is still common when characterizing exposures and effects of pesticide stresses. To demonstrate the possible effect of landscape structure in the risk assessment three scenarios were constructed based on the default scenario already described containing randomly allotted primary vole habitat patches ("natural grass" = NG). The natural grassland is a habitat type particularly suitable to the voles because it supplies the animals with food and cover throughout the year. We explored three landscape scenarios as follows: (1) The NG close to the orchards scenario (NGc), where the natural grassland was located around the orchards where pesticide was applied; (2) The natural grass not around orchards scenario (NGa), where the natural grassland was placed away from the orchards; and (3) the 0% natural grass scenario (NGz), where no natural grassland occurred in the landscape.

Landscape Structure Results

The NGc scenario resulted in the lowest impact of the landscape scenarios with a population depression of 3%, but the proportion of voles affected by the pesticide was also highest here (Fig. 5c, d). This seeming paradox arises because natural grassland in this scenario produced a connected set of suitable habitat fragments capable of sustaining a larger population size around the orchards than in the other scenarios. There were thus sufficient healthy males in the nearby natural grassland to provide viable sperm for females in orchards. This means there were still quite high abundances of voles in the orchards despite these being the sites of exposure of gestating females (Fig. 5d), and after spraying these populations recovered rapidly to baseline levels (Fig. 5c).

Compared with the NGc scenario the NGz scenario had the highest population depression and lowest recovery level of the landscape structure scenarios.

The natural grassland was removed from the landscape completely, thereby reducing connectivity between optimal habitats (which here are primarily the orchards). The affected vole frequency was lower, because of the reduced vole abundance around the orchards, but the impact was higher due to the reduced level of source nontreated populations in the landscape. Accordingly local extinction occurred on a larger scale resulting in the lowest level of recovery.

The NGa was used as a control for the NCc scenario, maintaining the area of grassland but locating it away from the orchards. Voles living in those grasslands were unaffected by the spraying, thus the proportion of affected voles was lower than in the default scenario (Fig. 5d), but the population depression was greater (Fig. 5c) because of a lack of healthy males in grasslands adjacent to orchards to provide viable sperm for females in the orchards. The lack of correlation between three different endpoints, namely, the total proportion of males affected, the proportion of these directly affected, and the baseline population size illustrates the nontrivial nature of the relationships between the factors considered (Table 1).

2.5 Example 4: Impacts of Pesticide Bans and Reductions at Landscape Scales

Jepsen et al. [21] utilized ALMaSS to evaluate the impact of a total pesticide ban on the abundance and distribution of five species: *Alauda arvensis* (skylark), *Microtus agrestis* (field vole), *Bembidion lampros* (beetle), *Oedothorax fuscus* (linyphiid spider), and *Capreolus capreolus* (roe deer). While it would be temptingly simple to create a scenario where, on the one hand, we had conventional agriculture and on the other the same thing but with no pesticides, this may be a rather too simple approach. Instead, a more holistic consideration of the problem is required. The debate surrounding the safe use of pesticides in Denmark prompted the establishment of a state-funded Pesticide Committee in 1999. This committee initiated a nation-wide evaluation of the economic and agronomic consequences of a partial or complete ban on pesticide usage in Danish agriculture, the conclusion of which was published by Jacobsen and Frandsen [56].

The results suggest that a total pesticide ban will have wide-reaching consequences for land use and also crop choices. For instance, under the EU CAP regulations relating to arable area payments at the time, farmers could claim payments and make a profit by sowing a crop they would never harvest. In other areas land would shift from arable to dairy production. In those areas where arable production remained there would be a reduction in areas of pesticide-intensive crops for harvest. In particular, a significant rise in the area of oil seed rape was indicated since this is cheap to sow and provides a good weed-suppressing cover. Jepsen et al. [21] simulated this outcome by comparing the distribution and abundance of the five species between agricultural practice as in 2003 and a scenario in which all crops were grown organically and where agricultural land altered its composition from 64 to 29% cereals, oil seed rape increased from 11 to 17% of the arable area, and where roughage (rotational grass, peas, etc.) increased from 19 to 59%, with the remaining areas being set aside. These simulations used the landscape of Fig. 3.

As expected due to reduced incidence of crop-management related stressors (insecticides and soil cultivation), beetle and spider numbers generally increased over the whole landscape. Field vole numbers also increased marginally and uniformly because of the increase in connectivity due to increasing the area of grass relative to arable fields. The skylark however, contrary to initial expectations, decreased in population size across the landscape with marked decreases in previously good habitats. These decreases were an integration of a number of positive and negative influences. The reduction in pesticides and subsequent increase in invertebrate food worked positively; however, the lack of tramlines caused by late-season pesticide applications meant that the food was less abundant. In addition, the grass areas would be grown for silage and would have very narrow windows of breeding opportunity before cutting and/or grazing resulted in them being useless as breeding habitat.

The response of the roe deer was also complex with a distinct spatial pattern to the changes. These local population changes were in response to changing crop locations relative to suitable wooded habitats, primarily hedgerows. In those areas where both hedgerows and suitable crops coincided, the deer could move out from woods and forage; in other areas, the lack of shelter meant that the improved forage was not utilized [21].

A similar interaction between pesticide changes and farm management was found when evaluating the impact on skylark population sizes of taxation measures to alter pesticide use [57]. The effects of using pesticides were compared with spraying nothing. The real effect of not spraying would be to not open tramlines, preventing skylark foraging and breeding access, because the farmer would not drive onto the field. Not spraying would also alter the crops grown. When these effects were taken into account the mean 4% impact of pesticides predicted in an earlier study [30] was reduced to a barely significant 1% impact [31]. However, in both studies other structural changes in the landscape management were capable of altering skylark populations by 20–50%. We conclude that a common sense, holistic, approach to simulation is needed so that "knock-on effects," such as changes in crop area allocations, are taken into account in policy evaluation.

2.6 Two Further Examples of Predictive, Fully Fledged ABMs

The development of the ALMaSS framework took 10 years, including program debugging and verifications. The development of a typical animal model with the ALMaSS framework, including testing, usually takes 1–2 years. The analysis of more theoretical scenarios of an existing animal model, however, can be performed rather quickly, typically within a few months. Historically, and due to reasons of page limitations in scientific journals, the extensive testing of ALMaSS so far has not been fully documented. Therefore, we here briefly describe two further fully fledged ABMs that were developed for ecological applications and where testing, verification, and validation have already been documented. These examples also show that basing a model on fitness-seeking behavior can make ABMs complex, but highly predictive. The trout model was explicitly developed for management support. The shorebird model has a more academic background but currently is being tailored to address a range of real-world applications.

2.6.1 Shorebird Models

The shorebird models of Goss-Custard et al. predict the impact of land reclamation, resource harvesting, and recreation on the winter mortality of shorebirds and waterfowl. The ABMs had to predict the effect of new environmental conditions for which no empirical rules or data were available [58–65].

In these models, the habitat is divided into discrete patches, which vary in their exposure and their quantity and type of food. During each time step birds choose where and on what to feed, or whether to roost. Time steps typically represent 1–6 h. The bird's state variables include foraging efficiency, dominance, location, diet, assimilation rate, metabolic rate, and amount of body reserves. Key environmental inputs to the models are the timings of ebb and flow and temperature. The submodels describing the bird's decision where to move, what to eat, and how much time to spend in feeding are based on principles mainly from optimal foraging theory. The individuals are assumed to always try and maximize fitness, i.e., their own chance of survival.

Model predictions were compared with many observed patterns during several iterations of the modeling cycle. The modeling cycle includes defining the model's purpose, choosing a model structure, and implementing and analyzing the model [16]. At the end of this process, patch selection, prey choice, and the proportion of time spent in feeding were accurately predicted for many species and sites. In one case, the increase in winter mortality due to land reclamation was known from observations. The model was parameterized for the preimpact situation, and then run for the situation after the land reclamation and the increase in winter mortality were determined. The match of observed and predicted increase in winter mortality was strikingly good [66].

2.6.2 Stream Fish Models

Railsback and coworkers developed a suite of stream fish ABMs (mainly cutthroat trout *Oncorhynchus clarki* [67–73]; see also the precursor model of Van Winkle et al. [74]). The models were developed to predict the effects of river management on fish populations. Fish adapt to changes in flow caused by dams and water diversions by moving to different habitat. Thus, to predict how fish populations react to new flow regimes it was necessary to know how fish select habitat. The trout model of Railsback and Harvey [70] uses daily time steps, with stream habitat represented as rectangular cells. The section of a stream represented in the model would usually comprise about 200 m consisting of about 100 cells (the number of cells varies because of varying water levels). Within a day, individual fish carry out the following actions: spawn, move, feed, and grow. Mortality could occur within each of these steps and model runs cover a time span of years or decades.

In the model, trout based their daily decision on the projection of current habitat conditions for 90 days into the future [67]. Railsback and Harvey [71] show that this "state-based, predictive" theory of habitat selection is, in contrast to alternative theories, capable of reproducing a set of six patterns observed in reality ("pattern-oriented modeling," [16,75]). In a management application, the trout IBM was used to predict the population-level consequences of stream turbidity (Harvey and Railsback, unpublished manuscript): over a wide range of parameter values, the negative effects of turbidity on growth (and consequently, reproduction) outweighed the positive effects on predation risk.

3 Advantages and Drawbacks of the ABM Approach

3.1 *Advantages*

Assuming that we have the option to make an ABM, what are the key advantages of this approach in ecotoxicology? The most important characteristic of ABMs is that we deal explicitly with spatiotemporal factors, and this coupled with the simple fact that toxicants are rarely distributed evenly in space and time in the real world is a major step forward in realism.

However, this is only half of the story. ABMs integrate the information in heterogeneous environments with the behavior of the agents, since ABMs pose a mechanistic approach. This is clearly demonstrated by the skylark and mechanical weeding example where integration of the management, weather, and skylark ecology and behavior provided the necessary understanding of the system to prescribe nondamaging weeding practices. This integration also allows the consideration of multiple stressors (example 2). Here again, the fact that the ABM integrated the impacts of different stressors with the animal ecology and behavior gave rise to important population-level responses. While consideration of multiple stressors might not be straightforward from a regulatory perspective, it is an area where ABMs could make a major contribution.

Probably the best example of the integrational power of ABMs is the vole example (example 3), which shows the use of an ABM as a virtual laboratory allowing a very wide range of factors to be modified separately or in unison and their impacts compared. This example also illustrates the point about flexibility in ABMs. The problem definition in the vole example required incorporation of individual-based genetic transfer of information due to the epigenetic impact of the pesticide, which in isolation could have been achieved using traditional population genetic approaches. However, this was further complicated by the behavioral ecology and individual-level impact of the pesticide. These factors include strong territorial behavior, high fecundity, and local habitat-dependent dispersal in a structurally complex and variably permeable (to dispersing voles) landscape, together with spatiotemporal variation in the distribution of the stressor and variable phenotypic and toxicological responses at the individual level.

It is hard to imagine a non-ABM approach that could integrate all of these aspects in a natural way and yet still provide a simple intuitive experimental system for manipulation and testing. This type of "virtual laboratory" approach has a huge potential in increasing our understanding of biological systems and their responses to toxic stressors. In fact, these approaches are already being used to tackle theoretical population ecology problems in spatially heterogeneous environments [76].

When used to evaluate policy changes, ABM results may often contraindicate a reductionist approach (as shown with the ALMaSS examples earlier). In the real world where so many factors interact it would be common sense to consider the changes in farm management that would result from any policy change, and the use of ABMs should be no different. Although ABMs can become very large and complex they are not capable of simulating systems to such a degree that a single model can encompass all ecological and socioeconomic aspects. However, integration of a range of multidisciplinary models so that inputs to ABMs are as realistic as possible is achievable. For example, Dalgaard et al. [77] linked socioeconomic, nitrogen-budgeting, hydrological, and ecological models together to assess land management scenarios. The flexibility of the complex ABM approach facilitates this process.

Information-rich systems such as the Army Risk Assessment Modeling System (ARAMS) [78] would be ideal candidates to take advantage of agent-based technology. This system already has a wildlife exposure module that uses a simple area use factor to determine exposure, but could be augmented with realistic animal movements and responses to remediation measures.

Another often overlooked advantage of an ABM approach is that the mechanistic detail forces the researcher to consider the system of study from another angle, and perhaps in greater detail than hitherto undertaken. This has the very real benefit of providing a framework for storing current knowledge and identifying areas where research is needed because information is currently lacking.

3.2 ABMs Versus More Aggregated Population Models

When considering the advantages and drawbacks of ABMs for ecotoxicological research we are thinking primarily of population-level effects. A common point of contention is whether ABMs are better than simple population models. This point comes up repeatedly at conferences (e.g., see [79]) and therefore we devote a little space to it here.

The question of whether the one type of model is better than the other misses the real point of models, which is to create a representation of a system that allows investigation of the properties of the system and, in some cases, prediction of future outcomes. There is nothing innately better about an ABM than, for example, a matrix model of population growth; the two types of model are different and meant for different purposes. A matrix model [80] is a mathematical representation of the current state of the population. Unless its parameters are allowed to vary, it cannot be used for prediction, but only for projection as to whether the population will grow or decline. An ABM, on the other hand, can make predictions because its components alter their states and behaviors in response to changing input variables.

This does not mean that the ABM is better than a matrix model. The ABM cannot be parameterized using the same parameters as the matrix model; it cannot be constructed as quickly as a mathematical model, and it is always more difficult to understand. Choice of model type depends on the resources available and the purpose of the analysis, and it is even less clear cut as we move up the continuum of increasing realism from scalar population models to spatially structured models such as metapopulation models. Here, the purposes of the two model types may overlap, but several factors affect choice of model type. There may be constraints of data availability that dictate a simple model structure, or other constraints such as on development time, available computational power, or even technical ability, which would dictate a simpler model. If such constraints are not important, then there is a common sense link between the accuracy of a model and the degree to which it represents reality (i.e., its realism), but at some point the generality of the model will be reduced as we make the model too specific. Tradeoffs exist between the accuracy of the model, the resources required to build it, and the desired generality [81]. There is no one solution to this problem; each application must be evaluated in its own right. The criteria, however, used for choosing a certain model should be made explicit in any application.

3.3 *Drawbacks*

3.3.1 Presumed Drawbacks

Some commonly heard arguments against increasing realism and therefore complexity in models, and by extension to increasing realism in risk assessment are as follows:

Increasing realism decreases generality. This argument probably has its roots with Levins [81], although it is a common general principle. To determine whether this is a drawback or not depends on how general we want our model to be. If our question is specific then a general model is likely to be imprecise (e.g., the use of TER and fixed threshold values for all species in pesticide regulation to predict risk in example 1). In ecotoxicology "general" models are unsatisfactory because there is no general target/nontarget organism, mode of action, or route of exposure. When constructing ABMs generality is not the aim per se; here we usually try to capture the essence of a specific system or class of systems, rather than a generality. However, generalities can be achieved if we evaluate our specific model over a sufficiently wide range of conditions. In principle, the exploration of carefully defined scenarios in ABMs could provide a sensitivity analysis of the probabilities of adverse effects as well as general rules. For example, in the vole (example 3) interactions between the different landscape structural factors could be evaluated in order to create general rules about pesticide impacts and habitat connectivity.

Adding detail makes the creation and testing of general ecological principles difficult. Not to be confused with a criticism of adding *unnecessary* detail, this is related to the generality argument, but is fundamentally flawed in that it assumes that we need generalities, that is, simplifications, before developing and testing theories. Surely theories are best derived from patterns emerging from as many varied and detailed observations as possible [16]. So given enough examples of specific systems (such as realistic ABMs) to experiment with, greater insight into general theories or even new paradigms may develop. This goes to the heart of the promise of complexity science and ought not to be perfunctorily dismissed.

Detailed models are unnecessarily complex. Naturally adding detail to a model without good reason would be foolish, because every additional detail causes additional work. So, as for other models the principle of parsimony holds for ABMs. We might use patterns to get ideas about optimal model complexity (see [75]), but ultimately it is the task of model analysis to see how much a model can be simplified while keeping its potential to serve its purpose. However, if we consider complexity in the same way, complexity has a price in terms of increased work in adding model details, but a distinct benefit in terms of richness, which we can utilize for testing, validation, and prediction [19].

Increasing realism leads to a loss of precision. This argument is based upon a traditional statistical approach to modeling. In a mathematical model the error in the prediction is related to the error terms in the parameter inputs in a predictable manner, and this can be propagated or compounded in complex models. While true of a mathematical construct this concept does not necessarily hold for complex systems in which checks and balances stabilize the outputs. It is especially untrue of models constructed using a pattern-oriented approach (see later), whereby error propagation is constrained by the form of model testing [82]. In fact, biological systems in general have sloppy parameter spaces, and focus should, therefore, be on predictions rather than parameter values and their errors [83]. This is incidentally also one of the reasons why these models do not result in deterministic chaos, which is another commonly held, but misinformed belief.

3.3.2 Real Drawbacks

There are, however, a number of much more significant drawbacks when considering building ABMs. The drawbacks of constructing and using an ABM approach, especially a comprehensive approach like ALMaSS, can be summed by the phrase "When you can change anything you have to consider everything." In considering "everything" you need both to be able to generate plausible mechanisms for interactions that must all be defined and to locate or generate data to support the parameterizing of these. In building or modifying the model the interactions must be considered again since what on the face of it may be a simple change can, in fact, have far-reaching consequences. The same is true of building a scenario after the model is finished; simply accepting default values may be counterproductive, for example, applying a reductionist approach to pesticide limitation as in examples 2 and 4.

The difficulties of model construction are already mentioned earlier. The complexity of the system means that the technical demands placed on the developer are higher than those typically placed on the ecological modeler. These demands are comparable to the technical skills required by other specialist branches of natural sciences such as biostatistics or molecular ecology, the difference being that there are few schools of computational biology, and so suitably qualified staff may be hard to find. This may be a major drawback to actually implementing an ABM approach.

Perhaps the biggest drawback to the increased use of ABM models in scientific disciplines in general is simply the fact that they are new. This means that ABMs lack some important characteristics compared with other modeling approaches, these being a rigorous theoretical basis and a standardized approach to construction, testing, and communication of models. In fact, the emergence of theory is a rapidly developing area under the auspices of complexity science. Complexity science aims to describe, explain, and control the collective objects and phenomena emerging at a particular spatiotemporal scale from the simpler interactions of their components at a finer scale. The search for a general theory to simplify understanding of complex systems is, however, elusive. For example, one general theory that might have been useful to describe the emergent patterns of multiagent systems is the theory of self-organized criticality [84]. However, this general theory seems not to have fulfilled its original promise and is perhaps better viewed as a way of sketching the essential structure of a system [85]. Seen in this light, ABMs might fulfill the role of filling in the mechanistic details in system functioning while the search for unifying principles continues at a higher level of organization.

Development of methods for communication and testing of ABMs has started, but is still in its infancy. There is a widely held view that models of this complexity are difficult, if not impossible, to validate. However, one emerging approach to validation is pattern-oriented modeling [75], which includes as a main element inverse modeling for parameterization [82, 86] whereby multiple field data patterns are used to simultaneously filter combinations of parameter values and model structures in order to achieve the twin aims of testing the behavior of the agents in the model and of reducing parameter uncertainty. The greater the number of real-world patterns that can be simulated concurrently, the greater the confidence in the model, and typically the smaller the possible parameter space. Pattern-oriented modeling is a new approach and so examples are few and far between (e.g., [42, 87, 88]), and as yet no structured protocols exist for carrying out an analysis. However, the basic approach is well described [89] and would be easily adaptable to an ecotoxicological problem, especially where large-scale field data are available from monitoring studies or field trials. So rather than being seen as a drawback, the novelty of pattern-oriented modeling could be seen as a challenge and an opportunity to develop the science and use of ABMs further.

Difficulty in communication of ABMs is a major drawback to their acceptance and general accessibility to nonspecialists. This seems paradoxical to some extent since good ABM construction practice is to use the ecological system to be modeled as the primary metaphor [16]. It follows then that explaining the model to ecologists ought to be relatively simple. This can indeed be the case at a superficial level, but description of the detailed choices made in construction and parameterization is far from simple. The two most critical sources of model documentation are the written model description and the source code; however, for ABMs these documents can be very large and are not usually easy to read. One approach suggested is to standardize the description such that once a reader has encountered a number of such descriptions familiarity increases transparency. This is the concept behind the ODD protocol (overview, design concepts, and details) of Grimm et al. [24] and Polhill et al. [90].

The idea of the ODD protocol is to define a fixed sequence in which different levels and elements of a model are described to allow the reader a quick overview of what the model is and what it does, that is, its structure and processes, without having to consider any detail at first. Then, important concepts underlying the design are discussed, for example, how adaptive behavior was represented, and how and why stochasticity was included. Finally, details of the model's implementation are provided. It can be useful, or even necessary, to present the actual code by which a certain process was represented. Thus, the separation of "overview" and "detail" takes into account that some readers are more interested in the overall structure and rationale of the model, for example, the ecotoxicologist, while others want to know the details of the model's implementation, for example, if they have to assess the model as a reviewer for a scientific journal or a regulatory authority.

ODD seems to gain ground in the literature but still is in its infancy and under development [14]. It can be difficult to apply it to ABM frameworks such as ALMaSS or FEARLUS [90] because the distinction between a specific model and the framework is not always easy to draw.

4 The Future of ABMs in Ecotoxicology

The examples of ABMs in ecotoxicology demonstrate the utility of the ABM approach and highlight that the system response is not easily predictable in advance due to the complex nature of the systems under study. If we do not include multiple stressors we can underestimate risks (example 2), and without evaluating the landscape structure and details of the toxicology of the stressor we also risk inaccurate prediction of the population impact (example 3). Even socioeconomic factors cannot be ignored in any but the most experimental of scenarios (example 4). It seems that almost all factors are important, and that is probably the cause for concern.

All is not lost however. If ABMs can be used to demonstrate that these effects are important, they can also be used to investigate the way these factors interact and thus increase our understanding of the system. In doing so and adding to the examples here, one could imagine an ABM/ecotoxicology utopia where series of representative landscapes were continually updated as agricultural practices change, and farmers responded to socioeconomic drivers and altered their management in response to these and weather variables. Aquatic and terrestrial environments would be combined in such a simulation, and surface and ground water flow of pesticides and fertilizers would be modeled. Entire suites of nontarget species could be modeled in these landscapes and whenever a new pesticide or policy change was to be tested it could be done against a well-documented comprehensive simulation of a real system with all the complexities of multiple stressors, varying crop coverage and farmer behavior, and landscape structure.

This would be a far cry from testing whether a TER value was less than 5, and while it might sound far fetched the technology to accomplish it already exists. Models of all basic subcomponents of the system exist, and hardware is easily capable of running such a system. For instance, ALMaSS can be run on a standard PC with one processing core while research computing facilities now exist with computers having >11,000 parallel processor cores [91]. What would be needed would be the resources and the will to construct and maintain such a model. On the other hand, it is important to keep in mind also that simpler ABMs and matrix and differential equation models all have their place. Ideally, such simpler models will be more or less directly linked to more complex ABMs such as the ALMaSS models to achieve a kind of "theoretical validation" of the complex model.

Even without embarking on such a project, the fact that it can now be feasibly imagined suggests that the future of ABMs in ecotoxicology is rosy, and naturally much can be achieved with the models we already have. It is our hope then that, as in other scientific disciplines, ABM development in ecotoxicology is going to be swift and exciting.

References

1. Folcik VA, An GC, Orosz CG (2007) The basic immune simulator: An agent-based model to study the interactions between innate and adaptive immunity. Theor Biol Med Model 4: 39. Available at http://www.tbiomed.com/content/4/1/39
2. Fossett M, Senft R (2004) SIMSEG and generative models: A typology of model-generated segregation patterns. Proceedings of the Agent 2004 Conference on Social Dynamics: Interaction, Reflexivity and Emergence, Chicago, IL, 39–78. Available at http://www.agent2005.anl.gov/Agent2004.pdf
3. Redfish (2008) http://www.redfish.com/stadium/
4. Chen X, Zhan FB (2008) Agent-based modelling and simulation of urban evacuation: Relative effectiveness of simultaneous and staged evacuation strategies. J Oper Res Soc 59: 25–33
5. Anwar SM, Jeanneret CA, Parrott L, Marceau DJ (2007) Conceptualization and implementation of a multi-agent model to simulate whale-watching tours in the St. Lawrence Estuary in Quebec, Canada. Environ Model Softw 22: 1775–1787
6. Mikler AR, Venkatachalam S, Ramisetty-Mikler S (2007) Decisions under uncertainty: A computational framework for quantification of policies addressing infectious disease epidemics. Stoch Environ Res Risk A 21: 533–543

7. Muller G, Grebaut P, Gouteux JP (2004) An agent-based model of sleeping sickness: Simulation trials of a forest focus in southern Cameroon. CR Biol 327: 1–11
8. Brede M, Boschetti F, McDonald D (2008) Strategies for resource exploitation. Ecol Complex 5: 22–29
9. Blaum N, Wichmann MC (2007) Short-term transformation of matrix into hospitable habitat facilitates gene flow and mitigates fragmentation. J Anim Ecol 76: 1116–1127
10. Mathevet R, Bousquet F, Le Page C, Antona M (2003) Agent-based simulations of interactions between duck population, farming decisions and leasing of hunting rights in the Camargue (Southern France). Ecol Model 165: 107–126
11. Satake A, Leslie HM, Iwasa Y, Levin SA (2007) Coupled ecological-social dynamics in a forested landscape: Spatial interactions and information flow. J Theor Biol 246: 695–707
12. DeAngelis DL, Gross LJ (1992) Individual-based models and approaches in ecology: Populations, communities and ecosystems. Chapman and Hall, New York
13. Louzoun Y, Solomon S, Atlan H, Cohen IR (2001) Modeling complexity in biology. Phys A 297: 242–252
14. Grimm V (2008) Individual-based models. In: Jørgensen SE, Fath BD (eds), Ecological Models, Vol. [3] of *Encyclopedia of Ecology*, Elsevier, Oxford, 5: 1959–1968
15. Grimm V (1999) Ten years of individual-based modelling in ecology: What have we learned and what could we learn in the future? Ecol Model 115: 129–148
16. Grimm V, Railsback SF (2005) Individual-based modelling and ecology. Princeton University Press, Princeton, NJ
17. DeAngelis DL, Mooij WM (2005) Individual-based modeling of ecological and evolutionary processes. Annu Rev Ecol Evol Syst 36: 147–168
18. Van den Brink P, Baveco JM, Verboom J, Heimbach F (2007) An individual-based approach to model spatial population dynamics of invertebrates in aquatic ecosystems after pesticide contamination. Environ Toxicol Chem 26: 2226–2236
19. DeAngelis DL, Mooij WM (2003) In praise of mechanistically-rich models. In: Canham CD, Cole JJ, Lauenroth WK (eds) Models in ecosystem science. University Press, Princeton, NJ
20. Holland EP, Aegerter JN, Dytham C, Smith GC (2007) Landscape as a model: The importance of geometry. PloS Comput Biol 3: 1979–1992
21. Jepsen JU, Baveco JM, Topping, CJ, Verboom J, Vos CC (2005) Evaluating the effect of corridors and landscape heterogeneity on dispersal probability: A comparison of three spatially explicit modelling approaches. Ecol Model 181: 445–459
22. DeAngelis DL, Cox DK, Coutant CC (1980) Cannibalism and size dispersal in young-of-the-year largemouth bass – Experiment and model. Ecol Model 8: 133–148
23. Topping CJ, Rehder MJ, Mayoh BH (1999) Viola: A new visual programming language designed for the rapid development of interacting agent systems. Acta Biotheor 47: 129–140
24. Grimm V, Berger U, Bastiansen F, Eliassen S, Ginot V, Giske J, Goss-Custard J, Grand T, Heinz SK, Huse G, Huth A, Jepsen JU, Jorgensen C, Mooij WM, Muller B, Pe'er G, Piou C, Railsback SF, Robbins AM, Robbins MM, Rossmanith E, Ruger N, Strand E, Souissi S, Stillman RA, Vabo R, Visser U, DeAngelis DL (2006) A standard protocol for describing individual-based and agent-based models. Ecol Model 198: 115–126
25. Caron-Lormier G, Humphry RW, Bohan DA, Hawes C, Thorbek P (2008) Asynchronous and synchronous updating in individual-based models. Ecol Model 212: 522–527
26. Crooks AT (2007) The repast simulation/modelling system for geospatial simulation. Centre for Advanced Spatial Analysis (University College London), London, UK. Working Paper 123. Available at http://www.casa.ucl.ac.uk/working_papers/paper123.pdf
27. Wilensky U (1999) NetLogo. Center for Connected Learning and Computer-Based Modeling, Northwestern University, Evanston, IL. Available at http://ccl.northwestern.edu/netlogo
28. Swarm (2006) Swarm: A platform for agent-based models. Available at http://www.swarm.org/
29. Topping CJ, Hansen TS, Jensen TS, Jepsen JU, Nikolajsen F, Odderskær P (2003) ALMaSS, an agent-based model for animals in temperate European landscapes. Ecol Model 167: 65–82
30. Topping CJ, Odderskær P (2004) Modeling the influence of temporal and spatial factors on the assessment of impacts of pesticides on skylarks. Environ Toxicol Chem 23: 509–520

31. Topping CJ, Sibly RM, Akcakaya HR, Smith GC, Crocker DR (2005) Risk assessment of UK skylark populations using life-history and individual-based landscape models. Ecotoxicology 14: 925–936
32. Bilde T, Topping C (2004) Life history traits interact with landscape composition to influence population dynamics of a terrestrial arthropod: A simulation study. Ecoscience 11: 64–73
33. Thorbek P, Topping CJ (2005) The influence of landscape diversity and heterogeneity on spatial dynamics of agrobiont linyphiid spiders: An individual-based model. Biocontrol 50: 1–33
34. Jepsen JU, Topping CJ (2004) Modelling roe deer (*Capreolus capreolus*) in a gradient of forest fragmentation: Behavioural plasticity and choice of cover. Can J Zool 82: 1528–1541
35. Fuller RJ, Gregory RD, Gibbons DW, Marchant JH, Wilson JD, Baillie SR, Carter N (1995) Population declines and range contractions among lowland farmland birds. Conserv Biol 9: 1425–1441
36. Chamberlain DE, Fuller RJ, Bunce RGH, Duckworth JC, Shrubb M (2000) Changes in the abundance of farmland birds in relation to the timing of agricultural intensification in England and Wales. J Appl Ecol 37: 771–788
37. Odderskær P, Topping CJ, Petersen MB, Rasmussen J, Dalgaard T, Erlandsen M (2006) Ukrudtsstriglingens effekter på dyr, planter og ressourceforbrug. Miljøstyrelsen, Bekæmpelsesmiddelforskning fra Miljøstyrelsen 105. Available at http://www2.mst.dk/Udgiv/publikationer/2006/87-7052-343-6/pdf/87-7052-344-4.pdf
38. Schläpfer A (1988) Populationsökologie der Feldlerche Alauda arvensis in der intensiv genutzten Agrarlandschaft. Der Ornitologische Beobachter 85: 309–371
39. Jenny M (1990) Populationsdynamic der Feldlerche *Alauda arvensis* in einer intensiv genutzten Agrarlandschaft des Sweizerischen Mittellandes. Der Ornitilogischer Beobachter 87: 153–163
40. Daunicht WD (1998) Zum Einfluss der Feinstruktur in der Vegatation auf die Habitatwahl, Habitatnutzung, Siedlungsdichte und Populationsdynamik von Feldlerchen (Alauda arvensis) in großparzelligem Ackerland. Phd Thesis. University of Bern, Bern
41. Navntoft S, Petersen BS, Esbjerg P, Jensen A, Johnsen I, Kristensen K, Petersen PH, Ørum JE (2007) Effects of mechanical weed control in spring cereals – Flora, fauna and economy. Danish Environmental Protection Agency, Pesticides Research No. 114
42. Wiegand K, Saltz D, Ward D, Levin SA (2008) The role of size inequality in self-thinning: A pattern-oriented simulation model for and savannas. Ecol Model 210: 431–445
43. Thorbek P, Bilde T (2004) Reduced numbers of generalist arthropod predators after crop management. J Appl Ecol 41: 526–538
44. Den Boer PJ (1990) The survival value of dispersal in terrestrial arthropods. Biol Conserv 54: 175–192
45. Errington PL (1934) Vulnerability of bob-white populations to predation. Ecology 15: 110–127
46. Dalkvist T, Topping CJ, Forbes VE (2009) Population-level impacts of pesticide-induced chronic effects on individuals depend more on ecology than toxicology. Ecotoxicol Environ Safety, 10.1016/j.ecoenv.2008.10.002
47. Anway MD, Cupp AS, Uzumcu M, Skinner MK (2005) Epigenetic transgenerational actions of endocrine disruptors and mate fertility. Science 308: 1466–1469
48. Anway MD, Leathers C, Skinner MK (2006) Endocrine disruptor vinclozolin induced epigenetic transgenerational adult-onset disease. Endocrinology 147: 5515–5523
49. Hansson L (1977) Spatial dynamics of field voles *Microtus agrestis* in heterogeneous landscapes. Oikos 29: 593–644
50. Evans DM, Redpath SM, Elston DA, Evans SA, Mitchell RJ, Dennis P (2006) To graze or not to graze? Sheep, voles, forestry and nature conservation in the British uplands. J Appl Ecol 43: 499–505
51. Smidt NM, Olsen H, Bildsøe M, Sluydts V, Leirs H (2005) Effects of grazing intensity on small mammal population ecology in wet meadows. Basic Appl Ecol 6: 57–66
52. Jensen TS, Hansen TS (2001) Effekten af husdyrgræsning på småpattedyr. In: Pedersen L B, Buttenschøn R, Jensen T S (eds) Græsning på ekstensivt drevne naturarealer – Effekter på stofkredsløb og naturindhold. Skov & Landskab, Hørsholm, Park- og Landskabsserien 34: 107–121

53. Bell G, Lechowicz MJ, Appenzeller A, Chandler M, DeBlois E, Jackson L, Mackenzie B, Preziosi R, Schallenberg M, Tinker N (1993) The spatial structure of the physical environment. Oecologia 96: 114–121
54. Clifford PA, Barchers DE, Ludwig DF, Sielken RL, Klingensmith JS, Graham RV, Banton MI (1995) An approach to quantifying spatial components of exposure for ecological risk assessment. Environ Toxicol Chem 14: 895–906
55. Purucker ST, Welsh CJE, Stewart RN, Starzec P (2007) Use of habitat-contamination spatial correlation to determine when to perform a spatially explicit ecological risk assessment. Ecol Model 204: 180–192
56. Jacobsen LB, Frandsen SE (1999) Analyse af de sektor-og samfundsøkonomiske konsekvenser af en reduktion af forbruget af pesticider i dansk landbrug. The Ministry of Food, Agriculture and Fisheries, Danish Research Institute of Food Economics, Denmark, 104
57. Topping CJ (2005) The impact on skylark numbers of reductions in pesticide usage in Denmark. Predictions using a landscape-scale individual based model. National Environmental Research Institute, Denmark, NERI Technical Report 527: 33. Available at http://technical-reports.dmu.dk
58. Goss-Custard JD, Durell SEALD (1990) Bird behaviour and environmental planning: Approaches in the study of wader populations. Ibis 132: 273–282
59. Goss-Custard JD, Caldow RWG, Clarke RT, Durell SEALD, Sutherland WJ (1995) Deriving population parameters from individual variations in foraging behaviour. I. Empirical game-theory distribution model of oystercatchers *Haematopus ostralegus* feeding on mussels *Mytilus edulis*. J Anim Ecol 64: 265–276
60. Sutherland WJ (1996) From individual behaviour to population ecology. Oxford University Press, Oxford
61. Goss-Custard JD, Sutherland WJ (1997) Individual behaviour, populations and conservation. In: JR Krebs, NB Davies (eds) Behavioural ecology: An evolutionary approach. Blackwell Science, Oxford
62. Stillman RA, Goss-Custard JD, West AD, Durell S, Caldow RWG, McGrorty S, Clarke RT (2000) Predicting mortality in novel environments: Tests and sensitivity of a behaviour-based model. J Appl Ecol 37: 564–588
63. Stillman RA, Goss-Custard JD, West AD, Durell S, McGrorty S, Caldow RWG, Norris KJ, Johnstone IG, Ens BJ, Van der Meer J, Triplet P (2001) Predicting shorebird mortality and population size under different regimes of shellfishery management. J Appl Ecol 38: 857–868
64. Stillman RA, Caldow RWG, Durell SEALD, West AD, McGrorty S, Goss-Custard JD, Perez-Hurtado A, Castro M, Estrella SM, Masero JA, Rodríguez-Pascual FH, Triplet P, Loquet N, Desprez M, Fritz H, Clausen P, Ebbinge BS, Norris K, Mattison E (2005) Coast bird diversity – maintaining migratory coastal bird diversity: Management through individual-based predictive population modelling. Centre for Ecology and Hydrology for the Commission of the European Communities, UK
65. Stillman RA, West AD, Goss-Custard JD, McGrorty S, Frost NJ, Morrisey DJ, Kenny AJ, Drewitt A (2005) Predicting site quality for shorebird communities: A case study on the Humber estuary, UK. Mar Ecol Prog Series 305: 203–217
66. Goss-Custard JD, Burton NHK, Clark NA, Ferns PN, McGrorty S, Reading CJ, Rehfisch MM, Stillman RA, Townend I, West AD, Worrall DH (2006) Test of a behavior-based individual-based model: Response of shorebird mortality to habitat loss. Ecol Appl 16: 2215–2222
67. Railsback SF, Lamberson RH, Harvey BC, Duffy WE (1999) Movement rules for individual-based models of stream fish. Ecol Model 123: 73–89
68. Railsback SF (2001) Getting "results": the pattern-oriented approach to analyzing natural systems with individual-based models. Nat Resour Model 14: 465–474
69. Railsback SF (2001) Concepts from complex adaptive systems as a framework for individual-based modelling. Ecol Model 139: 47–62
70. Railsback SF, Harvey BC (2001) Individual-based model formulation for cutthroat trout, Little Jonas Creek, California. General Technical Report PSW-GTR-182. Pacific Southwest Research Station, Forest Service, U.S. Department of Agriculture, Albany, CA

71. Railsback SF, Harvey BC (2002) Analysis of habitat selection rules using an individual-based model. Ecology 83: 1817–1830
72. Railsback SF, Harvey BC, Lamberson RH, Lee DE, Claasen NJ, Yoshihara S (2002) Population-level analysis and validation of an individual-based cutthroat trout model. Nat Resour Model 14: 465–474
73. Railsback SF, Stauffer HB, Harvey BC (2003) What can habitat preference models tell us? Tests using a virtual trout population. Ecol Appl 13: 1580
74. Van Winkle W, Jager HI, Railsback SF, Holcomb BD, Studley TK, Baldrige JE (1998) Individual-based model of sympatric populations of brown and rainbow trout for instream flow assessment: Model description and calibration. Ecol Model 110: 175–207
75. Grimm V, Revilla E, Berger U, Jeltsch F, Mooij WM, Railsback SF, Thulke HH, Weiner J, Wiegand T, DeAngelis DL (2005) Pattern-oriented modeling of agent-based complex systems: Lessons from ecology. Science 310: 987–991
76. Sibly RM, Nabe-Nielsen J, Forchhammer MC, Forbes VE, Topping CJ (2008) On population dynamics in heterogeneous landscapes. Ecol Lett 17: 10.1023/A:1027390600748
77. Dalgaard T, Kjeldsen T, Rasmussen BM, Fredshavn JR, Münier B, Schou JS, Dahl M, Wiborg IA, Nørmark P, Hansen JF (2004) ARLAS' scenariesystem. Et grundlag for helhedsorienterede konsekvensvurdringer af ændringer i arealanvendelsen. In: Hansen JF (ed) Arealanvendelse og landskabsudvikling Fremtidsperspektiver for natur, jordbrug, miljø og arealforvaltning. Danmarks Jordbrugsforskning, Markbrug, 110, 97–128 (in Danish with English summary). Available at: http://web.agrsci.dk/djfpublikation/djfpdf/djfma110.pdf
78. Dortch MS, Gerald JA (2004) Recent advances in the army risk assessment modeling system. In: Whelan G (ed) Brownfields, multimedia modeling and assessment. WIT Press, Southampton, UK. Available at: http://el.erdc.usace.army.mil/arams/pdfs/arams04-advance.pdf
79. Thorbek P, Forbes V, Heimbach F, Hommen U, Thulke HH, van den Brink P, Wogram J, Grimm V (2008) Ecological models in support of regulatory risk assessments of pesticides: Developing a strategy for the future. Integr Environ Assess Manag 5: 1
80. Caswell H (2001) Matrix population models: Construction, analysis, and interpretation. Sinauer Associates, Sunderland, MA
81. Levins R (1966) Strategy of model building in population biology. Am Sci 54: 421–431
82. Wiegand T, Jeltsch F, Hanski I, Grimm V (2003) Using pattern-oriented modeling for revealing hidden information: A key for reconciling ecological theory and application. Oikos 100: 209–222
83. Gutenkunst RN, Waterfall JJ, Casey FP, Brown KS, Myers CR, Sethna JP (2007) Universally sloppy parameter sensitivities in systems biology models. PloS Comput Biol 3: 1871–1878
84. Bak P (1997) How nature works: The science of self-organised criticality. University Press, Oxford
85. Frigg R (2003) Self-organised criticality – What it is and what it isn't. Stud His Philos Sci 34A: 613–632
86. Tarantola A (1987) Inverse problem theory: Methods for data fitting and model parameter estimation. Elsevier, New York
87. Lambert P, Rochard E (2007) Identification of the inland population dynamics of the European eel using pattern-oriented modelling. Ecol Model 206: 166–178
88. Rossmanith E, Blaum N, Grimm V, Jeltsch F (2007) Pattern-oriented modelling for estimating unknown pre-breeding survival rates: The case of the lesser spotted woodpecker (*Picoides minor*). Biol Conserv 135: 555–564
89. Kramer-Schadt S, Revilla E, Wiegand T, Grimm V (2007) Patterns for parameters in simulation models. Ecol Model 204: 553–556
90. Polhill GJ, Parker DC, Brown DG, Grimm V (2008) Using the ODD protocol for describing three agent-based social simulation models of land use change. J Artif Soc Sci Sim 11(2/3) <http://jasss.soc.surrey.ac.uk/11/2/3.html>
91. http://www.hector.ac.uk

Ecotoxicological Applications of Dynamic Energy Budget Theory

Sebastiaan A.L.M. Kooijman, Jan Baas, Daniel Bontje, Mieke Broerse, Cees A.M. van Gestel, and Tjalling Jager

Abstract The dynamic energy budget (DEB) theory for metabolic organisation specifies quantitatively the processes of uptake of substrate by organisms and its use for the purpose of maintenance, growth, maturation and reproduction. It applies to all organisms. Animals are special because they typically feed on other organisms. This couples the uptake of the different required substrates, and their energetics can, therefore, be captured realistically with a single reserve and a single structure compartment in biomass. Effects of chemical compounds (e.g. toxicants) are included by linking parameter values to internal concentrations. This involves a toxico-kinetic module that is linked to the DEB, in terms of uptake, elimination and (metabolic) transformation of the compounds. The core of the kinetic module is the simple one-compartment model, but extensions and modifications are required to link it to DEBs. We discuss how these extensions relate to each other and how they can be organised in a coherent framework that deals with effects of compounds with varying concentrations and with mixtures of chemicals. For the one-compartment model and its extensions, as well as for the standard DEB model for individual organisms, theory is available for the co-variation of parameter values among different applications, which facilitates model applications and extrapolations.

Keywords Dynamic energy budgets · Effects on processes · Kinetics · Metabolism · Transformation

1 Introduction

The societal interest in ecotoxicology is in the effects of chemical compounds on organisms, especially at the population and ecosystem level. Sometimes these effects are intentional, but more typically they concern adverse side-effects of other

S.A.L.M. Kooijman (✉)
Faculty Earth and Life Sciences
Vrije Universiteit, de Boelelaan 1085, 1081 HV Amsterdam, The Netherlands
e-mail: bas.kooijman@falw.vu.nl

J. Devillers (ed.), *Ecotoxicology Modeling*, Emerging Topics in Ecotoxicology: Principles, Approaches and Perspectives 2, DOI 10.1007/978-1-4419-0197-2_9,

(industrial) activities. The scientific interest in effects of chemical compounds is in perturbating the metabolic system, which can reveal its organisation. This approach supplements ideas originating from molecular biology, but now applied at the individual level, the sheer complexity of biochemical organisation hampers reliable predictions of the performance of individuals. Understanding the metabolic organisation from basic physical and chemical principles is the target of dynamic energy budget (DEB) theory [1,2]. In reverse, this theory can be used to quantify the effects of chemical compounds, i.e. *changes* of the metabolic performance of individuals. This chapter describes how DEB theory quantifies toxicity as a process.

The effects can usually be linked to the concentration of compounds inside the organism or inside certain tissues or organs of an organism. This makes that toxicokinetics is basic to effect studies. The physiological state of an organism, such as its size, its lipid content, and the importance of the various uptake and elimination routes (feeding, reproduction, excretion) interacts actively with toxicokinetics, so a more elaborate analysis of toxicokinetics should be linked to the metabolic organisation of the organism [3]. In the section on toxicokinetics, we start with familiar classic models that hardly include the physiology of the organism, and stepwise include modules that do make this link in terms of logical extensions of the classic models.

Three ranges of concentrations of any compound in an organism can be delineated: too-little, enough and too-much. The definition of the enough-range is that variations of concentrations within this range do not translate into variations of the physiological performance of the individual. Some of the ranges can have size zero, such as the too-little-range for cadmium. Effects are quantified in the context of DEB theory as changes of (metabolic) parameters as linked to changes in internal concentrations. These parameters can be the hazard rate (for lethal effects), the specific food uptake rate, the specific maintenance costs, etc. Changes in a single parameter can have many physiological consequences for the individual. DEB theory is used not only to specify the possible modes of action of a chemical compound, but also how the various physiological processes interact. An increase in the maintenance costs by some compound, for instance, reduces growth, and since food uptake is linked to body size, it indirectly reduces food uptake, and so affects reproduction (and development). The existence of the ranges too-little, enough and too-much of concentrations of any compound directly follows from a consistency argument, where no classification of compounds is accepted (e.g. toxic and non-toxic compounds), and many compounds (namely those that make up reserve) exist that do change in concentration in the organism, without affecting parameter values. An important consequence is the existence of an internal no effect concentration (NEC), as will be discussed later.

Effects at the population level are evaluated from those at the individual level, by considering populations as a set of interacting individuals [4–8]. Although DEB-based population dynamics can be complex, particular aspects of population performance, such as the population growth rate at constant environmental conditions, are not very complex. Moreover particular simplifying approximations are possible. The focus of toxicology is typically at time scales that are short relative to the life span of the individual. That of ecotoxicology, however, are much longer, involving the whole life history of organisms; effects on feeding, growth, reproduction and survival are essential and typically outside the scope of toxicology. This has strong implications for the best design of models. Although pharmacokinetics models frequently have many variables and parameters, such complex models are of little use for applications at the population level, where a strong need is felt for relatively simple models, but then applied to many species and in complex situations. DEB theory is especially designed for this task.

Chemical transformations are basic to metabolism, and transformations of toxicants are no exception. Compounds that dissociate in water should be considered as a mixture of ionic and molecular forms, and the pH affects that mixture. This makes that the effect of a single pure compound is of rather academic interest; we have to think in terms of the dynamic mixtures of compounds. Because of its strong links with chemical and physical principles, DEB theory has straightforward ways to deal with effects of mixtures. Some of this theory rests on the covariation of parameter values across species of organism and across chemical compounds. Theory on this covariation is implied by DEB theory, and is one of its most powerful aspects.

We first introduce some notions of DEB theory, and then discuss toxicokinetics and effects in the context of DEB theory.

2 The Standard DEB Model in a Nutshell

The standard DEB model concerns an isomorph, i.e. an organism that does not change in shape during growth, that feeds on a single food source (of constant chemical composition) and has a single reserve, a single structure and three life stages: embryo (which does not feed), juvenile (which does not allocate to reproduction) and adult (which allocates to reproduction, but not to maturation). This is in some respects the simplest model in the context of DEB theory, which is thought to be appropriate for most animals. Food is converted to reserve, and reserve to structure. Reserve does not require maintenance, but structure does, mainly to fuel its turnover (Fig. 1). Reserve can have active metabolic functions and serves the role of representing metabolic memory. Reserve and structure do not change in chemical composition (strong homeostasis). At constant food availability, reserve and structure increase in harmony, i.e. the ratio of their amounts, the reserve density, remains constant (weak homeostasis).

The shape (and so the change of shape) is important because food uptake is proportional to surface area, and maintenance mostly to (structural) volume. The handling time of food (including digestion and metabolic processing) is proportional to the mass of food "particles", during which food acquisition is ceased. The mobilisation rate of reserve to fuel the metabolic needs follows from the weak and strong homeostasis assumptions [9]; a mechanism is presented by Kooijman and Troost [10]. Allocation to growth and somatic maintenance (so to the soma) comprises a fixed fraction of mobilised reserve, and the remaining fraction is allocated to maturation (or reproduction) and maturity maintenance.

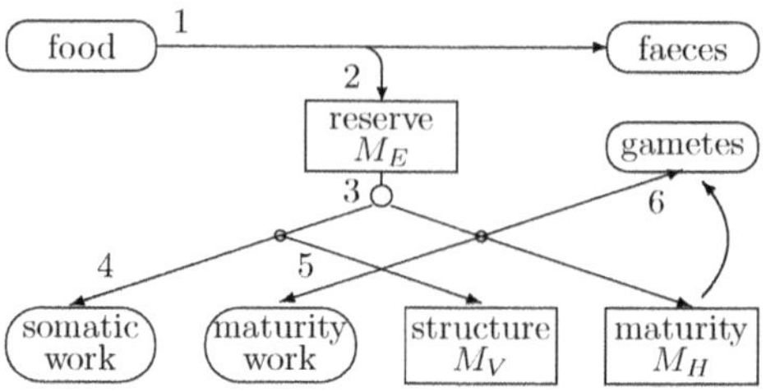

Fluxes of food and reserve	
1	feeding
2	assimilation
3	mobilisation
4	somatic maintenance
5	maturity maintenance
6	reproduction

Fig. 1 The standard DEB model with fluxes (moles per time) and pools (moles). Assimilation is zero during the embryo stage and becomes positive at the transition to the juvenile stage (birth) if food is available. Age is zero at the start of the embryo stage. Reproduction is zero during the juvenile stage and becomes positive at the transition to the adult stage (puberty), when further investment into maturation is ceased

The transition from the embryo to the juvenile stage (i.e. birth) occurs by initiating assimilation when the maturity exceeds a threshold value, and from the juvenile to the adult stage (i.e. puberty) by initiating allocation to reproduction and ceasing of allocation to maturation at another threshold value for maturity. Reserve that is allocated to reproduction is first collected in a buffer that is subjected to buffer handling rules (such as spawning once per season, or convert the buffer content into an egg as soon as there is enough).

Biomass consists of reserve and structure, and can, therefore, change in chemical composition (e.g. lipid content) in response to the nutritional condition; maturity has the status of information, not that of mass or energy. Apart from some minor details, the presented set of simple rules fully specify the dynamics of the individual, including all mass and energy fluxes, such as the uptake of dioxygen, the production of carbon dioxide, nitrogen waste and heat. It takes some time to see exactly how Sousa et al. [9] gives a nice evaluation. Aging is considered to result from a side effect of reactive oxygen species (ROS), and is so linked to the uptake of dioxygen [11]. A high food uptake results in a large amount of reserve, so a high use of reserve, a high uptake of dioxygen, an acceleration of aging and a reduction of life span. The induction of tumours is also linked to the occurrence of ROS, and other reactive molecules, such as mutagenic compounds. This gives a natural link between aging and tumour induction.

The standard DEB model has been extended into many directions for the various purposes. The allocation rule to the soma, for instance, can be refined to allocation to various body parts (e.g. organs), where growth of each body part is proportional to the allocated reserve flux minus what is required for maintenance of that part. Rather than using fixed fractions of the mobilised reserve, the fractions can be linked to the relative workload of the body part. This allows a dynamic adaptation of the body parts in interaction with their use. Tumours can be considered as body parts, and the "workload" of the tumour is the consumption of maintenance. This formulation produced realistic predictions of the effects of caloric restriction on tumour growth, and of the growth of tumours in young vs. old hosts [12]. This approach can be extended to various types of tumours, where tumour growth is not linked to that of the whole body, but that of a particular body part. Many tumours result from destruction of local cell-to-cell communication, rather than from genotoxic effects, but these different routes have similar dynamics.

Other types of extensions of the standard DEB model concern the inclusion of variations in chemical composition of food (with consequences for the transformation of food into reserve) and size-dependent selection of different food items. For example, many herbivores are carnivores when young. Animals are special because they feed on other organisms. Most other organisms take the food-compounds (energy source, carbon source, nitrate, phosphate) that they need independently from the environment, which necessitates the inclusion of more than one reserve; see Kooijman and Troost [10] for the evolutionary perspectives. Most microorganisms, on the one hand, grow and divide, and do not have the three life stages delineated by the standard model. This makes that their change in shape hardly matters and that surface areas can be taken proportional to volumes, which simplifies matters considerably. The partial differential equations that are required to described the physiologically structured population dynamics of isomorphs then collapse to a small set of ordinary differential equations. Plants, on the other hand, require at least two types of structure (roots and shoots) and have a complex adaptive morphology (i.e. surface area–volume relationships); their budgets are most complex to quantify.

3 Family of Toxicokinetic Models

Originally (before the 1950s) the focus of toxicology, i.e. the field that gave rise to ecotoxicology, was on medical applications of compounds in a pharmacological context. Subjects where given a particular dose, and the interest is in the redistribution inside the body, and in transformation and elimination. The aim is to reach the target organ and to achieve a particular effect that restores the health or well-being of the subject. A closely-related interest that developed simultaneously was health protection (disinfection and food protection products), with the purpose of killing certain species of pathogenic organism (especially microorganisms), or to reduce their impact.

After the 1950s, ecotoxicology began to flourish and gradually became more independent of toxicology, where the initial focus was in positive and (later) negative effects of biocides. This came with extensions of the interest in the various uptake and elimination routes that are of ecological relevance, and the environmental physical-chemistry of transport and transformation. The aim is to kill particular species of pest organism locally (insects and weeds), and to avoid effects on other, non-target, species (crop and beneficial species). After the 1970s, the interest further generalised to an environmental concern of avoiding effects of pollutants on organisms, with an increasing attention for (bio)degradation of compounds that are released into the environment, coupled to human activity [13].

This historic development and branching of the interest in toxico-kinetics came with a narrowing of the focus on a particular aspect of toxico-kinetics in the various applications that, we feel, is counter-productive from a scientific point of view. The purpose of this section is an attempt to restore the coherence in the field, by emphasising the general eco-physiological context and the relationship of the various models with the core model for toxico-kinetics: the one-compartment model. The more subtle models account for the interaction with the metabolism of the organism, which involves its metabolic organisation.

We here focus on the logical coherence of toxico-kinetics, bio-availability and metabolism, including effects (=changes in metabolism). It is not meant to be a review. For recent reviews on toxico-kinetic models, see Barber [14] and Mackay and Fraser [15]. See Table 1 for a list of frequently used symbols.

3.1 One-Compartment Model

The core model in toxico-kinetics is the one-compartment model, see Fig. 2. It states that the uptake rate is proportional to the environmental concentration c, and the elimination rate is proportional to the internal concentration Q:

$$\frac{d}{dt} Q = k_e \left(P_{0d} c(t) - Q \right), \tag{1}$$

Table 1 List of frequently used symbols, with units and interpretation

Symbol	Units	Interpretation
t	d	Time
c	M	Concentration of compound in the environment
n_i	mol m^{-1}	Density of compound
Q	mol C-mol^{-1}	Concentration of compound in an organism
r	d^{-1}	Specific growth rate of structure
k_e	d^{-1}	Elimination rate
k_{01}, k_{10}	d^{-1}	Exchange rates between compartments
P_{0d}	mol C-mol^{-1} M^{-1}	Bioconcentration factor (BCF)
v_{ij}, v_e	m d^{-1}	Velocity, elimination
d_i	m^2 d^{-1}	Diffusivity
m_E, m_{ER}	C-mol C-mol^{-1}	Reserve density, reproduction buffer density

The unit C-mol represents the number of C-atoms in a organism as multiple of the number of Avogadro.

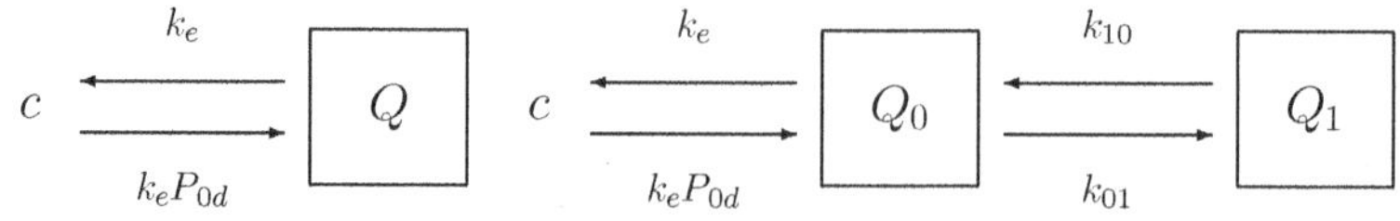

Fig. 2 The scheme of the one- and two-compartment models. The factor P_{0d} converts an external concentration into an internal one; all rates labelled k have dimension 'per time'. In the two-compartment model P_{0d} does not have the interpretation of the bioconcentration factor

$$Q(t) = Q(0)\exp(-tk_e) + k_e P_{0d} \int_0^t c(s)\exp((s-t)k_e)\,ds, \tag{2}$$

$$= Q(0)\exp(-tk_e) + P_{0d}\,c\,(1 - \exp(-tk_e)) \quad \text{for } c \text{ is constant}, \tag{3}$$

where $P_{0d} = Q(\infty)/c$ is the BioConcentration Factor (BCF) and k_e the elimination rate. The product $k_e P_{0d}$ is known as the uptake rate, and is frequently indicated with k_u, which is misleading because its units are $d^{-1}\,m^3\,C\text{-mol}^{-1}$. Even if we would work with kg rather than C-mol, and the specific density of the organism equals $1\,kg\,dm^{-3}$, the BCF is not dimensionless [1]. It is typically more convenient to work with molalities in soils (mol kg^{-1}), and with molarities (mol l^{-1}) in water. Molalities give the uptake rate the units $d^{-1}\,g\,C\text{-mol}^{-1}$. Many workers use gram rather than mole to quantify the compound, but this choice is less practical to compare the toxicity of different compounds. The elimination rate k_e has dimension 'per time' and is independent of how the compound is quantified; contrary to the uptake rate, the elimination rate can be extracted from effect data and determines how fast effects build up in time, relative to the long-term effect level.

The concentrations c and Q must obviously exist, meaning that the environment and the organism are taken to be homogeneous. This condition can be relaxed without making the model more complex by allowing a spatial structure (such as organs), and an exchange between the parts that is fast relative to the exchange between the organism and the environment. Other implicit assumptions of the one-compartment model are that the organism does not change in size or in chemical composition, so changes in food availability must be negligible. The (bio)availability of the compound remains constant. So transformations can be excluded, and the environment is large relative to the organism and well-mixed. Sometimes, e.g. in the case of a large fish in a small aquarium, this is not true and the dynamics of the concentration in the organism and the environment should be considered simultaneously: the 1-1 compartment model [16]. These restrictions will be removed later.

Since rates generally depend on temperature, and temperature typically changes in time, the elimination rate k_e can change in time as well [17]. In the sequel, we will discuss rates of metabolism, and like all rates, these also depend on temperature, frequently according to the Arrhenius relationship [1].

3.2 Multi-Compartment Models

If transport inside the organism is not fast, relative to the exchange with the environment, multiple-compartment models should be considered [18, 19]. If exchange with the environment is only via compartment number 0, the change in the concentrations in the nested compartments number 0 and 1 is

$$\frac{d}{dt}Q_0 = k_e\,(P_{0d}\,c - Q_0) + k_{10}Q_1 - k_{01}Q_0; \quad \frac{d}{dt}Q_1 = k_{01}Q_0 - k_{10}Q_1, \tag{4}$$

where k_{01} and k_{10} are the exchange rates between the compartments, see Fig. 2. The partition coefficient between the compartments equals $P_{10} \equiv Q_1(\infty)/Q_0(\infty) = k_{01}/k_{10}$, while $P_{0d} \equiv Q_0(\infty)/c$ remains unchanged. In many cases whole body measurements are used. If M_0 and M_1 denote the masses of compartment 0 and 1, the whole body partition coefficient with the environment amounts to $P_{+d} = (M_0 + M_1 P_{10}) P_{0d}/M_+$, with $M_+ = M_0 + M_1$ is the whole body mass. This rather complex behaviour of the whole body partition coefficient can be a source of problems in fitting models to data. In many practical cases, it is not possible to identify the compartments and to measure the concentrations in these compartments directly. The use of multi-compartment models cannot be recommended in such cases.

This extension still classifies as a transport model, so in a clean environment ($c = 0$), the organism will loose all its load ($Q_0(\infty) = Q_1(\infty) = 0$). Quite a few data sets on the kinetics of ('heavy') metals in organisms show that once loaded, an organism never fully looses its load [20–22]. Such behaviour cannot be captured by multi-compartment models because this involves an extension to transformation of compounds (technically speaking, sequestered compounds belong to a different chemical species).

If $k_{01}, k_{10} \gg k_e$, we can assume that $Q_1(t) \simeq P_{10}\, Q_0(t)$, and the kinetics (4) reduces to (1), with Q replaced by Q_0. This situation is called time-scale separation.

The interpretation of the compartments can be a special tissue or organ, or, more abstract, reserve (compartment 1) and structure (compartment 0). The latter makes sense in the context of DEB theory, where both compartments are assumed to have a constant, but different, chemical composition, while reserve is relatively rich in lipids in many animal taxa. While (4) assumes that the size of all compartments does not change, we will relax on this below, when we allow more interactions with metabolism and energetics.

We can obviously include more compartments, and more complex interactions with the environment, but the number of parameters rapidly increases this way. In practice multi-compartments are used if the one-compartment model fits data badly. The introduction of more parameters generally improves the fit, but not necessarily for the right reasons. As a rule of thumb, it is only advisable to use more compartment models if data on the concentrations inside the compartments are available. If lack of fit of the one-compartment model is the only motivation, alternatives should be considered that are discussed later.

3.3 Film Models

Film models are conceptually related to multi-compartment models because both are extensions of the one-compartment model that include more detail in transport (so in physical factors), though in different but complementary ways. Film models are especially popular in environmental chemistry for following the transport of compounds from one environmental compartment (such as water) to another (such as air). Both compartments are assumed to be well-mixed, except for a narrow film at the interface of both compartments, where transport is by diffusion.

To follow the dynamics for the densities of the compound n (mol/m), we need to define a spatial axis perpendicular to the interface and choose the origin at the boundary between the bulk and the film (on each side of the interface). Let L_i be the depth of the film, d_i the diffusivity of the compound in the film, and v_{ij} the exchange velocity of the compound between the two media. As discussed in Kooijman et al. [16], the dynamics of the densities is given by partial differential equations (pdes) for medium $i = 1 - j$ and $j = 0$ or 1

$$0 = \frac{\partial}{\partial t} n_i(L,t) - d_i \frac{\partial^2}{\partial L^2} n_i(L,t) \quad \text{for } L \in (0, L_i), \tag{5}$$

with boundary conditions at $L = 0$ (i.e. the boundary between the film and the well-mixed medium) for $v_i = d_i / L_i$

$$0 = \frac{\partial}{\partial t} n_i(0,t) - v_i \frac{\partial}{\partial L} n_i(0,t), \tag{6}$$

and boundary conditions at $L = L_i$ (i.e. the interface between the media where the two films meet)

$$0 = v_{ji} n_j(L_j,t) - v_{ij} n_i(L_i,t) + d_i \frac{\partial}{\partial L} n_i(L_i,t), \tag{7}$$

The boundary condition at $L = 0$, and the diffusion process in the film is rather standard, but we believe that the boundary condition at $L = L_i$ is presented for the first time in Kooijman et al. [16]. Users of the popular film models typically skip the formulation of the pde and directly focus at steady-state situations; they typically use the concentration jump across the interface that belongs to the situation when there is no net transport across the interface. As long as there is transport, however, the concentration jump differs from this equilibrium value.

The depth of the films is typically assumed to be small and the transport in the films in steady state, which makes that the density profiles in the films are linear. This leads to the 1-1-compartment kinetics for the bulk densities

$$\frac{d}{dt} n_i(0) = k_{ij}(P_{ij} n_j(0) - n_i(0)); \quad k_{ij} = \frac{v_i / \mathcal{L}_i}{1 + P_{ij} v_i / v_j - v_i / v_{ij}}, \tag{8}$$

where $\mathcal{L}_i$ is the depth of the medium. This approximation only applies if $v_i v_j < v_{ij} v_j + v_{ji} v_i$ and the transport in the film is rate limiting. The 1-1-compartment kinetics also results, however, if the film depths reduce to zero and if the diffusivities are high. The rate from i to j then reduces to $k_{ij} = v_i \mathcal{L}_i^{-1} (1 - v_i / v_{ij})^{-1}$. In these two situations, transport in the film is no longer rate limiting.

The applicability of film models to toxico-kinetics in organisms is still an open question. It can be argued that a stagnant water film sticks to aquatic or soil organisms (and air to a terrestrial organism), and that the skin (or cuticula) is not well served by the internal redistribution system (blood) of the organism. If toxicokinetics is fully limited by transport in the film, and if it is not limited by the film, one-compartment kinetics results; only in the intermediary situation we can expect some deviations. Yet the discussion is not completely academic, since these details matter for how the elimination rate depends on the partition coefficient [16,23].

3.4 Uptake and Elimination Routes

We now consider extensions of the one-compartment model due to biological factors by accounting for various uptake and elimination routes. These routes depend on the type of organism, its habitat and properties of the compound. Animals that live in (wet) soil are in intense contact with the water film around soil particles, and their situation has similarities with that of aquatic animals. Direct transport through the skin can be important, which involves the surface area of an organism. Some parts of the skin are more permeable, especially that used by the respiratory system for dioxygen uptake and carbon dioxide excretion. The uptake rate might be linked to the respiratory rate, which depends on the energetics of the organism. Generally, the respiration rate scales with a weighted sum of surface area and volume, but the proportionality constants depend on the nutritional conditions of the organism [1]. For terrestrial animals, uptake via the lungs from air and via skin contact with the soil must be considered. Sometimes uptake is via drinking; the DEB theory quantifies drinking via the water balance for the individual and involves a.o. metabolism and transpiration. The details can be found in Kooijman [1].

A second important uptake route is via food and the gut epithelium. The feeding rate depends on food availability, food quality and the surface area of the organism [1].

The elimination can follow the same routes as uptake, but there are several additional routes to consider, namely via products of organisms. The first possibility is the route that excreted nitrogen waste follows (urination). Reproductive products (mostly eggs and sperm) can also be an important elimination route. Moulting (e.g. ecdysozoans, including the rejection of gut epithelium, e.g. collembolans) or the production of mucus (e.g. lophotrochozoans) or milk (e.g. female mammals) are other possible excretion routes. The DEB theory quantifies reproductive and other products as functions of the amounts of reserve and structure of the individual. In the standard DEB model, they work out to be cubic polynomials in body length, but the coefficients depend on the nutritional conditions (amount of reserves per structure) [1].

3.5 Changes in Body Size and Composition

The body size of an organism matters in the context of toxico-kinetics for several reasons [24]. As exchange is via surface area, and is proportional to concentration, surface area–volume interactions are involved. This problem also applies to compartment and film models, but gets a new dimension if we consider changes in body size, which are linked to the nutritional condition of organisms (lipid content), and so to (changes in) body composition. Small changes in size can have a substantial effect on the shape of accumulation curves.

If an organism does not change in shape during growth (so it remains isomorphic), surface area is proportional to volume$^{2/3}$, or to squared length. Moreover, dilution by growth should be taken into account, even at low growth rates. This modifies (1) to

$$\frac{\mathrm{d}}{\mathrm{d}t} Q = (P_{0d}\, c(t) - Q)\, v_e / L - Q r \quad \text{with} \quad r = \frac{\mathrm{d}}{\mathrm{d}t} \ln L^3, \tag{9}$$

where L is the length, and $v_e = L_m k_e$ is the elimination velocity for maximum length L_m. The last term, Qr, represents the dilution by growth. If it equals zero, we can replace v_e/L by the constant elimination rate k'_e, but its meaning still matters if we compare the kinetics in two organisms of different size. DEB theory specifies how the change in (cubed) length depends on the amount of reserve and structure of the organism, and how the change in reserve depends on these state variables and food availability. Food intake and maintenance play an important role in growth and together they control the maximum size an organism can reach, since food intake is proportional to a surface area and maintenance to structural volume. Wallace had this insight in 1865 already [25].

The DEB theory allows for particular changes in body composition, because reserve and structure can change in relative amounts and both have a constant composition. Food (substrate) is first transformed into reserve, and reserve is used for metabolic purposes, such as somatic and maturity maintenance, growth, maturation and reproduction. The change in reserve density for metabolic use is proportional to the reserve density per length, which makes that high growth rates come with high reserve densities, i.e. the ratio of the amounts of reserve and structure.

Reserves are in many animal taxa relatively enriched in lipids, which might have a strong influence on the kinetics of hydrophobic compounds. The body burden of eel in a ditch that is polluted with mercury or PCB might greatly exceed that of other fish partly because eel is relative rich in lipids. This illustrates the importance of lipid dynamics. Freshly laid eggs consist almost exclusively of reserve, which makes egg production a potentially important elimination route for lipophyllic compounds. The reserve allocation to reproduction is via a buffer that comes with species-specific buffer handling rules. Many aquatic species spawn once a year only (e.g. most bivalves and fish), which implies that the buffer size gradually increases between two spawning events and makes a sharp jump down at spawning. The body burden can also make a jump at spawning (up or down, depending on the properties of the compound).

The difference in lipid content between reserve and structure invites for the application of a nested two-compartment model, where the exchange with the environment is via structure. This links up nicely with food uptake, because reserve does not play a role in it, and food uptake is also proportional to squared structural length. An important difference with the nested two-compartment model is, however, that the size of the compartments typically changes in time, especially the reserve. When redistribution of the compound between the compartments reserve and structure is relatively fast, and the nested two-compartment model for the compound reduces to a one-compartment one, reserve dynamics still affects toxico-kinetics, because the lipid content is changing in time. The resulting dynamics for active uptake from food amounts to

$$\frac{d}{dt}Q = (P_{Vd}c_d + P_{VX}fc_X)\frac{v_e}{L} - Q(P_{VW}\frac{v_e}{L} + r) \quad \text{with } P_{VW} = 1 + P_{EV}(m_E + m_{ER}), \tag{10}$$

where c_d and c_X are the concentrations of the compound in the environment and in food, f is the scaled functional response, P_{Vd} and P_{VX} are the partition coefficients of the compound in structure and environment or food. The reserve density m_E and the reproduction buffer density m_{ER} now modify the partition coefficient between structure and biomass (i.e. reserve plus structure), via the partition coefficient between reserve and structure P_{EV}. DEB theory specifies how structural length L, the reserve density m_E and the reproduction buffer density m_{ER} change in time.

The reproduction buffer is not of importance in all species, and not always in males. If food density is constant, the reserve density m_E becomes constant. In those situations, the structure-biomass partition coefficient P_{VW} is constant as well. If also the dilution by growth can be neglected, i.e. $r = 0$ and L is constant, (10) still reduces to the one-compartment model (1).

Many accumulation–elimination experiments are done under starvation conditions; e.g. it is hardly feasible to feed mussels adequately in the laboratory. The reserve density decreases during the experiment, so the chemical composition is changing, which can affect the toxico-kinetics [26].

Some situations require more advanced modelling of the uptake and eliminations route, where, e.g. gut contents exchanges with the body in more complex ways, and defecation might be an elimination route.

3.6 Metabolism and Transformation

Both uptake and elimination can depend on the metabolic activity [27]. Respiration is frequently used as a quantifier for metabolic activity. This explains the popularity of body size scaling relationships for respiration [28], and the many attempts to relate many other quantities to respiration. In the context of DEB theory, however, and that of indirect calorimetry, respiration is a rather ambiguous term, because it can stand for the use of dioxygen, or the production of carbon dioxide or heat. These are not all proportional to each other, however. Moreover, all these three fluxes have contributions from various processes, such as assimilation, maintenance, growth, etc. Since the use of reserves fuels all non-assimilatory activities, this is an obvious quantifier to link to the rate at which compounds are transformed or taken up. For uptake of compounds via the respiratory system, the use of dioxygen might be a better quantity to link to uptake under aerobic conditions.

Respiration rates turn out to be cubic polynomials in structural length in DEB theory, which resemble the popular allometric functions numerically in great detail. The coefficients depend on the nutritional conditions in particular ways. Since elimination rates are inversely proportional to length because of surface area–volume interactions, as has been discussed, and the specific metabolic rate is very close to this relationship, it is by no means easy to evaluate the role of metabolism in direct uptake and elimination in undisturbed subjects.

The role of metabolism is easier to access for elimination via products and if the elimination rate is not proportional to the internal concentration, but has a maximum capacity. The classic example is the elimination of alcohol in human blood [29]. This type of kinetics can be described as

$$\frac{\mathrm{d}}{\mathrm{d}t}Q = k_{\mathrm{e}} P_{0d}\, c - k_{\mathrm{e}} Q/(K_Q + Q), \tag{11}$$

where K is a half saturation constant for the elimination process. It reduces to the one-compartment model (1) for small internal concentrations, relative to the half saturation constant, $K_Q \gg Q$. The elimination rate can now be linked to metabolic activity, and so to body size. If particular organs are involved, such as the liver in the case of alcohol, the DEB theory can be used to study adaptation processes to particular metabolic functions. In the case of alcohol, the uptake term should obviously be replaced by a more appropriate one that applies to the particular subject.

Many toxicants are metabolically modified. This especially applies to lipophyllic compounds, which are typically transformed into more hydrophyllic ones, which are more easily excreted but also metabolically more active. The rate of transformation can be linked to the metabolic rate, and so depends on body size and nutritional conditions. These metabolic products can be more toxic than the original lipophyllic compound.

4 Bio-Availability

Compounds are not only transformed in the organism, but also in the environment which affects their availability. Many have an ionic and a molecular form, which are taken up at different rates; the ionic species requiring counter ions, which complicates their uptake. Speciation depends on the concentration of compound and environmental properties, such as the pH. It can vary in time and also occurs inside the organism, but the internal pH usually varies within a narrow band only. Models for mixtures of chemicals can be used in this case (see section on effects). Internal concentration gradients could develop if transport inside the organism is slow; film models should be used in this case. A nice example of a case where concentration gradients result from transformation in combination with transport is the fluke *Fasciola*, which has an aerobic metabolism near its surface with the microaerobic environment inside its host, but an anaerobic metabolism in the core of its body [30].

Another problem, which occurs especially in soils, is that the transport through the medium can be slow enough for concentration gradients to develop around the organism. Film models should then be used again.

A major problem in the translation of laboratory toxicity tests to field situations is the formation of ligands with (mainly) organic compound that are typically abundant in the field, but not in the bioassay. Ligands reduce the availability substantially, and typically has a rather complex dynamics. Moreover compounds can be transformed by (photo)chemical transformation and by actions of (micro)organisms. This implies that the concentration of available compounds changes in time, and our methodology to assess effects of chemical compounds should be able to take this into account.

These processes of transformation require compound-specific modelling and this short section demonstrates that bio-availability issues interact with toxico-kinetics and effects of chemical compounds in dynamic ways, which calls for a dynamic approach to effects of chemicals [3, 31].

5 Effects at the Individual Level

Compound affect individuals via changes of parameter values as functions of the internal concentration [32, 33]. The parameter values are independent of the internal concentration in the 'enough'-range of the compound. This implies the existence of an internal no effect concentrations (NECs) at either end of the 'enough'-range; the upper end is typically of interest for ecotoxicological applications. Outside the 'enough'- range the value of the target parameter is approximately a linear function of the internal concentration as long as the changes in parameter value are small; the inverse slope is called the tolerance concentration (a large value means that the compound is not very toxic). Small changes in parameter values do not necessarily translate into small changes of some end-point, such as the cumulative number of offspring [34] or the body size [35] at the end of some standardised exposure period. DEB theory specifies how exactly changes in parameter values translate into the performance of the individual. Typical target parameters are the specific maintenance costs, or the yield of structure on reserve, or the maximum specific assimilation rate or the yield of reserve on food or the yield of offspring on reserve. For effects on survival, the hazard rate serves the role of target parameter, and the inverse 'tolerance concentration' for the hazard rate is called the killing rate. Mutagenic compounds can induce tumours [36], but also accelerate ageing by enhancing the effects of ROS. The partitioning fraction for mobilised reserve can be the target parameter for endocrine disruptors.

Using sound theory for how effects depend on internal concentrations, DEB-based theory can handle varying concentrations of toxicant [37–40], even pulse exposures [41]. DEB theory applies to all organisms, including bacteria that decompose organic pollutants. A proper description of this process should account for adaptation [42], co-metabolism [43] and the fact most bacteria occur in flocculated form in nature [44], which affect the availability of the compound.

The model of linear effects of internal concentrations on parameter values has been extended into several directions, such as adaptation to the compound, inclusion of the recent exposure history via receptor dynamics [45] and attempts to include particular molecular mechanisms [46].

5.1 Mixtures and NECs

Mixtures of compounds affect parameters values via addition of the effects of single compounds, plus an interaction term, which is proportional to the products of the internal concentrations of the compound [47]. This interaction term can be positive or negative; a construct that is the core of the analysis of variance (ANOVA) model and rests on a simple Taylor series approximation of a general non-linear multivariate function, which only applies for small changes of parameter values; the non-linearities of the effects should be taken into account for larger changes. These non-linearities might well be specific for the compound and the species and, therefore, lack generality. Notice that linear effects on parameter values translate into non-linear effects of the performance of the individual because the DEB models are non-linear. Also notice that each DEB parameter has a NEC value for any compound; the lowest value among all parameters might be considered as *the* NEC of a compound for the organism, but its estimation requires to study effects on all 13 parameters of the standard DEB model, in principle. Since this study can be demanding, it is in practice essential to talk about the NEC of a compound for an organism *for a particular DEB parameter*.

The NEC reflects the ability of the individual to avoid changes of performance. From a statistical point of view, this robust parameter has very nice properties [48–50]. The NEC is *not* meant to imply that some molecules of a compound do not have an effect, while other molecules do. The removal of a kidney in a healthy person can illustrate the NEC concept: the removal implies an effect at the sub-organismic level, but this effect generally does not translate into an effect at the individual level. The NEC, therefore, depends on the level of observation. We can delineate three cases of how compounds in the mixture combine for the NEC

- The presence of other compounds is of no relevance to the NEC of any particular compound
- The various compounds add, like they do for effects, and at the moment effects show up, the amounts of the compounds that show no effects remain constant
- The various compounds add, like they do for effects, and the amounts of the compounds that show no effects continue changing with the internal concentration

of the compounds; if a compound continues accumulation more than other compounds, its NEC increases while that of the other compounds in the mixture decreases

The third case is possibly most realistic, but also computationally the most complex. In many practical situations, the results are very similar to the second case, which can be used as an approximation. If compounds in a mixture are equally toxic, and so all have the same NEC, the second case is formally identical to this special case. This way of described effects of mixtures turns out to fit well with experimental data [47] and each pair of compounds have a single interaction parameter, which does not change in time. If there are k compounds in a mixture, there are $k(k-1)/2$ interaction parameters, just like in ANOVA.

A further reflection on the NEC might clarify the concept. Any compound affects (in principle) all DEB parameters (including the hazard rate), but the NEC for the various parameters differ. This makes that, if the internal concentration increases, the parameter with the lowest NEC first starts to change, but other parameters follow later. In a mixture of compounds, this can readily lead to a rather complex situation where in a narrow range of (internal) concentrations of compounds in a mixture several parameters start changing. Even in absence of the above-mentioned chemical interactions of compounds on a single parameter, interactions via the energy budget occur, which are hard to distinguish from the chemical interactions on a single parameter. Chemical interactions are typically rare, but interactions via the budget always occur.

5.2 Hormesis

Hormesis, the phenomenon that low concentrations of a toxicant seem to have a stimulating rather than an inhibiting effect on some endpoint, can result from interactions of the compound with a secondary stress, such as resulting from very high levels of food availability. If a compound decreases the yield of structure on reserve, it reduces growth and delays birth (if an embryo is exposed) and puberty (in the case of juveniles), but also reduces the size at birth. A reduction of growth indirectly reduces reproduction, because food uptake is linked to size. Since it also reduces size at birth, the overall effect can be a hormesis effect on reproduction (in terms of number of offspring per time) [51]. Indirect effects on reproduction differ from direct effects by not only reducing, but also delaying reproduction. This has important population dynamical consequences.

5.3 Co-Variation of Parameter Values

A very powerful property of the standard DEB model and the one-compartment model is that they imply rules for how parameter values co-vary among species and compounds [9, 23, 52]; this variation directly translates into how expected effects vary. These expectations can be used to fill gaps in knowledge about parameter values, but cannot replace the need for this knowledge. Evolutionary adaptations and differences in mode of action of compounds can cause deviation from expected parameter values for species and compounds, respectively.

The reasoning behind the scaling relationships for the standard model rests on the assumption that parameters that relate to the local biochemical environment in an organism are independent of the maximum body size of a species, but parameters that relate to the physical design of an organism depends on the maximum size. Strange enough, this simple assumption fully specifies the covariation of parameter values. The application is best illustrated with the maximum length L_m an endotherm can reach in the standard DEB model. This length is a simple function of three parameters, $L_m = \kappa\{p_{Am}\}/[p_M]$, where κ is the fraction of mobilised reserve that is allocated to the soma, $\{p_{Am}\}$ is the surface-area specific maximum assimilation energy flux and $[p_M]$ is the volume-specific somatic maintenance cost. Since κ and $[p_M]$ depend on the local biochemical environment, they are independent of maximum length, which implies that $\{p_{Am}\}$ must be proportional to maximum length. All other parameters can be converted in simple ways to quantities that depend on the local biochemical environment; these transformations then defined how they depend on maximum length. When we divide the maturity at birth and puberty by the cubed maximum length, we arrive at a maturity-density, which reflects the local biochemical environment and should not depend on maximum length. So the maturity at birth and puberty are proportional to the cubed maximum length. Many quantities, such as the use of dioxygen by an individual, can be written as functions of parameter values and amounts of reserve and structure. So the maximum respiration rate of a species is a function of parameter values, while we know of each parameter how it depends on maximum length. It can be shown that maximum respiration rate scales between a squared and a cubed maximum length, and the weight-specific respiration with weight to the power $-1/4$, a well-known result since Kleiber [53].

The reasoning behind the scaling relationships for the one-compartment model rests on the assumption that transport to and from the compartment is skewly symmetric [16]. The ratio of the concentrations in the compartment and the environment at equilibrium is a ratio of uptake and the elimination rates, just like the maximum length of the individual in the standard DEB model. This implies, see [16], that the uptake rate is proportional to the square root of the partition coefficient, and the elimination rate is inversely proportional to the square root of the partition coefficient. Film models are extensions of the one-compartment model, which behaves at the interface between the environments basically in the same way as an one-compartment model; only around this interface they differ because film models account for concentration gradients. This deviation can be taken into account with the result that elimination rates are (almost) independent of the partition coefficient for low values of the partition coefficient and inversely proportional to it at high values. Effects parameters can be included into the scaling reasoning is similar ways, with the result that the NEC is inversely proportional to the partition coefficient, and the tolerance concentration or the killing rate is proportional to the partition coefficient.

The octanol–water partition coefficient is frequently taken as a substitute for the body-water partition coefficient, with the advantage that reliable computational methods exist to evaluate this partition coefficient from the chemical structure of the molecule. In practice, however, octanol is not an ideal chemical model for organisms that change the chemical composition of their bodies. This makes that the co-variation of NECs, elimination and killing rates show less scatter and can be expected on the basis of the variation between each of these three parameters and the octanol–water partition coefficient [23].

We are unaware of any descriptive model for toxico-kinetics and/or metabolic organisation for which theory on the co-variation of parameter values is available, and we doubt that it even possible to derive such theory for descriptive models. Co-variation theory is not available for the so-called net-production models [54], for instance, where maintenance needs are first subtracted from assimilation before allocation to storage, growth or reproduction. The fact that the predicted relationships of now over 30 eco-physiological quantities, such as length of the embryonic and juvenile periods, maximum reproduction rate, maximum growth rate, maximum population growth rate, vary with the maximum body size of species in ways that match empirical patterns provides strong support of the DEB theory. The one-compartment and standard DEB models share the property that the independent variable (the partition coefficient in the case of toxicokinetics and the maximum length in the case of budgets) can be written as a ratio of an incoming flux (of toxicant and food, respectively) and an outgoing flux (excretion and maintenance, respectively). This shared property seems to be crucial for the core theory.

One of the many possible applications of the scaling relationships is in the effects of mixtures of compounds with similar modes of action, such as the polychlorinated hydrocarbons. Suppose we know the concentrations and the partition coefficients of the compounds in the mixture. We then link the elimination rates, the NECs and the tolerance concentrations to the partition coefficients in the way described, and estimate the three proportionality constants for the results of a bioassays with the mixture.

The sound theoretical basis for effects of toxicants in combination with rules for the co-variation of parameter values offers the possibility for extrapolation, from one individual to another, from one species of organism to another, and, sometimes, from one type of compound to another [23]. These crosslinks partly reduce the need for a huge experimental effort that should be invested in more advanced forms of environmental risk assessment, such as discussed in Brack et al. [13]. Moreover, the theory simplifies to parameter poor models under particular conditions. It has been demonstrated that many popular empirical models turn out to be special cases of the general theory [1]. This might help in particular applications.

6 Effects at the Population and Ecosystem Level

At high food levels, organisms grow and reproduce fast and the maintenance costs comprise only a tiny fraction of the budget of the individuals. If a compound increases the maintenance cost for individuals, say by a factor two, these effects are hardly felt by a fast-growing population. Fast growth never lasts long in nature, due to the depletion of food resources. At carrying capacity, where the generation of food resources just matches the maintenance needs of a population (this is the maintenance needs of the collection of individuals plus a low reproductive output that cancels the mortality), maintenance costs comprise the dominant factor of the budget of individuals. If a toxic compounds now increases the maintenance cost by a factor two, it in fact reduces the carrying capacity by a factor two. This simple argument shows that the effects of toxicants on populations is dynamic, even if the concentration of the compound would be constant [55]. It also shows that no single quantifier for toxicity can exist at the population level.

If a toxic compound increases the cost of growth or reproduction, the effects hardly depend on the growth rate of the population, so on the food level, which shows that the mode of action is important for how effects on individuals translate to those on populations. It might be difficult to tell the various modes of action apart on the basis of the results from a standardised toxicity bioassay. The reason why the mode of action is still important is in the biological significance of the observed effect, which must be found at the population level. Details in the reproduction strategy of populations turn out to be important for how effects on reproduction translate to the population level [56].

Although bioassays with meso-cosms have the charm of being close to the actual interest of effects of toxicants to be avoided, the experimental control is extremely weak, which results in a huge scatter of trajectories of experimental meso-cosms. The result is that the effects have to be huge to recognise them as effects [57]. Moreover, the expected long term behaviour of chemically perturbated ecosystems is very complex, as shown by bifurcation analysis [58].

The specific population growth rate integrates the various performances of individuals naturally, and can rather easily be evaluated [59]. A delay of the onset of reproduction can be at least as important as a reduction of the reproduction for the fate of the population.

7 Concluding Remarks

We argued that models for effects of chemical compounds should have three modules:

- Dynamic energy budgets for how organisms generally deal with resource uptake and allocation
- Toxico-kinetics for how organisms acquire the compounds
- Chances of budget parameters as function of the internal concentrations

We discussed the basics for each of these modules: the standard DEB model, the one-compartment model and the linear change in target parameters. We also indicated were and how these models can be extended, from simple to more complex, to include particular phenomena. We discussed how budgets affect both the kinetics and the effects and, therefore, have a central role in effects models. Practice teaches that the restriction of realistic modelling is not in the model formulation as such, but in the useful application of these models to data. More complex models have more parameters and many of these parameters are by no means easy to extract from available experimental data. They require knowledge of physiological and ecological processes that are typically outside the scope of typical (eco)toxicological research. Kooijman et al. [60] discusses why any particular application of DEB theory requires only a limited set of parameter values, and how these values can be obtained from simple observation on growth and reproduction at several levels of food availability.

The practical need to fill in gaps in knowledge about parameter values is the reason why due attention has been given to theory for the co-variation of parameter values; this theory naturally follows from the logical structure of the one-compartment model and the standard DEB model. Extensions of both models can modify the co-variation of parameters, as has been discussed.

Contrary to descriptive models, models with strong links with underlying processes can be used for a variety of extrapolation purposes, from acute to chronic exposure, from one species to another, from one compound to another, from individuals to populations, from laboratory to field situations [31]. Such extrapolations are typically required in environmental risk assessment, where NECs should play a key role [61]. The use of models to predict exposure in the environment is frequent, but to predict effects is still rare. The complexity of the response of organisms to changes in their chemical environment doubtlessly contributed to this. Yet we think that 30 years of applications of DEB theory to quantify effects of compounds on organisms have demonstrated that the theory is both effective and realistic. Many of the computations behind the models in this chapter can be done with the freely downloadable software package DEBtool: http://www.bio.vu.nl/thb/deb/deblab/

Acknowledgements This research has been supported financially by the European Union (European Commission, FP6 Contract No. 003956 and No 511237-GOCE).

References

1. Kooijman SALM (2000) Dynamic energy and mass budgets in biological systems. Cambridge University Press, New York
2. Kooijman SALM (2001) Quantitative aspects of metabolic organization; A discussion of concepts. Phil Trans R Soc B 356:331–349
3. Kooijman SALM (1997) Process-oriented descriptions of toxic effects. In: Schüürmann G, Markert B (eds) Ecotoxicology, pp. 483–519. Spektrum Akademischer Verlag, Heidelberg
4. Kooijman SALM, Andersen T, Kooi BW (2004) Dynamic energy budget representations of stoichiometric constraints to population models. Ecol 85:1230–1243
5. Kooijman SALM, Grasman J, Kooi BW (2007) A new class of non-linear stochastic population models with mass conservation. Math Biosci 210:378–394
6. Kooi BW, Kelpin FDL (2003) Structured population dynamics, a modeling perspective. Comm theor Biol 8:125–168

7. Kooijman SALM, Kooi BW, Hallam TG (1999) The application of mass and energy conservation laws in physiologically structured population models of heterotrophic organisms. J theor Biol 197:371–392
8. Nisbet RM, Muller EB, Brooks AJ, Hosseini P (1997). Models relating individual and population response to contaminants. Environ Mod Assess 2:7–12
9. Sousa T, Domingos T, Kooijman SALM (2008) From empirical patterns to theory: A formal metabolic theory of life. Phil Trans R Soc B 363:2453–2464
10. Kooijman SALM, Troost TA (2007) Quantitative steps in the evolution of metabolic organisation as specified by the dynamic energy budget theory. Biol Rev 82:1–30
11. Leeuwen IMM van, Kelpin FDL, Kooijman SALM (2002) A mathematical model that accounts for the effects of caloric restriction on body weight and longevity. Biogerontol 3:373–381
12. Leeuwen IMM van, Zonneveld C, Kooijman SALM (2003) The embedded tumor: Host physiology is important for the interpretation of tumor growth. Brit J Cancer 89:2254–2263
13. Brack W, Bakker J, Deerenberg C, Deckere E de, Gils J van, Hein M, Jurajda P, Kooijman S, Lamoree M, Lek S, López de Alda MJ, Marcomini A, Muñoz I, Rattei S, Segner H, Thomas K, Ohe P van der, Westrich B, Zwart D de, Schmitt-Jansen M (2005) Models for assessing and forecasting the impact of environmental key pollutants on freshwater and marine ecosystems and biodiversity. Environ Sci Pollut Res 12:252–256
14. Barber MC (2003) A review and comparison of models for predicting dynamic chemical bioconcentration in fish. Environ Toxicol Chem 22(9):1963–1992
15. Mackay D, Fraser A (2000) Bioaccumulation of persistent organic chemicals: mechanisms and models. Environ Pollut 110:375–391
16. Kooijman SALM, Jager T, Kooi BW (2004) The relationship between elimination rates and partition coefficients of chemical compounds. Chemosphere 57:745–753
17. Janssen MPM, Bergema WF (1991) The effect of temperature on cadmium kinetics and oxygen consumption in soil arthropods. Environ Toxicol Chem 10:1493–1501
18. Godfrey K (1983) Compartmental models and their application. Academic Press, London
19. Jacquez JA (1972) Compartmental analysis in biology and medicine. Elsevier, Amsterdam
20. Spurgeon DJ, Hopkin SP (1999) Comparisons of metal accumulation and excretion kinetics in earthworms (*Eisenia fetida*) exposed to contaminated field and laboratory soils. Appl Soil Ecol 11:227–243
21. Sheppard SC, Evenden WG, Cronwell TC(1997) Depuration and uptake kinetics of I, Cs, Mn, Zn and Cd by the earthworm (*Lumbricus terrestris*) in radiotracerspiked litter. Environ Toxicol Chem 16:2106–2112
22. Vijver MG, Vink JPM, Jager T, Wolterbeek HT, Straalen NM van, Gestel CAM van (2005) Elimination and uptake kinetics of Zn and Cd in the earthworm *Lumbricus rubellus* exposed to contaminated floodplain soil. Soil Biol Biochem 10:1843–1851
23. Kooijman SALM, Baas J, Bontje D, Broerse M, Jager T, Gestel CAM van, Hattum B van (2007) Scaling relationships based on partition coefficients and body sizes have similarities and interactions. SAR QSAR Environ Res 18:315–330
24. Kooijman SALM, Haren RJF van (1990) Animal energy budgets affect the kinetics of xenobiotics. Chemosphere 21:681–693
25. Finch CE (1994) Longevity, senescence, and the genome. University of Chicago Press, Chicago
26. Haren RJF van, Schepers HE, Kooijman SALM (1994) Dynamic energy budgets affect kinetics of xenobiotics in the marine mussel *Mytilus edulis*. Chemosphere 29:163–189
27. Molen GW van der, Kooijman SALM, Wittsiepe J, Schrey P, Flesch-Janys D, Slob W (2000) Estimation of dioxin and furan elimination rates from cross-sectional data using a pharmacokinetic model. J Exp Anal Environ Epidemiol 10:579–585
28. Peters RH (1983) The ecological implications of body size. Cambridge University Press, Cambridge
29. Wagner JG (1958) The kinetics of alcohol elimination in man. Acta Pharmacol Toxicol 14: 265–289
30. Tielens AGM (1982) The energy metabolism of the juvenile liver fluke, *Fasciola hepatica*, during its development in the vertebrate host. PhD thesis, Utrecht University, The Netherlands

31. Jager T, Heugens EHW, Kooijman SALM (2006) Making sense of ecotoxicological test results: towards process-based models. Ecotoxicol 15:305–314
32. Jager T, Crommentuijn T, Gestel CAM van, Kooijman SALM (2004) Simultaneous modelling of multiple endpoints in life-cycle toxicity tests. Environ Sci Technol 38:2894–2900
33. Péry ARR, Flammarion P, Vollat B, Bedaux JJM, Kooijman SALM, Garric J (2002) Using a biology-based model (debtox) to analyse bioassays in ecotoxicology: Opportunities and recommendations. Environ Toxicol Chem 21:459–465
34. Kooijman SALM, Bedaux JJM (1996) Analysis of toxicity tests on *Daphnia* survival and reproduction. Water Res 30:1711–1723
35. Kooijman SALM, Bedaux JJM (1996) Analysis of toxicity tests on fish growth. Water Res 30:1633–1644
36. Leeuwen IMM van, Zonneveld C (2001) From exposure to effect: A comparison of modeling approaches to chemical carcinogenesis. Mut Res 489:17–45
37. Péry ARR, Bedaux JJM, Zonneveld C, Kooijman SALM (2001) Analysis of bioassays with time-varying concentrations. Water Res 35:3825–3832
38. Alda Alvarez O, Jager T, Nunez Colao B, Kammenga JE (2006) Temporal dynamics of effect concentrations. Environ Sci Technol pages 2478–2484
39. Klepper O, Bedaux JJM (1997) A robust method for nonlinear parameter estimation illustrated on a toxicological model. Nonlin Anal 30:1677–1686
40. Klepper O, Bedaux JJM (1997) Nonlinear parameter estimation for toxicological threshold models. Ecol Modell 102:315–324
41. Pieters BJ, Jager T, Kraak MHS, Admiraal W (2006) Modeling responses of *Daphnia magna* to pesticide pulse exposure under varying food conditions: Intrinsic versus apparent sensitivity. Ecotoxicol 15:601–608
42. Brandt BW, Kelpin FDL, Leeuwen IMM van, Kooijman SALM (2004) Modelling microbial adaptation to changing availability of substrates. Water Res 38:1003–1013
43. Brandt BW, Leeuwen IMM van, Kooijman SALM (2003) A general model for multiple substrate biodegradation. Application to co-metabolism of non structurally analogous compounds. Water Res 37:4843–4854
44. Brandt BW, Kooijman SALM (2000) Two parameters account for the flocculated growth of microbes in biodegradation assays. Biotech Bioeng 70:677–684
45. Jager T, Kooijman SALM (2005) Modeling receptor kinetics in the analysis of survival data for organophosphorus pesticides. Environ Sci Technol 39:8307–8314
46. Muller EB, Nisbet RM (1997) Modeling the effect of toxicants on the parameters of dynamic energy budget models. In: Dwyer FJ, Doane TR, Hinman ML (eds) Environmental Toxicology and Risk Assessment: Modeling and Risk Assessment, Vol 6. American Society for Testing and Materials, Philadelphia, PA
47. Baas J, Houte BPP van, Gestel CAM van, Kooijman SALM (2007) Modelling the effects of binary mixtures on survival in time. Environ Toxicol Chem 26:1320–1327
48. Andersen JS, Bedaux JJM, Kooijman SALM, Holst H (2000) The influence of design parameters on statistical inference in non-linear estimation; a simulation study based on survival data and hazard modelling. J Agri Biol Environ Stat 5:323–341
49. Baas J, Jager T, Kooijman SALM (2009) Estimation of no effect concentrations from exposure experiments when values scatter among individuals. Ecol Model 220:411–418
50. Kooijman SALM, Bedaux JJM (1996) Some statistical properties of estimates of no-effects concentrations. Water Res 30:1724–1728
51. Kooijman SALM (2009) What the egg can tell about its hen: embryo development on the basis of dynamic energy budgets. J Math Biol 58:377–394
52. Kooijman SALM (1986) Energy budgets can explain body size relations. J theor Biol 121: 269–282
53. Kleiber M (1932) Body size and metabolism. Hilgardia 6:315–353
54. Lika K, Kooijman SALM (2003) Life history implications of allocation to growth versus reproduction in dynamic energy budgets. Bull Math Biol 65:809–834
55. Kooijman SALM (1985) Toxicity at population level. In: Cairns J (ed) Multispecies toxicity testing, pp. 143–164. Pergamon Press, New York

56. Alda Alvarez O, Jager T, Kooijman SALM, Kammenga J (2005) Responses to stress of *Caenorhabditis elegans* populations with different reproductive strategies. Func Ecol 19: 656–664
57. Kooijman SALM (1988) Strategies in ecotoxicological research. Environ Aspects Appl Biol 17(1):11–17
58. Kooi BW, Bontje D, Voorn GAK van, Kooijman SALM (2008) Sublethal contaminants effects in a simple aquatic food chain. Ecol Modell 112:304–318
59. Kooijman SALM, Hanstveit AO, Nyholm N (1996) No-effect concentrations in alga growth inhibition tests. Water Res 30:1625–1632
60. Kooijman SALM, Sousa T, Pecquerie L, Meer J van der, Jager T (2008) From food-dependent statistics to metabolic parameters, a practical guide to the use of dynamic energy budget theory. Biol Rev 83:533–552
61. Kooijman SALM, Bedaux JJM, Slob W (1996) No-effect concentration as a basis for ecological risk assessment. Risk Anal 16:445–447

Matrix Population Models as Relevant Modeling Tools in Ecotoxicology

Sandrine Charles, Elise Billoir, Christelle Lopes, and Arnaud Chaumot

Abstract Nowadays, one of the big challenge in ecotoxicology is to understand how individually measured effects can be used as predictive indices at the population level. A particular interesting aspect is to evaluate how individual measures of fitness and survival under various toxic conditions can be used to estimate the asymptotic population growth rate known as one of the most robust endpoint in population risk assessment. Among others, matrix population models are now widely recognized as a convenient mathematical formalism dedicated to the characterization of the population demographic health. They offer the advantage of simplicity, not only in the modeling process of underlying biological phenomena, but also in the sensitivity analyses and the simulation running. On the basis of different biological systems among aquatic animal species (from fish to zooplankton), we illustrate the use of matrix population models to quantify environmental stress effects of toxic type. We also show how critical demographic parameters for the population dynamics can be highlighted by sensitivity analyses. The first example will focus on coupled effects of food amount and exposure concentration on chironomid population dynamics in laboratory. The second example will exemplify the use of energy-based models coupled with matrix population ones to properly describe toxic effects on daphnid populations. Last, we will show how to introduce a spatial dimension in Leslie type models to describe space-specific aspects of contaminant induced population dynamics alteration with the case of brown trout population modeling at the river network scale.

Keywords Matrix population models · Ecotoxicology · Individual and population level · DEBtox · Spatial scale

---

S. Charles (✉)
Université de Lyon, F-69000, Lyon; Université Lyon 1; CNRS, UMR5558, Laboratoire de Biométrie et Biologie Evolutive, F-69622, Villeurbanne, France
e-mail: scharles@biomserv.univ-lyon1.fr

J. Devillers (ed.), *Ecotoxicology Modeling*, Emerging Topics in Ecotoxicology: Principles, Approaches and Perspectives 2, DOI 10.1007/978-1-4419-0197-2_10,

1 Introduction

Since pioneering Truhaut's works [1], ecotoxicology is recognized as a whole scientific field which encompasses chemistry, toxicology and ecology. It aims at assessing and predicting the ecological consequences of environmental contaminants at the different levels of organization [2]. Toxic effects are actually mainly quantified through numerous biochemical and biological variables which make sense according to the level of organization where they are measured. The integration of these measures at different levels of organization became one of the major challenges for ecotoxicology today, especially to infer toxic effects at the population level from bioassays carried out at the sub- or the individual level [3]. Indeed, in investigating the concept of stress in ecology, Forbes and Calow [4, 5] pointed out the population growth rate as a robust endpoint for assessing the ecological risks of chemicals. Stark and Banks [6] also promoted demographic toxicology as an approach to evaluate the toxicity using life history parameters and other measures of population growth rate.

The complexity of relationships between toxic compounds and individuals, between physiological and life history traits, or between individuals and populations makes necessary the use of modeling approaches especially when ecotoxicological outcomes are expected to be predictive and not only descriptive. Various modeling methods exist to extrapolate from the individuals to the population in an ecotoxicological context. Some of these methods have been reviewed by Mooij et al. [7] who also specifically present the use of an individual-based model to depict *Daphnia* population dynamics in the lake Volkerak. In addition to the egg-ratio method which has been applied to *Daphnia galeata* in littoral and pelagic areas [8] and to the modeling approach using partial differential equations [9] in the case of a *Daphnia pulex* laboratory population, the most often method used in ecotoxicology is the Euler–Lotka equation [10] which was recently modified to model effects of four synthetic musks on the life cycle of the harpacticoid copepod *Nitocra spinipes* [11]. But since now more than twenty years, matrix population models, originated by Leslie [12, 13], have proved their efficiency first in including internal structure in populations [14], second in describing toxic effects on population dynamics [15]. Matrix population models are indeed particularly relevant when different age or stage classes can be distinguished in their susceptibility to the toxic compounds [3]. First used to analyze life table experiments [16, 17], several authors have successfully described various population dynamics under toxic exposure since then; for example, Lopes et al. [18] evaluated the methiocarb effect on *Chironomus riparius* laboratory populations; Smit et al. [19] proposed a population model for *Corophium volutator*, a marine amphipod used in sediment bioassays; Klok et al. [20] extrapolated effects of copper in the common earthworm *Dendrobaena octaedra* to the population level; Billoir et al. [21] used a combined approach including process-based effect models and matrix population ones to look at the effect of cadmium on *Daphnia magna* populations; and Ducrot et al. [22] used a similar approach to deal with zinc-spiked sediments effects in the gastropod *Valvata piscinalis*.

Another advantage can be granted to matrix population models with regard to the estimation of many characteristic population endpoints (the asymptotic population growth rate, the generation time, the stable population structure, the reproductive values, ...) and the use of convenient tools for sensitivity analyses [14]. Last, when a spatial dimension is needed to understand the effects of dispersion of toxic compounds on the population dynamics, matrix population models can easily be extended without further complexity [23]; hence, Chaumot et al. [24] successfully developed a multiregional matrix population model to explore how the demography of a hypothetical brown trout population living in a river network varied in response to different spatial scenarios of cadmium contamination.

First and foremost, our chapter aims to present, as simply as possible, the implementation of matrix population models, from the Leslie standard model to a density-dependent version, but also their possible extensions when a spatial dimension and/or a toxic compound has to be considered. Hence, Sect. 2 is dedicated to the theoretical framework underlying matrix population modeling, while Sects. 3–5 will introduce different concrete case studies with an increasing degree of complexity in models involved. Section 2 is thus intentionally detailed in a pedagogic manner so that each step of building a matrix population model can autonomously be conducted by a non specialist. On the contrary, models in Sects. 3–5 will deliberately little itemized to emphasize biological and ecotoxicogical conclusions; those who are interested in more details can refer to the cited literature.

2 The Latest on Matrix Population Models

In the last three decades, matrix population models have become one of the most popular tool for investigating the dynamics of age- or stage-classified populations [14]. First based on populations structured by chronological age [12, 13], they were rapidly extended to deal with any discrete state variable describing the internal structure of the population [25]. As the general framework remains the same in every case, Sect. 2 will focus on age-classified Leslie-type models in a non toxicological context first. Part 2.5 will be dedicated to the building of Leslie-type models when the effects of a toxic compound are introduced in the modeling framework.

2.1 Hypotheses and Notations of the Leslie Model and its Derivative

A Leslie-type model rests on three strong hypotheses:

1. The variable *age* is divided into a discrete set of ω age classes, the age-class i ($i = 1 \dots \omega$) gathering individuals of age bounded by $i - 1$ and i.
2. The time t is a discrete variable and the model describes the transition between the population density at time t and at time $t + 1$.
3. The time step exactly corresponds to one age class duration.

As illustrated in Fig. 1, the dynamics of a population can be schematized by a life cycle graph where nodes denote age classes and arrows denote transition between

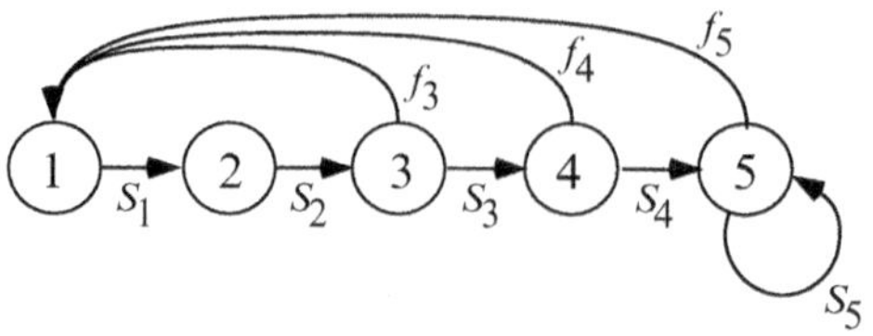

$$\mathbf{A} = \begin{pmatrix} 0 & 0 & f_3 & f_4 & f_5 \\ s_1 & 0 & 0 & 0 & 0 \\ 0 & s_2 & 0 & 0 & 0 \\ 0 & 0 & s_3 & 0 & 0 \\ 0 & 0 & 0 & s_4 & s_5 \end{pmatrix}$$

$$\begin{cases} n_1(t+1) = f_3 n_3(t) + f_4 n_4(t) + f_5 n_5(t) \\ n_2(t+1) = s_1 n_1(t) \\ n_3(t+1) = s_2 n_2(t) \\ n_4(t+1) = s_3 n_3(t) \\ n_5(t+1) = s_4 n_4(t) + s_5 n_5(t) \end{cases} \Leftrightarrow N(t+1) = \mathbf{A} N(t)$$

Fig. 1 Example of an age-structured life-cycle graph with $\omega = 5$ age classes. Fertilities (f_i) and survival rates (s_i) are age-specific. The corresponding age-structured population transition matrix **A** is also given

age classes. At time t, the population is described by the number of individuals in each age class. Let $n_i\,(t)$ be the number of individuals in age class i $(i = 1, \ldots, \omega)$ at time t and $N\,(t)$ the population vector at time t. Then:

$$N\,(t) = (n_1\,(t), \ldots, n_i\,(t), \ldots, n_\omega\,(t))^{\mathrm{T}}, \tag{1}$$

where superscript T denotes transposition.

From time t to time $t + 1$, the number of individuals in age class i is calculated from the numbers of individuals in the other age classes at time t, by means of proportionality coefficients, denoted by a_{ij}, corresponding to survival during the year and fecundity rates. This calculation can be translated into a linear combination containing these coefficients:

$$n_i\,(t+1) = \sum_{j=1}^{\omega} a_{ij} n_j\,(t) \qquad \forall\, i = 1 \ldots \omega. \tag{2}$$

Coefficients a_{ij}, which correspond to survival and fecundity rates, are directly derived from the life cycle graph as illustrated in Fig. 1. Therefore (2) can easily be converted into a matrix equation as follows:

$$N\,(t+1) = \mathbf{A}\, N\,(t), \tag{3}$$

with $\mathbf{A} = \left[a_{ij}\right]_{1 \le i, j \le \omega}$ the transition matrix.

Among a_{ij}, age-specific fertilities (f_i) appear on the first row of the matrix, while age-specific survival rates (s_i) appear on the lower subdiagonal of the matrix (Fig. 1). Coefficients f_i and s_i are called vital rates.

Because time is a discrete variable and because *birth-pulse* populations in which reproduction is limited to a short breeding season were exclusively considered, entries of the transition matrix are discrete coefficients (from one age class to the next). Their values will thus depend on the census time, i.e., on the choice for the beginning of the time step. Censuses can be carried out just *after* reproduction (thus called postbreeding censuses) or just *before* reproduction (prebreeding censuses). Under a prebreeding hypothesis, the first age class mortality is included into the fertility coefficients. In this chapter, only prebreeding censuses will be considered.

The transition matrix **A** has initially been defined for age-structured populations leading to the *classical* Leslie model [12, 13]. From time t to time $t+1$, all individuals go from the age class i to the age class $i+1$ and ω is the maximum age of the individuals ($s_5 = 0$ in Fig. 1).

However, when ω is unknown, the *standard* Leslie model allows to consider a last age class in which individuals can remain indefinitely: the last term in the diagonal remains not null ($s_5 \neq 0$ in Fig. 1).

A particular case sometimes appears in classical Leslie models when only the last age-class is able to reproduce ($f_3 = f_4 = 0$ in Fig. 1). In such a case, the transition matrix **A** is said *imprimitive*, which will induce a specific dynamic behavior.

2.2 *What about the Population Dynamics?*

As detailed in Caswell [14], several demographic endpoints can be deduced from the analysis of a Leslie model. Except in the case where the transition matrix **A** is imprimitive, under the theoretical assumption that vital rates remain constant with time, the population exponentially grows with a constant population growth rate λ during the asymptotic phase, after a transient phase corresponding to the first time steps from the initial condition (Fig. 2b). During this asymptotic phase, the age distribution (i.e., the proportion of the different age classes as given by a vector $\boldsymbol{w}$) becomes stable (Fig. 2c), and reproductive values (i.e., the contributions of each age class to the future generations, as given by a vector $\boldsymbol{v}$) are fixed. The net reproductive rate R_0 (i.e., the mean number of offspring by which a newborn individual will be replaced at the end of its life, or on another point of view, the rate by which the population increases from one generation to the next) also has a fixed value and can be used to calculate the generation time T (which also corresponds to the time required for the population to increase by a factor R_0): $T = \ln R_0 / \ln \lambda$. At last, the duration of the transient phase can be estimated by the speed of convergence (or the damping ratio ρ) toward the asymptotic phase.

As summarized in Table 1, all these population characteristics are directly related to mathematical outcomes obtained from the transition matrix **A**.

The particular case of an imprimitive matrix **A** leads the population size to oscillate (Fig. 2e). Consequently, the age distribution does not converge to a stable distribution (Fig. 2f). Nevertheless, Cull and Vogt [26] showed that a running average of the population vector $N(t)$, taken over the period of oscillation, converges to the vector v associated to λ.

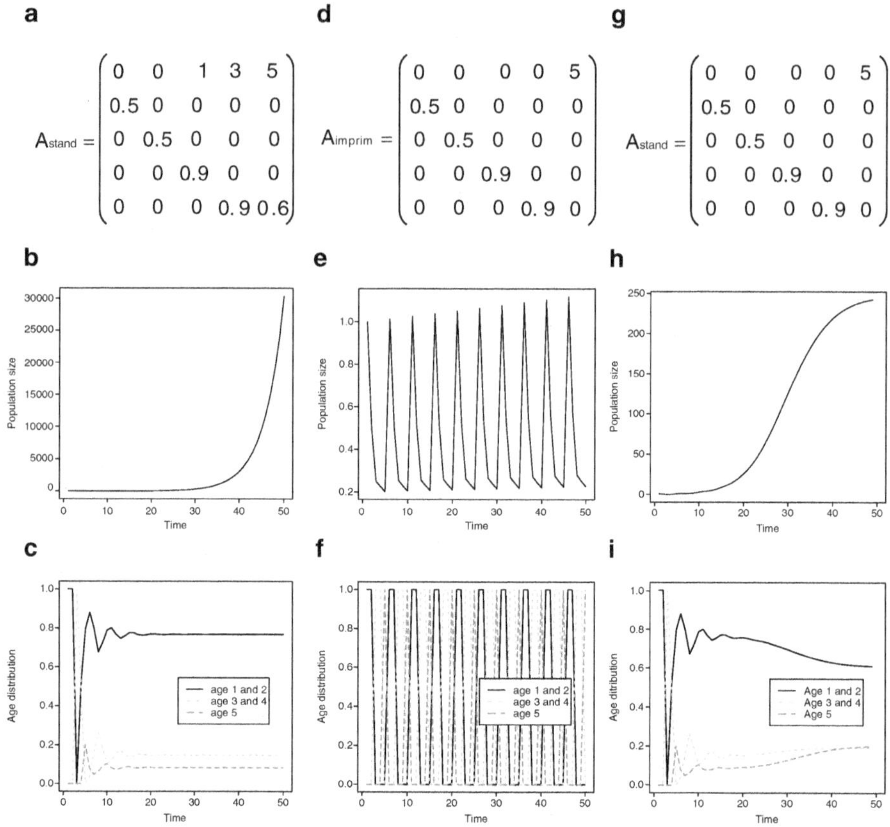

Fig. 2 Population dynamics simulations with a projection matrix (**a**) $\mathbf{A}_{\text{stand}}$ corresponding to a standard Leslie model, (**d**) $\mathbf{A}_{\text{imprim}}$ which is imprimitive, and (**g**) $\mathbf{A}_N$ which is density-dependent. Under (**a**) hypothesis, the population size grows exponentially (**b**) and the age distribution converges to a stable state (**c**). Under (**d**) hypothesis, the population size oscillates (**e**) leading also to oscillations for the age distribution (**f**). Under (**g**) hypothesis, both the population size (**h**) and the age distribution (**i**) reach an equilibrium state. In graphics (**c**), (**f**), and (**i**) *black* corresponds to age classes 1 and 2, *gray* to age classes 3 and 4, and *slate gray* to age class 5

2.3 Adding Density-dependence Hypotheses

Vital rates are likely to vary in response to changes in population density under the effect of interactions between individuals (e.g., intra-specific competition). In this case, vital rates f_i and s_i depend on age class sizes at time t. The transition equation (3) thus becomes:

$$N(t+1) = \mathbf{A}_{N(t)} N(t)\,, \quad (4)$$

by making the matrix entries functions of density.

Table 1 Relationships between classical demographic endpoints for a population, dynamics of which being ruled by a Leslie model (classical or standard), and their corresponding mathematical outcomes obtained from a transition matrix **A**

Symbol	Mathematical outcome	Demographic meaning
λ	Dominant eigenvalue of the transition matrix A	Population growth rate
w	Right eigenvector of A associated to λ	Stable age distribution
v	Left eigenvector of A associated to λ	Reproductive values
R_0	$R_0 = \sum_{i=1}^{\omega} f_i \prod_{k=1}^{i-1} s_k$	Net reproductive rate
T	$T = \dfrac{\ln R_0}{\ln \lambda} \Leftrightarrow R_0 = \lambda^T$	Generation time
ρ	$\rho = \lambda/\lvert\lambda_2\rvert$, where λ_2 is the eigenvalue of A with the second largest magnitude	Damping ratio

Because the model is nonlinear, the population growth is not exponential but is characterized by an equilibrium state $N^* = \left(n_1^*, \ldots, n_i^*, \ldots, n_\omega^*\right)^{\mathrm{T}}$ defined as follows:

$$N^* = \mathbf{A}_{N^*} N^*. \tag{5}$$

A linear approximation methodology [14] makes possible a local stability analysis of the equilibrium state N^*. Population endpoints are thus calculated from this equilibrium as the population size ($\sum_{i=1}^{\omega} n_i^*$) and the age distribution at equilibrium.

The *standard* Leslie model, the case with an imprimitive transition matrix **A**, and the density-dependent Leslie model are numerically illustrated in Figs. 2g–i. On these figures, the density-dependant function is expressed as follows:

$$f(N) = \mathrm{e}^{-0.005N}, \tag{6}$$

where N corresponds to the total number of individuals at time t.

2.4 Adding a Spatial Dimension

Modeling age-structured populations with Leslie-type models can also be applied to spatially fragmented populations [27]. Originated from Rogers [28, 29] in human demography and called multiregional models, they have been extended in ecology [30, 31] to become metapopulation models.

The dimension of such models describing real systems in detail (demography and dispersal) increases with the number of patches and the number of age classes considered. But, whatever the dimension, if the model remains linear with constant vital rates, the general theoretical framework previously presented still applies.

On a general point of view, a population structured in ω age classes and living in an environment fragmented into p patches is considered.

Let $n_{i,j}(t)$ be the number of individuals in age class i ($i = 1, \ldots, \omega$) and patch j ($j = 1, \ldots, p$) at time t, $N_i(t)$ the population vector of the age class i among all patches at time t and $N(t)$ the total population vector at time t:

$$N_i(t) = \left(n_{i,1}(t), \ldots, n_{i,j}(t), \ldots, n_{i,p}(t)\right)^{\mathrm{T}}, \tag{7}$$

$$N(t) = \left(N_1(t), \ldots, N_i(t), \ldots, N_\omega(t)\right)^{\mathrm{T}}. \tag{8}$$

Let $f_{i,j,k}$ be the fecundity of individuals in age class i and patch j reproducing in patch k. Let $s_{i,j,k}$ be the survival rate of individuals in age class i and patch j surviving and dispersing in patch k.

Fecundity and survival matrices can thus be defined for each age class i:

$$\mathbf{F}_i = \begin{pmatrix} f_{i,1,1} & \cdots & f_{i,j,1} & \cdots & f_{i,p,1} \\ \cdots & \cdots & & & \cdots \\ f_{i,1,k} & \cdots & f_{i,j,k} & \cdots & f_{i,p,k} \\ \cdots & & & \cdots & \cdots \\ f_{i,1,p} & \cdots & f_{i,j,p} & \cdots & f_{i,p,p} \end{pmatrix} \quad \text{and} \quad \mathbf{S}_i = \begin{pmatrix} s_{i,1,1} & \cdots & s_{i,j,1} & \cdots & s_{i,p,1} \\ \cdots & \cdots & & & \cdots \\ s_{i,1,k} & \cdots & s_{i,j,k} & \cdots & s_{i,p,k} \\ \cdots & & & \cdots & \cdots \\ s_{i,1,p} & \cdots & s_{i,j,p} & \cdots & s_{i,p,p} \end{pmatrix}. \tag{9}$$

Hence, in a standard Leslie approach, the transition matrix writes now as follows:

$$\mathbf{A} = \begin{pmatrix} \mathbf{0} & \cdots & & \mathbf{F}_i & \cdots & \mathbf{F}_\omega \\ \mathbf{S}_1 & \mathbf{0} & & \cdots & & \mathbf{0} \\ \mathbf{0} & \cdots & & & & \\ \cdots & \cdots & \mathbf{S}_i & \cdots & & \vdots \\ & & & \cdots & & \mathbf{0} \\ \mathbf{0} & & \cdots & \mathbf{0} & \mathbf{S}_{\omega-1} & \mathbf{S}_\omega \end{pmatrix}. \tag{10}$$

Figure 3 illustrates how it works in a particular case with $\omega = 5$ age classes and $p = 2$ patches. In this example, individuals from patch 1 are supposed to reproduce only in patch 1, while individuals in patch 2 can also reproduce in patch 1; only individuals in age classes 2–4 from patch 1 can disperse to patch 2. Here follow the corresponding matrices:

$$\mathbf{F}_3 = \begin{pmatrix} f_{3,1,1} & f_{3,2,1} \\ 0 & f_{3,2,2} \end{pmatrix}, \quad \mathbf{F}_4 = \begin{pmatrix} f_{4,1,1} & f_{4,2,1} \\ 0 & f_{4,2,2} \end{pmatrix}, \quad \mathbf{F}_5 = \begin{pmatrix} f_{5,1,1} & f_{5,2,1} \\ 0 & f_{5,2,2} \end{pmatrix}, \tag{11}$$

$$\mathbf{S}_1 = \begin{pmatrix} s_{1,1,1} & 0 \\ 0 & s_{1,2,2} \end{pmatrix}, \quad \mathbf{S}_2 = \begin{pmatrix} s_{2,1,1} & 0 \\ s_{2,1,2} & s_{2,2,2} \end{pmatrix}, \quad \mathbf{S}_3 = \begin{pmatrix} s_{3,1,1} & 0 \\ s_{3,1,2} & s_{3,2,2} \end{pmatrix},$$

$$\mathbf{S}_4 = \begin{pmatrix} s_{4,1,1} & 0 \\ s_{4,1,2} & s_{4,2,2} \end{pmatrix}, \tag{12}$$

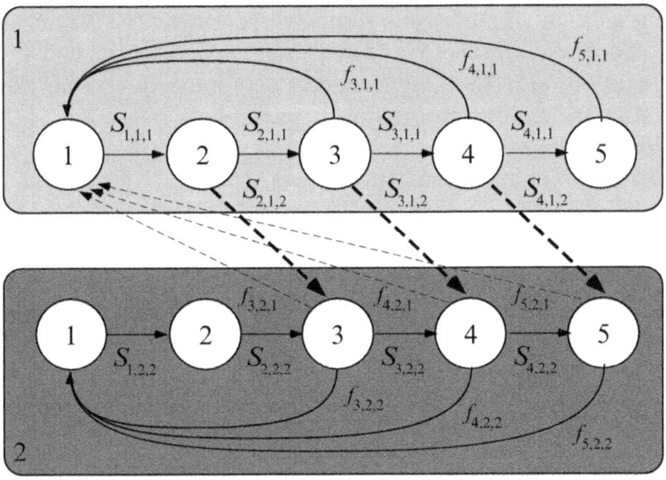

Fig. 3 Example of a life cycle graph in a spatialized environment with two patches. The population is subdivided into five classes. Individuals in patch 1 live and reproduce in patch 1, but some individuals from classes 2–4 can move in patch 2. Individuals in patch 2 live and reproduce in patch 2, but some individuals from classes 3–5 can reproduce in patch 1

$$\mathbf{A} = \left(\begin{array}{cc|cc|cc|cc|cc}
0 & 0 & 0 & 0 & f_{3,1,1} & f_{3,2,1} & f_{4,1,1} & f_{4,2,1} & f_{5,1,1} & f_{5,2,1} \\
0 & 0 & 0 & 0 & 0 & f_{3,2,2} & 0 & f_{4,2,2} & 0 & f_{5,2,2} \\
\hline
s_{1,1,1} & 0 & 0 & 0 & 0 & 0 & 0 & 0 & 0 & 0 \\
0 & s_{1,2,2} & 0 & 0 & 0 & 0 & 0 & 0 & 0 & 0 \\
\hline
0 & 0 & s_{2,1,1} & 0 & 0 & 0 & 0 & 0 & 0 & 0 \\
0 & 0 & s_{2,1,2} & s_{2,2,2} & 0 & 0 & 0 & 0 & 0 & 0 \\
\hline
0 & 0 & 0 & 0 & s_{3,1,1} & 0 & 0 & 0 & 0 & 0 \\
0 & 0 & 0 & 0 & s_{3,1,2} & s_{3,2,2} & 0 & 0 & 0 & 0 \\
\hline
0 & 0 & 0 & 0 & 0 & 0 & s_{4,1,1} & 0 & 0 & 0 \\
0 & 0 & 0 & 0 & 0 & 0 & s_{4,1,2} & s_{4,2,2} & 0 & 0
\end{array}\right). \quad (13)$$

2.5 *Introducing Effects of a Toxic Compound*

Toxic compounds are widely recognized to act on different life history stages of organisms, suggesting that population responses to a toxic stress might differ with age-structure of the population [3]. In this context, matrix population models are particularly well adapted to take into account toxic effects on the complete set of vital rates [20] to evaluate more precisely how contaminants influence the population growth rate [5].

To go further than with life table response experiment (LTRE) analyses, which measure the population-level effects of an environmental stress under two or more exposure conditions [16], Leslie-type models permit to express vital rates and consequently the population growth rate λ as a continuous function of the exposure concentration c.

On a practical point of view, the model writes:

$$N(t+1) = \mathbf{A}(c)\, N(t)\,, \tag{14}$$

with $\mathbf{A}(c) = \left[a_{ij}(c)\right]_{1 \le i,j \le \omega}$ the exposure concentration-dependent transition matrix.

The difficulty of this approach lies in the choice of effect models relating matrix entries $a_{ij}(c)$ to the exposure concentration c. This choice strongly depends on available data:

- From data collected in the literature, often obtained from bioassays in various laboratory conditions, it is possible to build concentration–response curves where the X-axis plots log[concentration] and the Y-axis plots response are expressed as a reduction coefficient, which applies on vital rates. Concentration–response curves follow a standard decreasing sigmoid shape, between 0 and 1, and are defined by two parameters: the curvature and the concentration that induces a response halfway between baseline and maximum (EC_{50}). A collection of classical regression models can be used to analyze such kind of data: polynomial, log-logistic, Probit or Weibull models [32].
- When data are collected at different exposure concentrations and different exposure times, typically the number of survivors, the body length of individuals or the number of offspring, process-based effect models can be constructed from the dynamic energy budget (DEB) theory and its applications in ecotoxicology – DEBtox – [33, 34]. The idea behind the DEB theory is that individuals manage their energy; this energy is taken up in the form of food and is then assimilated to be used for reproduction, growth, or maintenance. Under DEBtox mechanistic hypotheses, the energy management is assumed to be disturbed by a toxic compound when the exposure concentration exceeds a threshold concentration called the no effect concentration (NEC). Let us take one example to illustrate this point.

The internal concentration of the toxic compound in the organism at time t, $c_q(t)$, is assumed to be ruled by a one-compartment kinetics model:

$$\frac{dc_q(t)}{dt} = k_e\left(c - c_q(t)\right), \tag{15}$$

where k_e is the elimination rate (in time^{-1}) and c is the exposure concentration.

To model the effects of the toxic compound on vital rates, a stress function $\sigma\left(c_q\right)$ is used, which has a different dimension with regards to lethal or sublethal effects. Indeed, in the lethal effect model, $\sigma\left(c_q\right)$ is homogeneous to a rate (in time^{-1}), whereas in the sublethal effect model, $\sigma\left(c_q\right)$ is dimensionless.

When effects are lethal, the stress function $\sigma_{\mathrm{L}}\left(c_{\mathrm{q}}\right)$ is expressed as follows:

$$\sigma_{\mathrm{L}}(c_{\mathrm{q}}) = {c_{\mathrm{L}}}^{-1}(c_{\mathrm{q}} - \mathrm{NEC_L})_{+}, \tag{16}$$

with c_{L} the lethal tolerance concentration (in concentration $\times$ time unit) and $\mathrm{NEC_L}$ the no effect concentration under the lethal effect hypothesis.

When effects are sublethal, the stress function $\sigma_{\mathrm{S}}\left(c_{\mathrm{q}}\right)$ is expressed as follows:

$$\sigma_{\mathrm{S}}(c_{\mathrm{q}}) = {c_{\mathrm{S}}}^{-1}(c_{\mathrm{q}} - \mathrm{NEC_S})_{+}, \tag{17}$$

where c_{S} is the sublethal tolerance concentration (in concentration unit) and $\mathrm{NEC_S}$ the no effect concentration under the sublethal effect hypothesis.

$(c_{\mathrm{q}} - \mathrm{NEC}_{*})_{+}$ means $\max\left(0, c_{\mathrm{q}} - \mathrm{NEC}_{*}\right)$ with $* = \mathrm{L, S}$.

Hence, the toxic compound effects on survival can mathematically be expressed by the probability for an individual to survive until exposure time t at exposure concentration c:

$$q\,(t, c) = \exp\left(-\int_0^t h\,(\tau, c)\,\mathrm{d}\tau\right), \tag{18}$$

where $h\,(\tau, c)$ is the hazard rate at time τ and exposure concentration c:

$$h\,(\tau, c) = \begin{cases} m + \sigma\left(c_{\mathrm{q}}\,(\tau)\right) & \text{if } c > \mathrm{NEC_L} \text{ and } \tau > t_0 \\ m & \text{else} \end{cases}, \tag{19}$$

where m is the natural mortality rate and t_0 the time at which $c_{\mathrm{q}}\,(\tau)$ exceeds the $\mathrm{NEC_L}$:

$$t_0 = -k_{\mathrm{e}}^{-1} \ln\left(1 - \frac{\mathrm{NEC_L}}{c}\right). \tag{20}$$

Survival rates can be calculated as follows [14]:

$$s_i\,(c) = \frac{q\,(i+1, c)}{q\,(i, c)} \text{ in the case of a prebreeding census,} \tag{21}$$

$$s_i\,(c) = \frac{q\,(i, c)}{q\,(i-1, c)} \text{ in the case of a postbreeding census.} \tag{22}$$

As for the lethal effects of a toxic compound, sublethal effects can also be mathematically written and fecundity rates deduced; for more details see [21, 35–37] and Sect. 4 of this chapter.

From a more general point of view, the population dynamics may depend on various environmental factors which affect the vital rates, as for example food [18, 38] or temperature and water discharge [39]. See details in Sect. 3 of this chapter.

2.6 Perturbation Analyses

Originating from Caswell [40], perturbation analyses ask how population endpoints, and particularly the asymptotic population growth rate λ, change according to the vital rates. As stated by Caswell [41], a distinction has to be made between prospective (sensitivities and elasticities) and retrospective (LTRE) perturbation analyses. While the latter expresses observed variation in λ as a function of observed (co)variation in the vital rates, prospective perturbation analyses project on λ the consequences of future changes in one or more of the vital rates. Only the prospective perturbation analyses will be here considered.

The sensitivity of the asymptotic population growth rate λ to changes in the vital rates can easily be calculated for matrix population models [14, 41] from the transition equation (3) and population endpoints (λ, the stable age distribution w, and the reproductive values v). If only one entry, a_{ij}, changes, the sensitivity writes:

$$\frac{\partial \lambda}{\partial a_{ij}} = \frac{v_i w_j}{v^{\mathrm{T}} w}, \tag{23}$$

where v_i and w_j are the ith and the jth coordinate of v and w, respectively.

Sensitivities compare the absolute effects on λ of the same absolute changes in vital rates. A sensitivity matrix can be written as follows:

$$\mathbf{S} = \left(\frac{\partial \lambda}{\partial a_{ij}}\right)_{1 \le i,j \le \omega} = \frac{v w^{\mathrm{T}}}{v^{\mathrm{T}} w}. \tag{24}$$

In contrast to sensitivities, elasticities quantify the proportional change in λ resulting from an infinitesimal proportional change in matrix entries a_{ij} [42]:

$$\mathrm{e}_{ij} = \frac{\partial\left(\log \lambda\right)}{\partial\left(\log a_{ij}\right)} = \frac{a_{ij}}{\lambda}\frac{\partial \lambda}{\partial a_{ij}}. \tag{25}$$

Elasticities thus compare the relative effects on λ with the same relative changes in the values of the demographic parameters. An elasticity matrix can also be defined as:

$$\mathbf{E} = \frac{1}{\lambda}\mathbf{S} \circ \mathbf{A}, \tag{26}$$

where $\circ$ stands for the Hadamard product. Elasticities have mathematical properties such as:

$$\mathbf{E} \ge 0,\ a_{ij} = 0 \Rightarrow \mathrm{e}_{ij} = 0 \quad \text{and} \quad \sum_{i,j} \mathrm{e}_{ij} = 1. \tag{27}$$

If some factor x affects any of the entries a_{ij} then the total derivative of λ can be approximated by:

$$\frac{\mathrm{d}\lambda}{\mathrm{d}x} \approx \sum_{i,j} \frac{\partial \lambda}{\partial a_{ij}} \frac{\mathrm{d}a_{ij}}{\mathrm{d}x}. \tag{28}$$

In ecotoxicology, this quantity (28) is of broad interest as factor x can refer either to the toxic compound concentration or to the logarithm of the concentration.

Hence, the total sensitivity of λ to the toxic compound concentration can be decomposed into several contributions: for example, the one of fecundity rates and the one of survival rates, or with a spatial Leslie model, an interesting decomposition could be made per age class at the scale of the whole patch network. Applications of this decomposition will be illustrated in Sects. 3–5. For those who are interested, some peculiar tools have also been developed to perform sensitivity and elasticity analyses with density-dependent Leslie-type models [43, 44].

Here ends the purely theoretical part of the chapter. Following sections are dedicated to case studies in which models and their prospective results will be detailed in an ecotoxicological context.

3 A First Case Study with the Midge *Chironomus riparius*

As introduced in Sect. 2.5, matrix population models are useful tools to test population response to a stress. In particular, such models allow us to take into account different sensitivities to the stress between development stages, and individual level data can be included to calculate characteristic endpoints at the population level [14]. Matrix population models are here illustrated in their use to determine the combined effects of food limitation and pesticide exposure on a Chironomidae population. *Chironomus riparius* was chosen as a commonly used species in toxicity laboratory tests. It is widespread in river sediments and considered to be a good bioindicator of water quality. First, parameter inputs, when estimated from individual data using DEBtox models, are included into a matrix population model. Then, using experimental data obtained with *C. riparius*, lethal effects of a pesticide (methiocarb) on the asymptotic population growth rate of a laboratory population is estimated under two different food availability conditions. Lastly, a sensitivity analysis is performed to pinpoint critical age classes within the population for the purposes of the field management of populations. For more details, see [18] and [38].

3.1 The Biological Situation

C. riparius (Diptera: Chironomidae) is a nonbiting midge widely distributed in the northern hemisphere. Its life cycle (Fig. 4) comprises aquatic stages (eggs, larva, pupae) and aerial ones (adults). Individuals are synchronous and, in the field, there is usually only one life cycle per year with a winter diapause period during the fourth larval stage [45]. Under laboratory conditions, Charles et al. [38] have shown that food availability does not affect the duration of the egg stage (2 days), the first larval stages L_1 (2 days), the second larval stage L_2 (2 days), the third larval stage

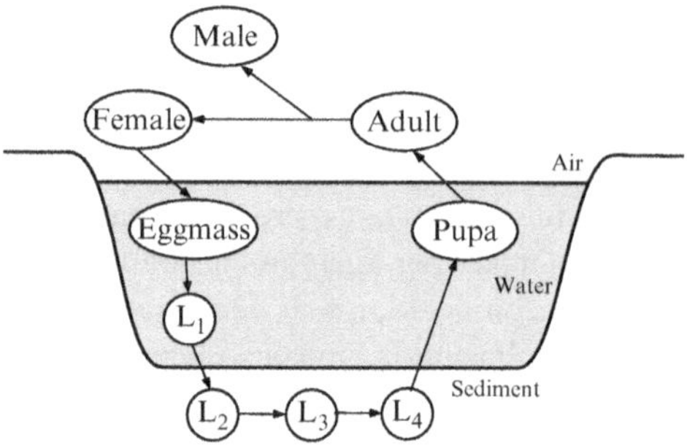

Fig. 4 The life cycle graph of *Chironomus riparius*

L_3 (3 days), the pupa stage (less than 1 day), and the adult stage (4 days). On the contrary, there was a strong effect of food amount on the duration of the fourth larval stage and on the adult fecundity. If food is not a limiting factor (*ad libitum* condition), the life cycle of *C. riparius* lasts about 17 days with stages occurring in a rapid succession.

The methiocarb was chosen for three main reasons: (1) it is more toxic than other similar chemicals [46]; (2) it has been found in field sediments at concentrations between 10 and 268 μg kg^{-1} (data from the Water Agency of Rhône-Méditerranée-Corse, France); and (3) only few studies are available concerning the effects of this pesticide on benthic organisms, except some works reporting an effect on individual survival rates only [47, 48].

To quantify effects of methiocarb on individual survival, data from the Cemagref (Lyon, France) were used. For L_2, L_3, and L_4 stages, the survival test data from [47] were used, in which organisms were exposed to six exposure concentrations (0, 25, 50, 280, 310, and 360 μg L^{-1}) under two food availability conditions: an *ad libitum* food condition (food amount equal to 1.4 mg per day per individual) and a food-limited condition (food amount equal to 0.2 mg per day per individual). Survival tests with eggs, L_1 and pupae stages were performed under seven exposure concentrations (0, 10, 20, 30, 40, 60, 80 μg L^{-1}). Food effect was not tested as it did not affect those stages (Péry, pers. comm.). For L_2, L_3, and L_4 stages, data were collected at several times, whereas for the other stages, data were obtained once at the end of the tests.

3.2 *The Model in Few Equations*

3.2.1 The Matrix Population Model

A *classical* Leslie model was used with a prebreeding census. Given the life cycle of *C. riparius* (Fig. 4), the pupa stage was combined with the fourth larval stage, and a daily time step was chosen for the model. Hence, the dimension of the transition matrix **A** was equal to the total duration of the life cycle. The exposure concentration had no effect on stage duration, while the food amount Q (in mg per day per individual) strongly affected the duration of the fourth larval stage D_4 (in days) and the adult fecundity f_k (in number of viable eggs per female) according to the following equations [38]:

$$D_4\left(Q\right) = 7.436 + \left(31.771 - 7.436\right)\mathrm{e}^{-6.596Q}, \tag{29}$$

$$f_k\left(Q\right) = 208.299\left(1 - \mathrm{e}^{-4.951Q}\right)R_{\mathrm{p}_k}, \tag{30}$$

where R_{P_k} is the reproduction probability at day k of the adult life.

As a consequence, the dimension of the transition matrix $\mathbf{A}$ depended on the food amount Q. Two matrix population models were thus developed: one for the *ad libitum* food condition and one for the food-limited condition. As survival rates depended on the methiocarb concentration (c), the transition matrix was finally a function of both Q and c.

The two population models can be written under the following general form:

$$N\left(t+1\right) = \mathbf{A}\left(Q, c\right)N\left(t\right), \tag{31}$$

with $\mathbf{A}\left(Q, c\right)$ the transition matrix for a food amount Q and an exposure concentration c:

$$\mathbf{A}_{(Q,c)} =$$

$$\begin{pmatrix}
0 & 0 & 0 & 0 & 0 & 0 & 0 & 0 & 0 & 0 & 0 & 0 & f_1(Q) & f_2(Q) & f_3(Q) & f_4(Q) \\
s_1(c) & 0 & 0 & 0 & 0 & 0 & 0 & 0 & 0 & 0 & 0 & 0 & 0 & 0 & 0 & 0 \\
0 & s_2(c) & 0 & 0 & 0 & 0 & 0 & 0 & 0 & 0 & 0 & 0 & 0 & 0 & 0 & 0 \\
0 & 0 & s_3(c) & 0 & 0 & 0 & 0 & 0 & 0 & 0 & 0 & 0 & 0 & 0 & 0 & 0 \\
0 & 0 & 0 & s_4(c) & 0 & 0 & 0 & 0 & 0 & 0 & 0 & 0 & 0 & 0 & 0 & 0 \\
0 & 0 & 0 & 0 & s_5(c) & 0 & 0 & 0 & 0 & 0 & 0 & 0 & 0 & 0 & 0 & 0 \\
0 & 0 & 0 & 0 & 0 & s_6(c) & 0 & 0 & 0 & 0 & 0 & 0 & 0 & 0 & 0 & 0 \\
0 & 0 & 0 & 0 & 0 & 0 & s_7(c) & 0 & 0 & 0 & 0 & 0 & 0 & 0 & 0 & 0 \\
0 & 0 & 0 & 0 & 0 & 0 & 0 & s_8(c) & 0 & 0 & 0 & 0 & 0 & 0 & 0 & 0 \\
0 & 0 & 0 & 0 & 0 & 0 & 0 & 0 & s_9(c) & 0 & 0 & 0 & 0 & 0 & 0 & 0 \\
0 & 0 & 0 & 0 & 0 & 0 & 0 & 0 & 0 & s_{10}(c) & 0 & 0 & 0 & 0 & 0 & 0 \\
0 & 0 & 0 & 0 & 0 & 0 & 0 & 0 & 0 & 0 & \ldots & 0 & 0 & 0 & 0 & 0 \\
0 & 0 & 0 & 0 & 0 & 0 & 0 & 0 & 0 & 0 & 0 & s_z(c) & 0 & 0 & 0 & 0 \\
0 & 0 & 0 & 0 & 0 & 0 & 0 & 0 & 0 & 0 & 0 & 0 & s_{z+1} & 0 & 0 & 0 \\
0 & 0 & 0 & 0 & 0 & 0 & 0 & 0 & 0 & 0 & 0 & 0 & 0 & s_{z+2} & 0 & 0 \\
0 & 0 & 0 & 0 & 0 & 0 & 0 & 0 & 0 & 0 & 0 & 0 & 0 & 0 & s_{z+3} & 0
\end{pmatrix}$$

Eggs (columns 1–2), L_1 (columns 3–4), L_2 (columns 5–6), L_3 (columns 7–9), L_4 (columns 10–12), Adults (columns 13–16)

where

$s_i\left(c\right)$ is the larval survival rate from day i to day $i+1$ at a given exposure concentration c. The z subscript in $s_z\left(c\right)$ refers to the last day of the fourth larval stage. Remind that the pupa stage lasts less than one day and that it has been combined with the fourth larval stage. Thus, $s_z\left(c\right) = s_{\mathrm{p}}\left(c\right) \times \widehat{s}_z\left(c\right)$, with $s_{\mathrm{p}}\left(c\right)$ the pupa survival rate and $\widehat{s}_z\left(c\right)$ the L_4 survival rate.

s_{z+k} is the adult survival rate from day k to day $k+1$ of adult life ($k = 1, \ldots, 3$). s_{z+k} was estimated to 1 whatever the exposure concentration. Adults were supposed to die just after reproduction what explains the null value of the last diagonal term in the transition matrix.

$f_k(Q)$ is the adult fecundity at day k of adult life ($k = 1, \ldots, 4$) as estimated by (30).

Under *ad libitum* food conditions, the duration of the fourth larval stage was about 7 days ((29), with $Q = 1.4$). The first adult fecundity f_1 was estimated at 208.1 viable eggs per female (30), while $f_k = 0$ for $k \geq 2$. Indeed, all females only reproduce the first day of their adult life [38]. Postreproductive age classes were thus removed from the model and the transition matrix $\mathbf{A}_{\text{al}}$ was an imprimitive matrix (see Sect. 2.1) of dimension 17.

Under the tested food-limited condition, the duration of the fourth larval stage was about 14 days ((29), with $Q = 0.2$) and adult fecundities were estimated in number of viable eggs per female at: 64.149 for f_1, 11.521 for f_2, 3.927 for f_3, and 1.309 for f_4 (30). The corresponding transition matrix $\mathbf{A}_{\text{lim}}$ was a primitive matrix of dimension 27.

To completely define the transition matrix, $s_i(c)$ and $\tilde{s}_z(c)$ need to be clarified. This was done by fitting individual effect models on laboratory experimental data.

3.2.2 Individual Effect Models

For egg, L_1 and pupae stages, a decreasing logistic concentration–response relationship was used to link survival rates $s_i(c)$ and the exposure concentration c:

$$s_i(c) = s_{0i} \times \frac{1 + \exp(\alpha)}{\exp(\alpha) + \exp(\beta c)}, \tag{32}$$

where c is the exposure concentration (in $\mu\text{g L}^{-1}$); s_{0i} is the natural survival rate without contaminant (per day); the second term is the survival reduction function for a given exposure concentration c. The parameter β is a curvature parameter, while the LC_{50}, i.e., the lethal concentration for 50% of the individuals, is related to the parameter α by $\alpha = \ln\left(e^{\beta \, LC_{50}} - 2\right)$.

Values of s_{0i} have been estimated by Charles et al. [38] at 0.836 for eggs and L_1, and at 1 for pupae. Parameters α and β have been estimated by a nonlinear fit of (32) on experimental data using the least squares criterion. Estimated values and their standard errors are given in Table 2. These parameter estimates remain valid whatever the food conditions.

Table 2 Parameter estimates of model (30) for egg and L_1 stages (bold) and for pupa (plain)

Parameters	Estimate	(sd)[a]
α	**8.478**	**(1.603)**
	8.749	(0.082)
β	**0.282**	**(0.053)**
	0.499	(0.004)

[a] The total number of experimental points is $n = 35$

Table 3 DEBtox parameter estimates (15) to (18) for *C. riparius* subjected to methiocarb, for L_2, L_3, L_4 stages and both food availability conditions. *Ad libitum* (in bold) and limited (in plain) food conditions; in brackets, the corresponding standard error ($n = 72$)

Stage	m (per day)	NEC_S (μg L^{-1})	c_L^{-1} (μg^{-1} L per day)	k_e (per day)
L_2	**0.051 (0.070)**	**$3.73\ 10^{-7}$ (5.667)**	**0.023 (0.006)**	**3.474 (2.167)**
	0.070 (0.021)	9.085 (4.152)	0.029 (0.006)	3.530 (1.759)
L_3	**0.038 (0.009)**	**236 (12.2)**	**0.014 (0.003)**	**3.171 (0.451)**
	0.021 (0.006)	176 (39.4)	0.010 (0.003)	3.340 (0.907)
L_4	**0.033 (0.008)**	**255 (6.155)**	**0.022 (0.004)**	**4.234 (0.582)**
	0.029 (0.007)	256 (6.530)	0.017 (0.003)	10.17 (4.828)

For L_2, L_3, and L_4 stages, survival rates were functions of both exposure concentration and exposure time. Data were thus analyzed by using DEBtox models as described in Sect. 2.5. The DEBtox software [33] was used to achieve the four estimates of m, c_L^{-1}, k_e, and NEC_L as involved in (15)–(20). Results are provided in Table 3.

3.3 *Main Results*

3.3.1 Methiocarb Effects on *C. riparius*

Methiocarb effects on *C. riparius* population dynamics was quantified, under each food availability condition, through the asymptotic population growth rate λ (Table 1).

Without contaminant, the asymptotic population growth rate was equal to $\lambda_\text{al} = 1.28$, under the *ad libitum* food condition, and $\lambda_\text{lim} = 1.14$, under the food-limited condition, what corresponds to a hypothetical population size daily increase of 28% and 14%, respectively, in accordance with Charles et al. [38]. Such values of λ were in fact consistent with the opportunistic characteristics of *C. riparius*, which is able to colonize organically-enriched aquatic habitats very quickly [49].

For methiocarb concentrations varying from 0 to 120 μg L^{-1}, the decrease in λ was simulated from (31). As shown in Fig. 5, the methiocarb has a major impact on the asymptotic population growth rate λ, which rapidly reaches the critical value 1 when the methiocarb concentration exceeds a threshold value of around 22 μg L^{-1} under the *ad libitum* food condition and of around 21 μg L^{-1} under the food-limited condition. Once exceeded these thresholds, the population became potentially extinct ($\lambda < 1$), under each food availability condition. Under the food-limited condition, the transition matrix $\mathbf{A}_\text{lim}$ was primitive, meaning that the impact of contaminant on stable stage distribution and reproductive values made sense. As shown in Fig. 6a, the exposure concentration allows the population to grow ($\lambda > 1$), while the methiocarb concentration does not affect the stable age distribution, except at 20 μg L^{-1}, where the L_4 proportion roughly increases to the detriment of egg and

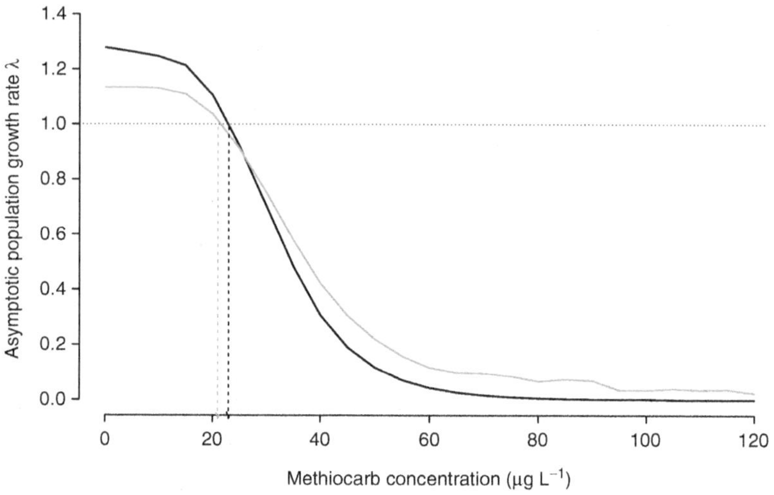

Fig. 5 Effects of the methiocarb concentration on the asymptotic population growth rate λ of *Chironomus riparius*, under *ad libitum* (in *black*) or limited (in *gray*) food conditions

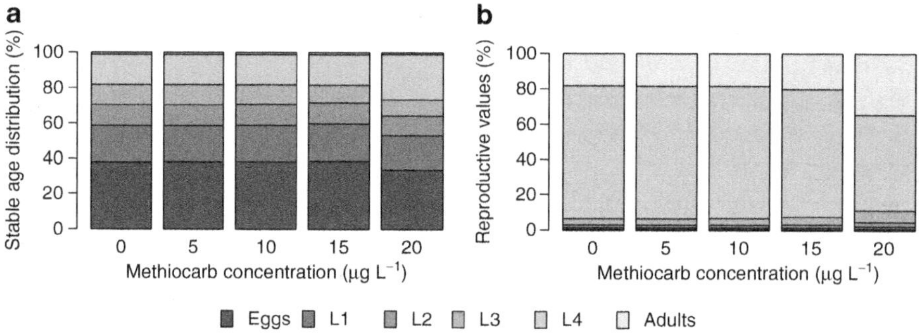

Fig. 6 (**a**) Stable stage distribution and (**b**) reproductive values of a laboratory *Chironomus riparius* population subjected to methiocarb under the food limited condition

L_3 stages. Furthermore, Fig. 6b clearly shows that the L_4 stage mainly contributes to the asymptotic population dynamics with the highest reproductive values. From $20\,\mu g\,L^{-1}$ of methiocarb, reproductive values change with the methiocarb concentration: the L_4 reproductive value become smaller in favor of the others.

3.3.2 Perturbation Analysis

The influence of each transition matrix input on λ was decomposed according to the perturbation analysis detailed in Sect. 2.6 – (26). The methiocarb was shown to only affect survival rates [47]. Consequently, only subdiagonal terms of the transition matrix $\mathbf{A}\,(Q, c)$ contributed to the sensitivity of λ to the exposure concentration.

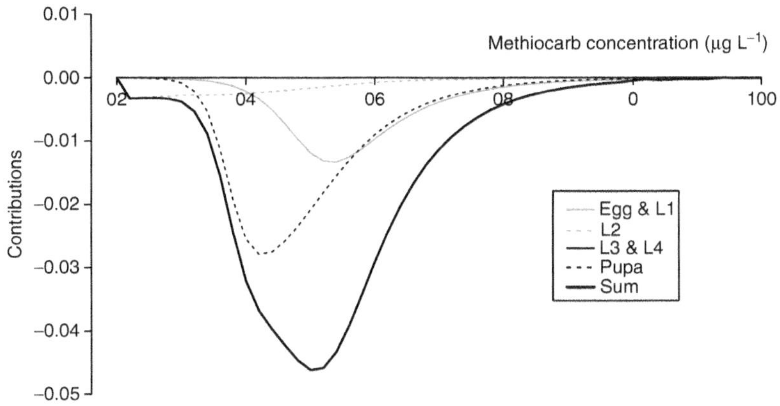

Fig. 7 Perturbation analysis results obtained by decomposition of the population response of a laboratory *Chironomus riparius* population subjected to methiocarb under the *ad libitum* food condition

As results under both food availability conditions were similar, only those corresponding to the *ad libitum* food conditions are presented in Fig. 7. Egg and L_1 stages have a moderate impact on λ in the middle-range concentrations, whereas L_2 stage only weakly contributes at low concentrations. Contributions for L_3, L_4, or adult stages can only be detected at very high concentrations (NEC $\geq$ 176 $\mu g\,L^{-1}$). On the contrary, a major contribution of the reduction in pupa stage survival can be observed at mid-range concentrations, which accounts for the emergence of strong individual effects during the brief period of pupae. From this sensitivity analysis, it appears that only egg, L_1 and pupa stages influenced the asymptotic population growth rate as a result of the methiocarb impact.

3.4 Summary

The case study presented in this section shows how a matrix population model can describe the dynamics of a theoretical Chironomidae population under laboratory conditions exposed to methiocarb and variable food availability conditions. In particular, this work shows how nested modeling approaches, classically used in both ecotoxicology and ecology, can help in understanding responses at the population level, by extrapolating effects from the individual level. Indeed, the population response to a contaminant was estimated in terms of the change in asymptotic population growth rate λ. Under two food availability conditions, methiocarb had a rapid effect on Chironomidae population dynamics: beyond of a threshold exposure concentration around 21–22 $\mu g\,L^{-1}$, the population became potentially extinct ($\lambda < 1$). The perturbation analysis permitted to decompose the population response as described by Caswell [14, 15] and to highlight critical age classes for the population dynamics, namely egg, L_I and pupa stages, survival rates of which strongly affect λ in the case of *C. riparius*. The pesticide here studied only affected survival rates, whereas other contaminants have been proved to also impact growth and/or fecundity. The next section will show how lethal and sublethal effects can simultaneously be taken into account in matrix population models.

4 A Second Case Study with the Cladoceran *D. magna*

This section will focus on the result analysis of a standardized chronic bioassay, the 21-day daphnid reproduction test [50,51]. From this common aquatic bioassay, both lethal and sublethal effects are measured at different times. At the individual level, the DEBtox approach [33] provides a set of mechanistic models of survival, reproduction, and growth as continuous functions of time and exposure concentration. By combining DEBtox theory and matrix population models, effects of the toxic compound at the individual level (reduced fecundity, growth, and survival) were extrapolated to the population one by taking into account age dependence of the contamination. The asymptotic population growth rate was obtained continuously vs. exposure concentration and accompanied by a confidence band. Perturbation analyses were also performed to highlight critical demographic parameters in the evolution of the asymptotic population growth rate as a function of contaminant concentration. The experimental data used in this case study corresponded to zinc (Zn) exposition of a *D. magna* population. For more details, see [21,35–37].

4.1 *The Biological Situation*

D. magna is a small cladoceran crustacean, which plays an important role in water purification. Daphnids eat by filtering water and retaining food particles. They spend their whole life in the same environment, where they develop through a succession of instars. As long as environmental conditions remain favorable, they reproduce predominantly by parthenogenesis. If only the parthenogenic mode of reproduction is considered, the life cycle is very simple. Daphnids can be divided into two age groups: juveniles, which are not yet able to reproduce, and adults.

D. magna is one of the most widely used animals in aquatic toxicity. In terms of sensitivity to toxic compounds, they are usually thought to be representative of other zooplankters [52], although this has been contested [53]. These animals have many properties that make them suitable for laboratory testing, such as their small size, high fecundity, short life span, parthenogenic reproduction, and ease of handling [54]. The 21-day daphnid reproduction test (ISO 10706) [51] is one of the two ISO normalized chronic tests on freshwater animals. In this routine test, organisms are exposed from their birth to several constant concentrations of toxic compounds, and several endpoints (e.g., survival, reproduction, and growth) are measured. For the sake of ecological relevance, all the effects observed on individuals were gathered to assess the toxic effects on the population dynamics.

Data collected in the Cemagref (Lyon, France) were used. Young daphnids (<24 h old) were exposed to constant concentrations of Zn (0, 0.074, 0.22, 0.66 mg L^{-1}) for 21 days under *ad libitum* food conditions. Mortality and number of offspring were daily recorded. Individual growth was measured at day 7, 14, and 21. See Billoir et al. [37] for details.

4.2 The Model in Few Equations

4.2.1 Effect Models at the Individual Levels

To deal with lethal effects of the contaminant, models described in Sect. 2 (15)–(20) were used, with $q(t,c)$ the probability for an individual to survive until exposure time t at exposure concentration c. As far as sublethal effects were concerned, five ways that the contaminant acts on *D. magna* energy management were proposed in the DEBtox framework [33, 34]: an increase in maintenance costs, an increase in growth costs, a decrease in assimilation, an increase in egg production costs, or a sur-mortality during oogenesis. These five assumptions led to different equations for the modeling of growth and reproduction, the two sublethal endpoints measured. Here only the maintenance cost increase assumption was considered, which was one of the two best ones (with the assimilation decrease) in case of a Zn contamination [37]. Hence, the growth and reproduction processes were modeled by the following equations, respectively [33, 36]:

$$\frac{\mathrm{d}l(t,c)}{\mathrm{d}t} = \gamma\left(1 - l(t,c)\left(1 + \sigma_S\left(c_q(t,c)\right)\right)\right), \tag{33}$$

$$R(t,c) = \frac{R_M}{1 - l_p{}^3}\left(1 + \sigma_S\left(c_q(t,c)\right)\right) \times \left(l^2(t,c)\frac{\left(1 + \sigma_S\left(c_q(t,c)\right)\right)^{-1} + l(t,c)}{2} - l_p{}^3\right) \quad \text{when } l(t,c) > l_p, \tag{34}$$

where the variables are the scaled length $l(t,c)$ (in mm), the internal concentration $c_q(t,c)$ (in mg L^{-1}) and the reproduction rate $R(t,c)$ (i.e., the number of offspring per mother per time unit). The covariables are the time t (in day) and the exposure concentration c (in mg L^{-1}). The four models ruling q (18), c_q (15), l (33), and R (34) as functions of t and c have to be considered together, respectively, because they share parameters and some of them are interdependent. Parameters involved in equations are: the maximum body length L_m (in mm), the scaled body length at puberty l_p (dimensionless), the von Bertalanffy growth rate γ (per day), the maximum reproduction rate R_M (in number of offspring per day per mother), the sublethal no effect concentration NEC_S (in mg L^{-1}), the sublethal tolerance concentration

Table 4 Descriptive statistics of empirical posterior distributions of DEBtox parameters

Parameter	Mean	(sd)	2.5th percentile	Median	97.5th percentile
L_m (mm)	4.08	(0.0246)	4.031	4.08	4.129
l_p (dimensionless)	0.615	(0.006297)	0.6029	0.6149	0.6278
g (per day)	0.1211	(0.002438)	0.1164	0.127	0.1261
R_M (# per day)[a]	8.107	(0.1647)	7.799	8.099	8.452
NEC_S(mg L^{-1})	0.1482	(0.006017)	0.1362	0.1482	0.1599
c_S(mg L^{-1})	1.128	(0.03018)	1.071	1.127	1.19
k_e (per day)	1.082	(0.4598)	0.7178	0.96	2.277
m (per day)	0.009047	(0.001548)	0.006017	0.009057	0.01207
NEC_L(mg L^{-1})	0.3398	(0.06432)	0.228	0.3349	0.4786
c_L (mg L^{-1} day)	0.05081	(0.01425)	0.03288	0.04748	0.08927

[a]The symbol # stands here for number of offspring

c_S (in mg L^{-1}), the elimination rate k_e (per day), the natural mortality rate m (per day), the lethal no effect concentration NEC_L (in mg L^{-1}) and the lethal tolerance concentration c_L (in mg L^{-1} day).

To estimate DEBtox parameters, Bayesian inference was used as already proposed by Billoir et al. [37]. From arbitrary prior probability distribution for each parameter, Bayesian inference provides estimates as posterior distributions given the data. In the case of complex models like ours, this estimation process generates samples of the joint posterior distribution of all the parameters [55]. Survival, growth, and reproduction data were used simultaneously to estimate DEBtox parameters. Descriptive statistics of posterior distributions are provided in Table 4. After checking the convergence of the estimation process, posterior distribution means were considered as estimates.

4.2.2 Matrix Population Model

Age-specific models are the most adapted to data sets from normalized bioassays. Indeed, such data provide much more information about reproduction and survival as a function of age than about body length, which is measured only two or three times all along the experiment. DEBtox effect models provide a survival function (18), and the reproduction rate (33) can be written as a function of time and exposure concentration. From these equations, it is possible to calculate survival rates from the age class i to the next (s_i) and the fecundity rates f_i in the case of a prebreeding census [14]:

$$s_i(c) = \frac{q\,(i+1, c)}{q\,(i, c)}, \tag{21}$$

$$f_i(c) = \int_i^{i+1} q\,(1, c)\,R\,(t, c)\,\mathrm{d}t, \tag{35}$$

where $q(t, c)$ is the probability of surviving until time t at a concentration c (18), and $R(t, c)$ is the reproduction rate (34).

A matrix model with ten age classes and a time step of two days (0–2 days, 2–4 days,..., 18–20 days) was chosen because chronic bioassays for daphnids last 21 days, and because daphnids generally reproduce every 2 days under standard conditions. The period between two reproductive events may vary in experiments [56]. This was taken into account through the matrix in the next age class thanks to the continuity of the reproduction effect models against time. Daphnids can live for more than 21 days and keep on reproducing, so a *standard* Leslie model was used, by adding a term s_{10} to the diagonal of the matrix. A value of $s_{10} = 0.95 \times s_9$ was used as estimated in a previous study [21].

Let $N\,(t)$ be the population vector at time t and exposure concentration c:

$$N\,(t) = \left(n_1\,(t)\,,\ldots,n_i\,(t)\,,\ldots,n_{10}\,(t)\right)^{\mathrm{T}}, \tag{36}$$

where superscript T denotes transposition.

Hence, the transition equation writes:

$$N\,(t+1) = \mathbf{A}\,(c)\,N\,(t)\,, \tag{37}$$

with

$$\mathbf{A}(c) = \begin{pmatrix} 0 & 0 & f_3(c) & f_4(c) & f_5(c) & f_6(c) & f_7(c) & f_8(c) & f(c)_9 & f_{10}(c) \\ s_1(c) & 0 & 0 & 0 & 0 & 0 & 0 & 0 & 0 & 0 \\ 0 & s_2(c) & 0 & 0 & 0 & 0 & 0 & 0 & 0 & 0 \\ 0 & 0 & s_3(c) & 0 & 0 & 0 & 0 & 0 & 0 & 0 \\ 0 & 0 & 0 & s_4(c) & 0 & 0 & 0 & 0 & 0 & 0 \\ 0 & 0 & 0 & 0 & s_5(c) & 0 & 0 & 0 & 0 & 0 \\ 0 & 0 & 0 & 0 & 0 & s_6(c) & 0 & 0 & 0 & 0 \\ 0 & 0 & 0 & 0 & 0 & 0 & s_7(c) & 0 & 0 & 0 \\ 0 & 0 & 0 & 0 & 0 & 0 & 0 & s_8(c) & 0 & 0 \\ 0 & 0 & 0 & 0 & 0 & 0 & 0 & 0 & s_9(c) & s_{10}(c) \end{pmatrix} \tag{38}$$

Here reproduction was assumed to occur from age class 3, but the age at first reproduction may increase with the toxic exposure concentration. All vital rates depended on the exposure concentration, and λ has been calculated as a continuous function of the exposure concentration.

To evaluate a confidence interval for the asymptotic population growth rate λ, a new method based on bootstrapping was proposed. For each exposure concentration c, 10,000 DEBtox parameter sets were drawn in their joint posterior distribution. Then, from each parameter set, the vital rates were deduced from (21) and (35), as well as the corresponding asymptotic population growth rate λ from the dominant eigenvalue of $\mathbf{A}\,(c)$. Thus 10,000 values of λ were obtained and the 2.5th and 97.5th quantiles of this sample were considered as the limits of a 95% confidence interval for λ at exposure concentration c.

4.3 Main Results

4.3.1 The Effect of Pollution on Population Dynamics

The asymptotic population growth rate λ was used to quantify the effects of Zn on daphnid population dynamics. The decrease in λ with the exposure concentration of Zn is shown in Fig. 8. It was calculated as a percentage of its maximum λ value, because only a potential relative effect from a reference situation without contaminant is considered.

In this case, the no effect concentration associated to sublethal effects (NEC_S) was consistently lower than the one associated to lethal effects (NEC_L) (Table 4). Consequently, the population dynamics was expected to be impacted first by the Zn effects on fecundity and second by the Zn effects on survival. Hence, below the NEC_S ($0.1482\,\mathrm{mg\,L^{-1}}$), the toxic compound did not affect population dynamics. Above the NEC_S, λ roughly fell as the toxic compound affected reproduction. On the contrary, the additional impact on survival which appeared from NEC_L ($0.3398\,\mathrm{mg\,L^{-1}}$) remained fairly soft due to a very weak killing rate c_L ($0.05081\,\mathrm{mg\,L^{-1}}$ day). By extrapolating beyond the tested concentration range

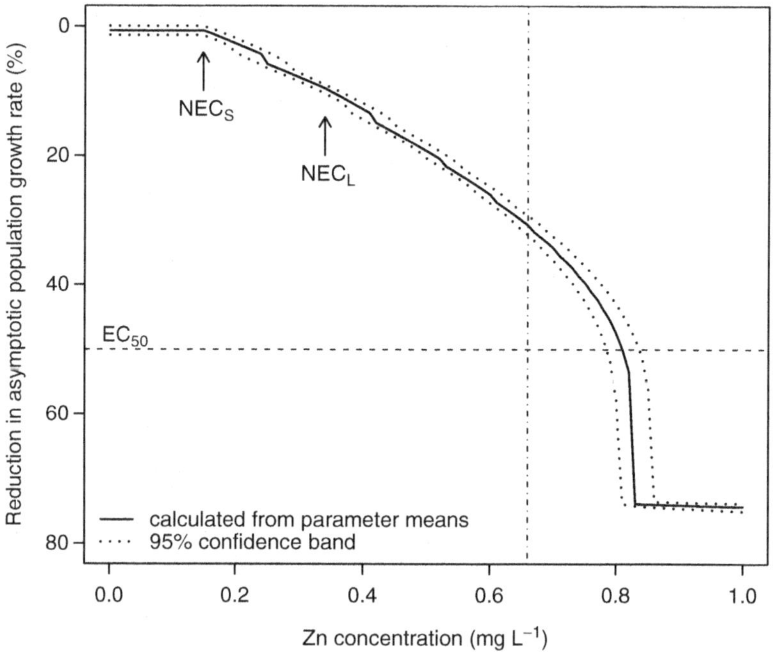

Fig. 8 *Daphnia magna* asymptotic population growth rate λ and its confidence band as a function of Zn exposure concentration ($\mathrm{mg\,L^{-1}}$) assuming an increase in maintenance costs as the main toxicological mode of action. The *vertical dashed line* corresponds to the strongest tested concentration. The *horizontal dashed line* corresponds to the critical threshold EC_{50}

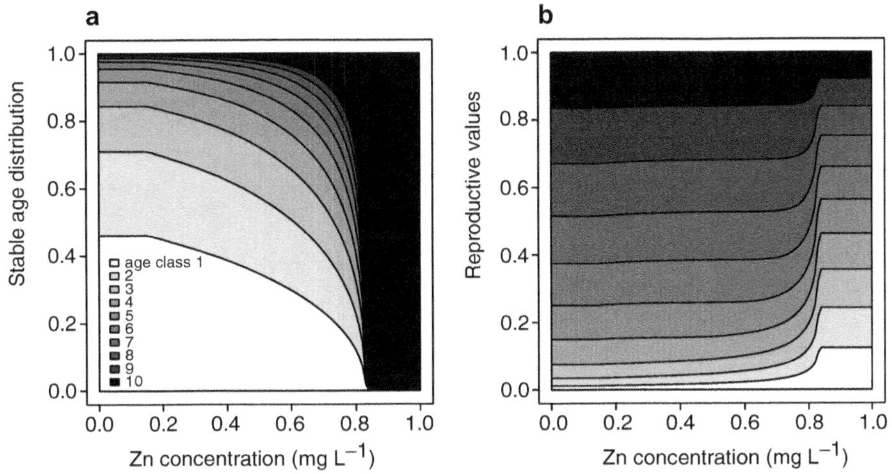

Fig. 9 (**a**) *Daphnia magna* stable age distribution and (**b**) reproductive values for each age class as a function of Zn exposure concentration (mg L^{-1}) assuming an increase in maintenance costs as the main toxicological mode of action

(0–0.66 mg L^{-1}), the DEBtox reproduction model predicted no more reproduction from $c = 0.83$ mg L^{-1} (whatever the age). Consequently, from this threshold exposure concentration equal to 0.83 mg L^{-1}, the population dynamics was only maintained thanks to survival and decreased slowly.

In Fig. 8, the confidence around λ was also plotted for all exposure concentrations in the range 0–1 mg L^{-1}. This result can be considered as a confidence band around the λ curve and so also be read horizontally. Hence, a prediction interval for the threshold concentration EC_{50} was calculated leading to a decrease of 50% in the asymptotic population growth rate, namely 0.78–0.85 mg L^{-1}.

Figure 9 shows the evolution of the stable age distribution and reproductive values with Zn exposure concentration. Without contamination, the first age class represented almost 50% of the population size. When exposure concentration increased, this proportion decreased for the benefit of the oldest age classes. From 0.83 mg L^{-1}, the organisms were no more able to reproduce; consequently, the population was asymptotically reduced to the last age class of daphnids only. Concerning reproductive values, reversed results were observed. The first age classes had weak contributions in the stable age distribution at weak Zn exposure concentrations. Their contributions increased later to the detriment of the oldest age classes.

4.3.2 Perturbation Analysis

Perturbation analyses highlight vital rates, which strongly contribute to the population response under stress condition (see Sect. 2.6 – (26)). As shown in Fig. 10a, contributions of fecundity and survival rates only begin from NEC_S (0.1482 mg L^{-1})

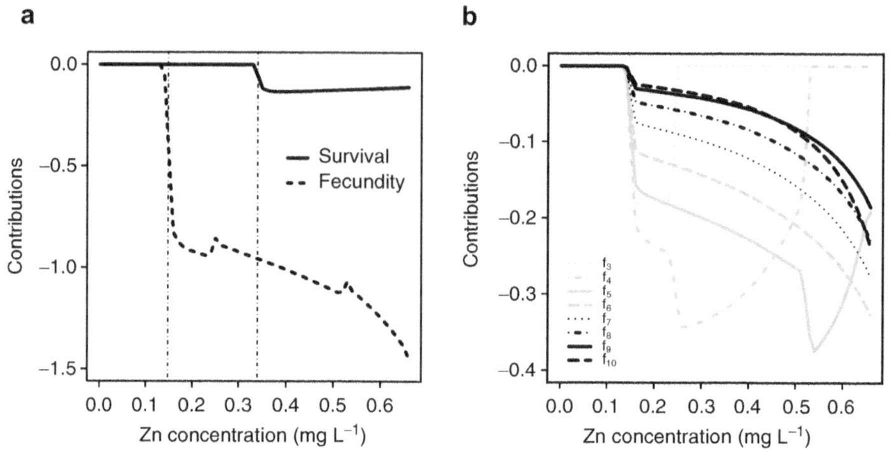

Fig. 10 Perturbation analysis of the asymptotic population growth rate λ of *Daphnia magna* exposed to Zn exposure concentration (see Sect. 2.6, (25)); contributions of fecundity and survival (**a**); detailed decomposition of the sensitivity of age-specific fecundities (**b**)

and NEC_L ($0.3398\,\mathrm{mg\,L^{-1}}$) concentrations from which vital rates are affected by the contaminant, respectively. The fecundities had much more important contributions than survival rates. In the DEBtox sublethal effect model (35), an increase in maintenance costs was assumed, that is both growth and reproduction were affected. Hence, as reproduction only began when daphnids reached length at puberty, the reproduction onset was delayed. As a consequence, the first age class to reproduce increased with the exposure concentration. A complete decomposition of contributions (Fig. 10b) revealed the marked contribution due to the fecundity of the first age class to reproduce which depended on the Zn concentration. Figure 10b shows the successive importance of f_3, f_4, f_5, and f_6 when concentration increases.

To conclude on this part, fecundity rates appeared as the most sensitive vital rates faced to contaminant, in particular the fecundity rate of the first age class to reproduce. These parameters had already been pointed out by Oli [57] as being the most sensitive in the population dynamics of "fast" populations. Oli studies involved mammals, but daphnid populations can also be considered to be a "fast" population, as they are characterized by $F/\alpha > 0.6$, α being the first age class to reproduce and F the mean fecundity.

4.4 Summary

Here again, the combination of DEBtox and matrix population models appeared very relevant in extrapolating individual results to the population level. All lethal and sublethal effects were integrated into the asymptotic rate of population increase, which is much more ecologically relevant than any statistically based parameter (e.g., NOEC and EC_x) derived from a single endpoint (mortality, growth, or reproduction). Moreover, our modeling approach provided the asymptotic rate of population increase as a continuous function of the exposure concentration and its associated confidence band from which a prediction interval for the threshold EC_{50} concentration was obtained. Once again, nested models proved their efficacy in assessing contaminant impacts on *D. magna* population dynamics and demography. Such models could, therefore, be of great use in guidelines for aquatic life safety and environmental health, even if many other stresses should be taken into account to properly extrapolate endpoints from laboratory to field. In Sects. 3 and 4, both case studies illustrated how to integrate effects of stress factors into matrix population models. In a similar way, density dependence, temperature, or predation could also be included. The next section will now exemplify how to deal with a spatial dimension.

5 When Space Matters: The Case of a Trout Population in a River Network

As underlined by the two previous case studies, transition matrix models offer the possibility to describe age-structured populations, and thus to consider differences in toxicological sensitivities between age classes or developmental stages. In ecological systems, spatial issues may compel to introduce a supplementary structure allowing a spatial description of exposed populations. This can be required to take into account simultaneously, on the one hand, a spatial heterogeneity of the contamination, and on the other hand, spatial processes involved in the population dynamics [58]. Here Leslie models are employed in such spatial contexts, by means of a dual structure for the description of the population: age and space [27]. Hence, space is underlined as modulating population effects of contamination through the use of results excerpted from the methodological development of a trout population model, which has been the object of a series of publications [24, 59–61]. In this section, instead of fully describing the building and the parametrization of the model, comparative results were chosen to show how spatialized matrix population models can supply insights in understanding how space issues could matter in the ecotoxicological risk at the population level.

5.1 *The Biological Situation*

The dynamics of a brown trout population (*Salmo trutta*) is depicted as resident of a river network symbolized by 15 connected patches corresponding to stretches of river from upstream to downstream levels (Fig. 11a). Biological parameterization as reported in Chaumot et al. [24] corresponds to natural populations from western France [62]. Since censuses are annual and first reproduction happens during the third year of life, three age classes are distinguished: young of the year, 1-year-old

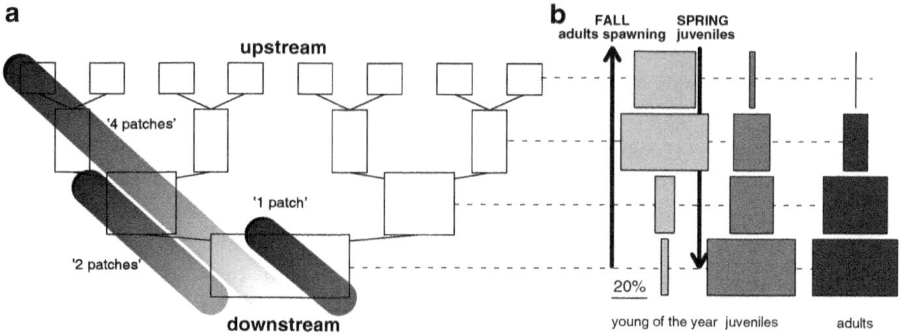

Fig. 11 Spatial dynamics of the brown trout population. (**a**) The river network considered for the study: 15 patches organized in four levels from upstream to downstream. *Shading bars* represent three spatial patterns of cadmium contamination due to three different discharge locations: upstream ("4 patches" are impacted with a dilution trend), third level ("2 patches" with dilution in the most downstream patch), and downstream ("1 patch"). (**b**) Spatial repartitions between the four levels of the river network under pristine conditions in fall, revealing the spatial segregation of the three age classes of the trout population (young of the year, 1-year-old juveniles and adults). Two major migratory events (*arrows*) occur each year (one downstream migration of 1-year-old juveniles in spring and one upstream spawning migration of adults in fall)

juveniles and adults. Breeding occurs in winter, mostly in upstream patches, while growing areas and feeding habitat of older age classes are spotted in the most downstream patches. This gives rise to a spatial segregation between the first and the two older age classes (revealed by the spatial distributions in Fig. 11b). This pattern articulated with age-specific differences in ecological requirements can emerge – thanks to two main spatial compensatory shifts in the occupancy of the river network during the life cycle of these salmonids. First, late in fall, adult breeders migrate upstream for spawning and then go back to their starting patch due to a "site-fidelity" behavior [63]. Second, in spring, 1-year-old juveniles conversely leave upstream areas and migrate down the river network in the sites, where they will spend the rest of their adult life. These migratory dynamics pattern leads us to the building of a "site-fidelity" model for the trout population dynamics in the network [24]. For methodological interests, a second version of the migratory model less consistent with biological observations was also developed, assuming a complete "population-mixing" instead of the downstream movement of juveniles each year in spring [59]. This population-mixing event randomly redistributes trout according to spatial occupancy reported in field studies. Here, only the structure of the "site-fidelity" model is reported below, but even less biological meaningful, the outputs of the "population-mixing" version are compared thereafter with the "site-fidelity" one, aiming at illustrating how Leslie matrix models handle the interplay between population spatial dynamics and spatial pattern of contamination.

Different scenarios of a cadmium (Cd) contamination are simulated assuming that they take place in one branch of the network with different patches of discharge; these scenarios take into account dilution at each confluence of the network (Fig. 11a). Either chronic or acute contamination are tested: chronic contaminations affect survival and fecundity rates, while 96h-acute events of pollution repeated each year at the same date only cause mortality. Effects are estimated by means of concentration–response curves established for published bioassays results with salmonids (see Sect. 2.5). Moreover, beside demographic perturbations, a second way of effects consisting in possible contaminant-induced alterations of migratory flows (avoidance or attraction) during the spawning migration of adults [60] was also tested.

5.2 The Model in Few Equations

The census is fixed in spring, after the emergence of the young of the year and before the downstream migration of 1-year-old juveniles. Let $n_{i,j}(t)$ be the number of trout females of age class i on patch j in year t ($j = 1 \dots 15$; $i = 1$ for young of the year; $i = 2$ for 1-year-old juveniles; $i = 3$ for adults). Thus 45 state-variables are gathered in the vector $N(t)$, depicting the population in spring of year t. This vector is a set of vectors $N_i(t)$ describing the internal structure of each age class as follows:

$$N(t) = (N_1(t), N_2(t), N_3(t))^{\mathrm{T}}, \tag{39}$$

$$\text{where } N_i(t) = \left(n_{i,1}(t), \dots, n_{i,j}(t), \dots, n_{i,15}(t)\right)^{\mathrm{T}}. \tag{40}$$

In first step, migrations were ignored. Considering that the census is positioned in spring and that spawning occurs late in fall, "fecundity transitions" between adult breeders in year t and young of the year in year $t+1$ must integrate 6 months of adult survival (square root of the annual survival rate), the fecundity rate, the fertilization rate, the sex ratio at birth, and the survival of young fishes during winter until the emergence in spring. Therefore, the transition matrix writes as follows:

$$\mathbf{A}(c) = \begin{bmatrix} \mathbf{O} & \mathbf{O} & \mathbf{F}(c)\sqrt{\mathbf{S}_3(c)} \\ \mathbf{S}_1(c) & \mathbf{O} & \mathbf{O} \\ \mathbf{O} & \mathbf{S}_2(c) & \mathbf{S}_3(c) \end{bmatrix}, \tag{41}$$

where $\mathbf{O}$ is a submatrix of 0 with 15 rows and 15 columns, and each non null element is a diagonal matrix of dimension 15 gathering patch-specific vital rates: $\mathbf{S}_i(c)$ survival matrix of age class i (the square root standing for 6 months of survival), $\mathbf{F}(c)$ fecundity matrix (including sex ratio and winter survival of newly born trout). The spatial heterogeneity of demographic rates throughout the river network (including contaminant effects) is thus translated in these patch-specific transition rates.

Considering the site-fidelity migratory pattern, downstream migrations of 1-year-old juveniles is introduced just after the census, with a conservative matrix $\mathbf{M_D}$ (of which the entry at row k and column l is the proportion of juvenile on patch l going to patch k). Because of the "site-fidelity" of adults, their absence from their patch is ignored during the spawning event. Moreover, considering that fecundity rates depend on the resident patch of adults and not on the spawning location, the spatial shift in fall is modeled exactly as a virtual upstream migration of eggs with a conservative matrix $\mathbf{M_U}$. The transition matrix becomes:

$$\mathbf{A}(c) = \begin{bmatrix} \mathbf{O} & \mathbf{O} & \mathbf{M_U}(c)\,\mathbf{F}(c)\,\sqrt{\mathbf{S}_3(c)} \\ \mathbf{S}_1(c) & \mathbf{O} & \mathbf{O} \\ \mathbf{O} & \mathbf{S}_2(c)\,\mathbf{M_D} & \mathbf{S}_3(c) \end{bmatrix}. \qquad (42)$$

$\mathbf{M_U}(c)$ is a function of the discharged Cd concentration in the river network because it integrates possible contaminant-induced perturbations of the migratory flows of spawners such as avoidance or attraction behaviors [60].

For each contamination scenario, a particular transition matrix is obtained corresponding to the general scheme presented in Sect. 2.4 (9). Asymptotic behavior of the model yields the asymptotic population growth rate λ, the stable age/spatial structure, the generation time, etc. allowing evaluating population response to the contamination in a spatial context.

5.3 *Results: Spatial Dynamics and Population Responses to Contamination*

Here, by means of precise examples, the interplay of spatial heterogeneities between contamination patterns and population dynamics is illustrated in the control of population responses to the pollution.

The first case points up how projection models can help to understand how the spatial contamination pattern influences the population effects. In Fig. 12a, considering "site-fidelity" behavior, it appears that for a given chronic Cd concentration, population impacts (in terms of reduction in asymptotic population growth rate) are less important when the discharge occurs more upstream, although more patches are contaminated (scenario "4 patches" vs. "2 patches"). The perturbation analysis (see Sect. 2.6 – (26)) reported on Fig. 12b, teaches us that the effect is mainly due to reductions in fecundity of adults in contaminated patches and that the contribution of survival reductions is quite insignificant. Thus, by considering on one hand the downstream position of adults (Fig. 11b) and on the other hand, the fact that the fecundity of adults (Fig. 12b) is the major limiting factors for the population renewal, more upstream discharges are understood as less impacting the population due to the dilution of the contamination. Moreover, the prominence of the effect of fecundity reduction also translates in other population endpoints: aging in the stable age structure, increase in the generation time [24]. Here the importance to integrate multiple population endpoints and not only to restrict the analysis to the widely employed asymptotic population growth rate is underlined in order to follow contaminant population impacts. In Fig. 13, the asymptotic spatial occupancy resulting from two different spatial patterns of contamination is presented ("1 patch" on Fig. 13a and "2 patches" on Fig. 13b). The comparison between the two situations reveals that population impacts on spatial occupancy are highly contrasted, while

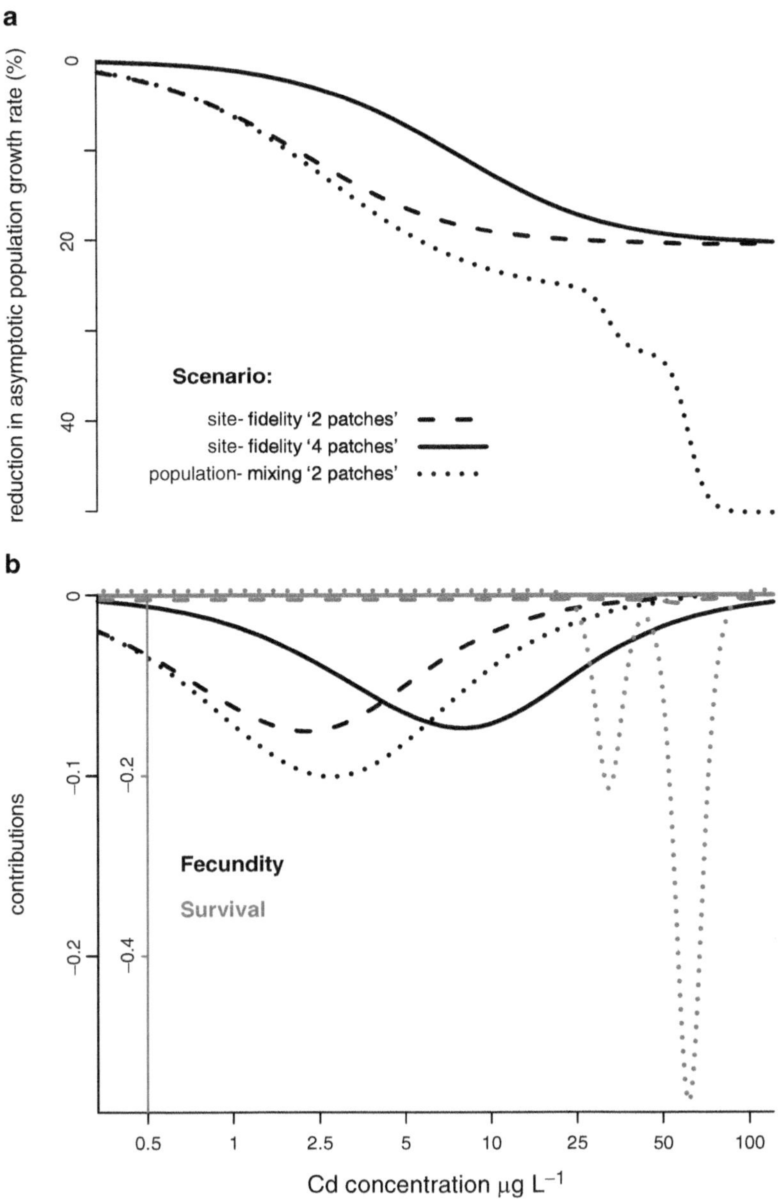

Fig. 12 (**a**) Effect on the asymptotic population growth rate of chronic contamination under three scenarios of chronic contamination: "2 patches" are contaminated (see Fig. 11) considering "site-fidelity" (*dashed line*) or "population-mixing" behavior (*dotted line*), or "4 patches" with the "site-fidelity" assumption (*solid line*). (**b**) Decomposition of sensitivity of the asymptotic population growth rate to cadmium concentration (see Sect. 2.6, (26)) between fecundity (*dark*) and survival (*gray*) contributions (same symbols for scenarios as on the *upper panel*). Scale for survival contributions is half-divided for clarity

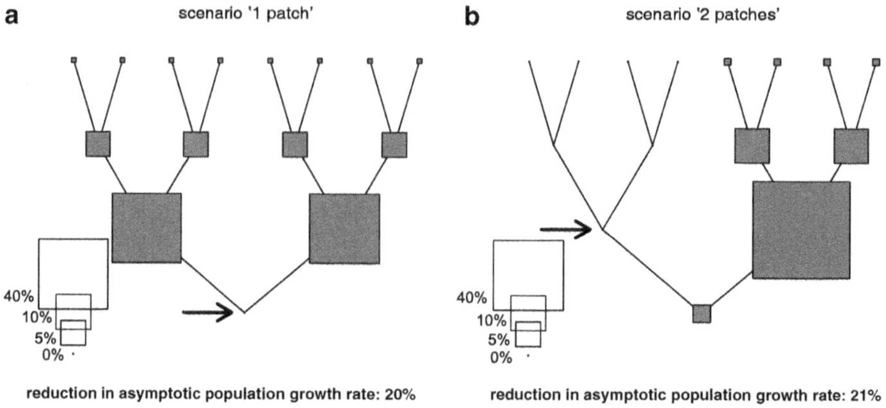

Fig. 13 Asymptotic adult distribution in the river network in fall under 2 spatial scenarios of chronic contamination with 70 Cd μg L^{-1} (see Fig. 11) and corresponding reduction in asymptotic population growth rate. Frequencies in the occupancy throughout the river network are quantified by the surfaces of the squares standing for each patch. *Arrows* locate the source of contamination in the network

quite the same reduction in the asymptotic population growth rate is calculated for both scenarios (about 20%). The assessment of population effects only based on potential growth could therefore be highly revised considering the fact that the trout population develops only one "side" of the river network in one of the two scenarios. Note that in scenario "2 patches," some patches are vacant even if uncontaminated. In view of the previous perturbation analysis (Fig. 12b), this "action at distance" [64] is well understood considering the importance of fecundity reductions in the population response together with the downstream location of breeders contributing to young trout production in upstream patches.

In case of acute pollution, the influence of the date in the year for Cd release supplies a supplementary illustration of the interplay between the spatial heterogeneity of contamination and the spatial dynamics of the trout population. Considering scenario "2 patches," the population impacts – in terms of asymptotic population growth rate – are more serious when the Cd discharge occurs after the downstream migration of 1-year-old juveniles (Fig. 14). This pattern is explained by the fact that fewer juveniles are exposed if pollution occurs before their migration due to the downstream location of the contamination. This is confirmed by the perturbation analysis [24], which reveals that reduction of survival of 1-year-old juveniles and adults mainly drives the population response in case of contamination after the downstream migration in spring.

For a same scenario of chronic contamination, the comparison of the decrease in the asymptotic population growth rate under the "site-fidelity" and "population-mixing" assumptions (Fig. 12) points up how the spatial behavior of the exposed species can contribute to the transmission of the effects to the population level. Strong negative peaks of 1-year-old juvenile and adult survival contributions (Fig. 12b) explained a supplementary drop of the potential population growth under

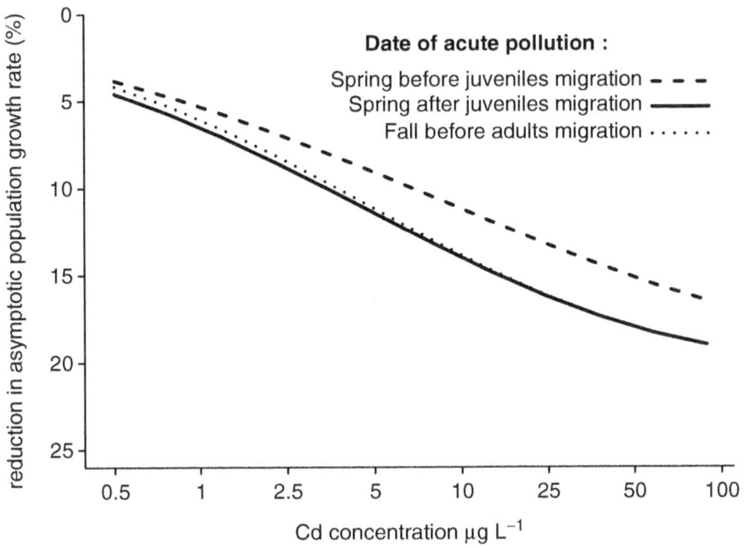

Fig. 14 Effect on the asymptotic population growth rate of acute contamination considering three different dates for cadmium release in the network

the mixing hypothesis (Fig. 12a). The difference in EC_{50} values fixed for the effects on fecundity and survival (1.3 and 30 $\mu g\,L^{-1}$, respectively) explained that these peaks appear for higher Cd concentrations. How to understand the weakness of survival contributions in the case of a "site-fidelity" strategy? Note that the breeders in contaminated patches have fecundity totally reduced for the concentrations affecting their survival (due to the difference in EC_{50}). In case of "site-fidelity," these trout never participate to the population renewal because each year they are situated in these polluted patches, what is not the case if a "population-mixing" occurs each spring. Therefore, survive or not does not matter for the population growth in the "site-fidelity" case. Thus, thanks to the perturbation analysis of the transition matrix, interactions between spatial dynamics and spatial pattern of contamination are disclosed and the emergence of the population responses is explained.

Finally, contamination may alter the population in a "second way" as space is concerned [60]: contamination can in fact induce migratory perturbations, what is reported for salmonids [65]. Spatial explicit modeling allows considering this potential impacts of migratory disruptions at the global population scale. If contaminant attraction or avoidance effects are introduced during the upstream migration of adults spawning, $\mathbf{M}_U\,(c)$ in (42), asymptotic spatial distributions are basically highly modified. Nevertheless, potential population growth is very less reduced [60]. This statement has to be totally revised if density-dependent regulations are introduced during the first year of survival because of limited carrying capacity of upstream patches. The analytical framework presented in Sect. 2.3 explained how the population could be described by an equilibrium size in such a nonlinear model. As reported in Chaumot et al. [60], migration perturbations generate imbalances in the spatial occupancy of young trout, what involves density-dependent mortality in this case. Such additional losses for the population can be very weighty. As an example, $2\,\mu g\,L^{-1}$ (chronic pollution, scenario "4 patches") decrease the total equilibrium population size by 2% via demographic rates reduction, while this drop can reach 37% in case of attraction behaviors and more surprisingly 32% in case of avoidance during breeder migration. Therefore, Leslie models integrating spatial dimension appear as relevant tools to consider this "second way" in the assessment of contaminant population effects in ecological context.

5.4 *Summary*

Caswell [15] underlines the importance of integrating the spatial dimension into ecotoxicological population models. Effectively, space can rule the emergence of population responses because, within the interplay between spatial heterogeneity of contamination and population spatial dynamics, an "action at distance" can emerge [58, 64, 66], some areas in the habitat could be key locations for the population, safety areas can influence the dynamics within impacted ones [58], or migratory and spatial behaviors can be altered by contamination and give rise to population dynamics disruptions. Here spatialized Leslie models and perturbation analysis illustrate the possibility to disclose how population response is built up via the identification of key vital rates, key age classes, and key locations in the global dynamics of a stressed population.

6 Conclusion

This chapter illustrated the way matrix population models can help in understanding toxic effects at the population level on the basis of individual level data analyses and integration of individual effect models relating life history traits to the exposure concentration. Lethal and sublethal individual effects considering sensitivity differences between developmental stages can be taken into account as well as a spatial dimension when individuals move from one location to another within a heterogeneously contaminated habitat. The power of matrix population models stems from easy writing, analysis, and interpretation, which are particularly suitable for dealing with complex biological phenomena. The demographic health of a population is readily accessible through characteristic endpoints, which are simply analytically handled from the transition matrix.

From an ecotoxicological point of view, matrix population models have the advantage of accounting for age-dependent sensitivities to toxic compounds as exhibited by many species. Hence, as stated in the introduction, they have proved their efficacy in numerous case studies involving various toxic compounds (heavy metals, pesticides, contaminated sediment, etc.) and different biological species (fishes, insects, crustaceans, gastropods, etc.). Moreover, results of a matrix population analysis consist in a set of demographic endpoints, which are functions of the vital rates, and thus of biological and environmental variables. In this context, perturbation analyses constitute a potent tool to identify key vital rates and key age classes, but also key locations when space matters.

Nevertheless, as in the case studies presented in Sects. 3–5, most of the matrix population models in ecotoxicology are based on a deterministic modeling approach by considering transition matrix entries constant over time and space. To tackle problems in a more realistic way, either an environmental stochasticity (i.e., when vital rates randomly vary over time), or a demographic stochasticity (i.e., when transitions and births become random events and are applied to a finite number of individuals) could be considered. Chaumot and Charles [61] proposed first few to introduce a demographic stochasticity in an ecotoxicological matrix population model. They showed that deterministic endpoints, such as equilibrium abundances, seem to seriously underestimate the endangering effect of pollution at the population level, and that the analysis of stochastic features such as the population extinction risk appears to be of broad interest to avoid this pitfall. Nevertheless, this work was only based on simulations under the software RAMAS Metapop® [67], whereas analytical results exist on stochastic matrix population models [68], thus securing the robustness of results.

Complementary to individual-based models [69] and simulations strategies [70], which aim at realism at first in the description of population dynamics [71], projection matrix models among state variable and top-down approaches – which accept a simplification of the system description – provide an appropriate way to understand the emergence of population impacts (understanding of broad interest for ecotoxicological risk assessment): first, because they use effects on individual life history traits as entries of the extrapolation scheme, parameters which are classical endpoints of bioassays; and second, because they allow an analytical treatment of the system dynamics, which guarantee the robustness of the statements, which can be formulated from such approaches. Finally, coupled to a quantitative analysis of toxic compound effects on life history traits through a statistical approach and the development of matrix population models, ecotoxicology can from now on aspire to propose a predictive approach at the various levels of biological organization, and thus to advance toward an integrated vision of the ecotoxicological risk assessment.

References

1. Truhaut R (1977) Ecotoxicology: Objectives, principles and perspectives. Ecotoxicol Environ Safety 1: 151–173
2. Levin SA, Harwell MA, Kelly JR, Kimball KD (1989) Ecotoxicology: Problems and approaches. Springer, New York
3. Emlen JM, Springman KR (2007) Developing methods to assess and predict the population level effects of environmental contaminants. Integr Environ Assess Manag 3: 157–165
4. Forbes VE, Calow P (1998) Is the per capita rate of increase a good measure of population-level effects in ecotoxicology? Environ Toxicol Chem 18: 1544–1556

5. Forbes VE, Calow P (2002) Population growth rate as a basis for ecological risk assessment of toxic chemicals. Philos Trans R Soc London Ser B 357: 1299–1306
6. Stark JD, Banks JE (2003) Population-level effects of pesticides and other toxicants on arthropods. Ann Rev Entomol 48: 505–519
7. Mooij WM, Hülsmann S, Vijverberg J, Veen A, Lammens EHRR (2003) Modeling *Daphnia* population dynamics and demography under natural conditions. Hydrobiologia 491: 19–34
8. Hülsmann S, Mehner T, Worischka S, Plewa M (1999) Is the difference in population dynamics of *Daphnia galeata* in littoral and pelagic areas of a long-term biomanipulated reservoir affected by age-0 fish predation? Hydrobiologia 408–409: 57–63
9. Nisbet RM, Gurney WSC, Murdoch WW, Mccauley E (1989) Structured population-models – A tool for linking effects at individual and population-level. Biol J Linn Soc 37: 79–99
10. Lotka AJ (1939) A contribution to the theory of self-renewing aggregates, with special reference to industrial replacement. Ann Math Stat 10: 1–25
11. Breitholtz M, Wollenberger L, Dinan L (2003) Effects of four synthetic musks on the life cycle of the harpacticoid copepod *Nitocra spinipes*. Aquat Toxicol 63: 103–118
12. Leslie PH (1945) On the use of matrices in certain population mathematics. Biometrika 33: 184–212
13. Leslie PH (1948) Some further notes on the use of matrices in poulation mathematics. Biometrika 35: 213–245
14. Caswell H (2001) Matrix population models – Construction, analysis, and interpretation. Sinauer Associates, Sunderlands, MA
15. Caswell H (1996) Demography meets ecotoxicology: Untangling the population level effects of toxic substances. In: MC Newmann, CH Jagoe (Eds), Ecotoxicology: A hierarchical treatment. Lewis Publishers, Boca Raton, FL
16. Caswell H (1996) Analysis of life table response experiments.2. Alternative parameterizations for size- and stage-structured models. Ecol Model 88: 73–82
17. Levin L, Caswell H, Bridges T, DiBacca C, Cabrera D, Plaia G (1996) Demographic responses of estuarine polychaetes to pollutants: Life table response experiments. Ecol Appl 6: 1295–1313
18. Lopes C, Péry ARR, Chaumot A, Charles S (2005) Ecotoxicology and population dynamics: Using DEBtox models in a Leslie modeling approach. Ecol Model 188: 30–40
19. Smit MGD, Kater BJ, Jak RG, van den Heuvel-Greve MJ (2006) Translating bioassay results to field population responses using a Leslie-matrix model for the marine amphipod *Corophium volutator*. Ecol Model 196: 515–526
20. Klok C, Holmstrup M, Damgaardt C (2007) Extending a combined dynamic energy budget matrix population model with a bayesian approach to assess variation in the intrinsic rate of population increase. An example in the earthworm *Dendrobaena octaedra*. Environ Toxicol Chem 26: 2383–2388
21. Billoir E, Péry ARR, Charles S (2007) Integrating the lethal and sublethal effects of toxic compounds into the population dynamics of *Daphnia magna*: A combination of the DEBtox and matrix population models. Ecol Model 203: 204–214
22. Ducrot V, Péry ARR, Mons R, Charles S, Garric J (2007) Dynamic engergy budgets as a basis to model population-level effects of zinc-spiked sediments in the gastropod *Valvata piscinalis*. Environ Toxicol Chem 26: 1774–1783
23. Lebreton J-D, Gonzales-Davila G (1993) An introduction to models of subdivided populations. J Biol Syst 1: 389–423
24. Chaumot A, Charles S, Flammarion P, Auger P (2003) Ecotoxicology and spatial modeling in population dynamics: An illustration with brown trout. Environ Toxicol Chem 22: 958–969
25. Lefkovitch LP (1965) The study of population growth in organisms grouped by stages. Biometrics 21: 1–18
26. Cull P, Vogt A (1973) Mathematical analysis of the asymptotic behavior of the Leslie population matrix model. Bull Math Biol 35: 645–661
27. Lebreton JD (1996) Demographic models for subdivided populations: The renewal equation approach. Theor Popul Biol 49: 291–313

28. Rogers A (1966) The multiregional matrix growth operator and the stable interregional age structure. Demography 3: 537–544
29. Rogers A (1985) Regional population projection models. Sage Publications, Beverly Hills, CA
30. Levin SA (1989) Applied mathematical ecology. Springer-Verlag, Berlin
31. Hanski I, Gilpin ME (1997) Metapopulation dynamics: Ecology, genetics and evolution. Academic Press, San Diego
32. Isnard P, Flammarion P, Roman G, Babut M, Bastien P, Bintein S, Essermeant L, Ferard JF, Gallotti-Schmitt S, Saouter E, Saroli M, Thiebaud H, Tomassone R, Vindimian E (2001) Statistical analysis of regulatory ecotoxicity tests. Chemosphere 45: 659–669
33. Kooijman SALM, Bedaux JJM (1996) The analysis of aquatic toxicity data. VU University Press, Amsterdam
34. Kooijman SALM (2000) Dynamic energy and mass budgets in biological systems. Cambridge University Press, Great Britain
35. Billoir E, da Silva Ferrão-Filho A, Delignette-Muller ML, Charles S (2009) DEBtox theory and matrix population models as helpful tools in understanding the interaction between toxic cyanobacteria and zooplankton. J Theor Biol 258: 380–388
36. Billoir E, Delignette-Muller ML, Péry ARR, Geffard O, Charles S (2008) Statistical cautions when estimating DEBtox parameters. J Theor Biol 254: 55–64
37. Billoir E, Delignette-Muller ML, Péry ARR, Charles S (2008) On the use of Bayesian inference to estimate DEBtox parameters. Environ Sci Technol 42: 8978–8984
38. Charles S, Ferreol M, Chaumot A, Péry ARR (2004) Food availability effect on population dynamics of the midge *Chironomus riparius*: A Leslie modeling approach. Ecol Model 175: 217–229
39. Charles S, Mallet J-P, Persat H (2006) Population dynamics of grayling: Modelling temperature and discharge effects. Math Model Nat Phenom 1: 33–48
40. Caswell H (1978) A general formula for the sensitivity of population growth rate to changes in life history parameters. Theor Popul Biol 14: 215–230
41. Caswell H (2000) Prospective and retrospective perturbation analyses: Their roles in conservation biology. Ecology 81: 619–627
42. De Kroon H, Van Groenendael J, Ehrlen J (2000) Elasticities: A review of methods and model limitations. Ecology 81: 607–618
43. Yearsley JM, Fletcher D, Hunter C (2003) Sensitivity analysis of equilibrium population size in a density-dependent model for Short-tailed Shearwaters. Ecol Model 163: 119–129
44. Caswell H, Takada T (2004) Elasticity analysis of density-dependent matrix population models: The invasion exponent and its substitutes. Theor Popul Biol 65: 401–411
45. Goddeeris BR, Vermeulen AC, De Geest E, Jacobs H, Baert B, Ollevier F (2001) Diapause induction in the third and fourth instar of *Chironomus riparius* (Diptera) from Belgian lowland brooks. Arch Hydrobiol 150: 307–327
46. Marking LL, Chandler JH (1981) Toxicity of 6 bird control chemicals to aquatic organisms. Bull Environ Contam Toxicol 26: 705–716
47. Péry ARR, Ducrot V, Mons R, Miege C, Gahou J, Gorini D, Garric J (2003) Survival tests with *Chironomus riparius* exposed to spiked sediments can profit from DEBtox model. Water Res 37: 2691–2699
48. Péry ARR, Mons R, Ducrot V, Garric J (2004) Effects of methiocarb on *Chironomus riparius* survival and growth with and without tube-building. Bull Environ Contam Toxicol 72: 358–364
49. Armitage PD, Cranston PS, Pinder LCV (1995) The Chironomidae: Biology and ecology of non-biting midges. Chapman and Hall, London, UK
50. OECD (1998) OECD guidelines for testing of chemicals. *Daphnia magna* reproduction test. Organization for Economic Co-operation and Development, Paris
51. ISO (2000) 10706. Water quality – Determination of long term toxicity of substances to *Daphnia magna* Straus (Cladocera, Crustacea). International Organization for Standardization, Geneva, Switzerland
52. Anderson BG (1944) The toxicity threshold of various substances found in industrial wastes as determined by the use of *Daphnia magna*. Sewage Work J 16: 156–165

53. Koivisto S (1995) Is *Daphnia magna* an ecologically representative zooplankton species in toxicity tests? Environ Pollut 90: 263–267
54. Adema DMM (1978) *Daphnia magna* as a test animal in acute and chronic toxicity tests. Hydrobiologia 59: 125–134
55. Gilks WR, Richardson S, Spiegelhalter DJ (1996) Markov Chain Monte Carlo in practice. Chapman and Hall, New York
56. Nogueira AJA, Baird DJ, Soares AMVM (2004) Testing physiologically-based resource allocation rules in laboratory experiments with *Daphnia magna* Straus. Ann Limnol 40: 257–267
57. Oli MK (2004) The fast-slow continuum and mammalian life-history patterns: An empirical evaluation. Basic Appl Ecol 5: 449–463
58. Ares J (2003) Time and space issues in ecotoxicology: Population models, landscape pattern analysis, and long-range environmental chemistry. Environ Toxicol Chem 22: 945–957
59. Chaumot A, Charles S, Flammarion P, Garric J, Auger P (2002) Using aggregation methods to assess toxicant effects on population dynamics in spatial systems. Ecol Appl 12: 1771–1784
60. Chaumot A, Charles S, Flammarion P, Auger P (2003) Do migratory or demographic disruptions rule the population impact of pollution in spatial networks? Theor Popul Biol 64: 473–480
61. Chaumot A, Charles S (2006) Pollution, stochasticity and spatial heterogeneity in the dynamics of an age-structured population of brown trout living in a river network. In: R Akcakaya (Editor), Population-level ecotoxicological risk assessment: Case studies. Oxford University Press, New York
62. Baglinière JL, Maisse G, Lebail PY, Nihouarn A (1989) Population dynamics of brown trout, *Salmo trutta* L., in a tributary in Brittany (France): Spawning and juveniles. J Fish Biol 34: 97–110
63. Ovidio M (1999) Annual activity cycle of adult brown trout (*Salmo trutta* L.): A radio-telemetry study in a small stream of the Belgian Ardenne. Bull Fr Peche Pisc: 1–18
64. Spromberg JA, John BM, Landis WG (1998) Metapopulation dynamics: indirect effects and multiple distinct outcomes in ecological risk assessment. Environ Toxicol Chem 17: 1640–1649
65. Hansen F, Forbes VE, Forbes TL (1999) Using elasticity analysis of demographic models to link toxicant effects on individuals to the population level: an example. Funct Ecol 13: 157–162
66. Sherratt JA (1993) The amplitude of periodic plane waves depends on initial conditions in a variety of lambda - Omega systems. Nonlinearity 6: 1055–1066
67. Akçakaya HR (1994) RAMAS metapop: Viability analysis for stage-structured metapopulations. Version 1.0. Setauket, New York
68. Tuljapurkar S, Caswell H (1997) Structured-population models in marine, terrestrial, and freshwater systems. Springer, Berlin
69. Van den Brink P, Baveco H, Verboom J, Heimbach F (2007) An individual-based approach to model spatial population dynamics of invertebrates in aquatic ecosystems after pesticide contamination. Environ Toxicol Chem 26: 2226–2236
70. Van Kirk RW, Hill SL (2007) Demographic model predicts trout population response to selenium based on individual-level toxicity. Ecol Model 206: 407–420
71. Grimm V (1999) Ten years of individual-based modelling in ecology: what have we learned and what could we learn in the future? Ecol Model 115: 129–148

Bioaccumulation of Polar and Ionizable Compounds in Plants

Stefan Trapp

Abstract The uptake of neutral and ionizable organic compounds from soil into plants is studied using mathematical models. The phase equilibrium between soil and plant cells of neutral compounds is calculated from partition coefficients, while for ionizable compounds, the steady state of the Fick–Nernst–Planck flux equation is applied. The calculations indicate biomagnification of neutral, polar, and nonvolatile compounds in leaves and fruits of plants. For electrolytes, several additional effects impact bioaccumulation, namely dissociation, ion trap effect, and electrical attraction or repulsion. For ionizable compounds, the effects of pK_a and pH partitioning are more important than lipophilicity. Generally, dissociation leads to reduced bioaccumulation in plants, but the calculations also predict a high potential for some combinations of environmental and physicochemical properties. Weak acids (pK_a 2–6) may accumulate in leaves and fruits of plants when the soil is acidic due to the ion trap effect. Weak bases (pK_a 6–10) have a very high potential for accumulation when the soil is alkaline. The model predictions are supported by various experimental findings. However, the bioaccumulation of weak bases from alkaline soils has not yet been validated by field studies.

Keywords Acids · Bases · Bioaccumulation · Ionic · pH · Plants · Model

1 Introduction

Living organisms are exposed to chemicals in the environment and may take up and concentrate them in their body. An example is the bioconcentration factor (BCF) of fish, which is the concentration of a chemical in fish divided by the concentration of the chemical in surrounding water:

$$\mathrm{BCF} = \frac{\text{Concentration of fish (mg/kg)}}{\text{Concentration in water (mg/L)}}. \tag{1}$$

S. Trapp (✉)
DTU Environment, Department of Environmental Engineering, Technical University of Denmark, Miljoevej, Building 113, DK – 2800 Kgs. Lyngby, Denmark
e-mail: stt@env.dtu.dk

J. Devillers (ed.), *Ecotoxicology Modeling*, Emerging Topics in Ecotoxicology: Principles, Approaches and Perspectives 2, DOI 10.1007/978-1-4419-0197-2_11,

Similar is the bioaccumulation factor (BAF), which is the concentration in an organism divided by the concentration in the surrounding medium:

$$\text{BAF} = \frac{\text{Concentration in organism (mg/kg)}}{\text{Concentration in surrounding medium (mg/L)}}. \quad (2)$$

The BCF was defined as the process by which chemical substances are adsorbed only through surfaces, whereas the BAF is due to all routes of exposure and includes dietary uptake [1]. Accordingly, "biomagnification" is a process in which the thermodynamic activity of the chemical in the body exceeds the activity in the diet.

Bioaccumulation in the food chain may lead to high doses of compounds in the diet of top predators and humans [2, 3] and is a highly undesired property of compounds [4]. In the European regulatory framework for chemical risk assessment, compounds with a BAF above 2,000 are considered as bioaccumulative and those with BAF above 5,000 as very bioaccumulative [4]. The same criterion (BAF of 5,000) is also used by other governments [5].

It is generally accepted that bioaccumulation is closely related to lipophilicity of a compound, measured as the partition coefficient between n-octanol and water, K_{OW}, or the partition coefficient between n-octanol and air, K_{OA} [6, 7]. Accordingly, a theoretical relation for aquatic biota was suggested [6, 8]. The lipid phase accumulates the compound similar to n-octanol, and therefore

$$\text{BCF} = L \times K_{\text{OW}}, \quad (3)$$

where L is the volumetric lipid content of an organism (L/L) and K_{OW} is the partition coefficient between n-octanol and water (mg/L octanol:mg/L water = L octanol/L water).

In general, most BAF estimation approaches describe the bioaccumulation behavior of organic substances solely by the octanol–water partitioning coefficient ($\log K_{\text{OW}}$). This may be correct for neutral lipophilic compounds. But there are other mechanisms that can lead to bioaccumulation, which are not connected to lipophilicity. One example is the accumulation due to uptake of water by plants from soil, with subsequent transport to leaves with the water stream, and subsequent accumulation in leaves when the water evaporates. Another example is the accumulation of weak electrolytes in living cells. Investigations show that for these dissociating compounds, other processes, such as pH-dependent speciation and electrical attraction, can be the decisive processes determining the accumulation in cells [9–11]. The $\log K_{\text{OW}}$ approach alone may lead to an under- or overestimation of the accumulation of ionizable substances. In a review using fish with 5,317 BCF values, about 20% of compounds had the potential for ionization. But for less than 40% of the tests, the pH of the water during the experiment was reported [1]. It seems that the critical role of pK_a and pH for the BCF of ionizable compounds is sometimes not sufficiently highlighted.

A mechanistic model described in this chapter will identify accumulation processes that are *not* related to lipophilic partitioning. The focus is on accumulation of compounds from soil in plants and, in particular, on ionizable compounds.

2 Electrolytes

"Electrolytes" is a common term for compounds with electrical charge. Common synonyms are "ionic compounds," "ionizable compounds," "dissociable compounds," "dissociating compounds," "electrolytic compounds," and "charged compounds." Electrolytes may be acids (valency -1, -2, etc.), bases (valency $+1$, $+2$, etc.), amphoters (valencies $+1$ and -1, $+1$ and -2, etc.), or zwitterions (valencies 0, $+1$, and -1, etc.). Weak electrolytes are compounds with weak acid- and base- groups, which dissociate only partly under usual environmental conditions (pH 4–10). Thus, weak electrolytes are commonly present in two or more different forms with very different properties, namely the neutral molecule and the ion, which can rapidly be transferred from one into the other if pH changes. "Very weak" electrolytes are named acids or bases that dissociate only to a minor degree at environmental pH (usually between pH 5 and 9), i.e., bases with $pK_a < 5$ and acids with $pK_a > 9$.

Different to the neutral molecule, ions can be attracted or rejected by electrical charges. Monovalent bases have a valency of $+1$ and are thus attracted by negative electrical potentials, while acids have a valency of -1 and would be attracted by positive electrical potentials. The neutral compound typically has a far higher lipophilicity than the corresponding ion. In average, $\log K_{OW}$ (ion) is equal to $\log K_{OW}$ (neutral) -3.5, which means the K_{OW} of ions is 3,162 times lower. For zwitterions, which have a permanent positive and negative charge, but a net charge of 0, the difference is smaller, $\log K_{OW}$ (zwitterion) $=$ $\log K_{OW}$ (neutral) -2.3 [12].

There is also a difference in vapor pressure. For ions, it is approximately 0. The vapor pressure of the total compound, p (Pa), can thus be calculated using the vapor pressure of the neutral molecule and multiplying with the fraction of neutral molecules.

The applicability domain of most QSAR regressions is limited to neutral compounds [4]. For ionizable compounds, the TGD suggests a correction of the physico-chemical properties ($\log K_{OW}$, Henry's law constant) by the neutral fraction of compound, F_n. For the BCF that means that

$$\log \mathrm{BCF} = 0.85 \times \log(F_n \times K_{OW}) - 0.70. \tag{4}$$

The reliability of this method for ionizable compounds was never critically evaluated. A recent survey by Ralph Kühne revealed that of his database with >10,000 compounds, at least 25% of compounds have structures that may dissociate, such as carboxylic acids, phenols, and amines. In fact, ionizable compounds are frequent and typical for many substance classes. Among pesticides, most herbicides are weak acids. Among pharmaceuticals, weak bases are frequent ("alkaloids"). Detergents are often anionic, cationic, or amphoteric. Metabolites of phase I reactions (oxidations) are usually acids, while the reduction of nitro-groups leads to amines. Given the widespread occurrence of weak electrolytes, it may surprise that very few models and regressions were developed for ionizable substances.

In the following, a dynamic model for plant uptake is developed – first the "standard approach" for neutral compounds. This is then modified to be applicable for electrolytic compounds.

3 Plant Uptake Models for Neutral Compounds

In this section, the uptake and accumulation of polar and ionizable compounds in plants is quantified. Based on physiological principles, the mass balance equations for the transport of compounds in the soil–plant–air system are derived and combined to mathematical models.

Plant uptake models for neutral compounds have been developed by several groups [13–16]. A series of crop-specific uptake models was derived, based on the PlantX model [17] – to mention are the one-compartment analytical solution of the latter [18], and the models for root vegetables [19], potatoes [20], and fruits [21,22].

3.1 How Plants Function

Figure 1 shows schematically how plants function: The large network of roots takes up water and solutes. In the pipe system of the xylem, these are translocated through the stem to the leaves. The leaves take up carbon dioxide from the atmosphere and simultaneously transpire the water. Carbohydrates produced in the leaves by photosynthesis are translocated in the phloem pipe system to the sinks (all growing parts, fruits, and storage organs).

In most ecosystems, plants transpire about two-thirds of the precipitation [23]. For humid conditions, this ranges from 300 to 600-L water per square meter per year. The water, which is taken up by the roots, does not stay there but is translocated in the xylem to the leaves and evaporates. Only 1–2% is taken up into the plant cells. Chemicals, which are dissolved in the "transpiration stream" (= the xylem sap), can be moved upward, too.

The water-use efficiency (growth of biomass per liter of transpired water) is typically at 20 g/L, which means that about 500-L water is transpired for 1 kg (dry) biomass. Dry matter content of leaves is typically between 5 and 20%. Thus, between 25 and 200-L water is transpired for 1-kg fresh weight leaves. Compounds that dissolve in the transpiration stream but are nonvolatile from leaves remain there. By this mechanism, an accumulation of compound in plants from soil may occur.

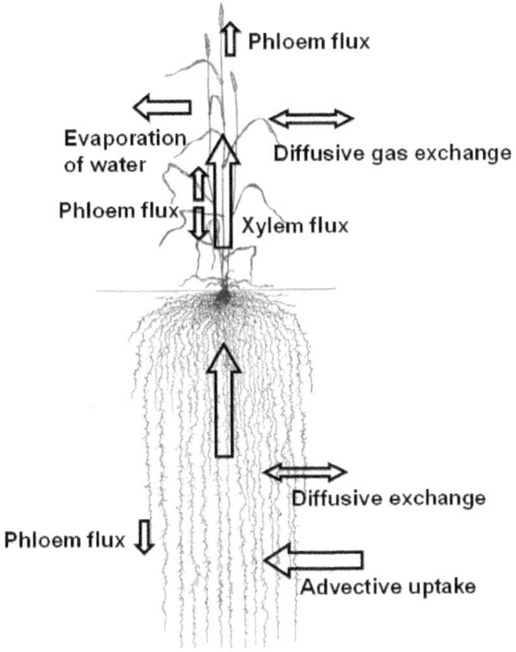

Fig. 1 "How plants function" – a sketch of plant organs and transport pathways

Different from lipophilic accumulation, the organism is not in chemical phase equilibrium with the surrounding medium – the chemical activity in leaves may increase above the activity in soil.

3.2 Equilibrium Partitioning

The diffusive flux between two compartments describes *Fick's first law of diffusion*:

$$J = P(a_o - a_i), \tag{5}$$

where J is the unit net flux of the neutral molecules from outside (o) to inside (i) of the compartment ($kg/m^2/s^1$), a is the activity (here: kg/m^3), and P is the permeability of the boundary (m/s).

It follows that diffusive exchange stops when $a_o = a_i$, that is, activities are equal. This does usually not mean equal concentrations. The activity ratio (which is, at low concentrations and low ionic strength, equal to the concentration ratio) in equilibrium is the *partition coefficient K* [often in the unit mg/L: (mg/L = L/L)].

An example is the partition coefficient between air and water, K_{AW}:

$$\frac{a_{Air}}{a_{Water}} = K_{AW} = \frac{p_S}{SRT}, \tag{6}$$

where K_{AW} is in the unit L/L (also known as dimensionless Henry's Law constant), p_S is the saturation vapor pressure (Pa), S is the solubility (kg/m^3), R is the universal gas constant (J/K/mol), and T is the absolute temperature (K). Analogous equilibrium partition coefficients can be defined for other phases.

3.2.1 Freely Dissolved Concentration of Neutral Organic Compounds in Soil

Soil is composed of water, the "soil matrix," and gas pores. Uptake into plants occurs only for the "bioavailable" fraction of compound, which is the concentration of freely dissolved compound [24].

The sorption to solids is described by the empirical Freundlich relation:

$$\frac{m_{\text{ads}}}{M_{\text{M}}} = K \times C_{\text{W}}^{1/n}, \tag{7}$$

where m_{ads} is the adsorbed amount of chemical (mg), M_{M} is the mass of sorbent, here the soil matrix (kg), K is the proportionality factor (Freundlich constant) (L water/kg soil), C_{W} is the equilibrium concentration in the aqueous solution (mg/L water), and n is a measure of nonlinearity of the relation. For small concentrations, values of n are close to 1. The Freundlich constant can then be seen as the slope of the linear adsorption/desorption isotherm. It is often called the distribution coefficient K_{d} between soil matrix and water:

$$\frac{m_{\text{ads}}}{M_{\text{M}}} = C_{\text{M}} = K_{\text{d}} \times C_{\text{W}}, \tag{8}$$

where C_{M} is the concentration sorbed to the soil matrix (mg/kg). The K_{d} of organic chemicals is related to the fraction of organic carbon in soil, OC:

$$K_{\text{d}} = \text{OC} \times K_{\text{OC}}. \tag{9}$$

K_{OC} (L/kg) is the partition coefficient between organic carbon and water and is described later.

Bulk soil. The natural bulk soil consists of soil matrix, soil solution, and soil gas. The concentration ratio of dry soil (= soil matrix, index M) to water has been described earlier. For a liter of dry soil (index Mvol), multiply with the dry soil density ρ_{dry} (kg/L).

$$C_{\text{Mvol}}(\text{mg/L})/C_{\text{W}}(\text{mg/L}) = K_{\text{d}} \times \rho_{\text{dry}} \quad (\text{mg/L} : \text{mg/L} = \text{L/L}). \tag{10}$$

For wet soil (C_{SoilVol}), add the volumetric pore water fraction P_{W} (L/L).

$$C_{\text{SoilVol}}/C_{\text{W}} = K_{\text{d}} \times \rho_{\text{dry}} + P_{\text{W}} \quad (\text{mg/L} : \text{mg/L} = \text{L/L}). \tag{11}$$

Now back to the unit mg/kg for the soil concentration (C_{Soil}). This is achieved by dividing by the wet soil density ρ_{wet} ($= \rho_{\text{dry}} + P_{\text{W}}$), unit kg/L:

$$C_{\text{Soil}}/C_{\text{W}} = (K_{\text{d}} \times \rho_{\text{dry}} + P_{\text{W}})/\rho_{\text{wet}} \quad (\text{mg/kg} : \text{mg/L} = \text{L/kg}). \tag{12}$$

Turning this around gives finally the concentration ratio between water (mg/L) and wet soil (mg/kg) K_{WS}:

$$C_{\text{W}}/C_{\text{Soil}} = K_{\text{WS}} = \rho_{\text{wet}}/(K_{\text{d}} \times \rho_{\text{dry}} + P_{\text{W}}) \quad (\text{mg/L} : \text{mg/kg} = \text{kg soil/L water}). \tag{13}$$

The relation describes the concentration ratio between soil water and bulk soil (wet soil) in phase equilibrium. Replacing the K_{d} by OC $\times$ K_{OC} gives an expression to calculate the dissolved concentration of a chemical in soil (mg/L) from the total concentration in soil (mg/kg, wet weight):

$$\frac{C_{\text{W}}}{C_{\text{Soil}}} = \frac{\rho_{\text{wet}}}{\text{OC} \times K_{\text{OC}} \times \rho_{\text{dry}} + P_{\text{W}}} = K_{\text{WS}} = \frac{1}{K_{\text{SW}}}. \tag{14}$$

As defined, C_{W} (mg/L) is the concentration of the chemical in soil water and C_{Soil} (mg/kg) is the concentration of the chemical in bulk (total) soil. ρ_{wet} is the density of the wet soil (kg/L), OC (also named f_{OC}) is the fraction of organic carbon (kg/kg), ρ_{dry} is the density of the dry soil, and P_{W} is the volume fractions of water in the soil (L/L).

The K_{OC} is the partition coefficient between organic carbon and water. For hydrophobic, neutral organic chemicals, the K_{OC} and can be estimated from:

$$\log K_{\text{OC}} = 0.81 \log K_{\text{OW}} + 0.1. \tag{15}$$

This is the regression suggested as default in the European chemical risk assessment tool EUSES [4]. Several other regressions are available, among them [25,26] are the following:

$$K_{\text{OC}} = 0.411 \times K_{\text{OW}}, \tag{16}$$

$$\log K_{\text{OC}} = 0.72 \log K_{\text{OW}} + 0.49. \tag{17}$$

The differences between these regression equations can be considerable, in particular in the extreme ranges (high or low K_{OW}).

3.2.2 Phase Equilibrium Between Roots and Water

The root concentration factor (RCF) was measured with pulverized ("mazerated") barley roots in shaking experiments with chemicals of different K_{OW} [27]:

$$\text{RCF} = \frac{\text{concentration in roots (mg/kg)}}{\text{concentration in water (mg/L)}}. \tag{18}$$

The RCF increased with K_{OW}. The fit curve between RCF and K_{OW} is as follows:

$$\log(\mathrm{RCF} - 0.82) = 0.77 \log K_{OW} - 1.52, \tag{19}$$

or

$$\mathrm{RCF} = 0.82 + 0.03\, K_{OW}^{0.77}, \tag{20}$$

where the RCF can be rewritten as K_{RW} (L/kg) to describe the equilibrium partitioning between root concentration C_R (mg/kg fresh weight) and water C_W (mg/L). The partitioning occurs into the water and the lipid phase of the root:

$$K_{RW} = W_R + L_R a K_{OW}^b, \tag{21}$$

where W and L are water and lipid content of the plant root, b for roots is 0.77, and $a = 1/\rho_{Octanol} = 1.22\,\mathrm{L/kg}$. Typical values for a carrot are 0.89 L/kg for W_R and 0.025 kg/kg for L_R [19]. With this parameterization, the equation gives practically identical values to the RCF.

3.3 *Dynamic Model for Uptake of Neutral Compounds into Roots*

3.3.1 Carrot Model

The "carrot model" calculates uptake into and loss from root with the transpiration water (Fig. 2). Diffusive uptake across the peel is neglected. It is assumed that the concentration in the xylem at the root tips, where the translocation stream enters the

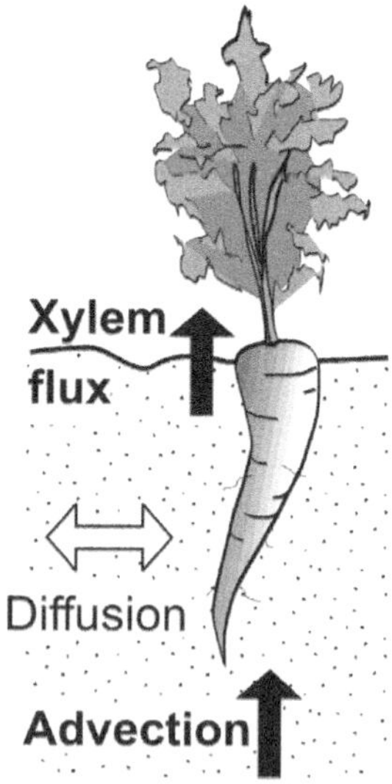

Fig. 2 Transport processes in a carrot root. *Full arrows*: considered by the model

roots, is in chemical phase equilibrium with the soil pore water. Since both solutions are water, the concentration is equal. At the outflow from the root, the concentration in xylem is in equilibrium with the root. In the mass balance, the change of chemical mass in roots is simply influx with water minus outflux with water.

$$\frac{dm_R}{dt} = C_W \times Q - C_{Xy} \times Q, \tag{22}$$

where m_R is the mass of chemical in roots, Q is the transpiration stream (L/d), and C_{Xy} is the concentration in the xylem (mg/L) at the outflow of the root. From chemical mass, concentration is received by dividing through the mass of the root M:

$$\frac{d(C_R \times M)}{dt} = \frac{dm_R}{dt} = C_W \times Q - C_{Xy} \times Q. \tag{23}$$

If growth is exponential, and the ratio Q/M (transpiration to plant mass) is constant, the growth by exponential dilution can be considered by a first-order growth rate k_{Growth} (per day). This rate is added to the rate for metabolism, k_M, to give the overall first-order loss rate constant k.

$$\frac{dC_R}{dt} = C_W \times \frac{Q}{M} - C_{Xy} \times \frac{Q}{M} - k \times C_R. \tag{24}$$

If the xylem sap is in equilibrium with the root, the concentration $C_{Xy} = C_R/K_{RW}$. Then,

$$\frac{dC_R}{dt} = C_W \times \frac{Q}{M} - C_R \times K_{RW} \times \frac{Q}{M} - k \times C_R. \tag{25}$$

Setting this to steady state ($dC_R/dt = 0$) gives for the concentration in the root C_R (mg/kg).

$$C_R = \frac{Q}{\frac{Q}{K_{RW}} + kM} C_W. \tag{26}$$

The ratio of the concentration in soil water, C_W, to that in bulk soil, C_{Soil}, is K_{WS}, and the BCF between root and bulk soil is as follows:

$$\text{BCF} = \frac{C_R}{C_{Soil}} = \frac{C_R}{C_W} \times K_{WS} = \frac{Q}{\frac{Q}{K_{RW}} + kM} \times K_{WS}. \tag{27}$$

Furthermore, the concentration in xylem sap when it leaves the root is as follows:

$$C_{Xy} = \frac{C_R}{K_{RW}} = \frac{Q}{\frac{Q}{K_{RW}} + kM} \times \frac{K_{WS}}{K_{RW}} \times C_{Soil}. \tag{28}$$

If there are not any sink processes (such as dilution by growth or metabolism inside the root), the root will reach phase equilibrium with the surrounding soil. The larger the term k_M compared with Q/K_{RW}, the more the result for C_R and C_{Xy} (outflow) differs from phase equilibrium. If k is only due to growth, then k_M is independent of chemical properties, while K_{RW} increases with log K_{OW}, so that Q/K_{RW} decreases. This means that for increasing K_{OW}, concentrations in root are below phase equilibrium, while for polar compounds, the concentration in root is near phase equilibrium, $C_R = K_{RW} \times K_{WS} \times C_S$.

The parameterization of the model has been done for 1-m^2 soil, with 1-kg roots, a transpiration of 1 L/d, and a root growth rate of 0.1 per day. The metabolism rate was set to 0 [19].

3.3.2 Other Approaches to Calculate the Chemical Concentration in Xylem Sap

The "transpiration stream concentration factor" (TSCF) is defined as the concentration ratio between xylem sap, C_{Xy}, and external solution (soil water), C_W.

$$\mathrm{TSCF} = C_{Xy}/C_W. \tag{29}$$

The TSCF is related to the K_{OW} [27] by a bell-shaped (Gaussian) curve:

$$\mathrm{TSCF} = 0.784 \times \exp\left\{\frac{-(\log K_{OW} - 1.78)^2}{2.44}\right\}. \tag{30}$$

For popular trees, a similar relation was found [28]:

$$\mathrm{TSCF} = 0.756 \times \exp\left\{\frac{-(\log K_{OW} - 2.50)^2}{2.58}\right\}. \tag{31}$$

These curves were found from laboratory experiments in hydroponic solution. Based on earlier model calculations, it was suggested that under "real" environmental conditions, that is, for plants growing in soil, the shape of the curve would be different [22]. The author argues that in hydroponic experiments, plants do not develop root hairs. Thus, these roots have a far lower surface area (factor 100) than when growing in soil. The small uptake of polar compounds predicted by the TSCF regressions is probably due to kinetic limitation of the diffusive uptake (resistance of the biomembrane to polar compounds). With higher surface, such a limitation would not occur.

3.3.3 Results of the Root Uptake Model

Figure 3 shows the calculated concentration in soil pore water, C_W (mg/L), for a bulk soil concentration of 1 mg/kg (equation 14: soil pore water concentration). For polar compounds, the concentration is >1 mg/L. This is because most of the polar compounds are present dissolved in soil pore water, but the soil pore water volume

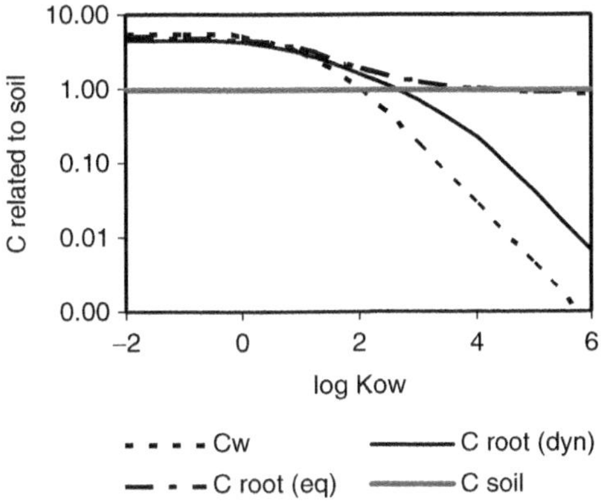

Fig. 3 Calculated concentration in soil pore water [C_w, (14)], equilibrium concentration in root (C_{root} eq), result of the dynamic root uptake model [C_{root} dyn, (27)] for a soil concentration of 1 mg/kg (C_{soil})

is only 0.35 L per liter of soil, and 1-L soil has 1.95-kg weight. 1.95 kg/L:0.35 L/L is 5.57 kg/L, and this is the maximum concentration ratio of soil pore water to bulk soil. At higher log K_{OW} ($\geq$2), the chemical sorbs to the soil organic carbon, and the concentration dissolved in pore water decreases tremendously. The figure also shows the phase equilibrium root to water, named C_{root} (eq) in the legend. This value is $K_{RW} \times K_{WS} \times C_{Soil}$, that is, $K_{RS} \times C_{Soil}$. For polar compounds (log $K_{OW} \leq 2$), the value is >1; this is because the root contains more water than the soil. For more lipophilic compounds, it approaches a value near 1. It is expected that, in reality, only fine roots or the peel of larger roots will approach this equilibrium concentration. The bulk (core) of larger roots will be below equilibrium, due to dilution by growth. This is represented in Fig. 3 by the curve C_{root} (dyn), which is the BCF $\times$ C_{Soil} described earlier (equation 27, dynamic root model). The more lipophilic the compound, the larger is the deviation from phase equilibrium, and the concentration in bulk root is very low. This pattern has been confirmed by experiments [29].

The regressions for the concentration ratio between xylem and soil solution (TSCF) derived by experiments [27, 28] and the model-based TSCF are shown in Fig. 4. Figure 5 shows the concentration ratio between xylem and bulk soil (TSCF $\times$ K_{WS}, equations 28–31: concentration in xylem). The TSCF, related to concentration in solution (Fig. 4), decreases for lipophilic compounds with log K_{OW} > 2 with all methods. But only the two regressions show the decrease for the polar compounds, with low TSCF for log K_{OW} < 0. The model predicts good translocation for the polar compounds. If the concentration in xylem is related to the concentration in bulk soil, the translocation decreases already for compounds with log K_{OW} > 1. The model predicts a very good uptake for the very polar compounds, which do not adsorb to organic carbon and are in soil exclusively present in soil solution. For polar compounds, the concentration in xylem is close to that in soil solution (Kws). This is confirmed by the new study of Dettenmaier et al. [78].

Note that the model does not consider the kinetics of permeability across biomembranes. As will be seen later, the uptake may be kinetically limited for very polar compounds (log $K_{OW} \leq -1$).

Fig. 4 Concentration ratio xylem solution to soil solution (TSCF) calculated with the model equation, compared with the TSCF regressions of Briggs et al. [27] and Burken and Schnoor [28]

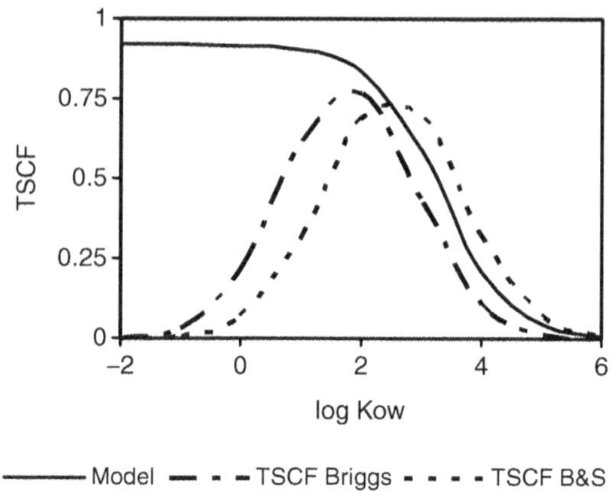

Fig. 5 Concentration ratio xylem solution to bulk soil (TSCF related to soil) calculated with the model (28), compared with the TSCF regressions of Briggs et al. [27] and Burken and Schnoor [28]. K_{WS} is equilibrium concentration in soil solution (14)

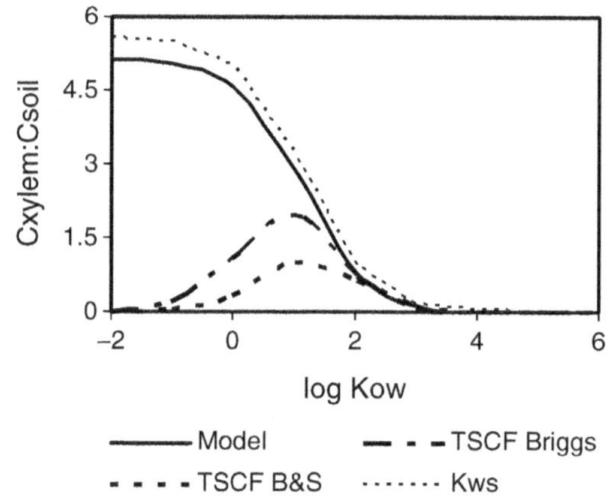

3.4 *Uptake of Neutral Organic Compounds into Shoots*

Uptake of chemical into shoots (stem and leaves) may be from soil or from air.

3.4.1 Transport of Compounds via Xylem from Soil

Uptake from soil is via the transpiration stream Q, and the change of mass is the product of transpiration Q (L/day) and the concentration in xylem, C_{Xy} (mg/L):

$$\frac{dm}{dt} = C_{Xy} \times Q. \tag{32}$$

Hereby, C_{Xy} can either be calculated using the TSCF regression or it is the outcome of the root model [19, 22]. We chose the latter here, while in the early models [17, 18], the TSCF was used. Additional transport processes from soil to shoots may be by gas phase [30, 31] or adsorbed to resuspended soil particles [32], but for the more polar compounds, these routes have less relevance.

3.4.2 Uptake from Air

Uptake from air can occur by dry gaseous deposition, wet deposition (rain), wet particulate deposition (particles in rain), and dry particulate deposition. The mass flux is the product of leaf area A (m^2), deposition velocity g (m/day), and concentration in air C_{Air} (mg/m^3):

$$\frac{dm}{dt} = A \times g \times C_{Air}. \tag{33}$$

A rough estimate for the overall deposition velocity g is a value of 1 mm/s [18] or 86.4 m/day.

3.4.3 Loss to Air

Diffusion is a two-way-process, and chemicals may also volatilize from leaf to air. The process is described similar to deposition:

$$\frac{dm}{dt} = A \times g \times \frac{\rho \times C_L}{K_{LA}}, \tag{34}$$

where K_{LA} is the partition coefficient of leaves to air (in the unit mg/m^3 leaves to mg/m^3 air), and ρ is the density of the leaves (kg/m^3). C_L has the unit mg/kg, and C_{Air} has the unit mg/m^3. If dry gaseous deposition (a diffusive process) is the only process considered, then g has the same value for volatilization as for deposition.

The partition coefficient K_{LA} is derived from the following:

$$K_{LA} = \frac{C_L}{C_A} = \frac{K_{LW}}{K_{AW}}, \tag{35}$$

where K_{LW} is the partition coefficient between leaves and water, and K_{AW} (L/L) is the partition coefficient between air and water (also known as dimensionless Henry's Law constant). Similar to the partition coefficient between roots and water,

$$K_{LW} = \frac{C_L}{C_W} = W + L \times a \times K_{OW}^{b}, \tag{36}$$

where W (L/kg) and L (kg/kg) are water and lipid content of the plant leaf, b for leaves is 0.95, $a = 1/\rho_{Octanol} = 1.22\,L/kg$.

3.4.4 Complete Uptake Model for Shoots

The complete mass balance for the shoots is as follows:

$$\frac{\mathrm{d}m_L}{\mathrm{d}t} = +Q \times C_{\mathrm{Xy}} + C_{\mathrm{Air}} \times g \times A - \frac{C_{\mathrm{L}} \times g \times A \times \rho}{K_{\mathrm{LA}}} - k_{\mathrm{M}} m_{\mathrm{L}}, \quad (37)$$

where the rate constant k_{M} describes metabolism.

Concentrations are derived as before by dividing by plant mass M. Growth is considered by adding a growth rate constant k_{Growth} to the metabolism rate k_{M}; the sum is the overall first-order rate k.

$$\frac{\mathrm{d}C_{\mathrm{L}}}{\mathrm{d}t} = +\frac{Q}{M_{\mathrm{L}}} \times \mathrm{TSCF} \times C_{\mathrm{W}} + \frac{C_{\mathrm{Air}} \times g \times A}{M_{\mathrm{L}}} - k \times C_{\mathrm{L}} - \frac{C_{\mathrm{L}} \times g \times A \times \rho}{K_{\mathrm{LA}} \times M_{\mathrm{L}}}. \quad (38)$$

The equation can be rewritten and gives the inhomogeneous linear differential equation:

$$\frac{\mathrm{d}C_{\mathrm{L}}}{\mathrm{d}t} = b - aC_{\mathrm{L}}, \quad (39)$$

with the standard solution:

$$C_{\mathrm{L}}(t) = C_{\mathrm{L}}(0) \times \mathrm{e}^{-at} + \frac{b}{a}(1 - \mathrm{e}^{-at}), \quad (40)$$

where loss rate a is as follows:

$$a = \frac{A \times g \times \rho}{K_{\mathrm{LA}} \times M_{\mathrm{L}}} + k \quad (41)$$

and source term b is as follows:

$$b = C_{\mathrm{Xy}} \times \frac{Q}{M_{\mathrm{L}}} + C_{\mathrm{Air}} \times g \times \frac{A}{M_{\mathrm{L}}}. \quad (42)$$

The steady-state solution ($t = \infty$) is as follows:

$$C_{\mathrm{L}}(t = \infty) = \frac{b}{a}. \quad (43)$$

In very similar form, the model was implemented in the *Technical Guidance Document* for chemical risk assessment [4]. The default parameterization of the model was taken from the original reference [18] and is shown in Table 1.

Table 1 Parameterization of the leafy vegetable model, normalized to 1 m^2 (data taken from the original publication [18])

Parameter	Symbol	Value	Unit
Shoot mass	M_L	1	Kg
Leaf area	A	5	m^2
Shoot density	ρ	500	kg/m^3
Transpiration	Q	1	L/day
Lipid content	L	0.02	kg/kg
Water content	W	0.8	L/kg
Conductance	G	10^{-3}	m/s
Loss rate (growth)	K	0.035	Per day
Time to harvest	T	60	Day

3.4.5 Results from the Model for Shoots

There are three chemical properties that have the largest influence on the results, and these are as follows:

1. Metabolism rate. In the model, the loss rate k is the sum of metabolism rate and growth rate. Metabolism cannot be predicted from physicochemical properties, and the rate k_M was set to 0.
2. Partition coefficient octanol–water. The K_{OW} determines the freely dissolved concentration in soil, the concentration in root, the concentration in xylem, and thus the translocation upward, and it is also relevant for the sorption to leaves. High K_{OW} values will lead to low uptake from soil, but high uptake from air (if the concentration in air is not 0).
3. Partition coefficient air–water. The K_{AW} has a very high impact on loss to air. Chemicals with a high K_{AW} value will volatilize rapidly from leaves to air.

The impact of K_{AW} and K_{OW} on the calculated concentration in shoots [(37)–(42), shoot model] is shown in Fig. 6. In this simulation, the bulk soil concentration was set to 1 mg/kg, and the metabolism rate and the concentration in air were set to 0. It thus shows the potential for an uptake from soil into shoots, for varying combinations ("chemical space") of K_{AW} and K_{OW}.

As can be seen, compounds with $\log K_{AW} \geq -3$ ($K_{AW} \geq 10^{-3}$ L/L) generally do not tend to accumulate from soil in leaves, independently of the lipophilicity. Also, compounds with a high $\log K_{OW}$($\log K_{OW} \geq 6$) show low accumulation from soil in leaves. Very nonvolatile compounds ($\log K_{AW} \leq 6$) accumulate best when the compound is at the same time polar ($\log K_{OW} \leq 0$). Polar and nonvolatile compounds possess, in fact, a very high bioaccumulation potential from soil in leaves. If the compounds are persistent, a maximum concentration factor of leaves to bulk soil of >100 is predicted.

A very similar result was obtained earlier by simulations with the Fruit Tree model [22]. The predicted accumulation of a persistent neutral organic compound in leaves of an apple tree was highest for polar, nonvolatile compounds. A high

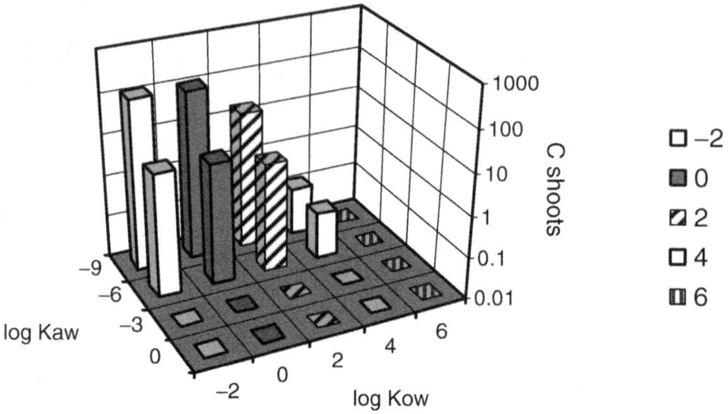

Fig. 6 Predicted concentration of persistent neutral organic compounds in shoots [(40): shoot model] after uptake from soil (C_{soil} is 1 mg/kg) for varying chemical properties (x-axis log K_{OW} and y-axis log K_{AW})

accumulation (up to factor 200 times the bulk soil concentration) was found for very polar (log $K_{OW} < 0$) and very nonvolatile (log $K_{AW} - 9$) compounds. Predicted accumulation in fruits was much lower (maximally factor 8) due to the lower flux of water into fruits [22]. The trend was opposite when uptake was from air. Then, very lipophilic, nonvolatile compounds were predicted to accumulate in fruits and leaves.

3.5 *Summary of Results for Neutral Compounds*

The simulations with the root and shoot model show that in both cases, it is the polar compounds that have the highest potential for an accumulation in plants. The maximum accumulation in leaves is far higher (>100) than in roots (>5). According to the model simulations, neutral polar compounds can thus bioconcentrate up to more than 100 times the concentration in soil, solely due to the translocation with water to leaves and the subsequent evaporation of the water, while the chemical remains if it is persistent and nonvolatile.

4 Electrolytes

Only a few plant uptake models are applicable to ionic or ionizable organic compounds [33]. To mention are the model of Kleier for phloem transport [34], the Satchivi model for pesticide spray application [35, 36], and the Fick–Nernst–Planck model by Trapp [33,37]. Here, the latter is connected with the root and plant uptake model presented in the last section.

Different from neutral organic compounds, dissociation and permeability of neutral and ionic compound across membrane play a key role in determining the uptake of electrolytes in organisms. Thus, pH and pK_a have a large effect on the bioconcentration of (weak) electrolytes. Also, electrical attraction or repulsion is a process not seen for neutral compounds. Thus, several processes have to be added to the plant uptake model(s) described earlier to predict the bioaccumulation potential of ionizable compounds.

4.1 Concentration and Activity

4.1.1 Dissociation

The activity ratio between ionic (index d) and neutral molecule (index n) is calculated by the *Henderson–Hasselbalch* equation [38]:

$$\log \frac{a_\mathrm{d}}{a_\mathrm{n}} = i\,(\mathrm{pH} - \mathrm{p}K_\mathrm{a}), \tag{44}$$

where a is the activity, d is the index for dissociated (synonym ionic), n for neutral, i is +1 for acids and −1 for bases, and pK_a is the negative logarithm ($\log_{10}$) of the dissociation constant. It follows for the fraction of neutral molecules F_n that

$$F_\mathrm{n} = \frac{1}{1 + 10^{i\,(\mathrm{pH}-\mathrm{p}K_\mathrm{a})}}. \tag{45}$$

The fraction of dissociated molecules F_d is $1 - F_n$.

4.1.2 From Concentration to Activity

In nondilute solutions, molecules interact with each other. The chemical potential is reduced due to these interactions, and the *activity a* is lower than the *concentration* C [39]:

$$a = \gamma \times C, \tag{46}$$

where γ is the activity coefficient (−). The activity coefficient of the ion, γ_d, can be calculated with the modified *Debye–Hückel* equation. Several approximations exist, among them the *Davies approximation* [40]:

$$\log \gamma_\mathrm{d} = -A \times z^2 \left(\frac{\sqrt{I}}{1 + \sqrt{I}} - 0.3 \times I \right) \quad \text{for I} \le 0.5\ \mathrm{M}, \tag{47}$$

where A depends on ambient pressure and temperature, $A = 0.5$ for 15°–20° and 1 atm. With an ionic strength I of 0.3 M, γ_d is 0.74 for a monovalent and 0.30 for a bivalent ion.

For neutral compounds, too, the activity differs from the dissolved concentration at high ionic strength. The activity coefficient of the neutral compound, γ_n, is found by the *Setchenov* equation:

$$\gamma_n = 10^{kI}, \tag{48}$$

where k is the Setchenov coefficient that increases with the size of the molecule. For smaller molecules, $k = 0.2$ (taken as default), and γ_n in plant saps with $I = 0.3$ M is 1.23. This means, in water with high ionic strength, the activity of neutral molecules is higher than in salt-free water. This is the reason for the well-known "out-salt" effect of neutral organic chemicals in salt water.

4.1.3 Activity and Adsorption

Not only in pure water, but in all phases, activity is related to the truly dissolved concentration. If, for example, in soil or in plant cells, the molecule is partly adsorbed, the activity can still be calculated. The relation between the activity a (kg/m^3) of free (truly dissolved) molecules and the total concentration C_t (kg/m^3) can generally be defined by fractions f, which consider dissociation, ionic strength, and sorption to lipids, so that.

The total (measurable) concentration C_t of the compound comprises the neutral (n) and dissociated (d) molecules, both kinds can be free in solution or sorbed state:

$$C_t = W \times C_{\text{free,n}} + L \times C_{\text{ads,n}} + W \times C_{\text{free,d}} + L \times C_{\text{ads,d}}, \tag{49}$$

where W and L are the volumetric fractions of water and lipids (L/L).

With $L \times C_{\text{ads}} = K \times C_{\text{free}}$ we can write as follows:

$$C_t = W \times C_{\text{free,n}} + K_n \times C_{\text{free,n}} + W \times C_{\text{free,d}} + K_d \times C_{\text{free,d}}. \tag{50}$$

Furthermore, using Henderson–Hasselbalch's equation and $a = \gamma \times C_{\text{free}}$, we receive for the relation between the activity a_n of the neutral molecules and the total concentration the "activity capacity" f:

$$f_n = \frac{a_n}{C_t} = \frac{1}{\frac{W}{\gamma_n} + \frac{K_n}{\gamma_n} + \left(\frac{10^{i(\text{pH}-\text{p}K_a)}}{\gamma_d}\right) \times (W + K_d)}. \tag{51}$$

The respective relation for the ions, with $a_d = a_n \times 10^{i(\text{pH}-\text{p}K_a)}$ is $f_d = a_d/C_t = f_n 10^{i(\text{pH}-\text{p}K_a)}$.

Note that the symbol K_d here describes the adsorption of the ionic (dissociated) molecules and is not related to the K_d (distribution coefficient) in Sect. 3.2.1, which describes the adsorption to the soil matrix.

4.2 Diffusive Exchange of Electrolytes Across Membranes

4.2.1 Neutral Compounds

The diffusive flux of neutral molecules across membranes, J_n, is described by *Fick's first law of diffusion* [analog to (5)]:

$$J_n = P_n(a_{n,o} - a_{n,i}), \tag{52}$$

where J is the unit net flux of the neutral molecules n from outside (o) to inside (i) of the membrane (kg/m^2/s), P_n is the permeability of the membrane (m/s) for neutral molecules, and a is the activity of the compound (kg/m^3).

4.2.2 Ions

The unit net flux of the dissociated (ionic) molecule species across electrically charged membranes, J_d, is described by the *Nernst–Planck* equation. An analytical solution for constant electrical fields is as follows [41–43]:

$$J_d = P_d \frac{N}{e^N - 1}(a_{d,o} - a_{d,i}e^N), \tag{53}$$

where P_d is the permeability of the membrane (m/s) for dissociated molecules, $N = zEF/RT$, where z is the electric charge (synonym valency, for acids −, for bases +), F is the Faraday constant (96,484.4 C/mol), E is the membrane potential (V), R is the universal gas constant (8.314 J/mol/K), and T is the absolute temperature (K).

4.2.3 Total Compound

The total flux J of the compound across the membrane is the sum of the fluxes of the neutral molecule and the ion, J_n and J_d:

$$J = P_n(a_{n,o} - a_{n,i}) + P_d \frac{N}{e^N - 1}(a_{d,o} - a_{d,i}e^N). \tag{54}$$

4.3 Diffusive Equilibrium for Ionizable Compounds

4.3.1 Definition of Equilibrium

Let us generally define the endpoint of diffusion as the equilibrium between compartments (i.e., the state with the highest entropy). The driving force for diffusive exchange is the *activity* gradient. It follows that diffusive exchange stops when $a_o = a_i$, that is, activities are equal. For neutral compounds, it follows from (52) that

$$J_n = P_n(a_{n,o} - a_{n,i}) = 0 \rightarrow a_{n,o} = a_{n,i},$$

where o denotes outside and i inside the compartment. For concentrations, using $a = f \times C$

$$C_{t,o} f_{n,o} = C_{t,i} f_{n,i}. \tag{55}$$

It follows that the equilibrium partition coefficient $K^{\mathrm{Eq,n}}$ of neutral compounds is the inverse ratio of the activity capacity values f:

$$\frac{C_{t,i}}{C_{t,o}} = \frac{f_{n,o}}{f_{n,i}} = K_{io}^{\mathrm{Eq,n}}. \tag{56}$$

For ions, too, the flux stops when equilibrium is reached. But diffusion is calculated with the *Nernst–Plank* equation, thus [identical to (53)]

$$J_d = P_d \frac{N}{e^N - 1}(a_{d,o} - a_{d,i}e^N).$$

The endpoint of diffusion is reached, with $N = zEF/RT$, when

$$\frac{a_i}{a_o} = e^{\frac{-zEF}{RT}} = K_{io}^{\mathrm{Eq,d}}. \tag{57}$$

This is the well-known *Nernst ratio* [44]. Because of the exponential relation, the theoretical accumulation can be quite high, in particular, for high electrical potentials, and for polyvalent bases ($z \geq +2$). For example, with a field of -120 mV (-0.12 V), K^{Eq} is 115 for $z = +1$, but 13,373 for $z = +2$.

4.3.2 Equilibrium in Binary Systems

Diffusion of both neutral compound and ion is calculated with (54). With $a = f \times C$, the flux into the compartment is as follows:

$$J_i = P_n f_{n,o} C_o + P_d \frac{N}{e^N - 1} f_{d,o} C_o \tag{58}$$

and the flux out is

$$J_o = P_n f_{n,i} C_i + P_d \frac{N}{e^N - 1} f_{d,i} e^N C_i. \tag{59}$$

In equilibrium, influx and outflux are equal, and the equilibrium of electrolytes results in:

$$K_{io}^{\mathrm{Eq}} = \frac{C_i}{C_o} = \frac{f_{n,o} \times P_n + f_{d,o} \times P_d \times \frac{N}{(e^N - 1)}}{f_{n,i} \times P_n + f_{d,i} \times P_d \times e^N \times \frac{N}{(e^N - 1)}}. \tag{60}$$

For dissociating compounds, the equilibrium concentration ratio is a complex function of the fractions in solution, f, the permeabilities for diffusive exchange, P, and of valency z, and charge E (because $N = \frac{zEF}{RT}$). Of course, the concentration ratios derived by the steady-state solution of the diffusive flux equation are not partition coefficients. In their mathematical handling, however, they resemble those.

4.4 *Electrolytes in Soil and Plant*

4.4.1 Electrolytes in Soil

The soil pH varies usually between 4 and 10, with most soils being slightly acidic to neutral (pH 6–7). The ratio between neutral and dissociated compound is calculated as described by the Henderson–Hasselbalch equation.

K_{OC}

The K_{OC} of electrolytes is calculated using special regressions [45], namely:

$$\log K_{OC} = 0.11 \log K_{OW} + 1.54 \quad \text{for the anion,} \tag{61}$$

$$\log K_{OC} = 0.47 \log K_{OW} + 1.95 \quad \text{for the cation,} \tag{62}$$

$$\log K_{OC} = 0.54 \log K_{OW} + 1.11 \quad \text{for the acid, neutral molecule.} \tag{63}$$

$$\log K_{OC} = 0.33 \log K_{OW} + 1.82 \quad \text{for the base, neutral molecule.} \tag{64}$$

As can be seen, cations show the strongest sorption, for a given $\log K_{OW}$.

Concentration in Soil Pore Water

For a liter of dry soil (index Mvol), we had [in Sect. 3.2.1, analog to (10)]

$$C_{\mathrm{Mvol}}/C_{\mathrm{W}} = K_{\mathrm{OC}} \times \mathrm{OC} \times \rho_{\mathrm{dry}}, \tag{65}$$

where concentration in soil matrix, C_{MVol}, and in soil pore water, C_{W}, were in the unit (mg/L). This changes for weak electrolytes to the following:

$$C_{\mathrm{Mvol}}/C_{\mathrm{W}} = (f_{\mathrm{n}} \times K_{\mathrm{OC,n}} + f_{\mathrm{d}} \times K_{\mathrm{OC,d}}) \times \mathrm{OC} \times \rho_{\mathrm{dry}}. \tag{66}$$

Consequently follows the concentration ratio of weak electrolytes, K_{WS}, between water (mg/L water) and wet soil (mg/kg):

$$\frac{C_{\mathrm{W}}}{C_{\mathrm{Soil}}} = K_{\mathrm{WS}} = \frac{\rho_{\mathrm{wet}}}{(f_{\mathrm{n}} \times K_{\mathrm{OC,n}} + f_{\mathrm{d}} \times K_{\mathrm{OC,d}}) \times \mathrm{OC} \times \rho_{\mathrm{dry}} + P_{\mathrm{W}}} \quad (\mathrm{mg/L : mg/kg = kg\ soil/L\ water}). \tag{67}$$

Advective fluxes (namely, the uptake of water by roots) are related to the dissolved concentration C_W. Diffusive exchange, however, will be related to the *activity* a of a compound. We can write for the relation f_n between activity a_n (mg/L) and total concentration in bulk soil C_{soil} (mg/kg) as follows:

$$f_n = \frac{a_n}{C_{Soil}}$$
$$= \frac{\rho_{wet}}{\frac{P_W}{\gamma_n} + K_{OC} \times OC \times \frac{\rho_{dry}}{\gamma_n} + 10^{i(pH-pK_a)} \times \frac{P_W}{\gamma_d} + 10^{i(pH-pK_a)} \times K_{OC} \times OC \times \frac{\rho_{dry}}{\gamma_d}}. \quad (68)$$

4.4.2 Charge and pH of Plant Cells

A typical plant cell consists of various organelles, which are embedded in the cell and often separated by own biomembranes. Plant cells are surrounded by a cell wall and a biomembrane called plasmalemma. The charge at the outside biomembrane of plant cells is between −71 and −174 mV [46]. Inside is the cell sap, cytosol, which has neutral pH (pH 7–7.4). The largest fraction of an adult plant cell is occupied by vacuoles, about 90% of volume. Vacuoles are the "waste bucket" of the plant cells (which have no excretion system) and are acidic (pH 4–5.5). The vacuoles are surrounded by a membrane called tonoplast. The tonoplast is positively charged, relative to the cytosol, with net charges of 0 to +88 mV [46] and +20 mV in average [23]. The ionic strength inside cells varies from 0.2 to 0.6 mol/L [46]; 0.3 mol/L is typical. Specialized plant cells are the phloem cells with high alkalinity (pH 7.4–8.7) [46] and the (dead) xylem vessels. The xylem fluid is acidic; pH values around 5.5 have been measured. The pH in the root zones is lower than in the bulk soil solution; values of pH 5 are common [46].

To summarize, all living cells are electrically charged and are with different charges in different organelles. But only ions react on these charges. Similarly, the pH of different cells and organelles varies. This can impact weak electrolytes strongly, but not has no effect on neutral compounds.

4.4.3 Partition Coefficients for Electrolytes in Plant Cells

Ionizable compounds undergo considerably more processes than the neutral ones. Besides diffusive or advective uptake into cells and xylem with subsequent lipophilic sorption, electrical attraction (or repulsion) at the charged biomembranes and ion trapping due to dissociation have a key impact. Figure 7 describes the processes considered by the model for weak electrolytes in a single plant cell. Each cell is composed by cytosol and vacuoles, and both consist of an aqueous and a lipid fraction and are surrounded by a biomembrane.

The concentration ratio K_{io} between inside and outside of a biomembrane is calculated with the flux equilibrium derived before [(60), equilibrium electrolytes].

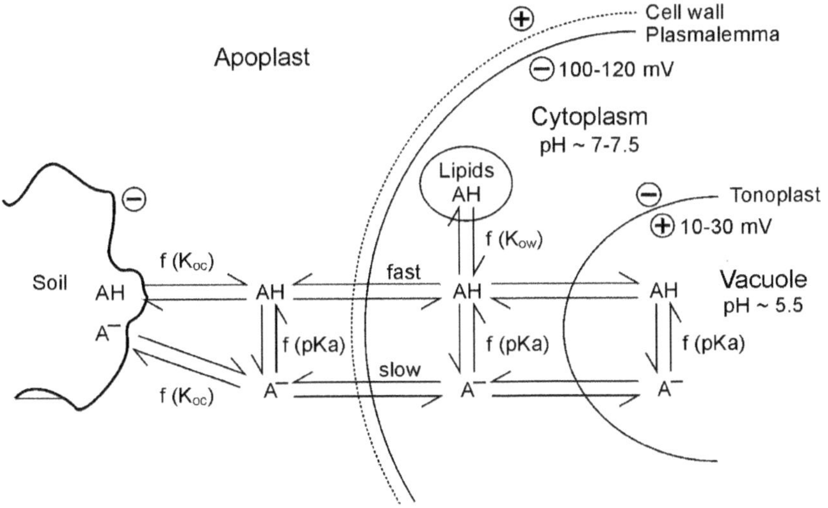

Fig. 7 Molecule species and model processes in the soil–solution–cell system shown for a weak acid. AH is the neutral molecule, A^- is the dissociated anion, and f() means "function of". Cited from [37]

For the partitioning between cytosol and soil, K_{cs}, soil is o outside and cytosol is i inside. For the partition coefficient between vacuole and cytosol, K_{vc}, cytosol is o outside and vacuole is i inside. Similarly, for xylem and phloem, cytosol is the outside compartment.

To derive the overall partition coefficient between xylem (and phloem; and vacuole) and soil solution, the partition coefficient xylem to cytosol is multiplied with the partition coefficient cytosol to soil solution:

$$K_{CS} = \frac{C_C}{C_S}; \quad \text{Cytosol to soil,} \tag{69}$$

$$K_{VS} = K_{VC} \times K_{CS} = \frac{C_V}{C_S}; \quad \text{Vacuole to soil,} \tag{70}$$

$$K_{XyS} = K_{XyC} \times K_{CS} = \frac{C_{Xy}}{C_S}; \quad \text{Xylem to soil,} \tag{71}$$

$$K_{PhS} = K_{PhC} \times K_{CS} = \frac{C_{Ph}}{C_S}; \quad \text{Phloem to soil.} \tag{72}$$

5 Plant Uptake Models for Electrolytes

The cell model was originally developed for single-celled algae [47] and later applied to roots [37], human cells [10], bacteria [11], and fish [48]. Here, it is coupled to the plant uptake model described in Sect. 3 [(22)–(28) and (37)–(42)].

5.1 Root Uptake Model for Electrolytes

The basic differential equation of the root model was (24):

$$\frac{dC_R}{dt} = C_w \times \frac{Q}{M} - C_{Xy} \times \frac{Q}{M} - k \times C_R. \qquad \text{[analog to (24)]}$$

This model assumes that the root tips were in phase equilibrium with the surrounding soil. This means that the concentration of chemical in the solution of the root tips ($C_{Xy,in}$) is equal to the concentration in soil solution (C_W) – or, in other words, the inflowing water has the same concentration as the external solution. For most neutral compounds (except the very polar ones that only slowly cross the biomembranes) this should be true, due to the high root surface and the rapid establishment of equilibrium. The same assumption was done for the Fruit Tree model approach, except only that fine roots and thick roots were separated [22].

The mass balance for roots can also be written as follows:

$$\frac{dC_R}{dt} = C_{Xy,in} \times \frac{Q}{M} - C_{Xy,out} \times \frac{Q}{M} - k \times C_R. \qquad (73)$$

For ionizable compounds, the concentration in xylem inflow is found using the flux-based equilibrium concentration ratios, so that

$$C_{Xy,in} = K_{XyC} K_{CS} C_S = K_{XyS} C_S, \qquad (74)$$

where C is the index for cytosol, S for bulk soil, and Xy for xylem. The concentration at the outflow of the xylem is in flux equilibrium to root. Root cells are actually composed of cytosol and vacuoles, so that

$$K_{XyR} = \frac{C_{Xy,out}}{\frac{C_c V_c + C_v V_v}{V_c + V_v}}. \qquad (75)$$

The new differential equation is as follows:

$$\frac{dC_R}{dt} = K_{XyS} C_S \times \frac{Q}{M} - K_{XyR} C_R \times \frac{Q}{M} - k \times C_R, \qquad (76)$$

which gives, in steady-state, for the bioconcentration in roots

$$\text{BCF} = \frac{C_R}{C_S} = \frac{K_{XyS} \times Q}{K_{XyR} Q + kM} \qquad (77)$$

and for the concentration in the xylem outflux

$$C_{Xy} = C_R \times K_{XyR}\,(\text{mg/L}). \qquad (78)$$

5.2 *Leaf Uptake Model for Electrolytes*

The mass balance of the shoots is + flux from soil ± exchange with air and is reformulated for ionizable compounds.

5.2.1 Transport of Electrolytes from Soil into Shoots

The flux of chemical from soil into shoots via xylem is the concentration in the xylem sap multiplied with the flow of water.

$$\frac{dm}{dt} = C_{Xy} \times Q. \tag{79}$$

C_{Xy} has been calculated in the previous Sect. 5.1.

5.2.2 Deposition from Air

Concentration in air is an input data, and the deposition from air is (as before for the neutral compounds) as follows:

$$\frac{dm}{dt} = A \times g \times C_{Air}. \tag{80}$$

5.2.3 Loss to Air

To describe loss to air of weak electrolytes is tricky: only the neutral fraction of the compound will volatilize (ions do not have a measurable vapor pressure). But how to find this fraction? The mass balance will calculate the total concentration in leaves. But the distribution between neutral and dissociated molecule will change within the cell compartments: cytosol, vacuole, and xylem have different properties. What happens? The xylem brings the solution upward into the leaves. It will require only diffusion through a few cells to reach either the holes of the stomata, or the cuticle. Both are apoplast (outside the living cells). It is the activity of the neutral molecules that drives volatilization, and we relate it to cytosol.

We defined before:

(a) The fraction of neutral compound in the cytosol [analog to (51), activity]:

$$f_{n,C} = a_{n,C}/C_{t,C} = \frac{1}{\frac{W}{\gamma_n} + \frac{K_n}{\gamma_n} + \left(\frac{10^{i\,(pH-pK_a)}}{\gamma_d}\right) \times (W + K_d)},$$

where K is the partition coefficient between lipids and solution, i.e., $K_n = L \times K^b_{OW,n}$ and $K_d = L \times K^b_{OW,d}$.

(b) The activity ratio between vacuoles and cytosol [analog to (60)]:

$$\frac{C_V}{C_C} = \frac{f_{n,C} \times P_n + f_{d,C} \times P_d \times \frac{N}{e^N - 1}}{f_{n,V} \times P_n + f_{d,V} \times P_d \times e^N \times \frac{N}{e^N - 1}} = K_{VC}.$$

(c) The volume ratio cytosol to vacuole, which is an input data (the ratio 1:9 is used).

The total concentration in leaves is as follows:

$$C_L = \frac{C_C V_C + C_V V_V}{V_C + V_V}. \tag{81}$$

With volume ratio $\frac{V_V}{V_C} = R_V$ and concentration ratio $\frac{C_V}{C_C} = K_{CV}$ follows:

$$\frac{C_C}{C_L} = K_{CL} = \frac{V_C + R_V V_C}{V_C + K_{VC} R_V V_C} = \frac{1 + R_V}{1 + K_{VC} R_V}. \tag{82}$$

The activity of the neutral molecule in the cytosol is as follows:

$$a_{n,C} = f_{n,C} \times C_C \quad \text{or} \quad a_{n,C} = f_{n,C} \times K_{CL} C_L. \tag{83}$$

What is missing is to relate this to the activity in air. For neutral compounds, the ratio of activity in air to the activity in water (cell solution) in equilibrium is [analog to (6)] as follows:

$$\frac{a_{n,Air}}{a_{n,W}} = K_{AW}. \qquad \text{[analog to (6)]}$$

Now the loss by diffusion to air is as follows:

$$\frac{dm}{dt} = A \times g \times a_{n,C} \times K_{AW}. \tag{84}$$

5.2.4 Comparison to the Method for Neutral Compounds

Compare this to the previous solution for neutral compounds (34).

$$\frac{dm}{dt} = A \times g \times \frac{\rho \times C_L}{K_{LA}} = A \times g \times \frac{\rho \times K_{AW} \times C_L}{K_{LW}}. \qquad \text{[analog to (34)]}$$

K_{LW} (36) was defined as follows:

$$K_{LW} = \frac{C_L}{C_W}, \quad \text{or} \quad C_W = \frac{C_L}{K_{LW}}, \qquad \text{[analog to (36)]}$$

where C_W is the concentration in the water phase in equilibrium with leaves. Per definition, $a = \gamma C_W$. Thus, if the activity coefficient γ would be neglected (or set to 1), the equation for the neutrals is as follows:

$$\frac{\mathrm{d}m}{\mathrm{d}t} = A \times g \times \frac{\rho \times C_L}{K_{LA}} = A \times g \times \rho \times K_{AW} \times C_W. \tag{85}$$

This equation is identical to the equation derived for the electrolytes (84), except that the unit of C_L here is in mg/kg; there $a_{n,L}$ is in mg/L.

5.2.5 Differential Equation for Concentrations of Weak Electrolytes in Shoots and Fruits

Shoots

The mass balance equation leads again to an inhomogeneous linear differential equation [analog to (39)].

$$\frac{\mathrm{d}C_L}{\mathrm{d}t} = -aC_L + b, \tag{86}$$

where loss rate a is [similar to 41)].

$$a = \frac{A \times g \times f_{n,C} \times K_{CL} \times K_{AW}}{V_L} + k \tag{87}$$

and source term b is [similar to (42)].

$$b = C_{Xy} \times \frac{Q}{V_L} + C_{Air} \times g \times \frac{A}{V_L}. \tag{88}$$

The steady-state solution ($t = \infty$) is [identical to (43)] as follows:

$$C_L(t = \infty) = \frac{b}{a}. \tag{89}$$

Fruits

For fruits, the approach is modified: the xylem flow into fruits is only $0.1 \times Q$ (flow to shoots), plus additional $0.1 \times Q$ for the phloem flow. The concentration in the phloem is calculated by using K_{PhC}, else as was done for xylem. The fruit surface area is 0.05 times the leaf area.

5.3 *Parameterization of the Plant Uptake Model for Electrolytes*

5.3.1 Cells

Cells have been described in Sect. 4.4.2. Tables 2 and 3 show the input data selected for the simulations. The data follows the suggestions in earlier work [18, 19, 37].

5.3.2 Permeabilities

Before a chemical can enter the cytoplasm, it must cross the cell wall and the plasmalemma. The cell wall may be considered as an unstirred aqueous layer with polysaccharides providing additional resistance. A permeability value of 0.25 mm/s was calculated earlier [37]. The cell wall permeability is neglected in this study, because for polar compounds it is very large compared with that of biomembranes and thus does not contribute to the overall permeability.

Table 2 Properties of the cell organelles as input data for the electrolyte plant model [37]

Parameter	Cytosol	Vacuole	Xylem	Phloem	Unit
Volume V	0.1	0.9	0.023	0.023	L
Surface area A	100	100	20	20	m^2
pH	7.0	5.5	5.5	8	(−)
Potential E	−0.12	−0.1	0	−0.12	V (to outside)
Ionic strength I	0.3	0.3	0.01	0.3	M
Water fraction W	0.943	0.943	1	1	L/L
Lipid fraction L	0.02	0.02	0	0	L/L

Table 3 Properties of roots, shoots, and fruits as input data for the electrolyte plant model

Parameter	Symbol	Value	Unit	Reference
Xylem flow to shoots	Q	1	L/day	[19]
Growth rate	K	0.1	Day	Typical value
Sorption parameter	B	0.85[a]	–	[27]
Growth rate roots	K	0.1	Day	[19]
Leaf area	A	5	M^2	[18]
Shoot volume	V_L	1	L	Typical value
Growth rate leaves	K	0.035	Day	[18]
Xylem flow to fruits	Q_F	0.1	L/day	Estimate
Phloem flow to fruits	Q_{Ph}	0.1	L/day	Estimate
Fruit volume	V_F	0.5	L	Estimate
Fruit surface area	A_F	0.25	m^2	Estimate

[a] A compromise of the two values 0.77 and 0.95 of Briggs et al. [27]

The permeability of biomembranes P_n (m/s) for neutral organic compounds is calculated from the compound lipophilicity. From diffusion velocities and membrane thickness, the following equation was derived [33]:

$$\log P_n = \log K_{OW} - 6.7. \tag{90}$$

Similar regressions have been suggested by other authors, based on measurements or model fits:

$$\log P_n = 1.2 \times \log K_{OW} - 5.85 \quad [34], \tag{91}$$

$$\log P_n = 0.33 \times \log K_{OW} - 8.0 \quad [49], \tag{92}$$

$$\log P_n = 1.2 \times \log K_{OW} - 7.5 \quad [50, 51]. \tag{93}$$

For the permeability of the neutral molecule, P_n, the $\log K_{OW}$ of the neutral molecule is used, and for the permeability of the dissociated molecules, P_d, the $\log K_{OW}$ of the ion (which is 3.5 log units lowered). Therefore, the membrane permeability of ions P_d is always factor 3,162 times lower than the corresponding P_n.

6 Simulation Results for Weak Electrolytes

A number of chemical and environmental parameters have impact on transport and accumulation of weak electrolytes in the soil–air–plant system. Neutral compounds "feel" only changes in $\log K_{OW}$ and K_{AW}. The behavior of ionizable compounds is additionally affected strongly by pK_a and soil pH, which are therefore in focus here.

6.1 Equilibrium Constants

First, the equilibrium constants derived for a single cell [Sect. 4.3, (60), (69–72)] are shown for varying compound pK_a and soil pH.

6.1.1 Acids

Figure 8a–d displays the equilibrium concentration ratio of a moderately lipophilic ($\log K_{OW} = 4$) monovalent acid ($z = -1$) between cytosol and bulk soil, vacuole and bulk soil, xylem and bulk soil, and finally phloem and bulk soil. The pH of the soil and the pK_a of the acid were varied.

Uptake of strong acids (pK_a low, anions) is generally low. One reason is that the charge of anions ($z = -1$) leads to electrical repulsion from the biomembrane (potential E at the plasmalemma of cytosol is -120 mV).

There are some exceptions from this rule: If the soil pH is below the pH inside the cell (i.e., soil pH 5), and the acids have a pK_a near soil pH, then the "ion trap"

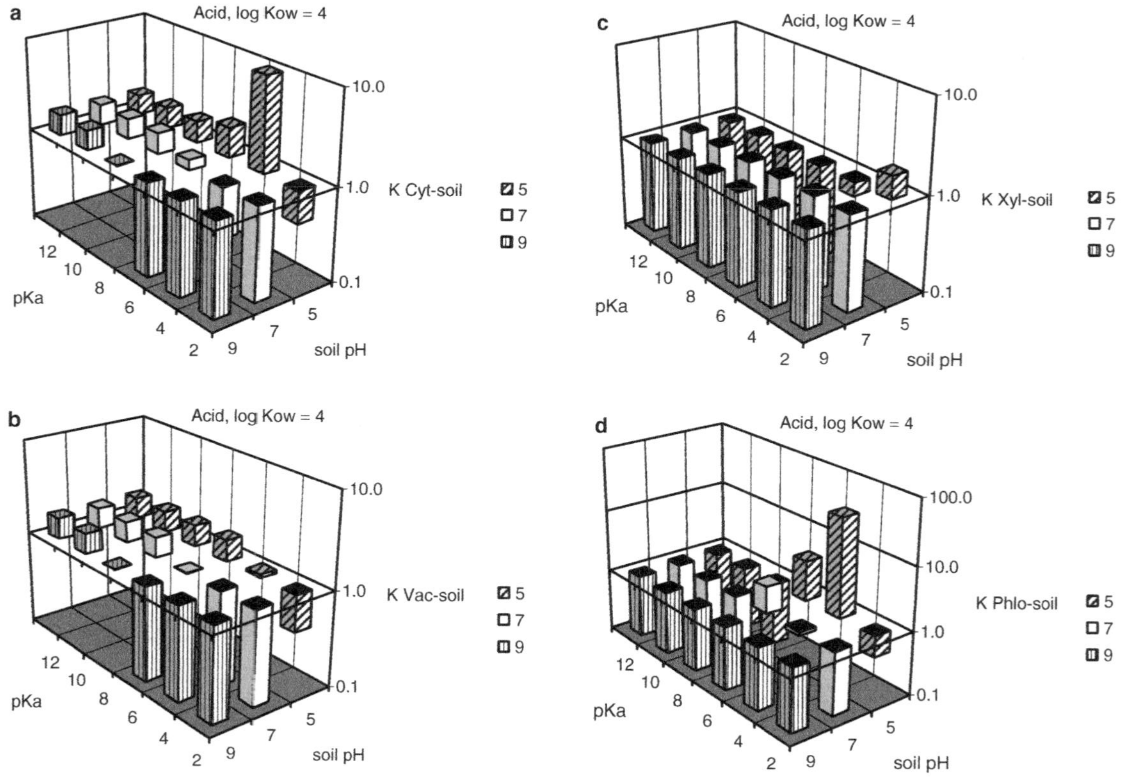

Fig. 8 Equilibrium concentration ratio to soil for an acid, $z = -1$, $\log K_{OW} = 4$, and soil pH of 5, 7, or 9 for (**a**) cytosol, (**b**) vacuole, (**c**) xylem sap, and (**d**) phloem sap

effect occurs: Outside the cell, in soil, the acid is present as neutral molecule, and the neutral molecule diffuses rapidly into the cell. Because the pH inside the cell is above pH outside, the weak acids dissociate. The anion diffuses only slowly across the cell membrane and thus is trapped inside. This affects concentrations in cytosol (Fig. 8a) when soil pH is 5 and acid pK_a is 4. The effect does not occur for vacuoles (Fig. 8b) and xylem (Fig. 8c), which are acidic (pH is 5.5). A very strong ion trap effect is predicted for phloem (Fig. 8d), which has the highest pH (pH 8). Subsequently, acids with pK_a between 4 and 6 show a high concentration ratio of phloem to soil, when the soil has a low pH.

Summarized, anions are not well taken up by the plant cells. This is due to the fact that plants cells have a negative electrical potential at the cell membrane, and this leads to a repulsion of the negatively charged anions. The lowest equilibrium constants show vacuoles and xylem due to the low pH of these compartments. A process that may lead to high accumulation of acidic compounds in cytosol and phloem is the ion trap. It occurs when pH of soil is below pH of cells, and when the pK_a is near the soil pH.

6.1.2 Bases

Impact of Soil pH and pK_a

Figure 9a, b displays the equilibrium concentration ratios to soil (equations 60 and 69) of a moderately lipophilic (log $K_{OW} = 4$) monovalent base ($z = +1$) for cytosol, vacuole, xylem, and phloem. The soil pH is 7 in Fig. 9a and in Fig. 9b.

The pattern is to some extent similar to that of acids, but opposite in trend with pK_a: for soil pH 7, small uptake is predicted, if the pK_a of the base is above soil pH. Then, the bases are dissociating. Strong bases show usually the lower accumulation than very weak bases, which are present as neutral molecules. However, several exceptions from this rule can be seen.

At pH 7, there is a small but noticeable accumulation of moderate bases with pK_a 6–8 in the xylem due to the ion trap. Furthermore, there is an accumulation of cations (p$K_a \geq 12$) in cytosol, vacuoles, and phloem due to electrical attraction by the negatively charged organelles.

The ion trap effect is stronger when the pH of soil is above the pH of the cell organelles, i.e., in alkaline soils with pH 9 (Fig. 9b). Then, a strong ion trap effect is predicted for xylem and vacuoles, leading to equilibrium concentration ratios of 7 (xylem) and 12 (vacuoles). With soil pH 9, the equilibrium concentration ratio is low for phloem, except for cations (p$K_a \geq 12$); it is high for xylem and pK_a 6–10 (peak at pK_a 8), but decreases strongly with lower pK_a; it is high also for vacuoles and pK_a around 8, but with only small decrease at lower pK_a, and it varies only little for cytosol, with K^{eq} between 1.1 and 2.2.

Phloem and xylem sap have no lipids; this explains the low concentration ratio to soil for the base with log $K_{OW} = 4$. The peak in the xylem is due to the ion trap, which occurs at high soil pH (pH 9).

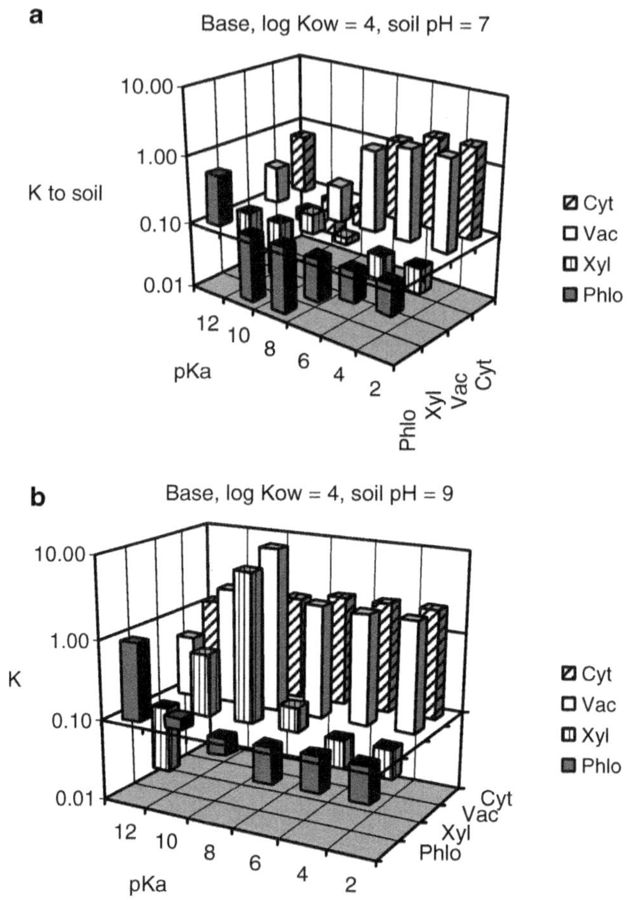

Fig. 9 Equilibrium concentration ratio to soil (**a**) of pH 7 and (**b**) of pH 9 derived from the steady-state flux equation for a base, $z = +1$, $\log K_{OW} = 4$, and cytosol Cyt, vacuole Vac, xylem sap Xyl, and phloem sap Phlo

Impact of Log K_{OW} on the Kinetics of Uptake

The model predicts throughout a good accumulation for cations ($pK_a \geq 12$) if the compartment has a negative potential (cytosol, vacuole, and phloem). This is due to the electrical attraction of the positively charged cation ($z = +1$). But beware! This is the result of the steady-state solution of the flux equation. If a dynamic mathematical solution is applied instead, polar cations show a very slow uptake into cells [10]. This is different from the predictions for the neutral compounds: For those, the model predicts that polar compounds are generally better taken up into roots and translocated to shoots and leaves (see Figs. 3–6).

This is not necessarily the case for the ionizable compounds, as is documented by the calculations done with the dynamic solution shown in Fig. 10. The polar cation

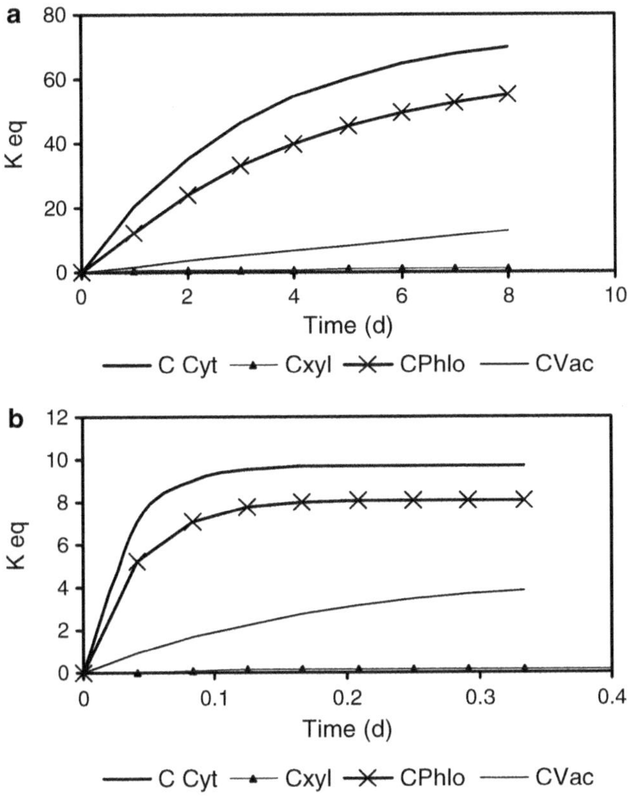

Fig. 10 Calculated concentration ratio versus time for soil pH 9 and a cation (pK_a = 12) with log K_{OW} of the neutral molecule of (**a**) 0 and (**b**) 2. This corresponds to a log K_{OW} of the cation of (**a**) −3.5 and (**b**) −1.5

(Fig. 10a) with log K_{OW} = 0 (of the neutral molecule) is taken up very slowly into the cell. After 8 days, steady-state equilibrium is not yet reached. In particular low is the uptake into the vascular system, that is, the xylem. The reason is that the log K_{OW} of the dissociated base is 3.5 log units lower than that of the neutral molecule, thus, the apparent log K_{OW} (also known as log D) is −3.5. With these low values, the membrane permeability is accordingly very low. For the transport of cations in real plants this means that the equilibrium will not be reached for polar cations. If the cation is more lipophilic (log K_{OW} neutral = 2 in Fig. 10b), the uptake is more rapid. Still, the uptake into vacuoles and xylem (which requires first uptake from soil into cytosol and then from cytosol into vacuoles and xylem) may be kinetically limited. For cations with log K_{OW} = 4 (of the neutral molecule; this corresponds to a log D of 0.5), the permeability across cell membranes is sufficiently rapid, and kinetic limitation is not expected. But the reader is reminded that for polar cations (and anions), there are kinetic limitations, and uptake and, in particular, the translocation may be overestimated by steady state.

For very lipophilic cations (log $K_{OW} \gg 2$), uptake decreases again, similar as it was for neutral compounds (Figs. 3–6). For very weak bases (pK_a low, always neutral), the optimum uptake is as it was for neutral compounds, for log K_{OW} – 1 to 2, while for stronger bases (p$K_a \gg 6$), uptake is probably optimal at log K_{OW} of the neutral molecule between 2.5 and 5.5, which corresponds to a log D (apparent log K_{OW}) between −1 and 2.

6.1.3 Potential for Uptake and Accumulation

The equilibrium concentration ratios [(60), (69–72)] indicate a potential for uptake. Acids generally are taken up less than neutral molecules, due to electrical repulsion and slow transfer across membranes. One process can lead to very high uptake and translocation, and that is the ion trap effect. It occurs when the pH in soil is below the pH inside the cells, and when the pK_a of the acid is close to soil pH. The highest pH inside plants is in phloem, therefore is the ion trap of acids, in particular, strong for phloem.

Cations are attracted by the electrical potential of living cells. Therefore, strong bases have generally a higher potential for uptake than strong acids. As pointed out, this process may be kinetically limited for polar bases. On the other hand, bases sorb stronger than acids to soil organic carbon and to negatively charged clay particles. This reduces their bioavailability and uptake. An ion trap of bases occurs when soil pH is above cell pH and is strongest for the acidic compartments vacuole and xylem. An optimum uptake is expected for moderately lipophilic bases with pK_a 8–10. Their equilibrium concentration ratios xylem to soil solution (TSCF) are even above 1 L/L (up to 10 L/L).

6.2 Predicted Concentrations in Soil and Roots

So far we had only calculated the flux-based equilibrium constants. The following section shows the results obtained using the dynamic root model for electrolytes (Sect. 5.1).

6.2.1 Acids in Roots and Xylem

Figure 11a–c shows calculated steady-state concentrations in root and xylem sap (= TSCF related to soil) for an acid (log $K_{OW} = 4$). At soil pH 5 (Fig. 11a), the ion trap occurs and leads to elevated uptake of acids with pK_a below 6. Interestingly, the concentration in xylem outflow is higher for the strong acids (low pK_a) than for the very weak acids (high pK_a) – this is due to the lower adsorption of anions to soil organic carbon and plant tissue. The model predicts a relatively good translocation to shoots of the dissociated acids with log K_{OW} (of the neutral molecule) of 4. If the

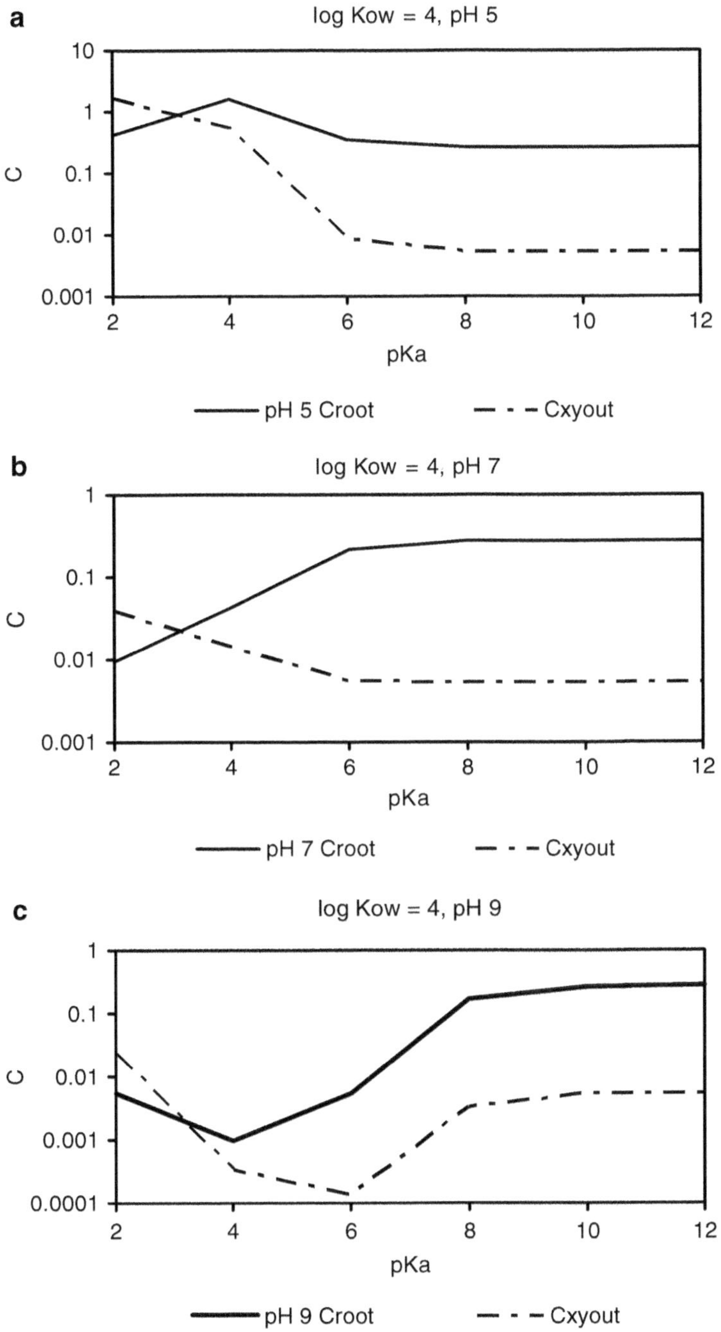

Fig. 11 C_{root} and concentration in xylem outflow from root (C_{Xyout}) for an acid ($\log K_{\mathrm{OW}} = 4$, $z = -1$) and (**a**) soil pH 5, (**b**) soil pH 7, and (**c**) soil pH 9. Varying pK_{a}. Concentration of acid in soil is 1 mg/kg

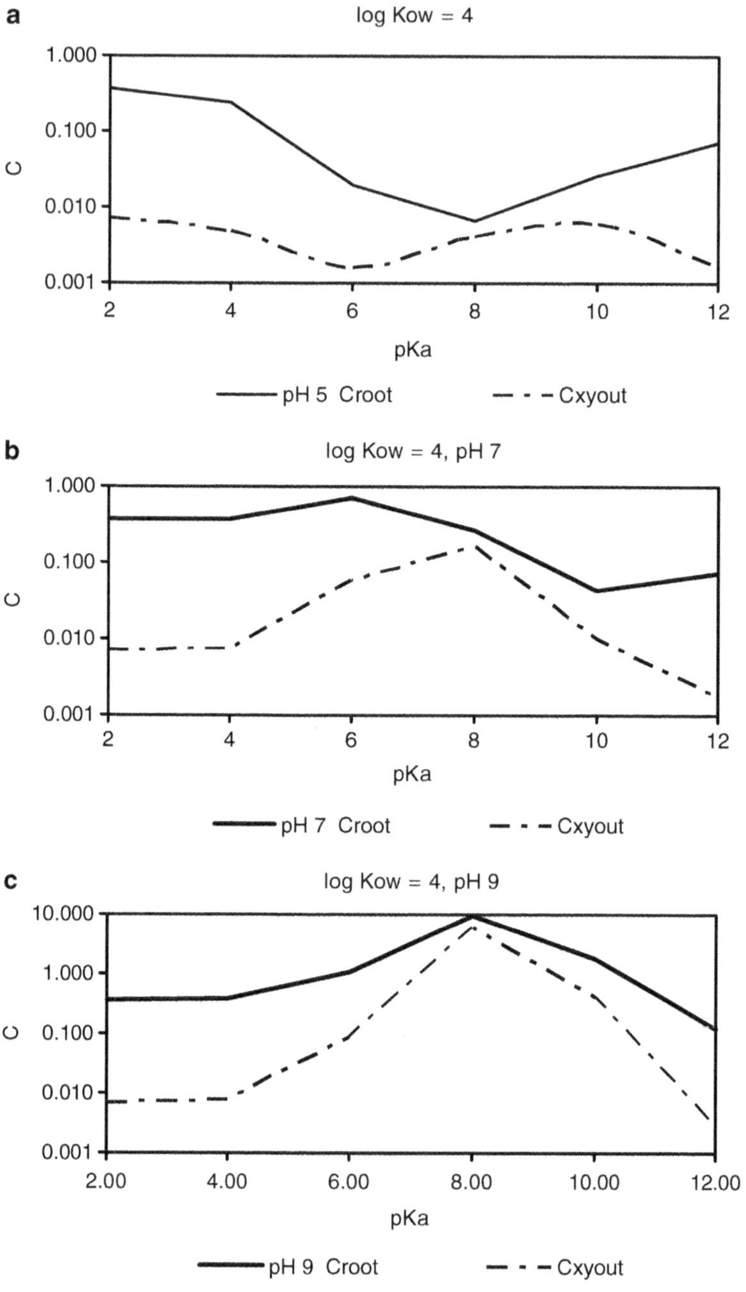

Fig. 12 Concentration in root (C_{root}) and concentration in xylem outflow from root (C_{Xyout}) for a base ($z = +1$, $\log K_{\mathrm{OW}} = 4$) and for (**a**) soil pH 5, (**b**) soil pH 7, and (**c**) soil pH 9. Varying pK_{a}. Concentration of base in soil is 1 mg/kg

acid does not dissociate (p$K_a \geq 6$), the translocation is very low: then, the concentration in xylem out of the roots, C_{xyout}, is below 0.01 mg/L for soil concentrations of 1 mg/kg.

At soil pH 7 (Fig. 11b), the model predicts that very weak acids (high pK_a, neutral) do better accumulate in root cells than strong acids. For the translocation to shoots (concentration in xylem out of the roots, C_{xyout}), the opposite trend is predicted. But note that in any case, translocation is small.

For soil pH 9 (Fig. 11c), uptake is very small, but somewhat higher when the acid is protonated, that is, neutral. Concentrations in xylem sap are very low, which means only small translocation upward to stem and leaves. An opposite ion trap occurs that keeps moderate acids (pK_a 4–6) out of the xylem.

6.2.2 Bases in Roots and Xylem

Figure 12a–c shows calculated steady-state concentrations in root and xylem sap (= TSCF related to soil) for a moderately lipophilic base ($\log K_{OW} = 4$) with varying pK_a and for soil with (a) pH 5, (b) pH 7, and (c) pH 9.

At low soil pH (pH 5, Fig. 12a), uptake into roots is best for very weak bases, which are neutral in soil and in root cells. An opposite ion trap effect keeps moderate bases (pK_a 6–10) out of the root cells – because soil pH is below the pH of the cell organelles. Stronger bases (cations) show higher concentrations than very weak bases (neutral at usual pH), because cations are electrically attracted by the negative electrical potential of root cells. Concentrations in xylem sap at the outflow from the roots (C_{xyout}) are generally very low, which means low translocation of very weak bases from soils with low pH to shoots.

At soil pH 7 (Fig. 12b), the ion trap turns around, and moderate bases (pK_a 6–10) accumulate to some extent in the xylem. This results in a more effective translocation to shoots.

With alkaline soils (pH is 9, Fig. 12c), the ion trap gets very strong and leads to effective accumulation of moderate bases (pK_a 6–10) both in root cells (and here mainly in vacuoles) and in xylem sap.

6.3 Predicted Concentrations in Shoots and Phloem

This section shows the results obtained with the model for electrolytes' uptake into shoots and fruits [(86)–(89)].

6.3.1 Acids

Figure 13 shows the calculated concentration in shoots and fruits for an acid with $\log K_{OW}$ of 4, varying pK_a and soil pH of 5, 7, and 9. In this scenario, air

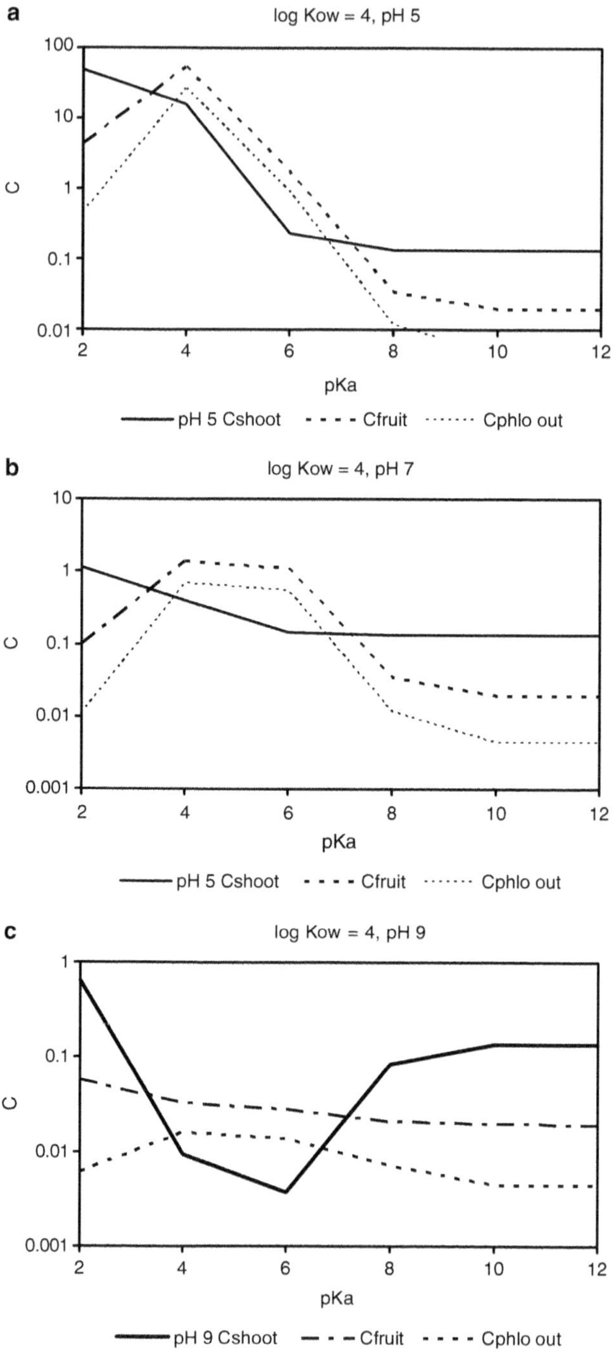

Fig. 13 C_{shoot}, C_{fruit}, and C_{Phloem} (out of root) for of an acid (log $K_{\mathrm{OW}} = 4$) with (**a**) soil pH 5, (**b**) soil pH 7, and (**c**) soil pH 9. Concentration in soil 1 mg/kg

concentration was set to 0, and uptake is exclusively from soil. But loss to air may occur, because the Henry's law constant is moderate (10^{-5} L/L). In the model, translocation into leaves (shoots) is only in the xylem (no phloem flux), while flow of water into fruits is $^1/_2$:$^1/_2$ via xylem and phloem.

For soil with low pH (pH 5, Fig. 13a), high concentrations in shoots are predicted for the strong acids (low pK_a). A major reason may be that anions do not volatilize from leaves (ions have no measurable vapor pressure), while the very weak acids remain neutral in cytosol and do escape to air. The K_{AW} was taken as 10^{-5} L/L. Even though not high, volatilization from leaves is a major fate process. Another reason is that anions are more polar than the corresponding neutral molecules and are less retained in soil and roots. The calculations were made for a compound with a log K_{OW} of 4 (of the neutral molecule species). At this log K_{OW}, the translocation of neutral compounds is already reduced, due to strong sorption to soil and to roots (see Figs. 3–6). The dissociated molecule has an apparent log K_{OW} (log D) of 0.5, which is in the optimum region for translocation (see Fig. 5). Additionally, with acidic soils, a strong ion trap effect occurs, which traps acids in the phloem. The maximum is for acids with pK_a at 4. Subsequently, these acids are translocated in the phloem to fruits. The predicted concentration in fruits is even higher than in shoots (stem and leaves). Very weak acids, with p$K_a > 7$, are predominantly neutral in soil and plant and are too lipophilic and too volatile for effective translocation in xylem and phloem and accumulation in leaves and fruits.

In neutral soils, pH 7 (Fig. 13b), translocation of acids to shoots is much less. The ion trap is weak, and acids with pK_a 4–6 are less efficiently translocated upward.

In alkaline soils, pH 9 (Fig. 13c), the opposite ion trap occurs, which keeps the acids with pK_a 4–6 out of phloem and fruits. Transport to aerial plant parts is generally low.

6.3.2 Bases

Figure 14a, b shows the calculated concentration in shoots, phloem, and fruits for a moderately lipophilic base (log $K_{OW} = 4$) of varying pK_a. The equilibrium constants, shown in the last section, dictate the pattern of accumulation. An ion trap occurs for bases with pK_a above 4 and below 10, with maximum effect at pK_a 8. This leads to notable accumulation in shoots and fruits. Different from acids, predicted concentrations in fruits are without exception lower than in shoots. This is because weak bases do not tend to accumulate in phloem, but in xylem.

The effect of the ion trap is stronger in alkaline soils, that is, with soil pH 9 (Fig. 14b). Accumulation of weak bases due to the ion trap effect is predicted for shoots with maximum above concentration factor 100 and for shoots above factor 10.

On the contrary, concentrations are in particularly low for uptake of bases from acidic soils (pH 5, not shown).

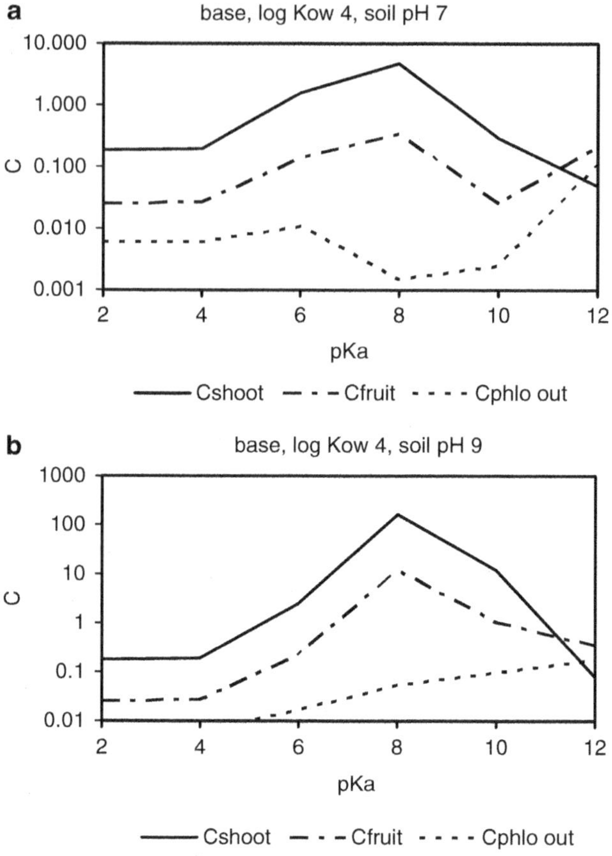

Fig. 14 C_{shoot}, C_{fruit}, and C_{Phloem} (out of root) for a base (log $K_{\mathrm{OW}} = 4$) and uptake from (**a**) soil pH 7 and (**b**) soil pH 9

6.3.3 Volatilization

The recent simulations were all done for a partition coefficient air to water (Henry's law constant, K_{AW}) of 10^{-5} L/L. Model calculations for concentrations in root, xylem, and phloem (at the outflow from the roots) do not vary with K_{AW} and air concentrations. But concentrations in shoots and fruits are strongly affected by a variation of these parameters. Figure 15 shows a simulation for soil pH 7 and an acid or base with log $K_{\mathrm{OW}} = 2$ and varying pK_{a}. As can be seen, with this high K_{AW}, concentrations in shoots and fruits are low for the very weak acids. Very weak acids (p$K_{\mathrm{a}} > 8$) are neutral at environmental pH and volatilize rapidly, when K_{AW} is high. Strong acids (low pK_{a}), on the other hand, dissociate in cytosol (pH 7) and do not escape to air. Therefore, concentrations in leaves are much higher. Moderate acids (pK_{a} 4–8) are kept out of the plant by the opposite ion trap.

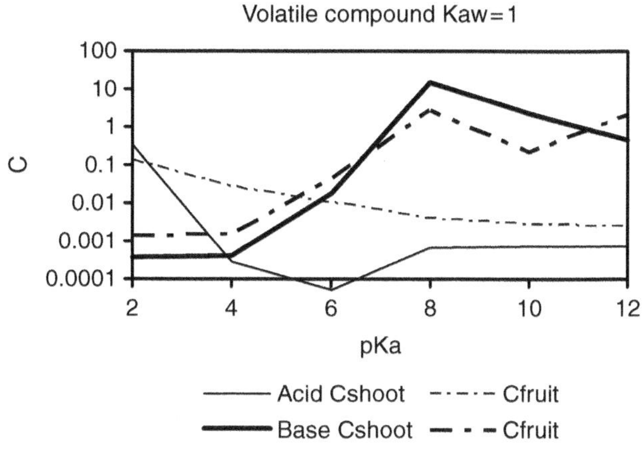

Fig. 15 Calculated concentration in shoots (C_{shoot}) and fruits (C_{fruit}) for uptake from soil with pH 7 and for a volatile acid or base ($\log K_{OW} = 2,\ K_{AW} = 1$)

For bases, the pattern is in part symmetrical: Strong bases (i.e., with high pK_a) do not tend to volatilize and may accumulate therefore in leaves and fruits. Very weak bases, on the other hand, escape to air. Moderate bases accumulate due to the ion trap (if the soil pH is not too low) in the xylem and show the highest translocation to leaves.

6.4 Comments to the Model Predictions

6.4.1 Limitations

The results shown in Sect. 6 were derived by model calculations, and no attempt was done so far to validate any of these predictions. How realistic and reliable can the predictions be? Before we use the model outcome for conclusions, the limitations and shortcomings should be brought into mind.

Constant soil concentration. For all simulations, concentrations in soil were assumed to be constant. This will not be the case in real environments. Compounds, once released to the environment, will degrade and dissipate. Degradation half-lives may be less than 1 day. But even with virtually persistent compounds, the root zone will quickly deplete from the compound when RCF and TSCF are high, because the compound is taken up by roots. Then, even though the uptake is potentially high, uptake decreases in the long run due to depletion of the root zone.

Inhomogeneity of environment. Similar, homogeneous environmental data were assumed. Soil is a very inhomogeneous medium, and relevant properties, such as humidity, organic carbon content, and in particular pH, may vary much even on small spatial scales. The same is true for plant data and micrometerological conditions.

Ageing. Ageing is the effect leading to reduced availability of compounds in soil with time. Experiments have shown that compounds in freshly spiked soil samples are available to plants and are taken up. From samples where the compounds have been applied years before, the uptake is much less. Often, K_{OC} values have been determined with freshly spiked soils and may therefore be too low.

Degradation in plants. Throughout the simulations, no degradation in plants was assumed. This is an unrealistic assumption, because plants are a living environment, and xenobiotics can react with enzymes [52, 53]. If metabolism of compounds occurs, the concentrations will decrease. This was shown for the example of cyanide [54]. Additionally, degradation in the root zone is usually much higher than in the neighborhood soil, due to a higher number of microbes in the root zone [55, 56]. Even inside plants, microbial degradation may play a role due to endophytic bacteria [57].

Steady state. The equilibrium concentration ratios as well as the concentrations in plants were calculated for the steady state. In reality, uptake may be limited kinetically, as was pointed out for polar bases. Thus, the predicted concentrations are an upper limit.

Uptake from air. In all simulations, uptake from air was 0, that is, the concentration of the chemical in air was set to 0. This was because the focus of this study is on bioaccumulation from soil. But, in particular, the lipophilic compounds, for which low uptake from soil into shoots and fruits is predicted, tend to accumulate to very high levels from air [58]. A low BCF shoot to soil does, therefore, not mean that the shoot is protected from contamination: it is only protected from contamination from soil via translocation. There may be uptake from air, and there may also be uptake from soil by other processes, such as particle resuspension and deposition on leaves [22, 32, 58].

Plant data. The data entered into the equations do not represent a special plant, but are rather typical values for cells and plants. The scenario is generic. For special crops, quite different parameters may be necessary, or even another model.

Feedback. Compounds taken up into plants may change the plant. For example, toxic effects may lead to reduced growth and transpiration [59]. But there may be also more subtle effects: for example, uptake of basic compounds into lysosomes (in plants: vacuoles) was accompanied by an increase in pH in these organelles [60]. A change of pH gradient in turn changes the uptake of ionizable compounds.

Additional diffusive resistances. For the calculation of the flux equilibrium, only membrane permeability was considered. Several other resistances to diffusive exchange may play a role. To mention are the cell wall resistance and boundary layer resistances. These two are independent of the lipophilicity of the molecule, while the biomembrane permeability increases with log K_{OW}. It may thus be expected that additional resistances play a role for lipophilic compounds.

Membrane permeability. The ion trap effect has been identified as one of the more relevant processes for accumulation of ionizable compounds in plants. This ion trap depends on the difference in membrane permeability for neutral and ionic compound. The membrane permeabilities were not measured, but estimated from log K_{OW}. The log K_{OW} of the ion was generally estimated to be 3.5 log-units lower. This value is not a constant. For example, delocalized cations show less difference in log K_{OW} and in their membrane permeability. Thus, the ion trap is weaker or nonexistent.

pKa shift at membranes. From the *Debye–Hückel theory* follows that the ionic strength I of solutions has impact on the pK_a. At $I = 0.3$M, the apparent pK_a of monovalent bases is 0.22 units lower, of acids higher. For bivalent bases and acids, the change is 0.62 units. This effect was not implemented in the model, but could be used to correct input data. It also means that the optimum ranges for pK_a can shift toward the extremes, that is, be higher than predicted for the bases and lower for acids.

The careful reader may find more flaws and shortcomings – for example, phloem flow only from roots to fruits, while it goes mainly from leaves to fruits or roots and others.

6.4.2 Potential for Uptake

Taken these shortcomings together it becomes clear that the main function of the model can not be precise predictions of real scenarios. The "news" drawn from the model simulations is the strong impact of pH and pK_a. This is not always taken into account. When experiments with ionizable compounds are done, the pH of the soil or solution should be controlled and reported.

Tables 4 and 5 categorize the impact of compound pK_a and soil pH on accumulation in roots, shoots, and leaves. For strong acids (anions), strongest accumulation occurs in shoots, and it is not dependent on soil pH, because strong acids are of anionic nature at all pH values. Moderate acids are best taken up at low pH and show

Table 4 Categorization of the BCF of acids of varying strength at varying soil pH

Compound	Soil pH 5	Neutral soil pH	Soil pH 9
Strong acids (p$K_a \leq 1$)	− Roots	− Roots	− Roots
	++ Shoots	++ Shoots	++ Shoots
	O Fruits	O Fruits	O Fruits
Moderate to weak acids ($2 \leq$ p$K_a \leq 6$)	O Roots	− Roots	−− Roots
	+ Shoots	O Shoots	−− Shoots
	++ Fruits	+ Fruits	− Fruits
Very weak acids (p$K_a \geq 8$)	O Roots	O Roots	O Roots
	O Shoots	O Shoots	O Shoots
	− Fruits	− Fruits	− Fruits

− − BCF $<$ 0.01; −BCF 0.01–0.1; O BCF 0.1–1; + BCF 1–10; ++BCF $>$ 10

Table 5 Categorization of the BCF of bases of varying strength at varying soil pH

Compound	Soil pH 5	Neutral soil pH	Soil pH 9
Strong base (p$K_a \geq 12$)	− Roots	− Roots	O Roots
	− Shoots	− Shoots	− Shoots
	O Fruits	O Fruits	O Fruits
Moderate to weak bases ($6 \leq pK_a \leq 10$)	−− Roots	− Roots	−− Roots
	O Shoots	+ Shoots	++ Shoots
	−− Fruits	O Fruits	+ Fruits
Very weak bases (p$K_a < 6$)	O Roots	O Roots	O Roots
	O Shoots	O Shoots	O Shoots
	− Fruits	− Fruits	− Fruits

where −− BCF < 0.01; −BCF 0.01–0.1; O BCF 0.1–1; + BCF 1–10; ++ BCF > 10

a particular high potential for accumulation in fruits, due to phloem transport. Their uptake at high soil pH is very low. Very weak acids also do not show a pH-dependent uptake, because they are neutral at all pH values.

Strong bases are less well taken up into plants (Table 5), because cations sorb much stronger to soil than anions. Moderate bases are taken up best at high soil pH, and most into shoots, due to good xylem translocation. Very weak bases, which are neutral at all pH, show the same behavior as the very weak acids.

For the moderate and weak acids, and for all types of bases, lipophilicity impacts uptake: increasing log K_{OW} generally reduces bioaccumulation in plants, due to increasing sorption to soil and increasing effect of growth dilution.

7 Comparison to Experimental Findings

7.1 Introduction

Several studies have been undertaken to test the models for neutral compounds [32, 61, 62]. Also, all publications that are the basics for the model for neutral compounds contain validation data [17, 19]. No such study is available for electrolytic compounds. This is the reason why the focus in this section is on findings for electrolytes. The comparison to experimental data should not be intermixed with a true validation study: No attempt was made to simulate the conditions during the experiments. Thus, only the tendency of the results can be compared. The question is whether the conclusions drawn in the previous section are mirrored in experimental results.

Chemical properties were taken from the original reference, or if not given these were estimated using the software ACD/Labs® (ACD/I-Lab, version 6.01, Toronto, Canada).

7.2 Polar Compounds

7.2.1 Uptake of Sulfolane and DIPA into Plants

Uptake of sulfolane (SU, tetrahydrothiophene 1,1-dioxide, $C_4H_8O_2S$) and di-isopropanolamine (DIPA) into wetland vegetation was measured in field and greenhouse studies [63]. SU is a neutral polar compound with log K_{OW} of −0.77 and K_{AW} of 2.14×10^{-4}. DIPA is a moderate to weak base with pK_a 9.14, log K_{OW} is −0.86, and K_{AW} is 3×10^{-9}. The pH of the spiked growth solution was not given, but was probably near neutral conditions.

The largest percentage of DIPA was recovered from the root tissue whereas for sulfolane the largest percentage was associated with the foliar portion of the plant. The measured RCF values for roots were 1–7 L/kg for DIPA and 0.3–1.4 L/kg for SU (related to initial concentration in solution). The BCF values for shoots were 1–2.5 for DIPA, but up to 160 for SU (related to initial concentration in solution). TSCF values were 0.1–0.9 for SU, but <0.01–0.02 for DIPA. This is in accordance with the model, which would (for neutral pH) predict low RCF, but high TSCF and BCF in shoots for a polar, nonvolatile neutral compound like SU. For a base with pK_a near 10, such as DIPA, low xylem translocation would be predicted for uptake from neutral solution, but some accumulation in cytosol and vacuole.

Another experiment with the chemical sulfolane was done for apples (Doucette W, personal communication). The concentration ratio fruit to soil was 2.8, while the Fruit Tree model would give a value of 8.2. For leaves, a BCF of 652 was found, and the model gave a BCF of 286 [22]. To the knowledge of the author, this is the highest BCF plant to soil that was ever measured, and it confirms well the model prediction that of all neutral compounds, polar nonvolatile chemicals are best translocated to leaves.

7.3 Uptake of Acids into Plants

7.3.1 Uptake of Sulphadimethoxine into Crop and Weed Plants

The compound sulphadimethoxine is a relatively polar weak acid (log K_{OW} 1.63, pK_a 6.70). Its properties are close to the optimum for uptake into cytosol and phloem.

Uptake and degradation by several plants was investigated by Migliore et al. [64–66]. In a study using agricultural crops in hydroponics, sulphadimethoxine was applied at an initial concentration of 300 mg/L in the nutrient solution. After 27 days, the contents were *Panicum miliaceum* root 2,070 mg/kg; leaf 110 mg/kg; *Pisum sativum* root 178 mg/kg, leaf 60 mg/kg; *Zea mays* roots 269 mg/kg; and leaf 12.5 mg/kg [64]. Severe toxic effects were observed. Similar tests were made with *Hordeum distichum* [65]. Bioaccumulation was higher from growth medium than from soil, but still high for the latter, with somewhat decreasing BAF with increasing

OC of the soils. In particular high accumulation of sulphadimethoxine was found for the common weeds *Amaranthus retroflexus*, *Plantago major*, and *Rumex acetosella* [66], with BAF plant to growth medium of 7.67, 20, and 3.3, respectively. The model predicts good uptake for polar compounds (Figs. 3–6) and also for moderate to weak acids (Figs. 8 and 11, Table 4), however, best from acidic soil or solution, and the optimum is at lower pK_a.

7.3.2 Uptake of Sulfonylurea Herbicides into Algae

The uptake of the sufonylurea herbicides metsulfuron-methyl, chlorsulfuron, and triasulfuron into *Chlorella* algae at varying pH was studied by Fahl et al. [67]. Metsulfuron-methyl ($\log K_{OW} = 1.70$) is a moderate acid with pK_a at 3.3; chlorsulfuron ($\log K_{OW} = 2.15$) has a pK_a at 3.6, and triasulfuron ($\log K_{OW} = 2.36$) at 4.5. These values are close to the optimum for uptake into cytosol (if solution pH is low, Fig. 8). The BCF of the single-celled algae and uptake from solution were directly calculated from the flux equilibrium in cytosol and vacuole, that is, as follows:

$$\mathrm{BCF} = \frac{K_{CW} V_C + K_{VC} \times K_{CW} \times V_V}{V_C + V_V}. \tag{94}$$

Table 6 shows the comparison to measured BCF. As predicted by the model, the BCF of the moderate acids increases with decreasing pH (6 to 5), even though at some measurements the BCF is highest at pH 5.3. The absolute BCF is predicted with maximum deviation of factor 4.

7.3.3 Uptake of the Herbicide Analogue 3,5-D by Curly Waterweed

The uptake of the herbicide analogue 3,5-dichlorophenoxy acetic acid (3,5-D) by curly waterweed (*Lagarosiphon major*) at varying pH was studied by De Carvalho et al. [68]. 3,5-D was used instead of 2,4-D (which is a growth-regulating herbicide) due to the very high toxicity of the latter. The physicochemical properties of both compounds are similar; 3,5-D has a $\log K_{OW}$ of 3.01 and a pK_a of 2.98 (estimated by ACD). The BCF was again directly predicted by the concentration ratios cytosol and vacuole to external solution. The measured BCF was highest (70) at the lowest pH

Table 6 Measured and predicted BCF of sulfonylurea herbicides in *Chlorella* algae at varying pH (measured data from Fahl et al. [67])

	pH = 6.0		pH = 5.3		pH = 5.0	
Compounds	Exp	Calc	Exp	Calc	Exp	Calc
Metsulfuron-methyl	1	4	14	17	17	33
Chlorsulfuron	8	4	36	17	21	22
Triasulfuron	9	5	54	32	30	39

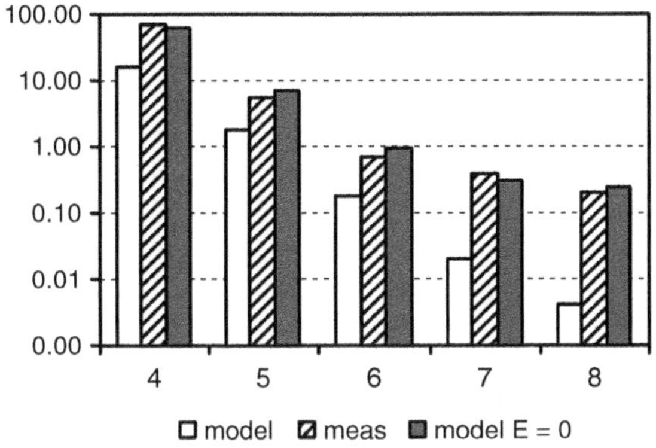

Fig. 16 Predicted ("model") and measured BCF of 3,5-D for the aquatic plant *Lagarosiphon* sp. [68]; "model $E = 0$" is the model prediction with membrane potential E set to 0

(pH 4) and decreased almost logarithmically with increasing pH (Fig. 16). The same trend was predicted by the model, even though predicted BCF values were generally too low, in particular for high pH. One reason for the difference at high pH may be that compounds partition partly into the "apparent free space" of the roots, which occupies 25–40% of the volume [69]. Because of this, the BCF can not fall below 0.25. This is currently not considered in the model, but can be added if required [37]. The BCF at low pH can be fitted by calibrating one or more model parameters (cell pH, volume ratio cytosol:vacuole, electrical potential of cytosol, permeabilities). As example, E (cytosol potential) was set to 0; this improves the prediction (Fig. 16). Probably, the default potential of -120 mV is too high in this case.

7.3.4 Uptake of Acids into Barley Shoots

Briggs et al. [70] determined the RCF and TSCF of organic acids with varying log K_{OW} between 0.06 and 4.51 and rather constant pK_a values between 2.9 and 3.1 for pH 4 and 7 in external solution [70]. At pH 7, RCF values ranged from 0.5 to 4.5. At pH 4, RCF ranged from 2.6 to 72. The largest increase (greater than factor 100) was for compounds with intermediate log K_{OW} (1.64 and 2.25). The TSCF at pH 7 was low, 0.04 to 0.05, while at pH 4, values between 0.1 and 4.2 were found. This tendency is in very good agreement with the model predictions for acids (Figs. 8 and 11). Briggs et al., too, used the ion trap process to interpret their results.

7.3.5 Phloem Translocation in *Ricinus communis*

Rigitano et al. [71] studied the phloem translocation of acids in *Ricinus communis*. Phenoxy acetic acids with pK_a around 3 and log K_{OW} between 1 and 3 were readily taken up and translocated in phloem to the apical leaf and upper stem of *Ricinus* plants, when injected into mature leaves. For moderate acids, the model predicts good phloem translocation.

7.3.6 Application of Herbicides

A number of systemic herbicides are used in crop protection. "Systemic" means that those compounds are translocated within plants. This can be via xylem or via phloem [72]. Phloem mobility is reached when dissociable acid groups are introduced into the molecule [34, 50, 51, 72]. Typical systemic herbicides are the acids 2,4-D (pK_a near 3), sulfonyl ureas (pK_a 3–5), the amphoter glyphosate (multiple pK_a values, pK_a acid at 2.2 and 6.2), fluroxypyr (pK_a 2.22), mecoprop (pK_a 3.19), imazethapyr (pK_a 2.90), or sethoxydim (pK_a 3.82), to mention some. For all these acids with low to moderate log K_{OW} and low pK_a, the model would predict low adsorption to tissue and good translocation in phloem to phloem sink (growing parts, fruits, roots). The opposite works, too: Typical contact insecticides, which are completely immobile in plants, are such compounds like DDT (neutral, log K_{OW} 6.17), lindane (neutral, log K_{OW} 3.8), or HCB (neutral, pK_a 5.8). For those, the model predicts null translocation and strong adsorption to plant tissue.

7.4 Uptake of Bases into Plants

7.4.1 Dodemorph and Tridemorph

Chamberlain et al. [73] studied the uptake into roots and translocation to shoots of the two bases dodemorph (log K_{OW} 5.2, pK_a 7.8) and tridemorph (log K_{OW} 5.6, pK_a 7.4) from solution at pH between 5 and 8. At pH 5, RCF values of dodemorph were <10 and of tridemorph about 20. With increasing pH, RCF increased to final values of 49 for dodemorph and 183 for tridemorph at pH 8. Similar, the TSCF increased from <1 for both bases at pH 5–24 for dodemorph and somewhat below 10 for tridemorph at pH 8. Both compounds have the pK_a in the optimum range for xylem translocation predicted by the model. To the knowledge of the author, the TSCF of dodemorph in this experiment was the highest TSCF ever determined. The experiments confirm the model predictions of good xylem translocation of moderate to weak bases at high external pH (Table 5, Figs. 12 and 14).

7.4.2 Substituted Phenethylamines and other Bases

Inoue et al. [74] measured uptake and translocation of structurally similar amines with varying log K_{OW} (−0.04 to 4.67), but rather constant pK_a (9.3–9.8), and of some bases with varying pK_a (4–8) into barley shoots from solution at pH 5–8. At pH 5, the RCF values ranged from 1.2 to 43.2 for the bases with pK_a > 9. Interestingly, the highest RCF was for the base with the lowest log K_{OW}. For the three other compounds with lower pK_a, the RCF values were from 0.7 to 2.9 at pH 4 of the solution and at 1.2–2.6 at pH 5. The RCF values of all bases increased with pH to values between 7.1 and 153 at pH 8. The lowest value at pH 8 (RCF 7.1) was for the compound with the low pK_a (5.03) and log K_{OW} 1.95. The highest value (RCF 153) was for the compound with pK_a 9.48 and log K_{OW} 1.26. One compound had a log K_{OW} of 4.67 (pK_a 9.28), but the RCF reached only 89.6. It can be concluded that the RCF was more depending on external pH and pK_a of the compound than on lipophilicity. This was also predicted by the model, even though the optimum pK_a determined in the experiments (between 9 and 12) seems to be higher than that found by the model (near 8, compare Figs. 9 and 14). Probably, this is due to the pK_a shift at high ionic strength at biomembranes.

The TSCF values did also increase with external pH. At pH 4 and 5, all TSCF values were below 1. At pH 8, TSCF ranged from 0.09 for a very polar compound (log K_{OW} – 0.04) to 23.2 for a compound with log K_{OW} 2.33 and pK_a 8.03. Only small increase was found for the weakest base (log K_{OW} 1.95, pK_a 5.03, TSCF from 0.10 at pH 4 to 0.17 at pH 8) and the most polar base (log K_{OW} – 0.04, pK_a 9.54, TSCF from 0.03 to 0.09).

This increase of RCF and TSCF of bases with increasing external pH is a convincing proof for the predictive power of the model approach. But the slow uptake and small increase for the most polar base also indicates that kinetic limitations occur in real life, which have not been considered accurately by the steady-state model approach. The model could also be solved for the dynamic case, and this deviation is a warning to consider kinetics in future developments. A kinetic solution was provided by Trapp [37] for TSCF, and that solution was quite accurate in predicting the TSCF of this polar base.

7.4.3 Veterinary Drugs, Field Study

Boxall et al. [75] measured the uptake of veterinary medicines from soil into plants. A light loamy sand soil (OC 0.4%, pH 6.3) was spiked with ten test chemicals. After harvest, only three compounds were detected in lettuce leaves, namely florfenicol, levamisole, and trimethoprim. Florfenicol is neutral and polar, with log K_{OW} at –0.04. Levamisol is also nondissociating and has a log K_{OW} of 1.84. Trimethoprim is a base with log K_{OW} at 0.91 and pK_a at 7.2. In carrot roots, diazinon, enrofloxacin, florfenicol, and trimethoprim were detected. Except for the last one, concentrations in peels were higher than in cores. Diazinon has a log K_{OW} of 3.81 and is practically neutral (pK_a base < 2.5); enrofloxazin is an amphoter with log K_{OW} at 0.7 and pK_a values at 6.0 and 8.8. All compounds detected in roots or shoots possess properties that favor accumulation, according to the model simulations.

Compounds that were not found at detectable levels in plant material were amoxicillin, oxytetracycline, phenylbutazone, sulfadizine, and tylosin. Amoxicillin degraded very quickly in soil (half-time < 1 day). Oxytetracycline is an amphoter with log K_{OW} – 0.9 and pK_a acid at 7.3 and pK_a base at 9.1 and 3.3 (outside the optimum range). Phenylbutazone is neutral (log K_{OW} 3.16). Tylosin is a base with log K_{OW} of 1.63 and pK_a of 7.1. Sulfazine is an acid with log K_{OW} at –0.09 at pK_a 6.48. These latter two compounds thus have properties that are near the optimum range for uptake into xylem or phloem. Some results concerning the uptake of the similar sulfonamide antibiotics (sulphadimethoxine) into certain plant species have been presented in the previous section.

It can be concluded that there was some agreement with the model predictions. However, the measured concentration factors were often lower than those predicted by the model. This was probably due to higher adsorption than estimated by the regression (eq. 62).

8 Outlook and Conclusions

8.1 Outlook

An issue in current research is the environmental fate of pharmaceuticals [75]. Table 7 contains a list of 12 high-volume drugs. Obviously, several of them have properties that make them candidates for uptake into plants. Clarity might bring field tests. For many of these compounds, the uptake depends critically on the soil pH.

8.2 When is Bioaccumulation from Soil into Plants Relevant at All?

Bioaccumulation as a problem for environment and human health is long known. However, accumulation from soil into plants has rarely been an issue, except on

Table 7 Properties of some high-volume pharmaceuticals (chemical properties estimated using ACD)

Drug	Use	Log K_{OW}	pK_a acid	pK_a base	Comment
Metformin	Antidiabetic	−2.31	None	13.86	Polar cation
Metoprolol	Antihypertensive	1.79	13.9	9.17	Base
Atenolol	Antihypertensive	0.1	13.88	9.16	Polar base
Verapamil	Antihypertensive	3.90	None	8.97	Base
Hydrochlorothiazide	Diuretics	−0.07		9.57; 8.95	Polar bivalent base
Furosemide	Diuretics	3.00	3.04	9.79	Amphotere
Simvastatin	Lipid lowering agent	4.42	13.49	None	Lipophilic, neutral
Isosorbide dinitrate	Antiangina pectoris drug	0.9	None	None	Neutral
Amitriptyline	Antipsychotic	4.92	None	9.18	Lipophilic base
Piracetam	Antidementia	−1.55	None	None	Polar neutral
Ibuprofen	Pain killer	3.7	5.2	None	Acid
Aspirin	Pain killer	1.2	3.5	None	Acid

polluted sites. Chemicals may reach humans via the diet, drinking water, inhalation of air, and other routes (e.g., direct contact, dermal uptake). Even though many chemicals are ubiquitous, the one or the other uptake pathway usually dominates, depending on the properties of the compound and the concentrations in the various media. When will bioaccumulation from soil into plants play a role?

The concentration ratio between air and water is described by the partition coefficient between air and water, K_{AW}(L/L), also known as dimensionless Henry's law constant. The concentration ratio between lipids and water is described by the K_{OW}(L/L). Subsequently, the ratio between lipids and air is described by the octanol–air partition coefficient, K_{OA}(L/L), which can approximately be calculated from the ratio K_{OW}/K_{AW}.

Compounds with a K_{AW} above 10^{-4} L/L would – in chemical equilibrium – be taken up in higher amounts via inhalation than by drinking water. Similarly, compounds with a log K_{OA} below 10^5 would in chemical equilibrium be taken up more rapidly with air than with lipids in the diet. Thus, unless a significant deviation from chemical equilibrium occurs, volatile compounds ($K_{AW} > 10^{-4}$, $K_{OA} < 10^5$) will preferably be inhaled, and diet is of no or low importance.

The uptake of persistent organic pollutants (POPs), such as polychlorinated dibenzo-*p*-dioxins and -furans PCDD/F, by humans is mainly via food ingestion [76]. Compounds with high log K_{OW} tend to accumulate in the food chain and will be mainly found in products from the aquatic food chain [7], such as fish, but also in milk and meat. Uptake into these animals may be from food crops, but the primary contamination source for the food crops is air, not soil [3, 22].

Thus, to be of relevance for human exposure via food crops, compounds in soil need to be polar and nonvolatile. These are neutral organic compounds with low vapor pressure (better: low K_{AW}) and low lipophilicity (low K_{OW}). But in particular, electrolytes fit into these scheme: due to dissociation, water solubility greatly increases, that is, lipophilicity decreases, and the vapor pressure of ions is near 0.

But it seems that high bioaccumulation of chemicals from soil into plants is a rare and not very likely process, at least it has not been described very often in the scientific literature.

There may be several reasons. First of all, chemicals have to be present in the soil. Deposition from air has been observed for many compounds, for example, polychlorinated biphenyls PCB or PCDD/F [58]. But it is only likely for persistent, semivolatile lipophilic compounds. These compounds will accumulate by the air–leaf or air–fruit pathway [7, 22]. An accumulation air–soil–fruit, however, will not occur (compare Sect. 3.4).

A possible source of contamination is irrigation with polluted water. This would indeed bring polar compounds into soil. Also, application of manure has been identified as a source of soil contamination, including veterinary drugs [75].

Another prerequisite for accumulation in plants is slow dissipation from the root zone. This excludes compounds that are rapidly degraded by microorganisms in the rhizosphere. Polar compounds have a higher bioavailability and thus are less likely to be persistent.

And indeed there is a high potential for bioaccumulation from soil in leaves and fruits for polar persistent compounds, as was proven by the work of Doucette et al. [63] for sulfolane. There is also a potential for organic acids to accumulate in fruits. This reminds to pesticides, in particular herbicides, which have frequently been detected in relatively high levels by food monitoring actions [77]. This may be due to spray application to leaf and fruit surfaces, but it may also be following application to soil with subsequent translocation.

There seems also – indicated by model predictions and laboratory experiments – a potential for accumulation of moderate bases of intermediate lipophilicity, and many pharmaceuticals are moderate bases. However, to the knowledge of the author, no study has yet shown that high bioaccumulation of moderate bases in plants occurs in field. This may be due to a lack of searching for those compounds and due to analytical difficulties. But it may also be because this bioaccumulation of moderate bases is only high at unusually high soil pH.

Finally, it shall be reminded that the difference between plant cells and animal cells is not very big in many aspects. Like plant cells, animal cells are surrounded by biomembranes with negative electrical potential, and some organelles (lysosomes) are acidic, and some (mitochondria) are alkaline. This means that the same processes identified in this study, ion trap and electrical attraction, which may lead to bioaccumulation of nonlipophilic compounds, may also lead to an accumulation of compounds in fish or other organisms.

Acknowledgments This work received financial support from the European Union Sixth Framework Programme of Research, Thematic Priority 6 (Global change and ecosystems), project 2-FUN (contract no. GOCE-CT-2007-036976) and project OSIRIS (contract no. GOCE-ET-2007-037017). Many thanks to our editor, James Devillers, for his initiative and engagement. Thanks to Hans-Christian Lützhøft, Wenjing Fu, Charlotte N. Legind and Antonio Franco for support.

References

1. Arnot JA, Gobas FAPC (2006) A review of bioconcentration factor (BCF) and bioaccumulation factor (BAF) assessments for organic chemicals in aquatic organisms. Environ Rev 14: 257–297
2. Travis CC, Hattemer-Frey HA, Arms AA (1988) Relationship between dietary intake of organic chemicals and their concentrations in human adipose tissue and breast milk. Arch Environ Contam Toxicol 17: 473–478
3. Czub G, McLachlan MS (2004) Bioaccumulation potential of persistent organic chemicals in humans. Environ Sci Technol 38: 2406–2412
4. EC (1996/2003). European Commission. Technical Guidance Document on Risk Assessment in support of Commission Directive 93/67/EEC on Risk Assessment for new notified substances, Commission Regulation (EC) No. 1488/94 on Risk Assessment for existing substances, and Directive 98/8/EC of the European Parliament and of the Council concerning the placing of biocidal products on the market; European Communities: Italy (1st edn 1996; 2nd edn. 2003). http://ecb.jrc.it/documents/TECHNICAL_GUIDANCE_DOCUMENT/EDITION_2/
5. US EPA United States Environmental Protection Agency (1976) Toxic substances control act. Washington, DC

6. Mackay D, Fraser A (2000) Bioaccumulation of persistent organic chemicals: Mechanisms and models. Environ Pollut 110: 375–391
7. Kelly BC, Ikonomou MG, Blair JD, Morin AE, Gobas FAPC (2007) Food web-specific biomagnification of persistent organic pollutants. Science 317: 236–239
8. Mackay D (1979) Finding fugacity feasible. Environ Sci Technol 13: 1218–1223
9. De Duve C, De Barsy T, Poole B, Trouet A, Tulkens P, Van Hoof F (1974) Commentary. Lysosomotropic agents. Biochem Pharmacol 23: 2495–2531
10. Trapp S, Horobin RW (2005) A predictive model for the selective accumulation of chemicals in tumor cells. Eur Biophys J 34: 959–966
11. Zarfl C, Matthies M, Klasmeier J (2008) A mechanistical model for the uptake of sulfonamides by bacteria. Chemosphere 70: 753–760
12. Hansch C, Leo A, Hoekman D (1995) Exploring QSAR: Fundamentals and applications in chemistry and biology. American Chemical Society, Washington DC
13. Versluijs CW, Koops R, Kreule P, Waitz MFW (1998) The accumulation of soil contaminants in crops, location-specific calculation based on the CSOIL module, Part 1: Evaluation and suggestion for model development. RIVM Report No 711 701 008, Bilthoven, NL
14. Trapp S, Matthies M, Scheunert I, Topp EM (1990) Modeling the bioconcentration of organic chemicals in plants. Environ Sci Technol 24: 1246–1251
15. Hung H, Mackay D (1997) A novel and simple model for the uptake of organic chemicals from soil. Chemosphere 35: 959–977
16. Paterson S, Mackay D, Mc Farlane C (1994) A model of organic chemical uptake by plants from soil and the atmosphere. Environ Sci Technol 28: 2259–2266
17. Trapp S, Mc Farlane JC, Matthies M (1994) Model for uptake of xenobiotics into plants – Validation with bromacil experiments. Environ Toxicol Chem 13: 413–422
18. Trapp S, Matthies M (1995) Generic one-compartment model for uptake of organic chemicals by foliar vegetation. Environ Sci Technol 29: 2333–2338; Erratum 30: 360
19. Trapp S (2002) Dynamic root uptake model for neutral lipophilic organics. Environ Toxicol Chem 21: 203–206
20. Trapp S, Cammarano A, Capri E, Reichenberg F, Mayer P (2007) Diffusion of PAH in potato and carrot slices and application for a potato model. Environ Sci Technol 41: 3103–3108
21. Trapp S, Rasmussen D, Samsøe-Petersen L (2003) Fruit Tree model for uptake of organic compounds from coil. SAR QSAR Environ Res 14: 17–26
22. Trapp S (2007) Fruit tree model for uptake of organic compounds from soil and air. SAR QSAR Environ Res 18: 367–387
23. Larcher W (1995) Physiological plant ecology, 3rd edn. Springer, Berlin
24. Reichenberg F, Mayer P (2006) Two complementary sides of bioavailability: Accessibility and chemical activity of organic contaminants. Environ Toxicol Chem 25: 1239–1245
25. Karickhoff SW (1981) Semi-empirical estimation of sorption of hydrophobic pollutants on natural sediments and soils. Chemosphere 10: 833–846
26. Schwarzenbach R, Westall J (1981) Transport of nonpolar organic compounds from surface water to groundwater: Laboratory sorption studies. Environ Sci Technol 15: 1360–1367
27. Briggs GG, Bromilow RH, Evans AA (1982) Relationship between lipophilicity and root uptake and translocation of non-ionised chemicals by barley. Pestic Sci 13: 495–504
28. Burken JG, Schnoor JL (1998) Predictive relationships for uptake of organic contaminants by hybrid poplar trees. Environ Sci Technol 32: 3379–3385
29. Wang M-J, Jones K (1994) Uptake of chlorobenzenes by carrots from spiked and sewage-sludge amended soil. Environ Sci Technol 28: 1260–1267
30. Trapp S, Matthies M (1997) Modeling volatilization of PCDD/F from soil and uptake into vegetation. Environ Sci Technol 31: 71–74
31. Jones KC, DuarteDavidson R (1997) Transfers of airborne PCDD/Fs to bulk deposition collectors and herbage. Environ Sci Technol 31: 2937–2943
32. Trapp S, Schwartz S (2000) Proposals to overcome limitations in the EU chemical risk assessment scheme. Chemosphere 41: 965–971
33. Trapp S (2004) Plant uptake and transport models for neutral and ionic chemicals. Environ Sci Pollut Res 11: 33–39

34. Kleier DA (1988) Phloem mobility of xenobiotics. Plant Physiol 86: 803–810
35. Satchivi NM, Stoller EW, Wax LM, Briskin DP (2000) A nonlinear dynamic simulation model for xenobiotic transport and whole plant allocation following foliar application I. Conceptual foundation for model development. Pest Biochem Physiol 68: 67–84
36. Satchivi NM, Stoller EW, Wax LM, Briskin DP (2000) A nonlinear dynamic simulation model for xenobiotic transport and whole plant allocation following foliar application. II. Model validation. Pest Biochem Physiol 68: 85–95
37. Trapp S (2000) Modeling uptake into roots and subsequent translocation of neutral and ionisable organic compounds. Pest Manag Sci 56: 767–778
38. Henderson LJ (1908) Concerning the relationship between the strength of acids and their capacity to preserve neutrality. J Physiol 21: 173–179
39. Debye P, Hückel E (1923) Zur Theorie der Elektrolyte. I. Gefrierpunktserniedrigung und verwandte Erscheinungen (The theory of electrolytes. I. Lowering of freezing point and related phenomena). Physikalische Zeitschrift 24: 185–206
40. Appelo CAJ, Postma D (1999) Geochemistry and groundwater pollution, 4th edn. Balkema, Rotterdam, NL
41. Goldman DE (1943) Potential, impedance and rectification in membranes. J Gen Physiol 127: 37–60
42. Hodgkin AL, Katz B (1949) The effect of sodium ions on the electrical activity of the giant axon of the squid. J Physiol 108: 37–77
43. Briggs GE, Hope AB, Robertson RN (1961) Electrolytes and plant cells. In: James WO (ed) Botanical monographs, Vol. 1. Blackwell Scientific, Oxford
44. Nernst W (1889) Die elektrische Wirksamkeit der Jonen. Z Physik Chem 4: 129–181
45. Franco A, Trapp S (2008) Estimation of the soil water partition coefficient normalized to organic carbon for ionizable organic chemicals. Environ Toxicol Chem 27(10): 1995–2004
46. Schopfer P, Brennicke A (1999) Pflanzenphysiologie, 5th edn. Springer, Berlin
47. Raven JA (1975) Transport of indolacetic acid in plant cells in relation to pH and electrical potential gradients, and its significance for polar IAA transport. New Phytol 74:163–172
48. Fu W, Franco A, Trapp S (2009) Methods for estimating the bioconcentration factor (BCF) of ionizable organic chemicals. Environ Toxicol Chem 28(7), in print
49. Hsu FC, Kleier DA (1990) Phloem mobility of xenobiotics. III. Sensitivity of unified model to plant parameters and application to patented chemical hybridizing agents. Weed Sci 38: 315–323
50. Grayson BT, Kleier DA (1990) Phloem mobility of xenobiotics. IV. Modelling of pesticide movement in plants. Pestic Sci 30: 67–79
51. Hsu FC, Kleier DA (1996) Phloem mobility of xenobiotics. VIII. A short review. J Exp Botany 47: 1265–1271
52. Sandermann H (1994) Higher plant metabolism of xenobiotics: The ‘green liver’ concept. Pharmacogenetics 4: 225–241
53. Komossa D, Langebartels C, Sandermann H Jr (1995) Metabolic processes for organic chemicals in plants. In: Trapp S, Mc Farlane C (eds), ‘Plant contamination – modeling and simulation of organic chemical processes’. Lewis Publisher, Boca Raton, FL
54. Larsen M, Ucisik A, Trapp S (2005) Uptake, metabolism, accumulation and toxicity of cyanide in willow trees. Environ Sci Technol 39: 2135–2142
55. Trapp S, Karlson U (2001) Aspects of phytoremediation of organic compounds. J Soils Sed 1: 37–43
56. Rein A, Fernqvist MM, Mayer P, Trapp S, Bittens M, Karlson U (2007) Degradation of PCB congeners by bacterial strains – Determination of kinetic parameters and modelling of rhizoremediation. Appl Microbiol Biotechnol 77: 469–481
57. Barac T, Taghavi S, Borremans B, Provoost A, Oeyen L, Colpaert JV, Vangronsveld J, van der Lelie D (2005) Engineered endophytic bacteria improve phytoremediation of water-soluble, volatile, organic pollutants. Nat Biotechnol 22: 583–588
58. McLachlan MS (1999) Framework for the interpretation of measurements of SOCs in plants. Environ Sci Technol 33: 1799–1804

59. Trapp S, Zambrano KC, Kusk KO, Karlson U (2000) A phytotoxicity test using transpiration of willows. Arch Environ Contam Toxicol 39: 154–160
60. Ishizaki J, Yokogawa K, Ichimura F, Ohkuma S (2000) Uptake of imipramine in rat liver lysosomes in vitro and its inhibition by basic drugs. J Pharmacol Exp Ther 294: 1088–1098
61. Rikken MGJ, Lijzen JPA, Cornelese AA (2001) Evaluation of model concepts on human exposure. Proposals for updating the most relevant exposure routes of CSOIL. RIVM report 711 701 022, Bilthoven, NL
62. Fryer ME, Collins CD (2003) Model intercomparison for the uptake of organic chemicals by plants. Environ Sci Technol 37: 1617–1624
63. Doucette WJ, Chard TJK, Moore BJ, Staudt WJ, Headley JV (2005) Uptake of sulfolane and diisopropanolamine (DIPA) by cattails (*Typha latifolia*). Microchemical J 81: 41–49
64. Migliori L, Brambilla G, Cozzolino S, Gaudio L (1995) Effects on plants of sulphadimethoxine used in intensive farming (*Panicum miliaceum, Pisum sativa* and *Zea Mays*). Agric Ecosyst Environ 52: 103–110
65. Migliori L, Brambilla G, Casoria P, Civitareale C, Cozzolino S, Gaudio L (1996) Effect of sulphadimethoxine contamination on barley (*Hordeum disticuhm* L., Poaceae, Liliopsidia). Agric Ecosys Environ 60: 121–128
66. Migliori L, Civitareale C, Brambilla G, Cozzolino S, Casoria P, Gaudio L (1997) Effect of sulphadimethoxine on cosmopolitan weeds (*Amaranthus retroflexus* L., *Plantago major* L. and *Rumex acetosella* L.). Agric Ecosys Environ 65: 163–168
67. Fahl GM, Kreft L, Altenburger R, Faust M, Boedeker W, Grimme LH (1995) pH-dependent sorption, bioconcentration and algal toxicity of sufonylurea herbicides. Aquat Toxicol 31: 175–187
68. De Carvalho RF, Bromilow RH, Greenwood R (2007) Uptake of pesticides from water by curly waterweed Lagarosiphon major and lesser duckweed *Lemna minor*. Pest Manag Sci 63: 789–797
69. Mc Farlane JC (1995) Anatomy and physiology of plant conductive systems. In Trapp S, Mc Farlane JC (eds) Plant contamination – Modeling and simulation of organic chemicals processes. Lewis Pubishers, Boca Raton, FL
70. Briggs GG, Rigitano RLO, Bromilow RH (1987) Physico-chemical factors affecting uptake by roots and translocation to shoots of weak acids in barley. Pestic Sci 19: 101–112
71. Rigitano R, Bromilow R, Briggs G, Chamberlain K (1987) Phloem translocation of weak acids in *Ricinus communis*. Pestic Sci 19: 113–133
72. Bromilow RLO, Chamberlain K (1995) Principles governing uptake and transport of chemicals. In Trapp S, Mc Farlane JC (eds) Plant contamination – Modeling and simulation of organic chemicals processes. Lewis Publishers, Boca Raton, FL
73. Chamberlain K, Patel S, Bromilow RH (1998) Uptake by roots and translocation to shoots of two morpholine fungicides in barley. Pestic Sci 54: 1–7
74. Inoue J, Chamberlain K, Bromilow RH (1998) Physico-chemical factors affecting the uptake by roots and translocation to shoots of amine bases in barley. Pestic Sci 54: 8–21
75. Boxall ABA, Johnson P, Smith EJ, Sinclair CJ, Stutt E, Levy LS (2006) Uptake of veterinary medicines from soils into plants. J Agric Food Chem 54: 2288–2297
76. Travis C, Arms A (1988) Bioconcentration in beef, milk and vegetation. Environ Sci Technol 22: 271–274
77. BVL Bundesamt für Verbraucherschutz und Lebensmittelsicherheit (2006) National report on pesticide residues in foodstuff. Available online at http://www.bvl.bund.de/berichtpsm (accessed March 17, 2006)
78. Dettenmaier EM, Doucette WJ, Bugbee B (2009) Chemical hydrophobicity and uptake by plant roots. Environ Sci Technol 43: 324–329

The Evolution and Future of Environmental Fugacity Models

Donald Mackay, Jon A. Arnot, Eva Webster, and Lüsa Reid

Abstract In this chapter we first review the concept of fugacity as a thermodynamic equilibrium criterion applied to chemical fate in environmental systems. We then discuss the evolution of fugacity-based models applied to the multimedia environmental distribution of chemicals and more specifically to bioaccumulation and food web models. It is shown that the combination of multimedia and bioaccumulation models can provide a comprehensive assessment of chemical fate, transport, and exposure to both humans and wildlife. A logical next step is to incorporate toxicity information to assess the likelihood of risk in the expectation that most regulatory effort will be focused on those chemicals that pose the highest risk. This capability already exists for many well-studied chemicals but we argue that there is a compelling incentive to extend this capability to other more challenging chemicals and environmental situations and indeed to all chemicals of commerce. Finally, we argue that deriving the full benefits of these applications of the fugacity concept to chemical fate and risk assessment requires continued effort to develop quantitative structure–activity relationships (QSARs) that can predict relevant chemical properties and programs to validate these models by reconciliation between modeled and monitoring data.

Keywords Mass balance modeling · Fugacity · QSARs · Chemical hazard assessment · Chemical risk assessment

1 Introduction: The Fugacity Concept

For the purposes of monitoring, modeling, and regulation, the obvious expression of the quantity of chemical present in compartments or phases such as in air, water, or fish is concentration with units such as ng/m^3, mg/L, μg/g, or mol/L. These concentrations do not directly convey any information about the relative equilibrium

D. Mackay (✉)
Centre for Environmental Modelling and Chemistry, Trent University, Peterborough, ON K9J 7B8, Canada
e-mail: dmackay@trentu.ca

J. Devillers (ed.), *Ecotoxicology Modeling*, Emerging Topics in Ecotoxicology: Principles, Approaches and Perspectives 2, DOI 10.1007/978-1-4419-0197-2_12,

status between phases. To obtain this information requires additional information in the form of equilibrium partition coefficients. Alternatively, by expressing the quantity present in terms of fugacity the equilibrium status between phases becomes immediately obvious since when phases reach equilibrium the thermodynamic criteria of fugacity, activity, or partial pressure are equal. When interpreting the results of multimedia mass balance models the use of fugacity conveys directly how close the system is to equilibrium and the direction of the diverse diffusive transfer processes. The use of partition coefficients in the various flux equations is thus avoided. It also transpires that the formulation of the mass balance equations in either algebraic or differential forms is much easier when using the fugacity formalism, and the results are more readily interpreted.

Mathematically, if two-phase concentrations are C_1 and C_2 and the partition coefficient is K_{12} then the relative equilibrium status is $C_1 : C_2K_{12}$ or $C_1/K_{12} : C_2$. In the fugacity formalism K_{12} is split into two phase-specific capacity terms such that K_{12} is Z_1/Z_2. The concentration C_1 is then $Z_1 f_1$ and C_2 is $Z_2 f_2$ where f_1 and f_2 are the fugacities that directly express the relative equilibrium status. The driving force for interphase diffusion is then $(f_1 - f_2)$ and at equilibrium f_1 and f_2 are equal. Fugacity is expressed in units of partial pressure, Pascals (Pa), and Z values (fugacity capacities) have units of mol/(m^3 Pa). Z values express the capacity or affinity of the phase for the chemical and depend on the phase composition, temperature, and physicochemical properties of the substance.

In the fugacity formalism mass transfer and reaction rate parameters or D values are defined such that the rates of transport or reaction are the product Df with units of mol/h. These D values can be regarded as fugacity rate coefficients.

Full details of methods of estimating Z and D values and fugacities are provided in the text by Mackay [1]. Our focus here is on how fugacity models have evolved over recent decades and on likely future developments.

2 Evolution of Multimedia Fugacity Models

The earliest or Level I models simulate the simple situation in which a chemical achieves equilibrium between a number of phases of different composition and volume. The prevailing fugacity is simply $M/\Sigma V_i Z_i$ where M is the total quantity of chemical (mol), V_i is volume (m^3), and Z_i is the corresponding phase Z value, [mol/(Pa m^3)]. Although very elementary and naive, this simulation is useful as a first indication of where a chemical is likely to partition. It is widely used as a first step in chemical fate assessments.

More realistic Level II models introduce the rate of chemical reaction or degradation and advection, but interphase equilibrium is still assumed. Level III models introduce intercompartmental transfer rates, thus equilibrium no longer applies. For Level III models it is then necessary to specify the chemical's mode-of-entry to the environment, that is, to air, water, or soil, or some combination of these media. Valuable insights obtained from these models include those of overall chemical persistence or residence time and potential for long-range transport (LRT) in air or water. Level IV models, which involve the solution of differential mass balance equations, can be used to describe the time-dependent or dynamic behavior of chemicals.

Figures 1–3 illustrate the results of Level I, II, and III models for pyrene using the chemical properties listed in Table 1. For the Level III model emission is 50% each to air and to water. A key feature of these models is that they identify the critical partitioning and degradation rate properties that control chemical fate.

From Fig. 1 it can be seen that most of the pyrene partitions to soil in the Level I modeled system. This reflects the high K_{OW} of pyrene and the much larger volume of soil than of sediment (by a factor of 36) in the standard equilibrium criterion (EQC) environment [2]. The Level II simulation shown in Fig. 2 gives a first estimate of chemical persistence, and since equilibrium is assumed in this case also, partitioning is still predominantly to soil. The model shows that less than half of the loss from the system is degradation in the soil; 24.2% is removed by advection in the air. Three chemical residence times are given: the total time, the reaction time, and the advection time. The total residence time is the "two-thirds" time for clearance of the chemical from the system. The reaction time is the "two-thirds" time for chemical removal by degradation alone and is generally considered to be the chemical persistence. The advection time considers only chemical removal by transport to a neighboring region. Thus, the persistence estimated by the Level II model for pyrene in a standard EQC environment is about 2 years. This estimate of persistence is refined using the Level III model. Figure 3 shows a persistence estimate for pyrene

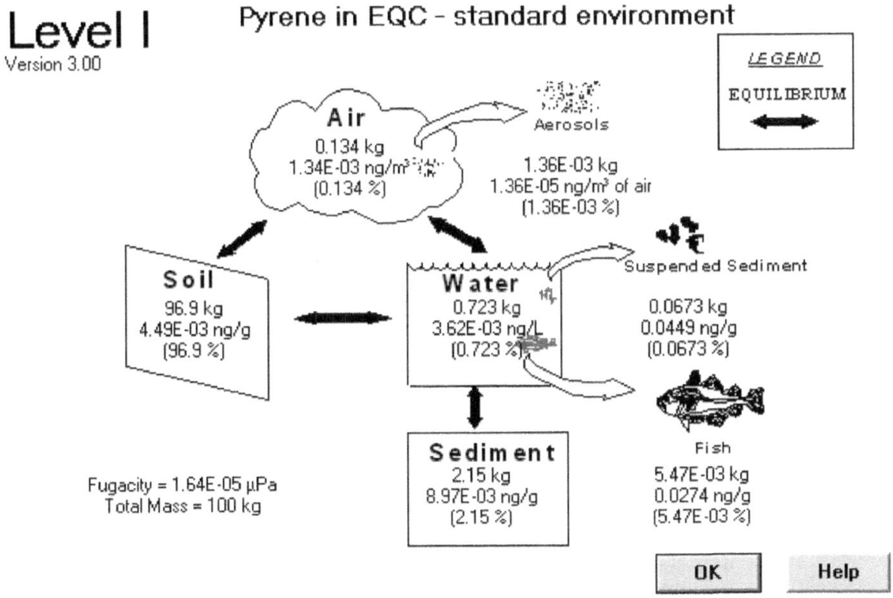

Fig. 1 Level I diagram for pyrene in the EQC environment

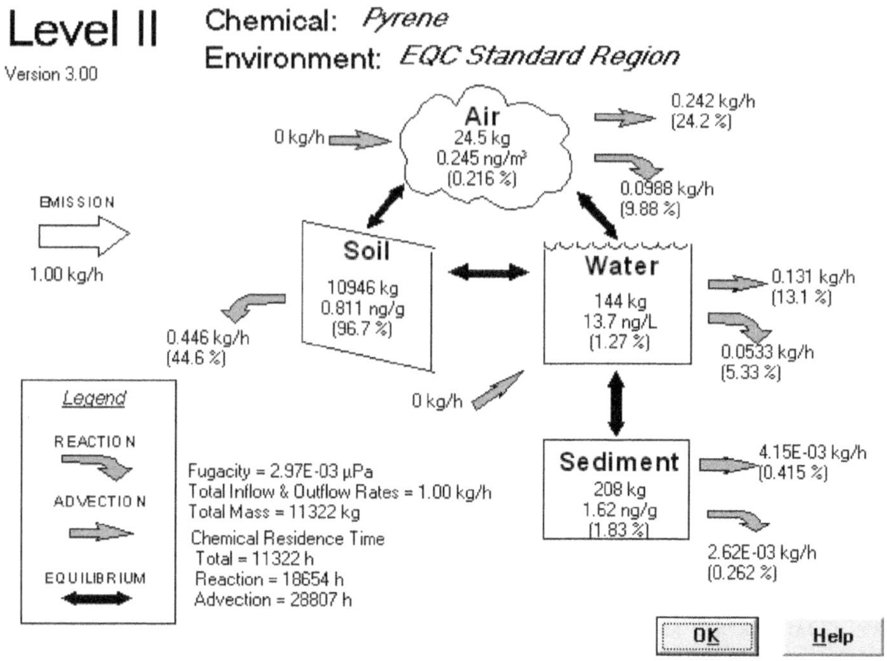

Fig. 2 Level II diagram for pyrene in the EQC environment

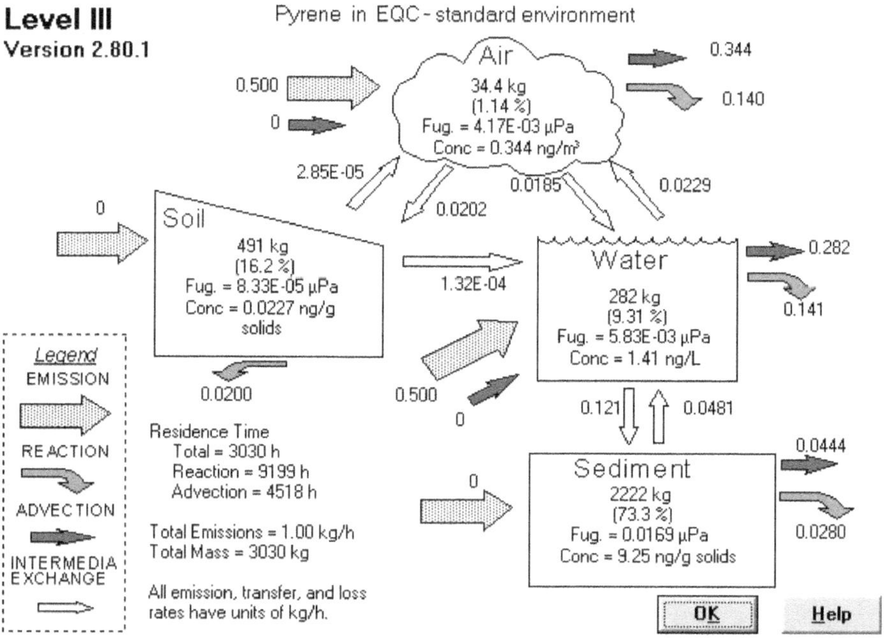

Fig. 3 Level III diagram for pyrene in the EQC environment with 50% of the emissions to each of air and water

Table 1 Properties of pyrene [32]

	Pyrene
CAS	129-00-0
Formula	$C_{16}H_{10}$
Molar mass (g/mol)	202.25
Melting point (°C)	150.62
Vapor pressure (Pa)	0.0006
Solubility (g/m^3)	0.132
Log K_{OW}	5.18
Half-lives (h)	
Air	170
Water	1,700
Soil	17,000
Sediment	55,000
Fish	50
Birds/mammals	17

of about 1 year, assuming equal emissions to air and water. Level III calculations do not assume that the chemical has achieved equilibrium between the different bulk compartments of the environment (air, water, soil, and sediment). This can be seen in Fig. 3 by examining the relative transfer rates between media. The majority of the pyrene in the system is now located in the sediment. This can be attributed to the emission to water and the relatively fast water-to-sediment transfer rate. Approximately 70% of the pyrene emitted to the air blows out of the modeled system and into the adjoining region while over half of the pyrene discharged to the water flows out of the system. The loss rates in air and water as a result of degradation processes are approximately equal and are about half the loss rate by water outflow (advection).

Table 2 lists a selection of fugacity models that have been applied to evaluative (hypothetical) and real environments. These and other fugacity models are available from the Centre for Environmental Modelling and Chemistry (formerly the Canadian Environmental Modelling Centre) website (http://www.trentu.ca/cemc).

3 Fugacity Models of Long-Range Transport

As Figs. 2 and 3 show, a useful feature of Level II and III models is that they can demonstrate the extent to which a chemical is lost from a region by atmospheric or water transport, that is, advective loss, as distinct from loss by degradation. It is possible to calculate the contribution of each loss mechanism to the overall persistence or residence time. When the advective residence time is short, that is, advection is rapid, the implication is that much of the chemical discharged into the region will flow to neighboring downwind or downstream regions. Whereas a local contamination problem is alleviated, the problem is merely transported to other regions where

Table 2 Fugacity models applied to evaluative and real environments

	Latest version/release date	Description	Publications describing/ using this model
Models for evaluative environments			
EQC	2.02/May 2003	The equilibrium criterion model uses chemical-physical properties to quantify chemical behavior in an evaluative environment. The environment is fixed to facilitate chemical-to-chemical comparison.	[2, 33, 34]
Level I	3.00/Sept. 2004	A model of the equilibrium distribution of a fixed quantity of conserved chemical, in a closed environment.	[1, 35]
Level II	3.00/Sept. 2005	A model of the equilibrium distribution of a nonconserved chemical discharged at a constant rate into an open environment at steady state.	[1, 35]
Level III	2.80/May 2004	A model of the steady-state distribution of a nonconserved chemical discharged at a constant rate into an open environment.	[1, 35]
RAIDAR	2.00/January 2010	The risk assessment identification and ranking model is a screening-level exposure and risk assessment model that brings together information on chemical partitioning, reactivity, environmental fate and transport, bioaccumulation, exposure, effect levels, and emission rates in a holistic framework.	[21, 22]
TaPL3	3.00/Sept. 2003	The transport and persistence Level III model is intended as an evaluative tool for the detailed assessment of chemicals for persistence and potential for long-range transport in either air, or water in a steady-state environment.	[10]
Models for real environments			
ChemCAN	6.00/Sept. 2003	A Level III model containing a database of 24 regions of Canada. By the addition of regional properties, it is easily applicable to other regions.	[36–38]
CalTOX	2.3/March 1997	A regional scale multimedia exposure model designed to assess the fate and human health impacts of contaminants. Human doses are derived as products of chemical concentrations in contact media and exposure factors for each media.	[39]

Most of these models and models listed in later tables are available from the Web site http://www.trentu.ca/cemc

the resulting contamination can be of concern, especially because there may be no direct control of sources. This general issue, which has ethical and international aspects, was first addressed in connection with SO_2 atmospheric transport from the United Kingdom and continental Europe to Scandinavia. It has become a major issue of concern as a result of the realization that levels of organic contaminants such as PCBs in the arctic environment and especially in arctic wildlife are remarkably high. Human exposure to these contaminants can be substantial because the resident population often consumes terrestrial and marine wildlife such as caribou and whale meat.

The Stockholm Convention has addressed this issue by regulating 12 substances or groups of substances that have been demonstrated to undergo LRT [3]. Scientific and modeling aspects of LRT have been addressed in a number of reports and books. Two general modeling approaches have emerged; multi-box Eulerian models and characteristic travel distance (CTD) Lagrangian models, both of which can employ the fugacity concept.

The most compelling evidence that significant LRT has occurred is provided by monitoring data in remote regions, for example, as summarized in Arctic Monitoring and Assessment Programme reports [4]. Multibox modeling can play a complementary role by demonstrating that monitoring data are consistent with our present understanding of LRT processes. Models can be used to identify and prioritize chemicals for persistence and LRT potential and provide estimates of the fraction of the mass of chemical released in one location that may reach a distant region as well as the rate of transport. Examples of this approach are Wania's arctic contamination potential (ACP) [5, 6], MacLeod's transfer efficiency [7], and the distant residence time (DRT) concept [8].

The CTD models are typically used to rank chemicals because of their simplicity and ease of interpretation. To calculate the CTD of a chemical, a one-region environment is simulated and then an expected "distance" that a chemical may be transported in a mobile phase (air or water) that is moving at a defined speed is calculated. The distance travelled by the chemical is related to several factors including the fugacity of the chemical in the transporting phase as well as the expected time that the chemical will exist in that phase (persistence) [9, 10].

Table 3 lists studies of LRT, many of which employ the fugacity concept.

4 Evolution of Bioaccumulation Fugacity Models

Bioaccumulation is the net result of competing rates of chemical uptake and elimination in an organism and can result in concentrations in organisms that are orders of magnitude greater than those in the air or water environment [11, 12]. Bioaccumulation includes uptake by respiration (bioconcentration) of chemical from the environment surrounding the organism (air or water) and dietary exposures. Dietary exposures can result in biomagnification, an increase in concentration

Table 3 Models and studies of long-range transport

	Latest model version/ release date	Description	Publications describing/ using this model
TaPL3	3.00/Sept. 2003	See Table 2.	[10, 40]
BETR-North America		A regionally segmented multicompartment, continental-scale, mass balance chemical fate model for North America.	[41]
BETR-World	409/2003 500/	A regionally segmented multicompartment, global-scale, mass balance chemical fate model.	[8, 42]
BETR-Global		A global-scale multimedia contaminant fate model that represents the global environment as a connected set of 288 multimedia regions on a 15° grid.	[43]
GloboPOP	1.10/2003	A zonally averaged multimedia model describing the global fate of persistent organic chemicals on the time scale of decades.	[5, 44]

from food to the consuming organism. Biotransfer factors are also used to express the food-to-organism increases in concentration, especially in an agricultural setting [13].

The fugacity concept proves to be particularly useful when simulating the uptake of chemical by organisms such as fish from their environment (e.g., water) and their food. The bioconcentration phenomenon is essentially a result of the chemical seeking equi-fugacity between the respiring organism and its environment. The concentration ratio or bioconcentration factor (BCF) is essentially Z_O/Z_E where Z_O applies to the organism and Z_E to the environment. Hydrophobic, bioaccumulative substances such as DDT and PCBs tend to have low values of Z_E and high values of Z_O and thus high BCFs.

Two general approaches have been used to assess and predict bioaccumulation: relatively simple regression models or QSARs and more complex mechanistic models that simulate all uptake and loss processes [11, 12].

Regressions for BCF–octanol water partition coefficient (BCF–K_{OW}) for fish and biotransfer factor–K_{OW} (BTF–K_{OW}) for agricultural species in the human food chain are still widely used for bioaccumulation and human exposure assessments. The use of simple regression equations implies that all chemicals with the same K_{OW} have the same BCF in fish or BTF in agricultural food webs. Biomagnification and biotransformation processes can, however, result in orders of magnitude difference in exposures, particularly for more hydrophobic chemicals, and these processes are not explicitly accounted for using simple regression equations. Laboratory-derived BCF data do not include dietary exposure, which is an important route of exposure for hydrophobic chemicals in the environment. Air-breathing organisms exchange chemical with the air for which the octanol–air partition coefficient (K_{OA}) is an important property and is not explicitly included in K_{OW}-based regressions for BTFs.

In response to these problems, bioconcentration models have been extended to address bioaccumulation by including food uptake and losses by metabolic conversion, respiration, fecal egestion, and growth dilution. It is relatively straightforward to apply these models to multiple organisms comprising food webs. Most effort has been devoted to aquatic organisms but recently there has been increasing attention to air-breathing organisms [14–17]. The major challenge has been to describe dietary rates and feeding preferences, especially during different seasons and life stages. Differences in species' physiology (body size, feeding rates) and characteristics (herbivores, carnivores, bioenergetics, feeding preferences) play a role in bioaccumulation processes and can be included in fugacity bioaccumulation models, resulting in more accurate simulations and predictions. Important considerations for using mechanistic bioaccumulation models include the principle of parsimony (Occam's Razor), parameterization, and reliable physical chemical property information (e.g., K_{OW}, K_{OA}, biotransformation rates). Sensitivity and uncertainty analyses can help direct priorities for accurate input data requirements.

These models have shown that uptake by edible vegetation from air and soil is fundamental to compiling reliable models of bioaccumulation in wildlife and humans. The uptake losses and translocation of chemicals in vegetation have proved to be challenging but fugacity models can provide insights into the important process of plant bioaccumulation.

As more information becomes available on the processes of uptake, release, and internal disposition of chemicals in fish and wildlife, the logical next step is to compile a more detailed model of chemical fate within the organism. The simple models discussed earlier generally treat the organism as a single compartment or "box." The more detailed models exploit the considerable experience in physiologically based pharmacokinetic (PBPK) models developed for medical and pharmaceutical purposes. These PBPK models can provide more information for accumulation in specific organs within the body and the rates of transport and transformation within the body and excretion processes. Most PBPK models are based on conventional concentration/rate constant/partition coefficient expressions [18, 19], but they can be rewritten in fugacity format [20]. Again, the fugacity formalism is advantageous because differences in the equilibrium status of chemical levels between blood and various organs and tissues are immediately apparent.

Table 4 lists a number of bioaccumulation models and studies.

Table 4 Bioaccumulation and PBPK models and studies

	Latest version/ release date	Description	Publications describing/ using this model
Fish	2.00/November 2004	A single organism bioaccumulation model treating the steady-state uptake and loss of an organic contaminant by a fish.	[1,35]
FoodWeb	2.00/March 2006	A mass balance model of contaminant flux through an aquatic food web.	[45]
Mysid	1.00/August 2007	A single organism bioaccumulation model treating the dynamic uptake and loss of an organic contaminant by the opossum shrimp (*Mysis relicta*).	[46]
AquaWeb	1.2/March 2007	A steady-state aquatic food web bioaccumulation model for estimating of chemical concentrations in organisms from chemical concentrations in the water and the sediment.	[47–51]
BAF-QSAR	1.5/May 2008	A model to estimate bioaccumulation factors for fish species in lower, middle, and upper trophic levels of aquatic food webs.	[52,53]
ACC-HUMAN		A nonsteady-state bioaccumulation model predicting human tissue levels from concentrations in air, soil, and water.	[54]
PBPK	1.0/January 2003	A physiologically based pharmacokinetic model describing the disposition of contaminants in an adult male human. It treats a parent chemical and, if desired, two metabolites that may be formed reversibly or irreversibly. Tissue concentrations for the chemical and any metabolites can be simulated for acute, occupational, and environmental exposure regimes.	[55]
PBPK/PBTK models		Some publications available outlining physiologically based pharmaco-/toxico-kinetic models for various species.	[18–20,56–58]
Terrestrial-based bioaccumulation models		Some publications available outlining terrestrial-based food web bioaccumulation models for various species.	[14,54,59–61]

5 Fugacity Models of Specific Compartments and Processes

The results of multimedia mass balance models often show the need to focus more attention on specific compartments such as soils to which a pesticide is applied or to water bodies that receive chemical discharges from direct discharges or from sewage treatment plants. Several such models have been developed, especially for sewage treatment plants, lakes, and rivers. For those evaluating chemical fate it is useful to have the capability of addressing in detail the likely chemical fate in these more local and site-specific conditions. The models may be used to explore remedial options and likely remediation times. As is discussed later such models are best regarded as individual "tools" available from a "tool box" of models.

Table 5 lists a number of these models.

Table 5 Fugacity models of specific compartments and processes

	Latest version/ release date	Description	Publications describing/using this model
AirWater	2.00/Nov. 2004	A model to calculate air–water exchange characteristics, including unsteady-state conditions, based on the physical chemical properties of the chemical and total air and water concentrations.	[1, 35]
BASL4	1.00/Apr. 2007	The biosolids-amended soil: Level IV model calculates the fate of chemicals introduced to soil in association with contaminated biosolids amendment.	[62, 63]
QWASI	3.10/Feb. 2007	The quantitative water air sediment interaction model assists in understanding chemical fate in lakes.	[64–69]
Sediment	2.00/Nov. 2004	A model to calculate the water–sediment exchange characteristics of a chemical based on its physical chemical properties and total water and sediment concentrations.	[1, 35]
Soil	3.00/Aug. 2005	A model for the simple assessment of the relative potential for reaction, degradation, and leaching of a pesticide applied to surface soil.	[1, 35]
STP	2.11/Mar. 2006	The sewage treatment plant model estimates the fate of a chemical present in the influent to a conventional activated sludge plant as it becomes subject to evaporation, biodegradation, sorption to sludges, and to loss in the final effluent.	[70, 71]

6 Evolution of More Comprehensive Multimedia and Bioaccumulation Fugacity Models

A logical step in modeling chemical fate, exposure, and even effects is to combine models that describe the fate of the chemical in the largely abiotic environment with bioaccumulation and food web models resulting in a more complete simulation of chemical behavior and exposure to humans and wildlife. Table 6 lists a selection of fugacity and non-fugacity models that combine fate, exposure, and effects and can be used for regulatory purposes. These models can be used to screen list of chemicals to identify those substances that are of greatest potential risk to humans and the environment for more comprehensive assessments using monitoring data. For example, the risk assessment, identification, and ranking (RAIDAR) model combines information on chemical partitioning, reactivity, environmental fate and transport, food web bioaccumulation, exposure, effect endpoint, and emission rate in a coherent mass balance evaluative framework [21, 22]. RAIDAR fate calculations are similar to those in the EQC model [2]; however, food web models representative of aquatic and terrestrial species such as vegetation, fish, wildlife, agricultural products, and humans are also included. RAIDAR is distinct from other models listed in Table 6 because the food web models assessing exposure to humans and ecological receptors include mechanistic expressions for chemical uptake and elimination. Thus, biomagnification and biotransformation processes in the food web can be included for the exposure assessment. An illustration of the RAIDAR model for chemical assessments is given in Sect. 6.1.

Table 6 Comprehensive models of chemical fate and bioaccumulation

	Latest version/ release date	Description	Publications describing/using this model
CalTOX	2.3/March 1997	See Table 2.	[39]
EUSES	2.0/2004	The European Union System for the Evaluation of Substances brings together exposure and effect assessments and risk characterization for environmental populations and humans, including occupational and consumer scenarios at local, regional, and continental scales.	[72, 73]
IMPACT 2002		The IMPACT 2002 model provides characterization factors for the midpoint categories: human toxicity, aquatic ecotoxicity, and terrestrial ecotoxicity for life-cycle impact assessments.	[74]
RAIDAR	2.00/January 2010	See Table 2.	[21, 22]

Combining the key elements of chemical exposure and effect at a screening level allows for a holistic approach for evaluating chemicals and may prove to be a valuable educational tool for regulators, scientists, and students. Combined model predictions can guide environmental monitoring programs by identifying the media in the environment (physical and biological) in which chemical concentrations and fugacities are expected to be the greatest. A holistic approach for chemical risk assessment (emissions, exposure, and effect) also provides the opportunity to identify the key processes and chemical properties that contribute the most uncertainty to the underlying risk calculation. Uncertainty and sensitivity analyses can be used to prioritize data gaps that often occur for the large numbers of chemicals requiring chemical assessment.

6.1 *An Illustrative Case Study for Chemical Exposure and Risk Assessment*

Figure 4 illustrates the output of RAIDAR fate calculations for pyrene using an arbitrary unit emission rate of 1 kg/h to air. This is a hypothetical rate of emission and the model user can choose either Level II or Level III fate calculations. As discussed earlier for Level II calculations, equilibrium between the environmental compartments of air, water, soil, and sediment is assumed; therefore, there is no need to

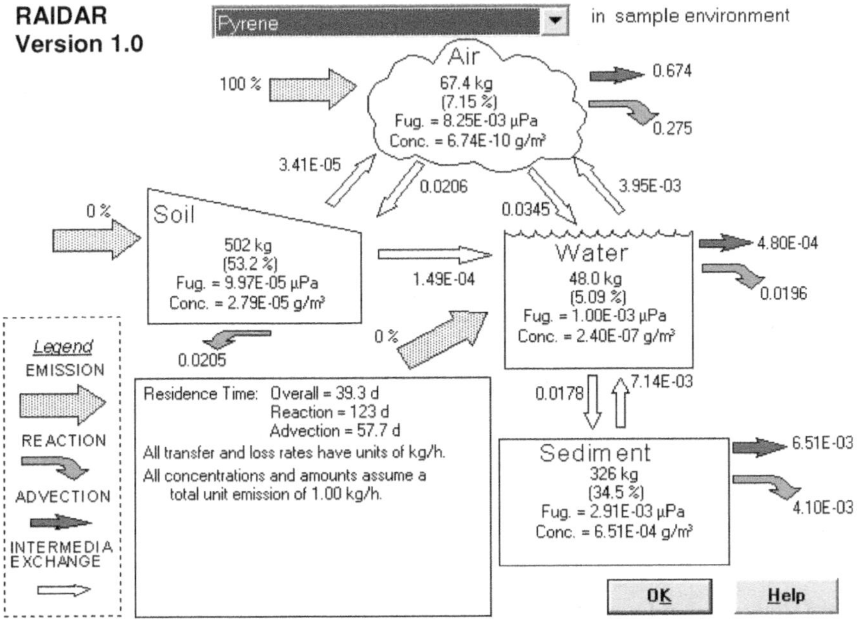

Fig. 4 Level III fate calculations for pyrene in the RAIDAR environment assuming 100% emissions to air

select a mode-of-entry for chemical release to the environment. For Level III calculations the predicted distribution of a substance in the physical compartments of the environment is determined from the specified mode-of-entry information. In this illustration it is assumed that 100% of the chemical is released to air.

The overall residence time in the evaluative regional environment (100,000 km^2) is 39.3 days. This overall residence time includes chemical transfers out of the region (advection) and chemical degradation (reaction) within the region. Approximately 67% of the pyrene that is released to air in the region is removed from the region by advection in air and the advection residence time is 57.7 days. The reaction residence time, or persistence, is 123 days. Thus, overall persistence in the system based solely on reaction is quite different from the overall residence time. This highlights the need to clearly determine the specific assessment objectives and the influence of model assumptions when comparing chemical persistence.

Based on predicted chemical concentrations and fugacities in the bulk physical compartments of air, water, soil, and sediment, chemical concentrations and fugacities are then calculated in the representative species in RAIDAR using mass balance food web models. Figure 5 displays fugacities for pyrene in the biological species in the model food webs. These fugacities are based on the assumed unit emission rate and include estimated biotransformation rates [23]. For certain persistent chemicals the fugacities are observed to increase in higher trophic level organisms (biomagnification). In this example, the fugacities decrease in higher trophic level organisms, a phenomenon known as trophic dilution. For example, the biomagnification factor (BMF) from the terrestrial herbivore to the terrestrial carnivore is 0.19 (BMF < 1). This is largely due to biotransformation within the predator organisms. Lack of biotransformation as slow growth usually leads to biomagnification.

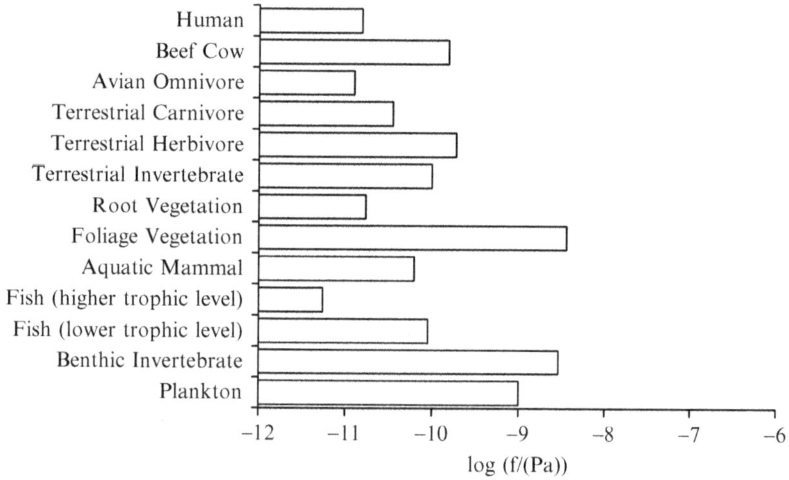

Fig. 5 Illustration of fugacities (f) for pyrene in some of the biological compartments in the RAIDAR evaluative environment

The next step is to include toxicity in the chemical assessment by selecting an effect level or concentration. In this illustration an acute narcotic toxic effect endpoint of 5 mmol/kg wet weight is selected [24]. The hazard assessment factor (HAF) is an intensive hazard property being a combined function of persistence, bioaccumulation, and the selected toxicity endpoint [22]. The HAF is the dimensionless ratio of the calculated unit concentration in an organism (C_U) to the toxic effect endpoint (C_T) assuming a hypothetical "unit" emission of 1 kg/h. The HAF provides a single value for comparing all chemicals of interest for the combined properties of persistence, bioaccumulation, and the selected toxicity endpoint. As illustrated previously the fugacities and unit concentrations (C_U) can be calculated for all representative RAIDAR species based on the assumed unit emission rate. In the present example for pyrene, "benthic invertebrates" are identified as the representative species with the greatest hazard quotient ($C_U/C_T = 3.0 \times 10^{-5}/5$) and thus the HAF is 6.0×10^{-6}. If biotransformation estimates were not included in the assessment for pyrene biomagnification in the food webs would occur resulting in the identification of "terrestrial carnivores" as the most vulnerable species and the HAF would be 4.5×10^{-3} (about 1,000 times larger).

The previously described calculations are independent of the actual quantity of chemical released to the environment, being based on assumed unit emission rates, and are therefore only hazard metrics. A screening level RAIDAR risk assessment factor (RAF) can be simply calculated from the HAF by multiplying by an estimate of the actual chemical emission rate [22]. For example, an estimated emission rate in Canada for pyrene is 10.7 kg/h [25], and the resulting RAF is 6.4×10^{-5}. The implication is that prevailing levels are well below levels at which pyrene is expected to cause toxic effects. This case study illustrates the need to consider all elements of a chemical's properties (persistence, P, bioaccumulation, B, and toxicity, T) and quantity discharged (Q) when evaluating chemicals for their potential risks to humans and the environment [22].

7 The Issue of Fidelity and Complexity

These models can become very complex by attempting to include numerous organisms and vegetation types. Further, there may be a need to include municipal and industrial waste treatment processes. It is also apparent that urban regions often experience higher levels of emissions than rural regions, thus urban regions often experience higher levels of contamination than rural regions and urban residents and wildlife may experience greater exposure. This can be addressed by replacing the single soil environment with urban, rural, or agricultural and pristine soils. Pesticides may be preferentially applied in an agricultural setting. It is increasingly apparent that for some chemicals used domestically and in consumer products, for purposes such as plasticizers or for reducing flammability, indoor exposure can greatly exceed outdoor exposure. The implication is that detailed simulation of environmental fate is largely irrelevant for humans who experience their greatest exposure indoors.

A tension thus develops between the need to increase model complexity to address all possible routes of exposure and the need to ensure that the model is robust, transparent, understandable, and is free from gross errors. The optimal answer may be to develop a suite of modeling "tools" that address a variety of aspects of chemical fate. This *tool box* can contain models of the types described earlier, as well as models addressing specific situations such as waste water treatment, indoor exposure, pesticide dissipation in an agricultural setting, and even less common conditions such as aquaculture. If this is to be accomplished the model-to-model transition should be as simple and as user-friendly as possible. The use of fugacity in this context offers the advantage that a common system of units applies, thus the fugacity output from one model becomes the input to the next model. The nature of the process in changing fugacity also becomes immediately apparent. For example, an effective waste water treatment plant may typically achieve a reduction in fugacity of a contaminant by a factor of 10, that is, essentially 90% removal. A bird such as an owl consuming a contaminated rodent should experience a fugacity increase as a result of biomagnification by a factor such as 30 if the contaminant is not biotransformed, but only by a factor of 3 or less if the bird has the metabolic capability to degrade the substance. In short, viewing the environmental fate of chemicals through the lens of fugacity can provide valuable insights into the many varied and complex processes that chemicals undergo in the environment.

8 The Future: A Speculation

Society through its many national and international regulatory agencies has become increasingly intolerant of inadvertent exposure to chemical substances. There are increasing demands for improved assessment of the risks of adverse effects to humans and wildlife and for more vigorous and effective measures to identify the chemicals of greatest concern and restrict their use accordingly. This is a demanding task, especially because there are believed to be some 100,000 chemicals requiring assessment. Fugacity modeling can, we believe, contribute to this process but many challenges remain. In this final section we speculate on some needs and directions.

8.1 Chemical Properties

Models of chemical fate, fugacity, or otherwise require information of sufficient accuracy on chemical properties such as vapor pressure, partition coefficients, and reactivity in a variety of media ranging from the atmosphere to the human liver. The availability of such data is very limited, especially for the less-studied substances and for mixtures [26]. There is thus an obvious incentive to develop and improve QSARs or QSPRs that can estimate these properties from chemical structure. Considerable progress has been made, but much remains to be done, especially for more complex molecules containing oxygen, nitrogen, sulfur, phosphorus, silicon, fluorine, and metal moieties. Present models do not always satisfactorily address ionizing and surface active substances or those of high molar mass such as dyes and pigments. A coordinated program of laboratory-based property determination and QSAR development is needed.

8.2 Ground-Truthing Models

There is concern that model-based predictions of chemical fate may be subject to systematic error because some important processes are omitted or poorly described. An example is the role of snow in scavenging the atmosphere and accumulating chemical seasonally in snowpacks or ice. Modeling is relatively inexpensive and easy compared with monitoring, and there has thus been a tendency for predictions to outstrip observations. What is clearly needed is a continuing program of "ground-truthing" models by comparison of modeling and monitoring data, especially including exposure. An example is the recent study by McKone et al. [27] of the fate and exposure of organo-phosphorus pesticides by agricultural workers in which the model predictions extended from application conditions, to environmental concentrations, to exposure, and to levels of metabolites in urine. Another is the assessment of fate and exposure to phthalate esters by both environmental routes and from consumer products [28]. Unless there is a continuing effort to ground truth models, there is a danger that exposure may be underestimated, with implications for adverse effects on human or ecosystem health. Conversely overestimation may result in unnecessary restrictions and economic penalties to industry and to society at large.

8.3 Fugacity and Toxicity

Some 70 years ago Ferguson showed that for nonselective or narcotic chemicals toxic effects were elicited at a relatively constant chemical activity in the organisms' "circum environment" of air or water [29]. The corresponding concentrations varied over many orders of magnitude. This concept is inherent in the concepts of critical body residue or body burden corresponding to toxic endpoints. Fugacities, like concentrations, vary greatly but both can be readily converted into activities and to body burdens providing a direct link from fugacities in the environment as predicted from multimedia models and activity levels in the exposed organism. Of course, many chemicals exert selective toxicity as a result of specific biochemical interactions, but if toxic potency can be estimated for specific modes of toxic action in the form of multiples of narcotic levels, this could provide a predictive capability for nonnarcotics. The potential of this approach has been suggested by Verharr et al. [30], McCarty et al. [24], and others [31].

If a robust link can be established between toxic levels of chemicals and their external and internal fugacities this has the potential to provide a coherent mechanism by which the proximity of environmental levels to those of concern from the viewpoint of toxic effects could be quantified and evaluated. Fugacity can then play an increasingly valuable role in assisting society to manage the multitude of chemicals of commerce on which our present standard of living depends, with assurance that levels of risk of adverse effects are acceptably low.

References

1. Mackay D (2001) Multimedia environmental models: The fugacity approach, 2nd edition. Lewis Publishers, Boca Raton, FL
2. Mackay D, Di Guardo A, Paterson S, Cowan C (1996) Evaluating the environmental fate of a variety of types of chemicals using the EQC model. Environ Toxicol Chem 15: 1627–1637
3. United Nations Environment Programme (1998) Report of the first session of the INC for an international legally binding instrument for implementing international action on certain persistent organic pollutants (POPs), in UNEP report (Vol. 15). International Institute for Sustainable Development (IISD), Canada
4. Arctic Monitoring and Assessment Programme (2004) AMAP assessment 2002: Persistent organic pollutants (POPs) in the Arctic. Arctic Monitoring and Assessment Programme (AMAP): Oslo, Norway
5. Wania F (2003) Assessing the potential of persistent organic chemicals for long-range transport and accumulation in polar regions. Environ Sci Technol 37: 1344–1351
6. Wania F (2006) Potential of degradable organic chemicals for absolute and relative enrichment in the arctic. Environ Sci Technol 40: 569–577
7. MacLeod M, Mackay D (2004) Modeling transport and deposition of contaminants to ecosystems of concern: A case study for the Laurentian Great Lakes. Environ Pollut 128: 241–250
8. Mackay D, Reid L (2008) Local and distant residence times of contaminants in multicompartment models, Part 1: Theoretical basis. Environ Pollut 156: 1196–1203
9. Bennett DH, Kastenberg WE, McKone TE (1999) General formulation of characteristic time for persistent chemicals in a multimedia environment. Environ Sci Technol 33: 503–509
10. Beyer A, Mackay D, Matthies M, Wania F, Webster E (2000) Assessing long-range transport potential of persistent organic pollutants. Environ Sci Technol 34: 699–703
11. Gobas FAPC, Morrison HA (2000) Bioconcentration and biomagnification in the aquatic environment. In: Boethling RS, Mackay D (eds) Handbook of property estimation methods for chemicals: Environmental and health sciences. CRC Press: Boca Raton, FL
12. Mackay D, Fraser A (2000) Bioaccumulation of persistent organic chemicals: Mechanisms and models. Environ Pollut 110: 375–391
13. Travis CC, Arms AD (1988) Bioconcentration of organics in beef, milk and vegetation. Environ Sci Technol 22: 271–274
14. Gobas FAPC, Kelly BC, Arnot JA (2003) Quantitative structure–activity relationships for predicting the bioaccumulation of POPs in terrestrial food webs. QSAR Comb Sci 22: 329–336
15. Kelly BC, Ikonomou M, Blair JD, Morin AE, Gobas FAPC (2007) Food web-specific biomagnification of persistent organic pollutants. Science 317: 236–239
16. Kelly BC, Gobas FAPC (2001) Bioaccumulation of persistent organic pollutants in lichen–caribou–wolf food chains of Canada's central and western arctic. Environ Sci Technol 35: 325–334
17. Czub G, McLachlan MS (2004) Bioaccumulation potential of persistent organic chemicals in humans. Environ Sci Technol 38: 2406–2412
18. Himmelstein KJ, Lutz RJ (1979) A review of the application of physiologically based pharmacokinetic modeling. J Pharmacokinet Biopharm 7: 127–137

19. Nichols JW, McKim JM, Lien GJ, Hoffman AD, Bertelsen SL (1991) Physiologically based toxicokinetic modeling of three chlorinated ethanes in rainbow trout (*Oncorhynchus mykiss*). Toxicol Appl Pharmacol 110: 374–389
20. Paterson S, Mackay D (1987) A steady-state fugacity-based pharmacokinetic model with simultaneous multiple exposure routes. Environ Toxicol Chem 6: 395–408
21. Arnot JA, Mackay D, Webster E, Southwood JM (2006) Screening level risk assessment model for chemical fate and effects in the environment. Environ Sci Technol 40: 2316–2323
22. Arnot JA, Mackay D (2008) Policies for chemical hazard and risk priority setting: Can persistence, bioaccumulation, toxicity and quantity information be combined? Environ Sci Technol 42: 4648–4654
23. Arnot JA, Mackay D, Parkerton TF, Bonnell M (2008) A database of fish biotransformation rates for organic chemicals. Environ Toxicol Chem 27: 2263–2270
24. McCarty LS, Mackay D (1993) Enhancing ecotoxicological modeling and assessment. Environ Sci Technol 27: 1719–1728
25. Environment Canada (2005) National pollutant release inventory, 2003. Environment Canada: Ottawa, ON
26. Environment Canada (2006) Existing substances program at environment Canada (CD-ROM). Ecological categorization of substances on the Domestic Substances List (DSL). Existing Substances Branch, Environment Canada: Ottawa, ON
27. McKone TE, Castorina R, Harnly ME, Kuwabara Y, Eskenazi B, Bradmanm A (2007) Merging models and biomonitoring data to characterize sources and pathways of human exposure to organophosphorus pesticides in the Salinas Valley of California. Environ Sci Technol 41: 3233–3240
28. Cousins IT, Mackay D (2003) Multimedia mass balance modeling of two phthalate esters by the regional population-based model (RPM). Handbook Environ Chem 3: 179–200
29. Ferguson J (1939) The use of chemical potentials as indices of toxicity. Proc R Soc Lond B Biol Sci 127: 387–404
30. Verhaar HJM, Van Leeuwen CJ, Hermens JLM (1992) Classifying environmental pollutants. I. Structure–activity relationships for prediction of aquatic toxicity. Chemosphere 25: 471–491
31. Maeder V, Escher BI, Scheringer M, Hungerbuhler K (2004) Toxic ratio as an indicator of the intrinsic toxicity in the assessment of persistent, bioaccumulative, and toxic chemicals. Environ Sci Technol 38: 3659–3666
32. Mackay D, Shiu WY, Ma KC, Lee SC (2006) Handbook of physical–chemical properties and environmental fate for organic chemicals, Vol I-IV, 2nd edition. CRC Press: Boca Raton, FL
33. Mackay D, Di Guardo A, Paterson S, Kicsi G, Cowan CE (1996) Assessing the fate of new and existing chemicals: A five stage process. Environ Toxicol Chem 15: 1618–1626
34. Mackay D, Di Guardo A, Paterson S, Kicsi G, Cowan CE, Kane M (1996) Assessment of chemical fate in the environment using evaluative, regional and local-scale models: Illustrative application to chlorobenzene and linear alkylbenzene sulfonates. Environ Toxicol Chem 15: 1638–1648
35. Mackay D (2001) Multimedia environmental models – The fugacity approach. Second edition, Boca Raton, FL, Lewis Publishers
36. Kawamoto K, MacLeod M, Mackay D (2001) Evaluation and comparison of mass balance models of chemical fate: Application of EUSES and ChemCAN to 68 chemicals in Japan. Chemosphere 44: 599–612
37. MacLeod M, Fraser A, Mackay D (2002) Evaluating and expressing the propagation of uncertainty in chemical fate and bioaccumulation models. Environ Toxicol Chem 21: 700–709
38. Webster E, Mackay D, Di Guardo A, Kane D, Woodfine D (2004) Regional differences in chemical fate model outcome. Chemosphere 55: 1361–1376
39. McKone TE (1993) CalTOX, a multimedia total exposure model for hazardous-waste sites. U.S. Department of Energy: Washington, DC
40. Gouin T, Mackay D, Jones KC, Harner T, Meijer SN (2004) Evidence for the "grasshopper" effect and fractionation during long-range atmospheric transport of organic contaminants. Environ Pollut 128: 139–148

41. MacLeod M, Woodfine D, Mackay D, McKone T, Bennett D, Maddalena R (2001) BETR North America: A regionally segmented multimedia contaminant fate model for North America. Environ Sci Pollut Res 8: 156–163
42. Toose L, Woodfine DG, MacLeod M, Mackay D, Gouin J (2004) BETR-World: A geographically explicit model of chemical fate: Application to transport of a-HCH to the arctic. Environ Pollut 128: 223–240
43. MacLeod M, Riley WJ, McKone TE (2005) Assessing the influence of climate variability on atmospheric concentrations of polychlorinated biphenyls using a global-scale mass balance model (BETR-global). Environ Sci Technol 39: 6749–6756
44. Armitage J, Cousins IT, Buck RC, Prevedouros J, Russell MH, Macleod M, Korzeniowski SH (2006) Modeling global-scale fate and transport of perfluorooctanoate emitted from direct sources. Environ Sci Technol 40: 6969–6975
45. Campfens J, Mackay D (1997) Fugacity-based model of PCB bioaccumulation in complex food webs. Environ Sci Technol 31: 577–583
46. Patwa Z, Christensen R, Lasenby DC, Webster E, Mackay D (2007) An exploration of the role of mysids in benthic–pelagic coupling and biomagnification using a dynamic bioaccumulation model. Environ Toxicol Chem 26: 186–194
47. Arnot JA, Gobas FAPC (2004) A food web bioaccumulation model for organic chemicals in aquatic ecosystems. Environ Toxicol Chem 23: 2343–2355
48. Nichols JW, Schultz IR, Fitzsimmons PN (2006) In vitro–in vivo extrapolation of quantitative hepatic biotransformation data for fish. I. A review of methods, and strategies for incorporating intrinsic clearance estimates into chemical kinetic models. Aquat Toxicol 78: 74–90
49. Nichols JW, Fitzsimmons PN, Burkhard LP (2007) In vitro–in vivo extrapolation of quantitative hepatic biotransformation data for fish. II. Modeled effects on chemical bioaccumulation. Environ Toxicol Chem 26: 1304–1319
50. Arnot JA, Mackay D, Bonnell M (2008) Estimating metabolic biotransformation rates in fish from laboratory data. Environ Toxicol Chem 27: 341–351
51. Barber MC (2008) Dietary uptake models used for modeling the bioaccumulation of organic contaminants in fish. Environ Toxicol Chem 27: 755–777
52. Arnot JA, Gobas FAPC (2003) A generic QSAR for assessing the bioaccumulation potential of organic chemicals in aquatic food webs. QSAR Comb Sci 22: 337–345
53. Han X, Nabb DL, Mingoia RT, Yang C-H (2007) Determination of xenobiotic intrinsic clearance in freshly isolated hepatocytes from rainbow trout (*Oncorhynchus mykiss*) and rat and its application in bioaccumulation assessment. Environ Sci Technol 41: 3269–3276
54. Czub G, McLachlan MS (2004) A food chain model to predict the levels of lipophilic organic contaminants in humans. Environ Toxicol Chem 23: 2356–2366
55. Cahill T, Cousins I, Mackay D (2003) Development and application of a generalized physiologically based pharmacokinetic model for multiple environmental contaminants. Environ Toxicol Chem 22: 26–34
56. Lawrence GS, Gobas FAPC (1997) A pharmacokinetic analysis of interspecies extrapolation in dioxin risk assessment. Chemosphere 35: 427–452
57. Hickie B, Mackay D, de Koning J (1999) Lifetime pharmacokinetic model for hydrophobic contaminants in marine mammals. Environ Toxicol Chem 18: 2622–2633
58. Kannan K, Haddad S, Beliveau M, Tardif R (2002) Physiological modeling and extrapolation of pharmacokinetic interactions from binary to more complex chemical mixtures. Environ Health Perspect 110: 989–994
59. McLachlan MS (1996) Bioaccumulation of hydrophobic chemicals in agricultural food chains. Environ Sci Technol 30: 252–259
60. Kelly BC, Gobas FAPC (2003) An arctic terrestrial food-chain bioaccumulation model for persistent organic pollutants. Environ Sci Technol 37: 2966–2974
61. Armitage JM, Gobas FAPC (2007) A terrestrial food-chain bioaccumulation model for POPs. Environ Sci Technol 41: 4019–4025
62. Hughes L, Webster E, Mackay D (2008) A model of the fate of chemicals in sludge-amended soils. Soil Sediments Contam 17: 564–585

63. Hughes L, Mackay D, Webster E, Armitage J, Gobas F. 2005. Development and application of models of chemical fate in Canada: Modelling the fate of substances in sludge-amended soils. Report to Environment Canada. CEMN Report No 200502, Trent University: Peterborough, ON
64. Mackay D, Joy M, Paterson S (1983) A quantitative water, air, sediment interaction (QWASI) fugacity model for describing the fate of chemicals in lakes. Chemosphere 12: 981–997
65. Mackay D, Paterson S, Joy M (1983) A quantitative water, air, sediment interaction (QWASI) fugacity model for describing the fate of chemicals in rivers. Chemosphere 12: 1193–1208
66. Mackay D, Diamond M (1989) Application of the QWASI (quantitative water air sediment interaction) fugacity model to the dynamics of organic and inorganic chemicals in lakes. Chemosphere 18: 1343–1365
67. Diamond ML, Poulton DJ, Mackay D, Stride FA (1994) Development of a mass-balance model of the fate of 17 chemicals in the bay of Quinte. J Great Lake Res 20: 643–666
68. Diamond ML, MacKay D, Poulton DJ, Stride FA (1996) Assessing chemical behavior and developing remedial actions using a mass balance model of chemical fate in the Bay of Quinte. Water Res 30: 405–421
69. Webster E, Lian L, Mackay D (2005) Application of the quantitative water air sediment interaction (QWASI) model to the great lakes. Report to the lakewide management plan (LaMP) committee. Canadian Environmental Modelling Centre, Report No 200501, Trent University: Peterborough, ON
70. Clark B, Henry JG, Mackay D (1995) Fugacity analysis and model of organic-chemical fate in a sewage-treatment plant. Environ Sci Technol 29: 1488–1494
71. Seth R, Webster E, Mackay D (2008) Continued development of a mass balance model of chemical fate in a sewage treatment plant. Water Res 42: 595–604
72. Vermeire TG, Jager DT, Bussian B, Devillers J, den Haan K, Hansen B, Lundberg I, Niessen H, Robertson S, Tyle H, van der Zandt PTJ (1997) European union system for the evaluation of substances (EUSES). Principles and structure. Chemosphere 34: 1823–1836
73. Vermeire TG, Rikken M, Attias L, Boccardi P, Boeije G, Brooke D, de Bruijn J, Comber M, Dolan B, Fischer S, Heinemeyer G, Koch V, Lijzen J, Muller B, Murray-Smith R, Tadeo J (2005) European union system for the evaluation of substances (EUSES): The second version. Chemosphere 59: 473–485
74. Pennington DW, Margni M, Payet J, Jolliet O (2006) Risk and regulatory hazard-based toxicological effect indicators in life-cycle assessment (LCA). Hum Ecol Risk Assess 12: 450–475

The Application of Structurally Dynamic Models in Ecology and Ecotoxicology

Sven E. Jørgensen

Abstract Structurally dynamic models (SDMs) are models that account for the changes in the model parameters due to the adaptation or the shift in species composition resulting from current changes in the forcing functions. The parameter changes are found by introduction of eco-exergy as goal function in the model. The set of parameters that give the highest eco-exergy by the prevailing conditions are currently selected. The theory behind the use of eco-exergy as goal function in ecological models is presented as a translation of Darwin's theory to thermodynamics. Two examples of SDMs are presented to illustrative the advantages and disadvantages of this model type.

Keywords Structurally dynamic models · Adaptation · Shifts in species composition · Exergy · Eco-exergy · Darwin finches · Copper

1 Introduction

Ecosystems differ from most other systems by being extremely adaptive, having the ability of self-organization and having a large number of feedback mechanisms. The real challenge to modeling is, therefore, how can we construct models that are able to account for this enormous adaptability. The model type *structurally dynamic model* (SDM) has been developed to meet this demand. The next section will present the characteristics, the advantages, and the disadvantages of this model type and where it is most recommendable to consider to apply SDM. The following section, Sect. 3, presents the theory behind SDM, followed by a section where an illustrative example of SDMs applied in ecology is presented. The example shows clearly the idea behind this model type and under which circumstances it is advantageous to use it. The fifth section presents an ecotoxicological example and the last section concludes on the application of SDM in ecology and particularly in ecotoxicology.

S.E. Jørgensen (✉)
Section for Environmental Chemistry, Institute A, Copenhagen University, University Park 2, 2100 Copenhagen Ø, Denmark
e-mail: msijapan@hotmail.com

J. Devillers (ed.), *Ecotoxicology Modeling*, Emerging Topics in Ecotoxicology: Principles, Approaches and Perspectives 2, DOI 10.1007/978-1-4419-0197-2_13,

This model type will most probably be used increasingly in the coming years in our endeavor to make better prognoses, because reliable prognoses can only be made by models with a correct description of ecosystem properties including the ability to change the structure and the properties of key species. If our models do not describe properly adaptation and possible shifts in species composition, the prognoses will inevitably be more or less incorrect. The SDMs attempt to overcome this crucial model problem.

Ecology deals with irreducible systems with many feedback mechanisms that will regulate simultaneously all the factors and rates, and they interact and are also functions of time, too, as pointed out by Straskraba [1, 2].

Table 1 shows the hierarchy of regulation mechanisms that are operating at the same time. An ecosystem has so many interacting components that it is impossible ever to be able to examine all these relationships and even if we could, it would not be possible to separate one relationship and examine it carefully to reveal its details, because the relationship is different in nature from that in a laboratory where the examined components are separated from the other ecosystem components.

Known phrases in system ecology are "everything is linked to everything" and "the whole is greater than the sum of the parts." It implies that it may be possible to examine the parts separately, but when the parts are put together, they will form a whole that behaves differently from the sum of the parts. A model seems the only useful tool when we are dealing with irreducible systems. However, we need several models simultaneously to capture a more complete image of reality. It seems our only possibility to deal with the very complex living systems. Our brain simply cannot overview what will happen in a system where, for instance, several interacting processes are working simultaneously.

The number of feedbacks and regulations is extremely high and makes it possible for the living organisms and populations to survive and reproduce in spite of changes in external conditions. These regulations correspond to levels 3 and 4 in Table 1. Numerous examples can be found in the literature. If the actual properties of the species are changed the regulation is named adaptation. Phytoplankton

Table 1 The hierarchy of regulating feedback mechanisms [3]

Level	Explanation of regulation process	Exemplified by phytoplankton growth
1	Rate by concentration in medium	Uptake of phosphorus in accordance with phosphorus concentration
2	Rate by needs	Uptake of phosphorus in accordance with intracellular concentration
3	Rate by other external factors	Chlorophyll concentration in accordance with previous solar radiation
4	Adaptation of properties	Change of optimal temperature for growth
5	Selection of other species	Shift to better-fitted species
6	Selection of other food web	Shift to better-fitted food web
7	Mutations, new sexual recombinations, and other shifts of genes	Emergence of new species or shifts of species properties

is, for instance, able to regulate its chlorophyll concentration according to the solar radiation. If more chlorophyll is needed because the radiation is insufficient to guarantee growth, more chlorophyll is produced by the phytoplankton. The digestion efficiency of the food for many animals depends on the abundance of the food. The same species may be of different sizes in different environments, depending on what is most beneficial for survival and growth. If nutrients are scarce, phytoplankton becomes smaller and vice versa. In this latter case the change in size is a result of a selection process, which is made possible because of the distribution in size.

The feedbacks are furthermore constantly changing, that is, the adaptation itself is adaptable in the sense that if a regulation is not sufficient, another regulation process higher in the hierarchy of feedbacks – see Table 1 – will take over. The change in size within the same species is, for instance, only limited. When this limitation has been reached, other species may take over; see levels 5 and 6 in Table 1. It implies that not only the processes and the components, but also the feedbacks can be replaced, if it is needed to achieve a better utilization of the available resources.

2 The Characteristics of SDM

An ecosystem is a very dynamic system. All its components and particularly the biological ones are steadily moving and their properties are steadily modified, which is why an ecosystem will never return to the same situation again.

Every point is furthermore different from any other point and therefore offering different conditions for the various life forms.

This enormous heterogeneity explains why there are so many species on earth. There is, so to say, an ecological niche for "everyone" and *everyone* may be able to find a niche where he is best fitted to utilize the resources.

Darwin's theory describes the competition among species and states that the species that are best fitted to the prevailing conditions in the ecosystem will survive. Darwin's theory can, in other words, describe the changes in ecological structure and the species composition, but cannot directly be applied quantitatively (example given) in ecological modeling; see, however, the next section.

All species in an ecosystem are confronted with the challenge: how is it possible to survive or even grow under the prevailing conditions? The prevailing conditions are considered as all factors influencing the species, that is, all external and internal factors including those originating from other species.

All natural external and internal factors of ecosystems are dynamic – the conditions are steadily changing, and there are always many species waiting in the wings, ready to take over, if they are better fitted to the emerging conditions than the species dominating under the present conditions. There is a wide spectrum of species representing different combinations of properties available for the ecosystem. The question is which of these species are best able to survive and grow under the present conditions and which species are best able to survive and grow under the conditions one time step further and two time steps further and so on? The necessity

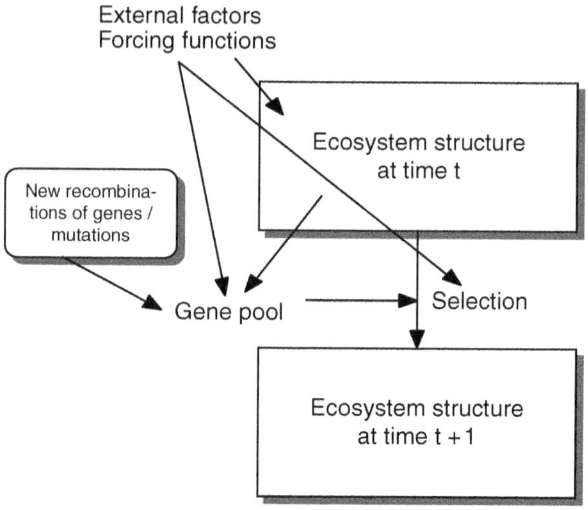

Fig. 1 Conceptualization of how the external factors steadily change the species composition. The possible shifts in species composition are determined by the gene pool, which is steadily changed due to mutations and new sexual recombinations of genes. The development is, however, more complex. This is indicated (1) by *arrows* from "structure" to "external factors" and "selection" to account for the possibility that the species can modify their own environment and thereby their own selection pressure and (2) an arrow from "structure" to "gene pool" to account for the possibilities that the species can to a certain extent change their own gene pool. Several mechanisms for this possibility can be found in the literature

in Monod's sense is given by the prevailing conditions – the species must have genes or maybe rather phenotypes (meaning properties) that match these conditions, to be able to survive. But the natural external factors and the genetic pool available for the test may change randomly or by "chance."

Steadily, new mutations (misprints are produced accidentally) and sexual recombinations (the genes are mixed and shuffled) emerge and give steadily new material to be tested by the question: which species are best fitted under the conditions prevailing just now?

These ideas are illustrated in Fig. 1. The external factors are steadily changed and some even relatively fast – partly at random, such as the meteorological or climatic factors. The species of the system are selected among the species available and represented by the genetic pool, which again is slowly but surely changed randomly or by *chance*. The selection in the Fig. 1 includes the level 4 of Table 1. It is a selection of the organisms that possess the properties best fitted to the prevailing organisms according to the frequency distribution. What is named ecological development is the change over time in nature caused by the dynamics of the external factors, giving the system sufficient time for the reactions.

Evolution, on the other hand, is related to the genetic pool. It is the result of the relation between the dynamics of the external factors and the dynamics of the genetic pool. The external factors steadily change the conditions for survival and the genetic pool steadily comes up with new solutions to the problem of survival.

The species are continuously tested against the prevailing conditions (external as well as internal factors) and the better they are fitted, the better they are able to maintain and even increase their biomass. The specific rate of population growth may even be used as a measure for the fitness (see, e.g., Stenseth [4]). But the property of fitness must, of course, be inheritable to have any effect on the species composition and the ecological structure of the ecosystem in the long run.

Natural selection has been criticized for being a tautology: fitness is measured by survival, and survival of the fittest therefore means survival of the survivors. However, the entire Darwinian theory including the aforementioned three assumptions cannot be conceived as a tautology, but may be interpreted as follows: the species offer different solutions to survival under given prevailing conditions, and the species that have the best combinations of properties to match the conditions have also the highest probability of survival and growth.

UNEP has developed two SDMs of lakes, which are directly accessible. Pamolare launched by UNEP can be downloaded from the homepage: http://www.unep.or.jp/ietc/pamolare. Pamolare 1 contains the following lake models: a one-layer model, a two-layer model, a SDM, and a drainage area model. Pamolare 2 contains a structurally dynamic shallow-lake model.

The use of SDM has particular interest in ecotoxicology, because the competing organisms have most often very different susceptibility to various toxic substances. Therefore, a clear selection of the fittest species in an ecotoxicological case study is, therefore, frequently observed.

SDMs can be constructed by two different methods: either by expert knowledge or by introduction of a goal function. If it is known how the properties of the species will change when the prevailing conditions are changed, it is, of course, possible to introduce this expert knowledge into the model, which is possible either by formulation of rules or by artificial intelligence. Rules may be exemplified by this example: when the phosphorus concentration is between x and y, then the growth rate of phytoplankton will be changing gradually from z to v. Examples of this type of SDMs are given in Patten [5]. Artificial intelligence is able to find the rules from interpretation by a computer of a suitable set of observations.

Several goal functions have been proposed, but only very few models that account for change in species composition or for the ability of the species to change their properties within some limits have been developed. Eco-exergy has been the most frequently applied goal function in SDM. It has successfully been used to develop SDMs in 19 cases; see Jørgensen and coworkers [6, 7]. As eco-exergy is not generally known it is necessary to introduce this thermodynamic variable in the next section.

3 Eco-Exergy as Goal Function in SDM

Exergy is defined as the work capacity the system can perform when brought into thermodynamic equilibrium with the environment.

We will name this form of exergy here as technological exergy. Technological exergy is not practical to use in the ecosystem context, because it presumes that the environment is the reference state, which means for an ecosystem the next ecosystem. As the energy embodied in the organic components and the biological structure and information contributes far most to the exergy content of an ecosystem, there seems to be no reason to assume a (minor) temperature and pressure difference between the ecosystem and the reference environment. Eco-exergy is defined as the work the ecosystem can perform relatively to the same ecosystem at the same temperature and pressure but at thermodynamic equilibrium, where there are no gradients and all components are inorganic at the highest possible oxidation state. Under these circumstances we can calculate the exergy, which has been denoted as eco-exergy to distinguish from the technological exergy, as coming entirely from the chemical energy:

$$\sum_{c} (\mu_c - \mu_{co}) N_i .$$

This represents the nonflow biochemical exergy. We can measure the concentrations in the ecosystem, but the concentrations in the reference state (thermodynamic equilibrium) could be based on the usual use of chemical equilibrium constants. Eco-exergy is a concept close to Gibb's free energy but opposite to Gibb's free energy, eco-exergy has a different reference state from case to case (from ecosystem to ecosystem) and it can furthermore be used far from thermodynamic equilibrium, while Gibb's free energy in accordance to its exact thermodynamic definition is a state function close to thermodynamic equilibrium. In addition, eco-exergy of organisms is mainly embodied in the information content and should, therefore, not be considered the same as the chemical energy of fossil fuel.

As $(\mu_c - \mu_{co})$ can be found from the definition of the chemical potential replacing activities with approximations by concentrations, we get the following expressions for the exergy:

$$\text{Ex} = RT \sum_{i=0}^{i=n} C_i \ \ln C_i / C_{i,o}, \tag{1}$$

where R is the gas constant (8.317 J/K/moles $=$ 0.082071 atm/K/moles), T is the temperature of the environment, while C_i is the concentration of the ith component expressed in a suitable unit, $C_{i,o}$ is the concentration of the ith component at thermodynamic equilibrium, and n is the number of components. $C_{i,o}$ is, of course, a very small concentration (except for $i = 0$, which is considered to cover the inorganic compounds), corresponding to a very low probability of forming complex organic compounds spontaneously in an inorganic soup at thermodynamic equilibrium. $C_{i,o}$ is even lower for the various organisms, because the probability of forming the organisms is very low with their embodied information, which implies that the genetic code should be correct.

By using this particular exergy based on the same system at thermodynamic equilibrium as reference, the eco-exergy becomes dependent only on the chemical potential of the numerous biochemical components.

It is possible to distinguish in (1) between the contribution to the eco-exergy from the information and from the biomass. We define p_i as c_i/A, where A is the total concentration of all components in the system.

With the introduction of this new variable, we get

$$\mathrm{Ex} = ART \sum_{i=1}^{n} p_i \ \ln p_i/p_{io} + A \ \ln A/A. \tag{2}$$

As $A \approx A_o$, eco-exergy becomes a product of the total concentration A (multiplied by RT) and Kullback measure:

$$K = \sum_{i=1}^{n} p_i \ \ln(p_i/p_{io}),$$

where p_i and p_{io} are probability distributions, a posteriori and a priori to an observation of the molecular detail of the system. It means that K expresses the amount of information that is gained as a result of the observations and that eco-exergy $= ARTK$. For different organisms that contribute to the eco-exergy of the ecosystem, the eco-exergy density contribution becomes $c_i RT \ \ln(p_i/p_{io})$, where c_i is the concentration of the considered organism. $RT \ln(p_i/p_{io})$, is found by calculation of the probability to form the considered organism at thermodynamic equilibrium, which would require that organic matter is formed and that the proteins (enzymes) controlling the life processes in the considered organism have the right amino acid sequence. These calculations can be seen in Jørgensen et al. [8] and Jørgensen and Svirezhev [6], and Jørgensen and Fath [9]. In the latter reference the latest information about the calculations of *RTK* that denoted β values for various organisms is presented; see Table 1. The β value for detritus is in this table $= 1.00$, which means that the eco-exergy density is found by multiplication of the concentration c_i by β as g detritus equivalents per unit of volume or area. As detritus has about 18.7 kJ/g, eco-exergy can be found as kJ by multiplication by 18.7. For human, the β value is 2,173, when the eco-exergy is expressed in detritus equivalent or 18.7 times as much or 40,635 kJ/unit of volume or area. The β value has not surprisingly increased as a result of the evolution. To mention a few values from Table 2: bacteria 8.5, protozoa 39, flatworms 120, ants 167, crustaceans 232, mollusks 310, fish 499, reptiles 833, birds 980, and mammals 2,127.

The evolution has in other words resulted in a more and more effective transfer of what we could call the classical work capacity to the work capacity of the information. A value of 2.0 means that the eco-exergy embodied in the organic matter and the information is equal. As the values are much bigger than 2.0 (except for virus, where the value is 1.01 – slightly more than 1.0) the information eco-exergy is the most significant part of the eco-exergy of organisms.

Biological systems have many possibilities for moving away from thermodynamic equilibrium, and it is important to know along which pathways among the possible ones a system will develop. This leads to the following hypothesis sometimes denoted as the ecological law of thermodynamics (ELT) [6–12]: If a system receives an input of exergy, then it will utilize this exergy to perform work.

Table 2 ß Values = exergy content relatively to the exergy of detritus [6]

Early organisms	Plants		Animals
Detritus		1.00	
Virus		1.01	
Minimal cell		5.8	
Bacteria		8.5	
Archaea		13.8	
Protists	Algae	20	
Yeast		17.8	
		33	Mesozoa, Placozoa
		39	Protozoa, amoeba
		43	Phasmida (stick insects)
Fungi, moulds		61	
		76	Nemertina
		91	Cnidaria (corals,sea anemones, jelly fish)
	Rhodophyta	92	
		97	Gastrotricha
Prolifera, sponges		98	
		109	Brachiopoda
		120	Platyhelminthes (flatworms)
		133	Nematoda (round worms)
		133	Annelida (leeches)
		143	Gnathostomulida
	Mustard weed	143	
		165	Kinorhyncha
	Seedless vascula plants	158	
		163	Rotifera (wheel animals)
		164	Entoprocta
	Moss	174	
		167	Insecta (beetles, fruit flies, bees, wasps, bugs, ants)
		191	Coleodiea (sea squirt)
		221	Lepidoptera (buffer flies)
		232	Crustaceans, Mollusca, bivalvia, gastropodea
		246	Chordata
	Rice	275	
	Gynosperms (including pinus)	314	
		322	Mosquito
	Flowering plants	393	
		499	Fish
		688	Amphibia
		833	Reptilia
		980	Aves (birds)
		2,127	Mammalia
		2,138	Monkeys
		2,145	Anthropoid apes
		2,173	*Homo sapiens*

The work performed is first applied to maintain the system (far) away from thermodynamic equilibrium whereby exergy is lost by transformation into heat at the temperature of the environment. If more exergy is available, then the system is moved further away from thermodynamic equilibrium, which is reflected in growth of gradients. If more than one pathway to depart from equilibrium is offered, then the one yielding the highest eco-exergy storage (denoted Ex) will tend to be selected, or expressed differently: among the many ways for ecosystems to move away from thermodynamic equilibrium, the one maximizing $\mathrm{dEx}/\mathrm{d}t$ under the prevailing conditions will have a propensity to be selected.

This hypothesis can be considered a translation of Darwin's theory into thermodynamics. It is supported by several ecological observations and case studies [3,6,7]. Survival implies maintenance of the biomass, and growth means increase of biomass and information. It costs eco-exergy to construct biomass and gain information, and therefore biomass and information possess eco-exergy. Survival and growth can, therefore, be measured by use of the thermodynamic concept eco-exergy, which may be understood as the work capacity the ecosystem possesses.

The idea of SDMs is to find continuously a new set of parameters (limited for practical reasons to the most crucial, i.e., sensitive parameters) that is better fitted for the prevailing conditions of the ecosystem. "Fitted" is defined in the Darwinian sense by the ability of the species to survive and grow, which may be measured as discussed earlier by eco-exergy. Figure 2 shows the proposed modeling procedure, which has been applied in the development of totally 19 SDMs.

For all SDMs developed with eco-exergy as the goal function, the changes obtained by the model were in accordance with actual observations. At least in models the applicability of the eco-exergy calculations has shown their more practical use, which can be explained by a robustness in the model calculations by the use of the β values that, of course, have uncertainties. It is noteworthy that Coffaro et al. [14], in their structural-dynamic model of the Lagoon of Venice, did not calibrate the model describing the spatial pattern of various macrophyte species such as *Ulva* and *Zostera*, but used exergy-index optimization to estimate parameters determining the spatial distribution of these species. They found good accordance between observations and model, as was able by this method *without* calibration, to explain more than 90% of the observed spatial distribution of various species of *Zostera* and *Ulva*.

Figure 3 illustrates the theoretical considerations behind the development of a SDM with eco-exergy as the goal function.

SDM is, of course, more cumbersome to apply than other models due to the eco-exergy optimization, which, for instance, may take place in the model every 5–30 days, but with a modern fast computer the additional computation is, of course, limited. The advantage of SDM is clearly that eventually structurally dynamic changes are considered and if that is the case, a SDM will inevitably give a more accurate result. It may also be needed to use SDM for the calibration, because changes in, for instance, phytoplankton composition from spring to summer to fall may imply that different parameters should be applied in the different seasons. The conclusion is, therefore, that it is recommended to use SDM, whenever it is known or even suspected that structurally changes will take place. SDM requires, however, good

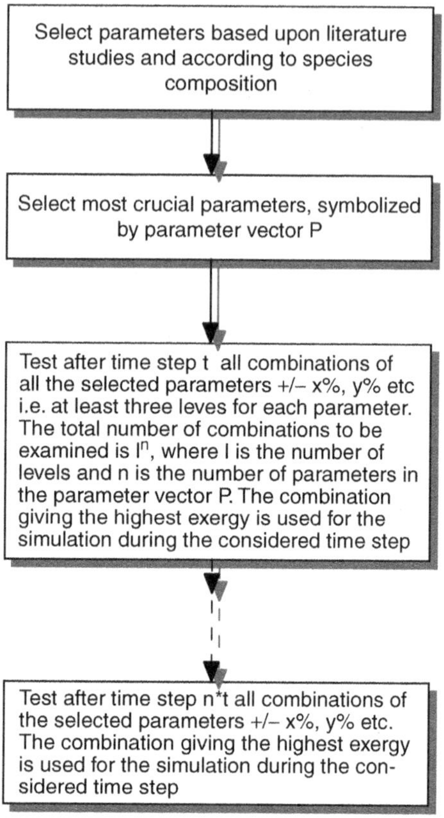

Fig. 2 The procedure used for the development of structurally dynamic models (reproduced from [13])

observations, in most cases also of some structural changes to give acceptable results. A SDM will not necessarily be more expensive to develop than other models, but the need for good observations and a good data set will often make the entire project more expensive.

4 An Illustrative SDM Example

The SDM of Darwin's finches by Jørgensen and Fath [7] is presented later as an illustrative example of SDM. The models reflect therefore – as all models – the available knowledge, which in this case is comprehensive and sufficient to validate even the ability of the model to describe the changes in the beak size as a result of climatic changes, causing changes in the amount, availability, and quality of the seeds that make up the main food item for the finches. The medium ground finches, *Geospiza fortis*, on the island Daphne Major were selected for this modeling case due to very detailed case-specific information found by Grant [15]. The model has three state variables: seed, Darwin's Finches adult, and Darwin's finches juvenile.

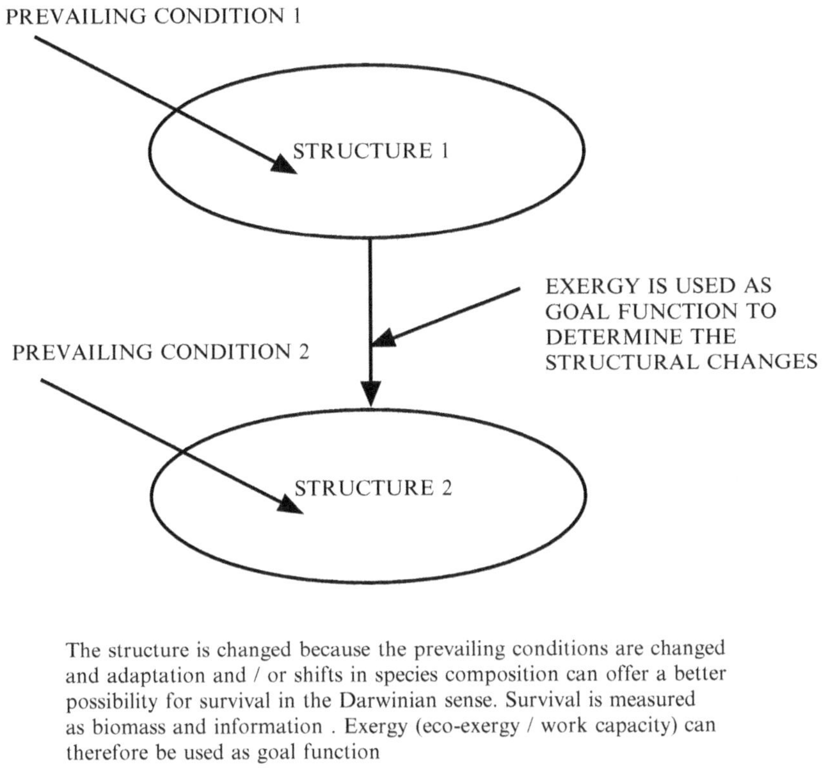

Fig. 3 The theoretical considerations behind SDMs developed with eco-exergy as the goal function are illustrated

The juvenile finches are promoted to adult finches 120 days after birth. The mortality of the adult finches is expressed as a normal mortality rate [15] + an additional mortality rate due to food shortage and an additional mortality rate caused by a disagreement between bill depth and the size and hardness of seeds.

The beak depth can vary between 3.5 and 10.3 cm [15] and the beak size = $\sqrt{DH}$, where D is the seed size and H is the seed hardness, which are both dependent on the precipitation, particularly in the months January to April. It is possible to determine a handling time for the finches for a given $\sqrt{DH}$ as function of the bill depth, which explains that the accordance between $\sqrt{DH}$ and the beak depth becomes an important survival factor. The relationship is used in the model to find a function called "diet," which is compared with $\sqrt{DH}$ to find how well the bill depth fits to the $\sqrt{DH}$ of the seed. This fitness function is based on information given in Grant [15] about the handling time. It influences, as mentioned earlier, the mortality of adult finches, but it has also impact on the number of eggs laid and the mortality of the juvenile finches. The growth rate and mortality of seeds is dependent on the precipitation, which is a forcing function known as a function of time. A function called shortage of food is calculated from the food required for the finches, which is known, and from the food available (the seed state variable). How the food shortage influences the mortality of juvenile finches and adult finches can be found in [15]. The seed biomass and the number of *G. fortis* as a function of time from 1975 to 1982 are known [15]. These numbers from 1975 to 1976 have been used to calibrate the following parameters: the coefficients determining the following:

1. The influence of the fitness function on (a) the mortality of adult finches, (b) the mortality of juvenile finches, and (c) the number of eggs laid.
2. The influence of food shortage on the mortality of adult and juvenile finches is known. The influence is, therefore, calibrated within a narrow range of values.
3. The influence of precipitation on the seed biomass (growth and mortality).

All other parameters are known from the literature.

The exergy density is calculated (estimated) as 275 × the concentration of seed + 980 × the concentration of Darwin's finches (see Table 2). Every 15 days it is found if a feasible change in the beak size, taking the generation time and the variations in the beak size into consideration, will give a higher exergy. If it is the case, then the beak size is changed accordingly. The modeled changes in the beak size were confirmed by the observations. The model results of the number of Darwin's

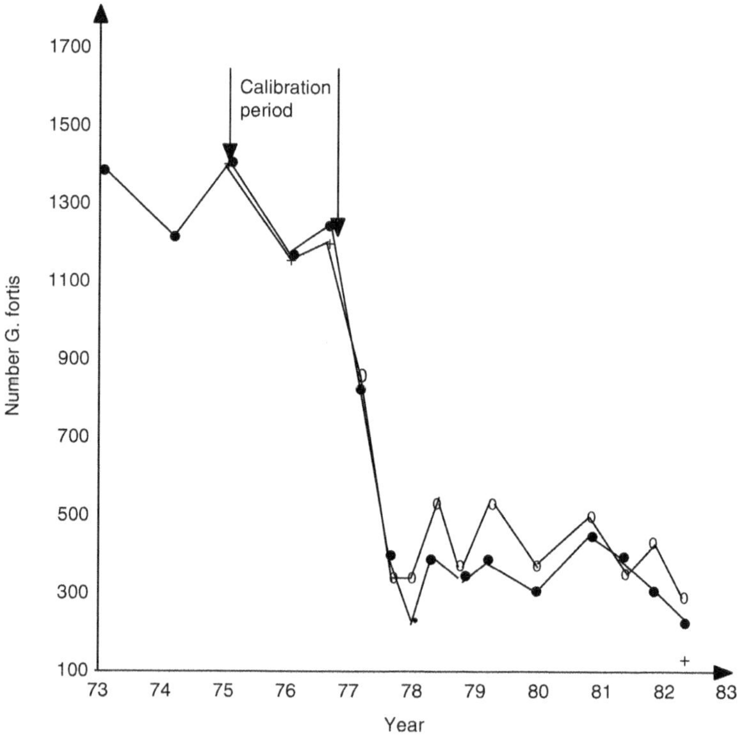

Fig. 4 The observed number of finches (*filled circles*) from 1973 to 1983, compared with the simulated result (*open circles*). 1975 and 1976 were used for calibration and 1977/1978 for the validation

finches are compared with the observations [15] in Fig. 4. The standard deviation between modeled and observed values was 11.6%, and the correlation coefficient, r^2, for modeled versus observed values is 0.977. The results of a nonstructural dynamic model would not be able to predict the changes in the beak size and would, therefore, give much too low values for the number of Darwin's finches because their beak would not adapt to the lower precipitation yielding harder and bigger seeds.

5 Ecotoxicological SDM Example

The conceptual diagram of the ecotoxicological model that is used to illustrate SDM is shown in Fig. 5, using the modeling software STELLA. Copper is an algaecide causing an increase in the mortality of phytoplankton [16] and a decrease in the phosphorus uptake and photosynthesis [17]. Copper is also reducing the carbon assimilation of bacteria [18]. The literature gives the change of the following three parameters in the model: growth rate of phytoplankton, mortality of phytoplankton,

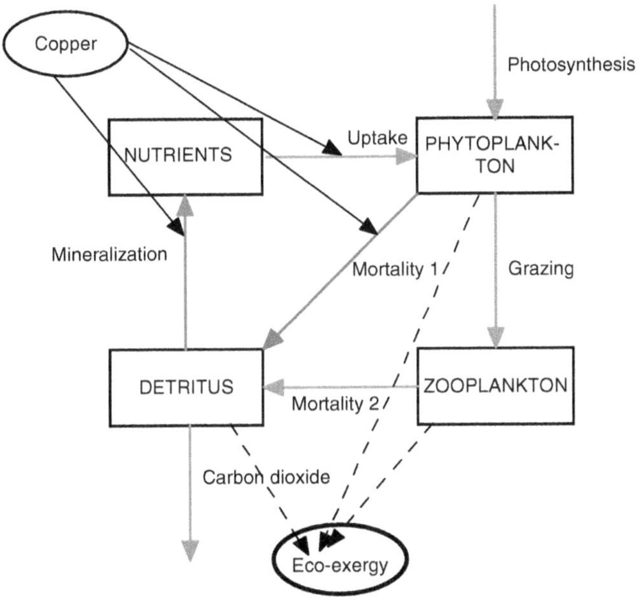

Fig. 5 Conceptual diagram of an ecotoxicologial model focusing on the influence of copper on the photosynthetic rate, phytoplankton mortality rate, and the mineralization rate. The boxes are the state variables; the *thick gray arrows* symbolize processes and the *thin black arrows* indicate the influence of copper on the processes and the calculation of eco-exergy from the state variables. Because of the change in these three rates, it is an advantage for the zooplankton and the entire ecosystem to decrease their size. The model is, therefore, made structurally dynamic by allowing zooplankton to change its size and thereby the specific grazing rate and the specific mortality rate according to the allometric principles. The size yielding the highest eco-exergy is currently found

and mineralization rate of detritus with increased copper concentration [16–19]. As a result the zooplankton is reduced in size [19], which according to the allometric principles means an increased specific grazing rate and specific mortality rate [19]. It has been observed that the size of zooplankton in a closed system (a pond for instance) is reduced to less than half the size at a copper concentration of 140 mg/m^3 compared with a copper concentration of less than 10 mg/m^3 [19]. In accordance with the allometric principles [20], it would result in more than doubling of the grazing rate and the mortality rate.

The model shown in Fig. 5 was made structurally dynamic by varying the size of zooplankton and using an allometric equation to determine the corresponding specific grazing rate and specific mortality rate. The equation expresses that the two specific rates are inversely proportional to the linear size [20]. In the range of different copper concentrations from 10 to 140 mg/m^3 are found by the model which zooplankton size yields the highest eco-exergy. In accordance with the presented SDM approach it is expected that the size yielding the highest eco-exergy would be selected. The results of the model runs are shown in Figs. 6–8. The specific grazing rate, the size yielding the highest eco-exergy, and the eco-exergy are plotted versus the copper concentration in these three figures.

As expected is the eco-exergy even at the zooplankton size yielding the highest eco-exergy decreasing with increase in copper concentration due to the toxic effect on phytoplankton and bacteria.

The selected size, see Fig. 7, at 140 mg/m^3 as also indicated in the literature is less than half, namely, about 40% of the size at 10 mg/m^3. The eco-exergy is decreasing from 198 kJ/l at 10 mg/m^3 to 8 kJ/l at 140 mg/m^3. The toxic effect of the

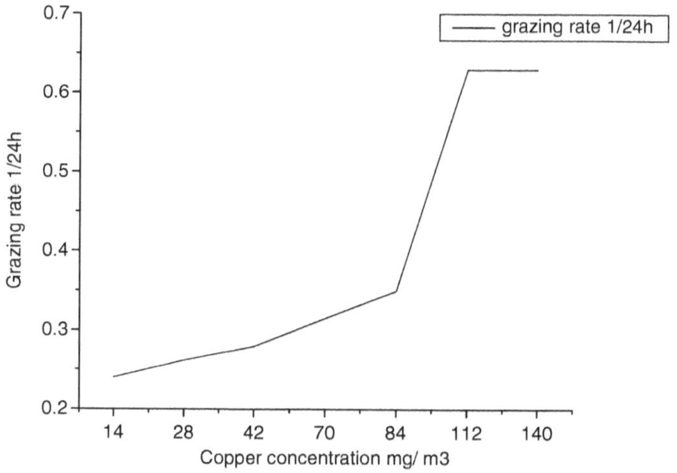

Fig. 6 The grazing rate that yields the highest eco-exergy is shown at different copper concentrations. The grazing rate is increasing more and more rapidly as the copper concentration is increasing but at a certain level, it is not possible to increase the eco-exergy further by changing the zooplankton parameters, because the amount of phytoplankton is becoming the limiting factor for zooplankton growth

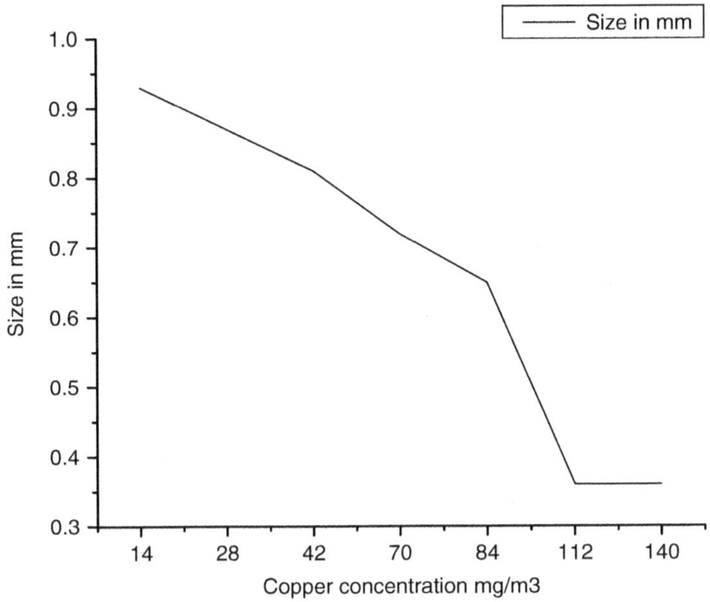

Fig. 7 The zooplankton size that yields the highest eco-exergy is plotted versus the copper concentration. The size is decreasing more and more rapidly as the copper concentration is increasing but at a certain level, it is not possible to increase the eco-exergy further by changing the zooplankton size, because the amount of phytoplankton is becoming the limiting factor for zooplankton growth

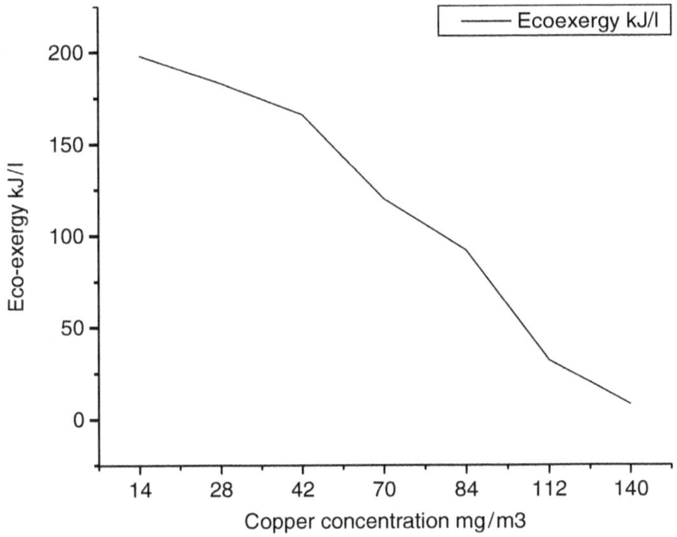

Fig. 8 The highest eco-exergy obtained when varying the zooplankton size is plotted versus the copper concentration. The eco-exergy is decreasing almost linearly with increasing copper concentration. The discrepancy from approximately a linear plot may be due to model uncertainty and discontinuous change of the copper concentration and the zooplankton size

copper is, in other words, resulting in an eco-exergy reduction to about 4% of the original eco-exergy level, which is a very significant toxic effect. If the zooplankton was not adaptable to the toxic effect by changing its size and thereby the parameters, the reduction in eco-exergy would have been even more pronounced already at a lower copper concentration. It is, therefore, important for the model results that the model is made structurally dynamic and thereby accounts for the change of parameters when the copper concentration is changed.

6 Conclusion

All organisms are able to change their properties to offer the best possibility for survival under the prevailing conditions. The generally applied bio-geo-chemical models do not consider this adaptation and they will, therefore, inevitably result in wrong prognoses. SDMs offer to solve this problem by changing currently the parameters to the values that yield the highest eco-exergy. The two presented illustrative examples have demonstrated how a SDM is working and how it is able to account for the adaptation. In the first example, the Darwin's finches are currently changing their beak size and in the ecotoxicological case study zooplankton is changing its size. In both cases the changes are approximately in accordance with the observations. It is, of course, an advantage that SDMs can predict approximately the changes of the species' properties, but it is an even more important advantage that the state variables are predicted closer to the observations by the SDMs than by bio-geo-chemical models. In accordance with the bio-geo-chemical model approach the Darwin finches would have died in the first presented example and also in the second example if the toxic effect of copper had been more pronounced, while the use of SDM in both cases gives reasonably approximate results.

References

1. Straskraba M (1979) Natural control mechanisms in models of aquatic ecosystems. Ecol Model 6: 305–322
2. Straskraba M (1980) The effects of physical variables on freshwater production: Analyses based on models. In: Le Cren ED, McConnell RH (eds) The Functioning of freshwater ecosystems (International Biological Programme 22), Cambridge University Press, Cambridge
3. Jørgensen SE (2002) Integration of ecosystem theories: A pattern, 3rd edition. Kluwer Academic, Dordrecht, The Netherlands
4. Stenseth NC (1986) Darwinian evolution in ecosystems: A survey of some ideas and difficulties together with some possible solutions. In: Casti JL, Karlquist A (eds) Complexity, language and life: Mathematical approaches, Springer Verlag, Berlin
5. Patten BC (1997) Synthesis of chaos and sustainability in a nonstationary linear dynamic model of the American black bear (*Ursus americanus* Pallas) in the Adirondack Mountains of New York. Ecol Model 100: 11–42
6. Jørgensen SE, Svirezhev YM (2005) Towards a thermodynamic theory for ecological systems, Elsevier, Amsterdam

7. Jørgensen SE, Fath B (2007) A new ecology: A systems approach, Elsevier, Amsterdam
8. Jørgensen SE, Patten BC, Straskraba M (2000) Ecosystems emerging. IV. Growth. Ecol Model 126: 249–284
9. Jørgensen SE, Fath BD (2004) Application of thermodynamic principles in ecology. Ecol Complexity 1: 267–280
10. Fath B, Jørgensen SE, Patten BC, Straskraba M (2004) Ecosystem growth and development. Biol Syst 77: 213–228
11. Jørgensen SE, Meyer HF (1977) Ecological buffer capacity. Ecol Model 3: 39–61
12. Jørgensen SE, Mejer HF (1979) A holistic approach to ecological modelling. Ecol Model 7: 169–189
13. Jørgensen SE (1992) Development of models able to account for changes in species composition. Ecol Model 62: 195–208
14. Coffaro G, Bocci M, Bendoricchio G (1997) Structural dynamic application to space variability of primary producers in shallow marine water. Ecol Model 102: 97–114
15. Grant PR (1986) Ecology and evolution of Darwin's finches. Reprinted in 1999. Princeton University Press, New Jersey
16. Kallqvist T, Meadows BS (1978) The toxic effects of copper on algae and rotifers from a soda lake. Water Res 12: 771–775
17. Hedtke SF (1984) Structure and function of copper-stressed aquatic microcosms. Aquat Toxicol 5: 227–244
18. Havens KE (1994) Structural and functional responses of a freshwater community to acute copper stress. Environ Pollut 86: 259–266
19. Fu-Liu X, Jørgensen SE, Tao S (1999) Ecological indicators for assessing the freshwater ecosystem health. Ecol Model 116: 77–106
20. Peters RH (1983) The ecological implications of body size, Cambridge University Press, Cambridge

Index

GPSR Compliance

The European Union's (EU) General Product Safety Regulation (GPSR) is a set of rules that requires consumer products to be safe and our obligations to ensure this.

If you have any concerns about our products, you can contact us on ProductSafety@springernature.com

In case Publisher is established outside the EU, the EU authorized representative is:

Springer Nature Customer Service Center GmbH
Europaplatz 3
69115 Heidelberg, Germany

Batch number: 10761599

Printed by Printforce, the Netherlands